Coastal Lowlands
geology and geotechnology

Coastal Lowlands

Geology and Geotechnology

Proceedings of the Symposium on Coastal Lowlands organized by The Royal Geological and Mining Society of the Netherlands (KNGMG), The Hague, 23 – 27 May 1987

Edited by

W.J.M. van der Linden
Institute of Earth Sciences, University of Utrecht, The Netherlands

S.A.P.L. Cloetingh
Institute of Earth Sciences, Free University, Amsterdam, The Netherlands

J.P.K. Kaasschieter
Nijverdal, The Netherlands

W.J.E. van de Graaff
KSEPL, Rijswijk (ZH), The Netherlands

J. Vandenberghe
Institute of Earth Sciences, Free University, Amsterdam, The Netherlands

J.A.M. van der Gun
TNO-DGV, Institute of Applied Geoscience, Delft, The Netherlands

Springer-Science+Business Media, B.V.

Library of Congress Cataloging in Publication Data

Coastal lowlands.

1. Geology--Netherlands--Congresses. 2. Coastal engineering--Netherlands--Congresses. 3. Engineering geology--Netherlands--Congresses. 4. Coastal engineering--Congresses. I. Linden, W. J. M. van der (Willem Jan Marie), 1931- . II. Nederlands Geologisch Mijnbouwkundig Genootschap.
QE273.C63 1989 554.92 88-27361

ISBN 978-90-481-4038-1 ISBN 978-94-017-1064-0 (eBook)
DOI 10.1007/978-94-017-1064-0

Originally published by Kluwer Academic Publishers in 1989
Softcover reprint of the hardcover 1st edition 1989

Preface

Coastal Lowlands by virtue of their position across the boundary of land and sea belong to the earth's most dynamic systems. This is true in the physical, i.e. geological and biological, as much as in the cultural and social sense. Although the nearness to the sea was and still is fraught with danger coastal lowlands have always attracted human interest, providing challenging opportunity, holding the promise of profitable enterprise. Coastal lowlands, especially where rivers enter the region, are the cradles of great civilisations and there, of old, populations reached highest densities. As an example, Dutch history is a tale of human struggle and endeavour with and against the sea. Dutch 'lowlanders' wrestled their land from the sea, in turn the sea forged a nation of independent fishermen, navigators, farmers and traders who built their towns and ships at the borders of the North and Zuyder Seas.

As lowlands subside and sea level rises, apparently these days at an increasing rate, concern about this environment world-wide is also rising. It certainly was appropriate and timely for the Royal Geological and Mining Society of the Netherlands when celebrating its 75th birthday to organize and call together a symposium, focussing attention on the geology and geotechnology of coastal lowlands; geology to better understand their formation and evolution, geotechnology to better manage and harvest resources as much as protect a unique and crucial environment.

We are indebted to H.E. Rondeel who carefully managed the financial side of this volume. F.B.J. Barends, H. de Boorder, S. Flint, R. Hillen, G.A.M. Kruse, P.M. Maurenbrecher, J. Oerlemans, O. van de Plassche, I. Shennan, D.J. Stewart, B.B.W. Thorborg and J.J. de Vries assisted with the editing of manuscripts.

Publication of these Proceedings has been made possible through contributions of the following sponsors: The Netherland's Ministry of Economic Affairs, the Ministry of Foreign Affairs and the Ministry of Transport and Public Works, The Royal Netherlands Academy of Sciences, The Royal Geological and Mining Society of the Netherlands (KNGMG), AKZO Zout Chemie and De Nederlandse Olie en Gas Exploratie en Productie Associatie (NOGEPA).

Utrecht, Summer 1988
The Editors

Preface

Contents

Geologie en Mijnbouw 68: 1–24 (1989)

KEYNOTE ADDRESS

Deltaic coastal wetlands

James M. Coleman & H.H. Roberts
School of Geoscience, Coastal Studies Institute, Louisiana State University, Bâton Rouge, Louisiana, U.S.A.

Received 15 October 1987; accepted in revised form 29 January 1988

Key words: Deltas, coastal wetlands, landloss, subsidence, sealevel

Abstract

Modern-day deltas exist in a wide variety of settings. Despite the various environmental contrasts, all actively prograding deltas have at least one common attribute: a river supplies clastic sediment to the coast and inner shelf more rapidly than it can be removed by marine processes. The most important processes controlling the geometry and landforms in deltas are climate, water and sediment discharge and its variability, river mouth processes, nearshore wave power, tides and tidal regime, nearshore currents, shelf slope, tectonics of the receiving basin, and receiving basin geometry.

Many present-day deltas are experiencing relatively large coastal landloss; this results from the complex interaction of many physical, chemical, and biological processes that operate in the natural environment and, in more recent times, the processes induced by man's utilization of this environment. All of these processes operate at different scales and magnitudes, in both time and space; some are amenable to manipulation by man, while others are essentially out of his control. Natural processes include sea level changes, subsidence and compaction, changes in deltaic sites of deposition, catastrophic events such as hurricanes, and biologically-induced factors. Man-induced factors include dams and levees, canal dredging, and fluid withdrawal.

Introduction

Since ancient times, river delta lowlands have been of fundamental importance to civilization. Owing to their early significance as agricultural lands, deltas received considerable attention from scholars such as Homer, Herodotus, Plato, and Aristotle. The term delta was first applied by the Greek historian Herodotus, approximately 450 B.C., to the triangular alluvial deposits at the mouth of the Nile River. In broader terms, deltas can be defined as those deposits, both subaerial and subaqueous, derived from riverborne sediments and dispersed by distributary channels. Because the different processes which control delta development vary considerably in relative intensity on a global scale, delta plain landforms span nearly the entire spectrum of coastal features and include distributary channels, river mouth bars, interdistributary bays, tidal flats, tidal ridges, beaches, dunes, dune fields, swamps, marshes, and evaporite flats.

River systems have been in existence throughout geologic times; the only major prerequisites are a partially elevated land mass, a depositional basin, rainfall, and chemical and physical degradation processes. River size and overall morphologies, however, have varied through time and are dependent on tectonic episodes, size of continents, basinal tectonics, climate, severity of weathering processes, sea level changes, and similar global processes. Today's modern river systems occur in a wide variety of geologic settings with associated environmental processes. A knowledge of these variations is helpful in defining present trends in coastal regions, as well as attempting to predict future trends in these important wetlands. Coleman (1976) showed, in a study of numerous modern worldwide deltas, that only a few major processes are responsible for the major variations seen in modern deltas. These processes are: climate, water and sediment discharge and its variability, sediment type, river mouth processes, nearshore wave power, tides and tidal regime, nearshore currents, shelf slope, and tectonics and geometry of receiving basin. This paper is a review of the variations displayed by modern day river systems and a discussion of the processes responsible for landloss as illustrated by the Mississippi River coastal wetlands.

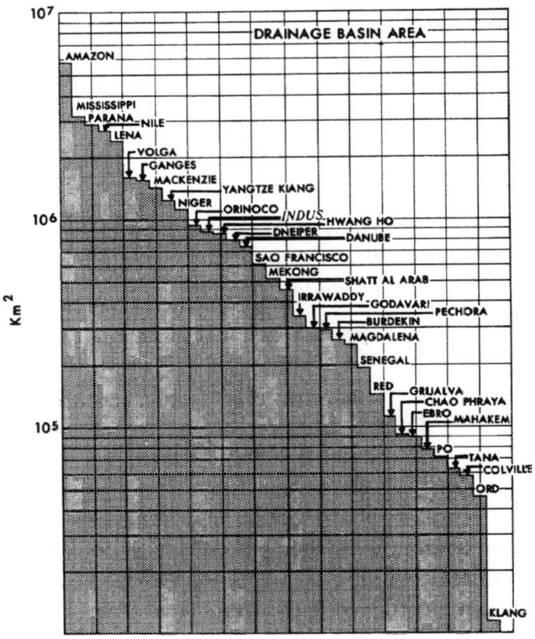

Fig. 1. Drainage basin areas of selected major river systems.

Delta attributes

Previous research has shown that deltaic facies associations are a function of numerous process variables. Attempts to incorporate some or all of these process variables into models for discriminating delta types have resulted in at least three classification themes. Fisher et al. (1969) proposed high constructive and high destructive delta types based on relative intensity of fluvial and marine processes. Coleman & Wright (1971) and Wright et al. (1974), using a broad range of parameters, quantified the process variables, then used statistical techniques to cluster deltas into discrete groupings. More recently, Elliott (1978) proposed a classification scheme based on the earlier work of Galloway (1975) wherein deltas were plotted on a ternary diagram to define general fields of fluvial, wave, and tide dominance. The most significant aspect of these studies is the recognition of the role of physical processes in producing specific and predictable responses.

Examination of a few major attributes of modern world river systems indicates that although a large number of variations exists, there are generalized trends and most exceptions can be logically explained. Figure 1 shows the drainage basin area of some 34 major world river systems. Note that these river systems span the climatic zones and represent deposition in a wide variety of depositional settings. Today's basin drainage areas span nearly three orders of magnitude, from less than 1×10^4 km^2 to greater than 4×10^6 km^2. Figure 2 illustrates the delta plain area of modern world deltas and shows a variation of approximately three orders of magnitude. Plotting only these two parameters gives the result shown in Figure 3, that is, as the drainage basin area increases, so does the delta plain area. However, there is a very wide variation in delta size for any given size of a drainage basin. Plotting of any two parameters shows similar results, general trends, but wide variation within those trends, illustrating that deltaic facies display variability because of numerous interacting parameters. For example, the San Francisco delta of Brazil is relatively small for the size of its drainage basin; this delta is characterized by extremely high wave action, and most of the fine-grained

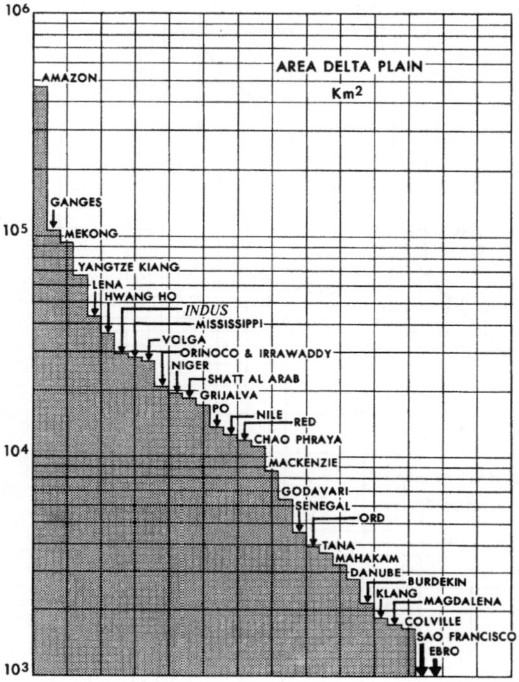

Fig. 2. Delta plain areas of selected major river systems (includes the subaqueous delta).

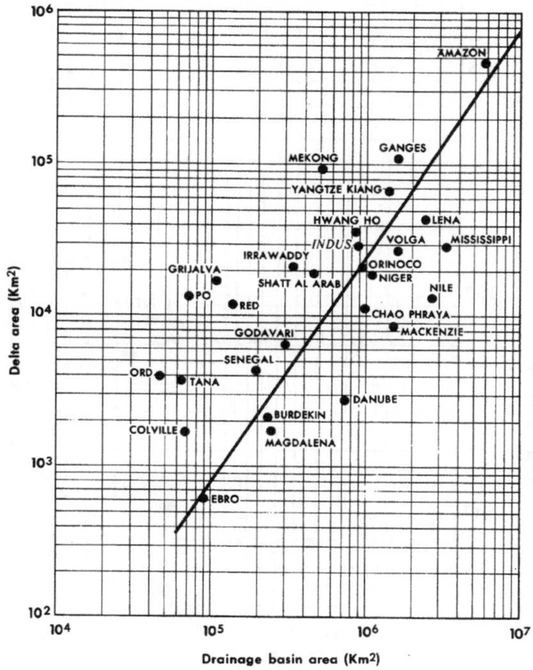

Fig. 3. Plot of drainage basin area against delta plain area. Sloping line is line of best fit.

sediment delivered to the basin is advected seaward by wave action and marine currents, while sands are concentrated at the shoreline as well as transported landward by eolian transgressive processes. In contrast, the Mekong delta of Vietnam is relatively large for the size of its drainage basin. The delta is rather stable (little subsidence) and is significantly influenced by tidal processes, which tend to laterally spread the deltaic facies associations.

Figure 4 illustrates river system discharge (m^3/sec) for several modern world deltas. Once again, there is nearly a three-fold magnitude in discharge among the rivers analyzed. Plotting discharge against delta area (Fig. 5) indicates that as discharge increases, delta area increases. Variations exist, but in the larger discharge river systems, this variation becomes minimal; it is the smaller discharge rivers which tend to show the highest variation primarily because of sediment load and sediment characteristics.

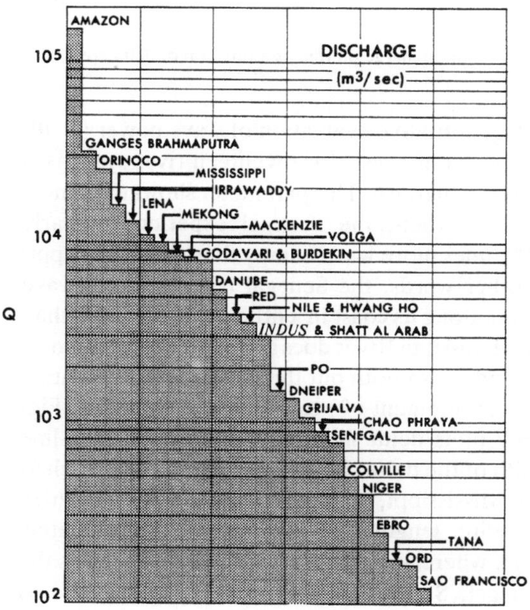

Fig. 4. Fluid discharge of selected major river systems.

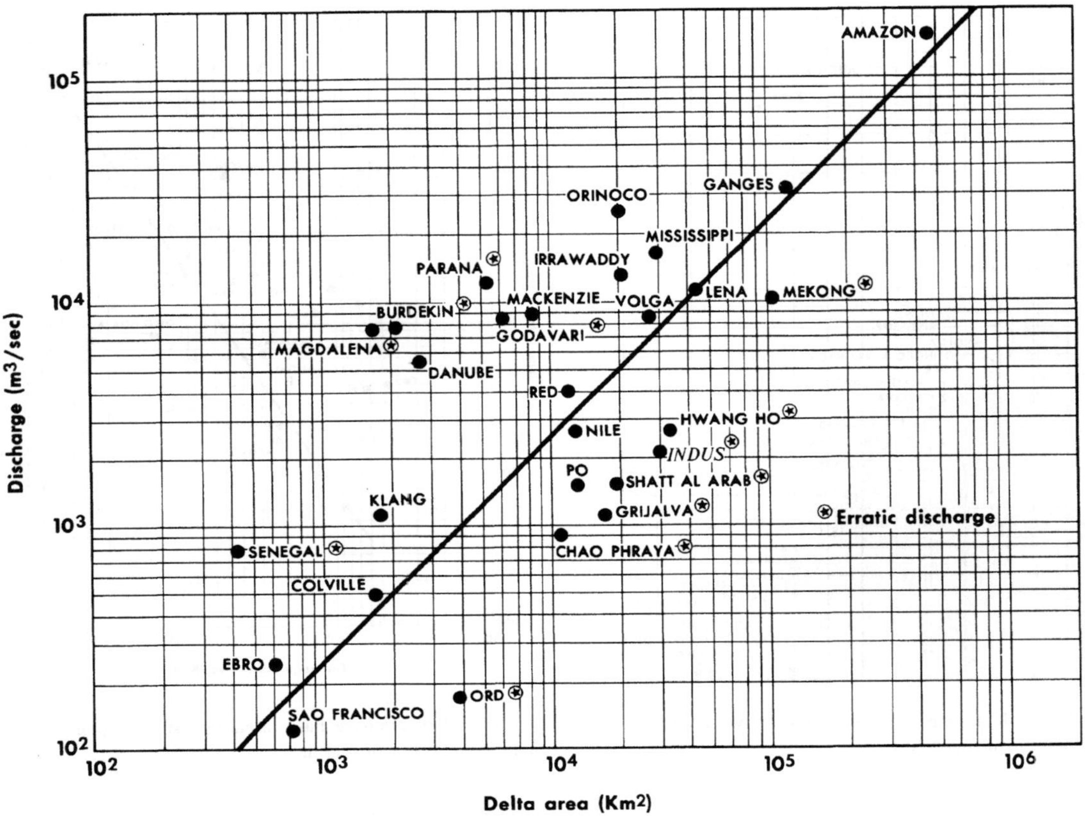

Fig. 5. Fluid discharge plotted as a function of delta area.

A quantitative evaluation of wave power ($\times 10^7$ ergs/sec/m of coast) for seventeen river systems is shown in Figure 6. This parameter shows extremely wide variation; the Senegal experiences nearly 1000 times more wave energy than the Mississippi. In other words, the Senegal receives more wave power along its coast in a little over two hours than the Mississippi River does all year. Such wave energy tends to smooth out the delta coast, preventing the development of protruding river mouths. Figure 7 plots delta shoreline length to straight-line width of the delta. Low wave energy deltas such as the Mississippi, Ganges, and Volga have high shoreline length to width ratios (approximately 4 : 1), whereas high wave energy deltas such as the Senegal, San Francisco, and Magdalena tend to show low ratios, generally 2 : 1 or less (Fig. 7). Although wave energy is highly dependent on the marine climate, one of the major controlling factors is the slope of the continental shelf fronting the delta. Figure 8 plots continental shelf angle against wave power. Those deltas having extremely low offshore slopes display relatively low shoreline wave energy, whereas those deltas displaying higher offshore slopes generally show much higher wave energy.

Tidal processes control the spatial relationships and geometries of deltaic facies. Three important characteristics of tidally dominated rivers can be identified: a) water-mass mixing by tidal activity destroys vertical density stratification, so that effects of buoyancy at river mouths are negligible; b) for part of the year tides account for the highest percentage of the sediment-transport energy, and

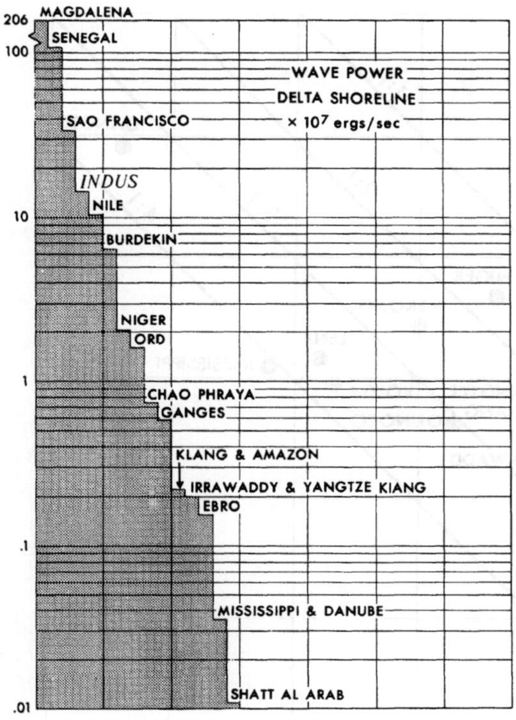

Fig. 6. Wave power ($\times 10^7$ ergs/sec) of selected delta shorelines.

flow both in and seaward of the river mouths is subjected to reversals over a tidal cycle; and c) the zone of marine-riverine interactions is greatly extended both vertically and horizontally. These effects result in widely differing geometries for the sand bodies that develop at the river mouth. Tidal processes are difficult to quantify, but Fig. 9 shows the average tidal range (in m) for 27 river deltas. The morphology of a low tide river delta such as the Nile or the Mississippi would be drastically altered in a short period of time if it was subjected to tidal inundation of nearly 6 m, as in the case of the Ord River delta of Australia.

Other factors are just as important as those discussed above, but many are very difficult to quantify. For example, the rate of subsidence controls the thickness of individual sand bodies and the stacking of the deltaic facies through time. However, this factor is virtually impossible to quantify because data on relative subsidence rates in modern world deltas do not exist. A more lengthy discussion of other factors that control deltaic facies development can be found in Coleman (1976), Coleman & Wright (1971), and Wright et al. (1974).

An alarming aspect of present-day deltaic lowlands is the rapid loss of subaerial wetlands. This is occurring on a worldwide basis; nearly all coastal deltaic plains have suffered extensive landloss within the past several centuries. The reasons are complex, yet an understanding of the major processes responsible for this loss is critical if mitigation measures are to be successful in slowing down this trend. Landloss is especially rapid in coastal Louisiana, the site of one of the world's major river deltas, the Mississippi River. In the third and fourth decades of this century, Russell (1936) and Fisk (1944) reported that Louisiana was losing its wetlands and that the state's coastal marshes were rapidly changing composition. Since that time, a considerable amount of research has been conducted to document and explain this wetland loss.

Wetland loss in the Mississippi River delta plain

The Mississippi River, the largest river system in North America, drains an area of 3,344,560 km^2 (Coleman, 1976) and has formed the largest coastal wetlands in North America. When De Soto found and named the Rio del Esperitu Santo, now the Mississippi River, in 1543, the Indians had been living in and utilizing its coastal zone for nearly 12,000 years (Gosselink, 1984). By the late 1800's industrial development had begun in the wetlands, and the construction of levees along the river accelerated this trend. The discovery of petroleum resulted in dredging of canals through the coastal wetlands for access to drilling sites. Geological and biological investigations of the delta began in the late 1800's (Lerch et al. 1892), but the most important studies of geomorphology and geology were concentrated in the middle to late 1900's. Significant studies include those by Russell (1936), Fisk (1944, 1952), and Kolb & Van Lopik (1958). Articles dealing with marsh ecology were published by Hathaway & Penfound (1936), Penfound & Hatha-

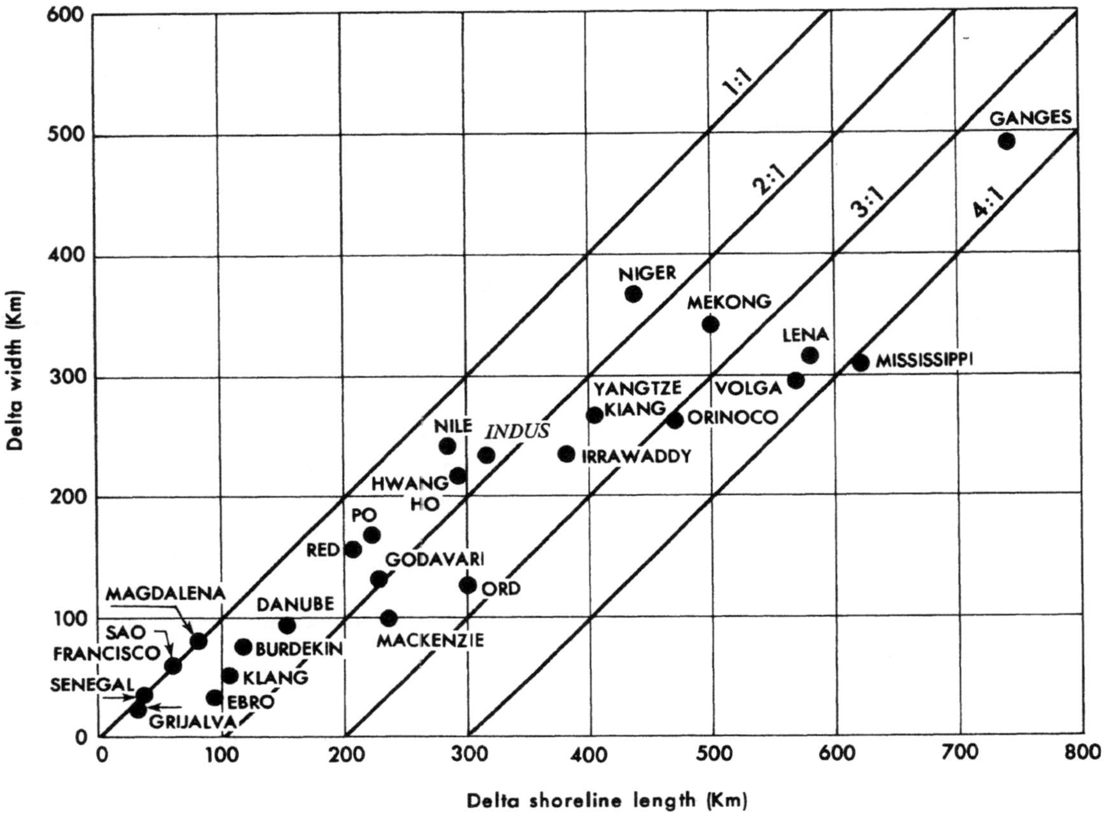

Fig. 7. Plot of delta shoreline length to straightline of delta width. Several ratio lines are shown for reference.

way (1938), and Brown (1944). More recent articles have extended the details of these early workers, and the advent of the Coastal Studies Institute and Sea Grant Program at Louisiana State University has expanded this field of study considerably.

The coastal area of Louisiana is a unique and valuable economic and environmental resource to the state and its citizens. Although early workers, especially Russell (1936) and Fisk (1944), called attention to the fact that Louisiana was losing its wetlands and that the coastal marshes were undergoing rapid change, little attention was paid to these predictions. Research by Gagliano & Van Beek (1970) and Gagliano et al. (1981) focused on this problem, and within a few years both the public and government agencies were acutely aware of the magnitude of the problem. Documented wetland loss rates averaging 0.8% per year (Gagliano et al., 1981; Turner et al., 1982) have caused major concerns about the future of the coastal parishes, state boundaries and hence oil and gas revenues, and renewable resources such as shrimp, crabs, and menhaden. Figure 10 shows the magnitude of this loss across Louisiana's wetlands. In some areas, especially the modern Mississippi River delta, the rates are exceptionally high; Figure 11 shows the accelerating rate of land loss in the Mississippi deltaic plain. Although various investigators differ as to the causes for rates of land loss, the numerous studies all indicate that the wetlands of Louisiana are being lost and undergoing rapid changes.

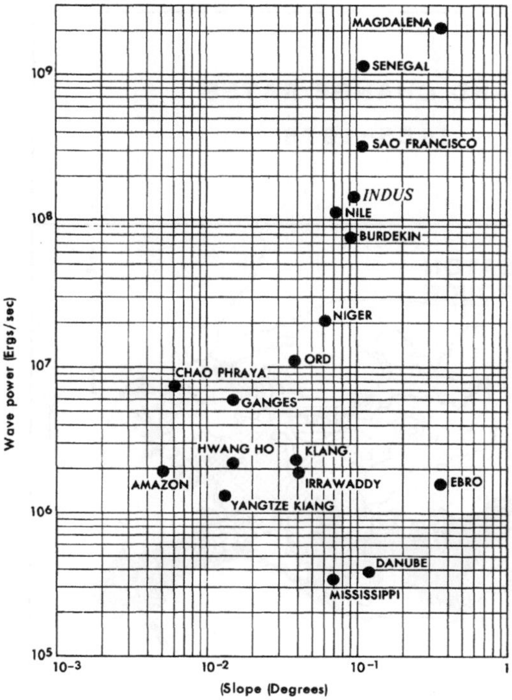

Fig. 8. Plot of wave power (ergs/sec) against offshore slope (degrees).

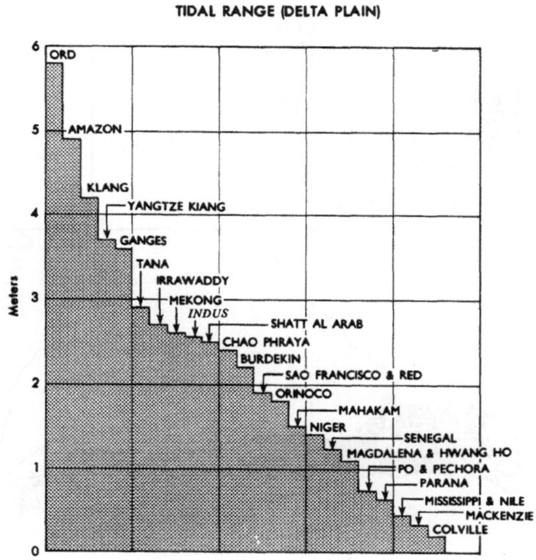

Fig. 9. Average tidal range of selected river deltas.

Causes of Louisiana's wetland loss

Rapid degradation of the wetlands is cause for considerable alarm. Wetland loss results from the complex interaction of many physical, chemical, and biological processes that operate in the natural environment and, in more recent times, the processes induced by man's utilization of this environment. All of these processes operate at different scales, in both time and space; some are amenable to manipulation by man, while others are essentially out of his control. In order to better understand some of these processes, it is helpful to view the development of the Gulf Coastal Plain over a long period of geological time so that present conditions can be placed in proper perspective.

The Mississippi River has had pronounced influence on the development of the northern Gulf of Mexico throughout a long period of geological time. Since the beginning of the Tertiary Period (some 65 million years ago), thermal cooling and the delivery of large volumes of sediment brought down by coastal rivers, especially the ancestral Mississippi Rivers, have created a major subsiding sedimentary basin, the Gulf Coast Geosyncline. Many of the ancient subsurface sedimentary sequences were laid down in localized depocentres, the prolific hydrocarbon-producing horizons that have formed the basis for Louisiana's recent economy. Geologists have documented that throughout this long period of geological time, there were major changes in the position of the shoreline and in the presence of large, extensive coastal plains that have been developed and subsequently lost by coastal inundation. Overall, however, the coastal plain has experienced a net gain in sediments, and the long-term pattern has been shoreline progradation and continued buildout of coastal environments. Causes of the 'cyclic' patterns of coastal retreat and loss are complex; they result from such processes as changes in the sediment yield, climate, sea level (both eustatic and subsidence), and patterns or sites of sediment deposition. In the Quaternary, changing sea levels associated with the

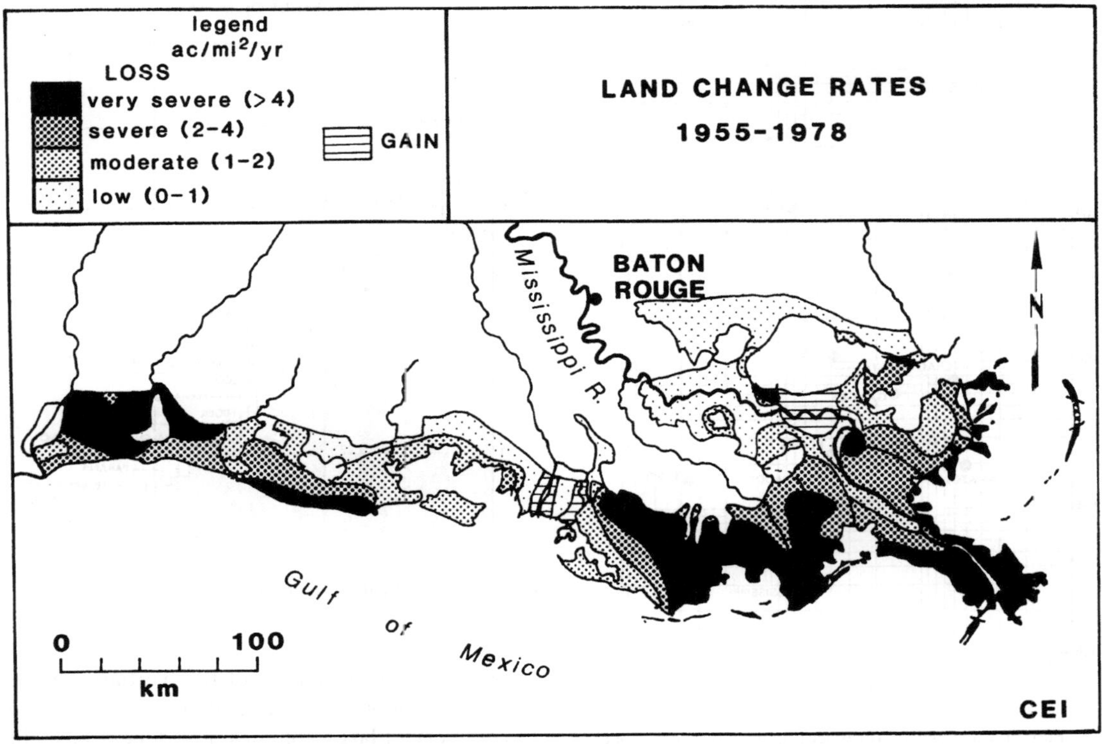

Fig. 10. Rates of land loss in the coastal wetlands of Louisiana (after Van Beek & Meyer-Arendt, 1982).

advance and retreat of inland glaciers have strongly influenced the near-surface sedimentary patterns of coastal wetland development. Numerous times during this period, extensive coastal advances and

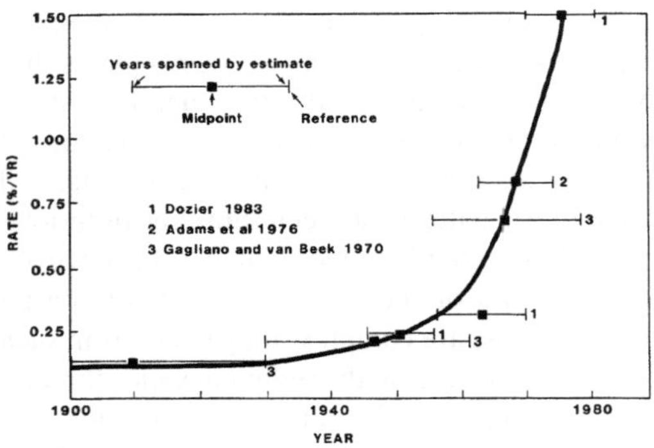

Fig. 11. Accelerated rate of land loss from 1900 to late 1980's. After Dozier, 1983.

retreats have taken place. Freshwater marsh deposits (representing older wetlands) have been documented from cores taken in offshore Louisiana and Texas in water depths of 200 m and at distances several hundred kilometres from the present shoreline.

In order to understand the changes in the coastal marshes that comprise our present-day wetlands, it is necessary to examine briefly some of the major factors, both natural and man-induced, that contribute to wetland loss. The following sections describe some of these major processes; it should be mentioned that there are second- and third- order processes not included in this compilation. However, we consider the factors listed below as the most important in contributing to wetland loss:

I. Geological factors
   A. Sea level changes
   B. Subsidence and compaction

C. Changes in deltaic sites of deposition

II. Catastrophic factors: (hurricanes)

III. Biological factors

IV. Man-induced factors

A. Dams and levees

B. Canal dredging

C. Fluid withdrawal

Geological factors

Sea level changes

Controls on sea level. The total volume of water in the ocean basins is believed to have remained fairly constant throughout the earth's evolution. The interplay of such processes as plate tectonics and climate has produced a variable and sometimes erratic record of sea level changes throughout geological time. Plate tectonics and climate control sea level position on a worldwide or 'eustatic' scale, whereas the regional influences of geology, climate, and hydrology interact to affect sea level on a local scale. Commonly, the local processes can override the global trend in sea level, resulting in regional sea level 'highs' or 'lows'.

Vail et al. (1977) have derived a global sea level curve showing relative high and low stands from the Precambrian (575 M.a. B.P.) to the present. Vail et al. (1977) curves indicate that the average position of sea level during geological time was higher than present sea level. Response of the oceans to climatic changes is the most important factor influencing short-term sea level positions. Five major sequences of glacial advance occurred during the 2.1 million years of Pleistocene time.

The most recent of the glacial advances (17,000 years B.P.) depressed sea level approximately 110 m below its present stand (Fig. 12; Nummedal, 1983). The subsequent rise in sea level following glacial retreat 15,000 years B.P. has been termed the Holocene transgression. Although the Holocene transgression is depicted as a smoothly increasing rise in sea level, Brooks et al. (1979) have shown that on a local scale, the rate of rise can be highly erratic (Fig. 13). Worldwide climate and local tectonic changes are probably responsible for the irregular Holocene sea level curve.

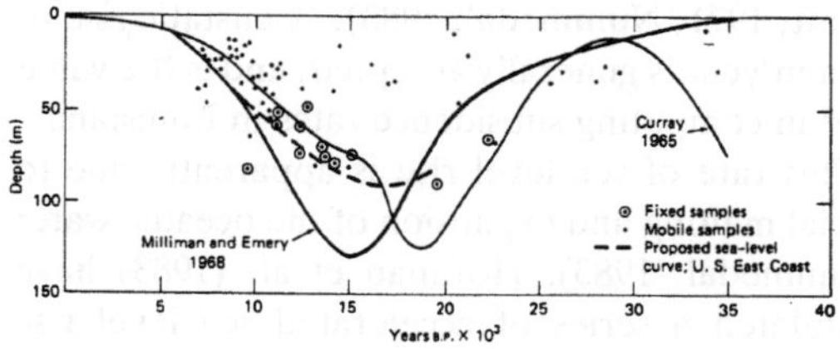

Fig. 12. Sea level curves for the Late Quaternary from sites on the east coast of the U.S. From Nummedal, 1983.

Coastal Louisiana is highly vulnerable to short-term changes in water level caused by hurricanes, cold-front passage, and flood waters. Increases in sea level produced by these processes may range from a few centimetres to several metres, and from a few hours to weeks in duration.

Recent sea level. Numerous attempts have been made to quantify the present rate of sea level rise, but, owing to the highly variable regional controls on sea level and the inability to acquire a reliable and representative data base, estimates of the eu-

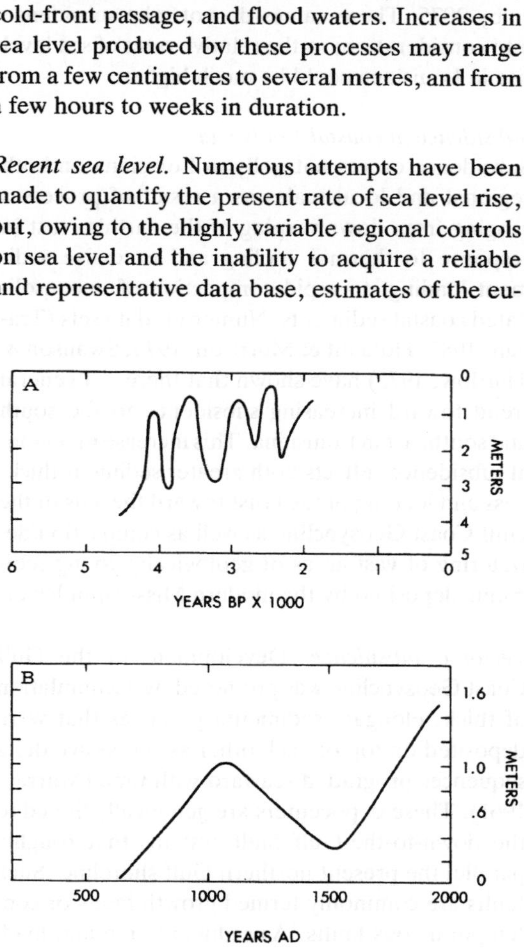

Fig. 13. Sea level fluctuations on (A) the South Carolina coast over the past 4000 years and (B) on the North Sea Coast of Germany since the year 650 AD. From Nummedal, 1983.

static sea level rise range from 1.2 to 3.0 mm/year (Kraft, 1971; Nummedal, 1983). A eustatic rise of 1.2 mm/year is generally accepted, and is the value used in computing subsidence rates in Louisiana.

This rate of sea level rise is apparently due to glacial melting and expansion of the oceanic water (Nummedal, 1983). Hoffman et al. (1983) have postulated a series of accelerated sea level rise scenarios based on climatic warming trends and projections of the greenhouse effect. Values for the medium sea level rise scenario outlined by the Environmental Protection Agency (Hoffman et al., 1983) suggest that sea level will be rising 3.3 mm/yr by the year 2025; 6.6 mm/yr by 2050; and 11.4 mm/yr by 2075. These values do not take subsidence into consideration; so the relative rates of sea level rise in Louisiana could be much higher.

Subsidence in coastal Louisiana

Subsidence occurs naturally in Louisiana on both regional and local scales as a result of processes ranging from downwarping of the earth's crust in response to thermal cooling and excessive sediment loading, to rapid compaction of unconsolidated coastal sediments. Numerous data sets (Trahan, 1982; Holdahl & Morrison, 1974; Swanson & Thurlow, 1973) have shown that there is a general trend toward increasing subsidence to the south and southeast in Louisiana. This increase in regional subsidence reflects both greater sediment thickness and loading of the crust toward the axis of the Gulf Coast Geosyncline as well as compaction/dewatering of vast areas of geologically young sediments deposited by the modern Mississippi River.

Tectonic subsidence. Development of the Gulf Coast Geosyncline was promoted by accumulation of thick, elongate sedimentary masses that were deposited on top of each other as successive delta sequences prograded seaward with time (Murray, 1961). These depocenters are genetically linked to the down-to-the-Gulf fault systems that roughly parallel the present northern Gulf shoreline. Such faults are commonly termed growth faults or contemporaneous faults. As sedimentation and loading continue, many of these faults remain active and thus add to the regional subsidence in Louisiana. Accurate rates of movement along growth faults over short times is not presently known, but offshore seismic data indicate that many of them are experiencing movement today. Much of the regional subsidence can be associated with fault compensation and with deformation of sediments under loading. However, lateral and vertical flowage of thick salt beds (Worzel & Burk, 1979) that underlie deposits of the ancestral Mississippi River, as well as the modern delta, also adds to the regional subsidence. Again, little quantitative data exist on the amount of subsidence that is attributable to salt withdrawal or salt solution at depth. Both crustal downwarping and salt mobilization are long-term components of regional subsidence.

Sediment loading and compaction. Shorter term processes that certainly add to sinking of the land, resulting in wetland loss, involve localized sediment loading, dewatering, and physical/chemical compaction of recently deposited sediments (younger than 6000 years) of the coastal plain. The dominantly fine grained and highly organic sediments of Louisiana's coastal plain are subjected to three processes that add to subsidence immediately after deposition (Terzaghi, 1943):

1. Primary consolidation – a reduction in the volume of the soil mass owing to dewatering under a sustained load. The load is transferred from the interstitial water to the soil particles.
2. Secondary compression – a decrease in soil volume associated with the rearrangement of constituent particles.
3. Oxidation of organic matter – reduction of soil volume as chemical reactions occur that cause organic matter to decompose into its mineral constituents.

These processes are fundamental properties of all sediment deposition. However, in areas where sedimentation rates are high, where the sediments contain high amounts of water, and where organic content is high, these processes are extremely active and contribute significantly to land loss. When viewed in a short period of time, for example the last 5000 years, it is apparent that sedimentation and accompanying compactional processes are not uniformly distributed across the coastal plain and

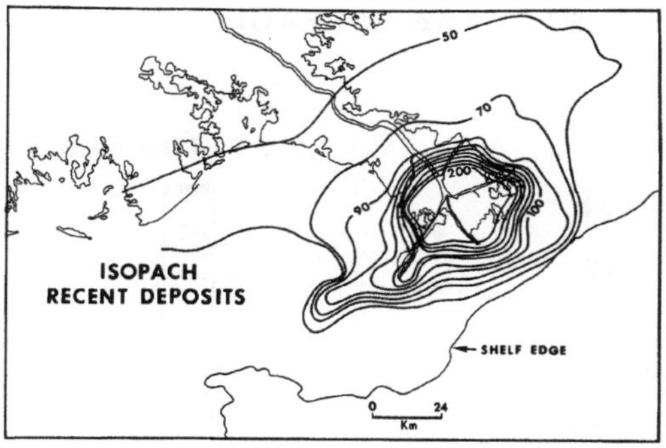

Fig. 14. Isopach of 'Recent' sediments in the modern Mississippi River delta. Contours in metres.

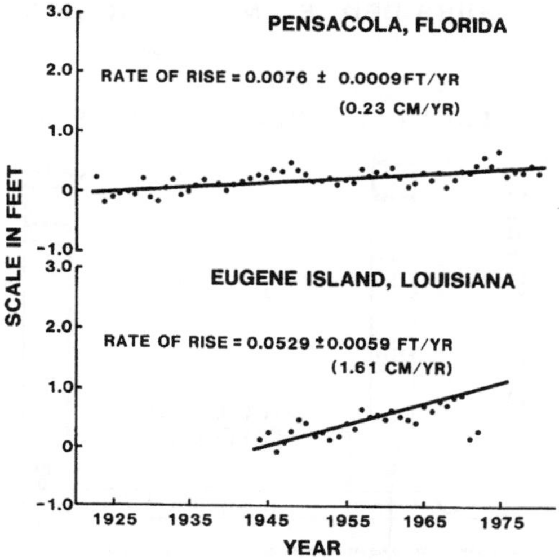

Fig. 15. Water level gauge records from Florida and Louisiana. Data courtesy of Louisiana Geological Survey.

shallow continental shelf. Switching of the site of deposition is the rule rather than the exception in coastal Louisiana. Although each depositional event (delta lobe) takes only about 1000 to 1500 years to complete (Kolb & Van Lopik, 1958; Frazier, 1967), Figure 14 illustrates that sediments in a single delta lobe can be over 100 m thick. A column of dominantly fine grained sediment of this magnitude, deposited in less than 1000 years, suggests that normal processes of compaction have not had time to consolidate the sediments, as would be the case under less rapid sedimentation conditions. Therefore, compaction in areas of thick deposits can be expected to be greater than in areas where sedimentation is slow and recent deposits are thin.

On a regional scale, this point of view can be supported by comparing long-term water level records. In dynamic areas of sedimentation such as the Mississippi delta, compaction causes the mean water level to increase at the gauge site relative to the rate of sea level rise caused by eustatic processes. Thus in areas experiencing high compaction and subsidence, water level rise over even short periods of time are significantly higher than in areas having less subsidence and compaction. Figure 15 compares a water level gauge record from the central coast of Louisiana with one from a much more stable area in western Florida. The major differences in the rate of water level rise can be attributed to subsidence and compaction of deltaic sediments associated with the Louisiana site. This rate of water level rise, 1.61 cm/yr, includes eustatic sea level rise. If eustatic sea level rise is subtracted, the rate of compaction and subsidence at the Louisiana site is 1.59 cm/yr. This figure is significant when compared to the vertical accretion rate of the marsh (Table 1).

On a local scale, the thickness of recently deposited sediments over a more consolidated Pleistocene surface can make a considerable difference in subsidence and compaction rates. Figure 16 illustrates subsidence data calculated from three soil borings across the central Louisiana coastal plain. The borings are located so that they cross the old

Table 1. Accretion Rates in Louisiana Coastal Marshes.

Marsh Type	Accretion Rate (cm/yr)	
	Mean	Range
Fresh – streamside	1.06	–
Fresh – inland	0.65	0.31–0.69
Intermediate – streamside	1.35	1.30–1.40
Intermediate – inland	0.64	0.38–1.06
Brackish – streamside	1.40	1.06–1.69
Brackish – inland	0.59	0.38–0.81
Salt – streamside	1.35	–
Salt – inland	0.75	0.56–0.94

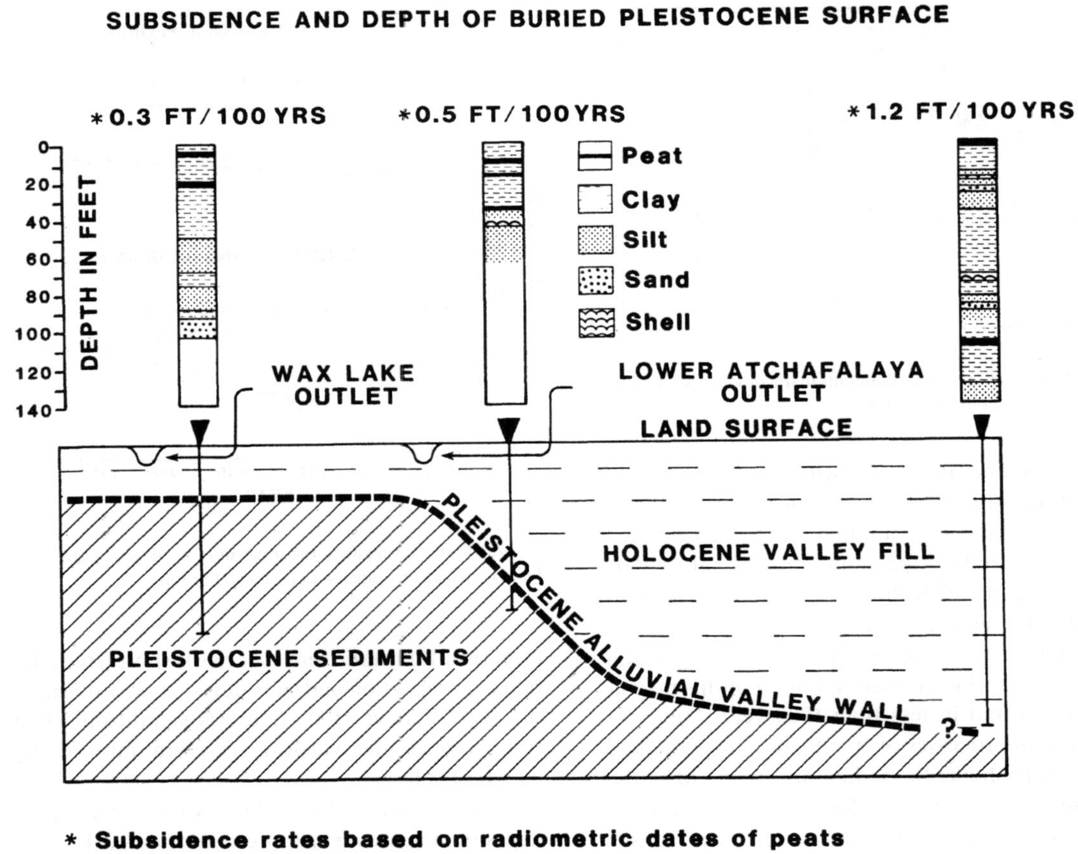

Fig. 16. Subsidence rates across the central Louisiana coastal plain. After Roberts, 1985.

Pleistocene alluvial valley wall, which was cut when sea level was lower than it is today. The combined effects of a relatively compacted Pleistocene surface and a thickening Recent sediment fill (Holocene Valley Fill) are seen in the subsidence rates calculated from radiometric dating of sediments from the cores. Subsidence in the area of thick Recent sediment fill is about four times as great as in the area of thin Recent sediment fill over the Pleistocene surface. This trend emphasizes the importance of local compactional processes and thickness of young deposits on subsidence, and hence land loss. Unfortunately, data are not available from many areas in the Louisiana coastal marshes. However, it is certain that thickness of the Holocene deposits varies considerably, and thus subsidence and compaction must also vary accordingly.

Changes in deltaic sites of deposition

Quaternary deltas. Sea level changes during the Pleistocene caused alternate entrenching and infilling of coastal river systems, especially the Mississippi River. When sea level was low, the river entrenched; and during periods of high sea level, the entrenched valleys were infilled. This infilliing process results in deposition of large volumes of sediment in the river valleys, thus depriving the coast of the low sea level sediment yield that forms coastal marshes. We are now in a high sea level

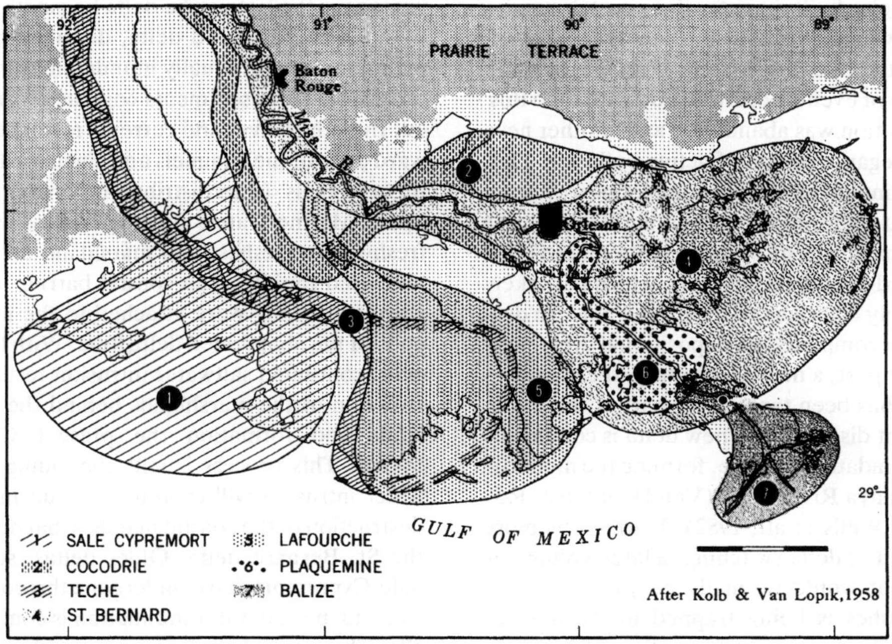

Fig. 17. Major delta lobes that have constructed the Holocene Mississippi River delta plain. Note the location of the most recent lobe in the Mississippi River delta complex, the Atchafalaya delta (A). Modified from Kolb & Van Lopik, 1958.

stage, and the present coast has been deprived of a large volume of sediment during the past 10 to 15,000 years. In addition to river entrenchment and infilling, changes in sea level also control the sites of deltaic deposition that form the broad coastal marshes. During falling sea level, the sites of delta deposition shift seaward, and deltaic sediments are deposited well out on the continental shelf. As sea level rises, the delta sites shift landward and coastal marshes suffer inundation. Seismic data and borings collected in offshore Louisiana allow depiction of the changing sites of delta deposition during the past 25,000 years. The presence of numerous delta lobes now buried beneath continental shelf deposits points out the role that sea level and subsidence play in controlling the total area of coastal marshes. From the last low sea level stand to the present, it is estimated that Louisiana's coastal marshes have decreased in area by approximately 40–50%. If submergence of the coast had not occurred along the Louisiana shoreline, many of these older deltaic lobes would still be present and the wetlands would be much more extensive.

Holocene delta sequences. The latest phase of the Quaternary cycle, characterized by relative stability of climates and relatively small changes in sea level, began approximately 5000 to 6000 years ago. This sequence involves the modern delta cycles described by Fisk & McFarlan (1955), Kolb & Van Lopik (1958), and Frazier (1967). Figure 17 illustrates the major Mississippi delta lobes that have developed during this period. The result of the building and subsequent abandonment of the Late Recent delta lobes was the construction of a modern deltaic coastal plain, which has a total area of 28,568 km^2, of which now only 23,900 km^2 is exposed (Coleman, 1976).

One of the earliest Mississippi deltas, the Sale-Cypremort, constructed a progradational deltaic lobe along the western flanks of the present Mississippi River deltaic plain. In approximately 1200 years, this delta built an extensive coastal marshland, then switched its course to another locus of deposition, the Cocodrie system (Fig. 17). A similar sequence of events occurred, and with time this site of deposition was abandoned and another new delta lobe began a period of active buildout. This process continued, each delta completing a cycle of buildout that required approximately 1000 to 1500 years. The most recent of these delta lobes is the Balize or bird-foot delta (Fig. 17), which has taken approximately 800 years to form. The modern delta has nearly completed its depositional cycle, and in the recent past, a new distributary, the Atchafalaya River, has been tapping off part of the water and sediment discharge. A new delta is commencing the progradational phase, forming the modern-day Atchafalaya River delta (Van Heerden & Roberts, 1980; Wells et al., 1982). During this most recent period of delta switching, a large volume of sediment that would normally be yielded to the coastal marshes is being trapped in the alluvial valley of this new river course, the Atchafalaya Basin. The consequence is that sediment which normally would nourish the coastal marshes is no longer reaching the coast, and these conditions severely impact the rate of coastal marsh loss.

Each deltaic progradational cycle, in which broad coastal marshes are formed, has been referred to as a constructional phase (Scruton, 1960). However, once the river begins to abandon its major site of deposition, the consolidated mass of deltaic sediments is immediately subjected to marine reworking and subsidence. Marine processes, such as waves and coastal currents, and subsidence result in inundation of the coastal deltaic marshes, and within a few thousand years the delta lobe has been transgressed by marine waters. Scruton (1960) referred to this stage of the delta cycle as the destructional phase. Thus over a relatively short period of geological time, land gain and land loss are a function of the stage of the delta cycle.

Reference to a map of the Mississippi delta lobes (Fig. 17) illustrates this point. The older St. Bernard delta lobe was actively prograding some 3500 years ago and remained active for approximately 1500 years, forming a broad coastal marshland along the eastern delta plain. Approximately 2000 years ago, the Lafourche delta began its progradation. The St. Bernard delta, deprived of its sediment load, was gradually inundated by marine waters as the delta sediments compacted and subsided. Initially, waves and currents reworked the seaward ends of the channels and small areas between the channels (interdistributary areas) began to open. Marine waters intruded into the formerly freshwater marshes, and marshland deterioration increased rapidly. The coastal barriers caused by wave reworking were attached to the more distal ends of the abandoned delta lobe. With time, however, continued saltwater intrusion resulted in destruction of the marshlands behind the barrier islands and formation the present Chandeleur Sound. This process is still continuing, and saltwater intrusion will eventually result in complete destruction of the coastal marshes that once capped the St. Bernard delta. Older deltas, such as the Sale-Cypremort, have undergone this complete cycle, and marine water inundation is complete (Fig. 17). Thus, to a large extent, marshland gain and loss are a function of stage in the deltaic cycle.

This cycle of marshland gain and loss caused by switching deltas was recognized by such early workers as R.J. Russell and H.N. Fisk, but was not condensed into a simple, understandable model. Later workers, especially Penland & Boyd (1981), formulated this concept into a relatively simple model (Fig. 18).

Bay fill cycles. The modern Balize delta has been formed in the past 800 years; and, because of its relatively young age, it offers an opportunity to evaluate the short-term processes responsible for delta building and deterioration. One of the major environments associated with this delta is the large bay fill or crevasse system that breaks off the main channels and infills the adjacent interdistributary bays. These sequences form the major coastal marshes in the modern delta. Figure 19 illustrates the bay fill sequences that have formed within the modern delta during the past few hundred years. Of six bay fills, four have been dated historically, and much of their development can be traced by historical maps.

Each bay fill forms initially as a break in the

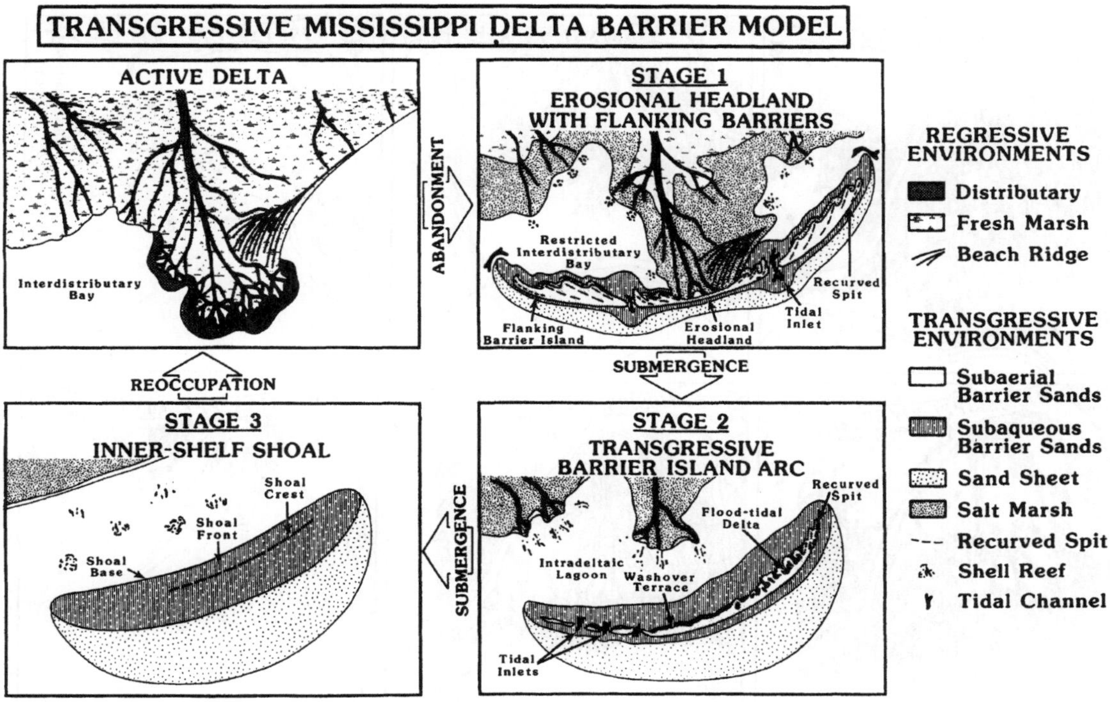

Fig. 18. Model of the evolution of a Mississippi River delta cycle. After Penland & Boyd, 1981.

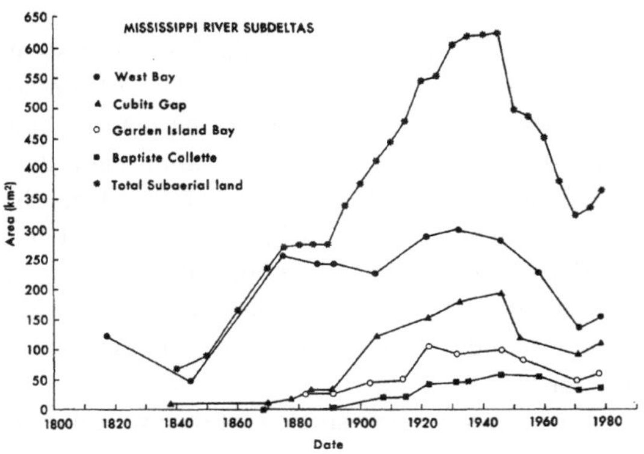

Fig. 19. Growth and deterioration of four bay fills within the Mississippi River delta. Growth rates range from 0.8 to 2.7 km^2/year; deterioration rates from 1.0 to 4.1 km^2/year. After Wells et al., 1982.

major distributary channel bank during flood stage, gradually increases in flow through successive floods, reaches a peak of maximum deposition, wanes, and becomes inactive (Coleman, 1976). As a result of compaction, the bay fill system is inundated by marine waters, reverting to an open bay environment, thus completing one cycle. The time framework varies as a function of many factors, but is generally within a 100 to 150 year period. Thus the bay fill cycle is very similar to the cycles of the major deltas, but develops in such a time framework that the details of the processes responsible for their formation can be adequately evaluated.

Historical maps of one of these bay fills, Cubits Gap, can be used to illustrate a cycle of subdelta building out and abandonment. Figure 20, which uses only a few of the numerous historic maps that are available for this area of the modern delta, shows the sequential development of the Cubits

Fig. 20. Historic development of the Cubits Gap bay as depicted from various maps. After Wells, et al., 1982.

Gap bay fill. The 1838 map was surveyed prior to the break and shows a narrow natural levee separating the Mississippi River from the shallow Bay Rondo. In 1862, a ditch that had been excavated by the daughters of Cubit, an oyster fisherman, was immediately responsible for a crevasse break. The original ditch was in the order of 120 m wide and extremely shallow. The flood of 1862 enlarged the opening, and by 1868 the width had increased to 740 m and the depth was in the order of several metres. By 1884, the map shows that initial buildout was begun by a complex series of distributary channels that had deposited relatively coarse sediment near the break. The map of 1905 shows that many of the major distributary channels had been developed and that rapid progradation had taken place in the eleven-year period since 1884. By 1922, a major part of the crevasse had been constructed; some small bays were already beginning to open, indicating that some parts of the bay fill system were being deprived of sediments. The 1946 map shows that sedimentation was taking place at the seaward ends of the distributary channels and that marshland loss was beginning to take place. In 1971, a large part of the crevasse system had been inundated by marine waters and marsh loss was significant. The only deposition was at the seawardmost ends of the distributaries and in the subaqueous delta.

Figure 21 shows on a single plot the cyclic nature of four of the historical Mississippi River bay fills; each cycle of buildout and growth was followed by deterioration. Projection of the present-day trends indicates that the cycle of each of the bay fills should be completed within the next 17 to 34 years (Wells et al., 1982). Thus the cycle of a bay fill can span a period of 115 to 175 years. Growth rates during the period of rapid buildout range from a

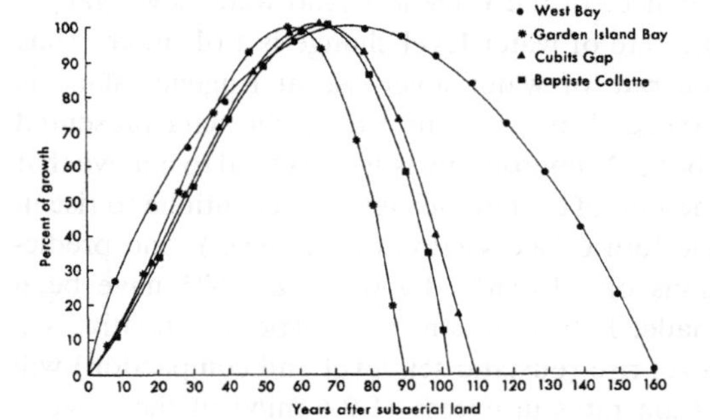

Fig. 21. Growth and deterioration curves of four bay fills from the Mississippi River delta as determined from historic map data.

low of 0.8 km^2 per year to a maximum of 2.7 km^2 per year. Deterioration rates range from 1.0 km^2 per year to a maximum of 4.1 km^2 per year.

This growth and deterioration cycle of the bay fills, although representing a relatively short period of time, is very similar to the delta cycle described earlier. The major point to be illustrated by these various cycles in delta growth is that at any one time, the state of the marshlands depends on the stage in the deltaic cycle. Because of the various ages of the delta lobes and the bay fills, wetland loss should vary considerably from site to site in the coastal wetlands.

Catastrophic factors: (hurricanes)

Hurricanes are tropical storms with sustained wind speeds of 120 km/hour. Those most likely to affect coastal Louisiana form in the Atlantic tropical latitudes, between 5° and 15° N, during August and September (Nummedal, 1982). Simpson & Lawrence (1971) have shown statistically that the Mississippi Delta and the upper Texas coast are the most likely areas for hurricane landfall. The severity of a hurricane's impact on the shoreline is controlled largely by the size, speed of movement, and path of the storm and by the slope of the continental shelf and orientation of the shoreline.

Circulation in a northern hemisphere hurricane is counterclockwise, and the most destructive forces of the hurricane are the winds and storm surge (Hayes, 1978). Storm surge is the term for superelevated water levels produced during a hurricane owing both to wind shear on the ocean and the very low barometric pressure. Nummedal et al. (1980) observed the effects of a 3.7-m storm surge at Gulf Shores, Alabama, produced during Hurricane Frederic. Hurricanes of this magnitude can easily submerge the low-lying coastal barriers. As a result of the cyclonic wind circulation, the combined erosive force of the wind and storm surge is most destructive along areas of the coast located to the right of the hurricane's eye. The dominant impact of hurricanes on coastal morphology is wind and wave erosion of the shoreline and coastal dunes. Elevated water levels expose a higher percentage of the shoreline to wave erosion, and facilitate wave overtopping of the barrier.

Penland & Boyd (1981) and Kahn & Roberts (1982) provide graphic examples of these processes. Dauphin Island, Alabama, is located to the right of the landfall point for Hurricane Frederic and was overtopped by a 3.6-m storm tide. Maximum seaward shoreline retreat was 40 m, but averaged 15 m along most of the barrier. Extensive erosion (25 m) occurred along the lagoonal shoreline and was attributed to increased current scour as water flowed across the barrier and into Mississippi Sound (Nummedal et al., 1980). Conversely, the Chandeleur Islands, off the southeast coast of Louisiana, are located to the left of the pathway of Frederic, and subsequently were exposed to a storm surge of only 1.3 m. These barriers were severely eroded (30-m retreat) and flattened. The Chandeleurs are a relatively sediment-starved chain of islands, and thus erosive events are generally more severe and leave longer lasting effects. Frederic reopened numerous channels across the barrier that had been formed by Hurricane Camille in 1969 (Nummedal et al., 1980; Kahn & Roberts, 1982). Much of the sand from the subaerial part of the beach was eroded and redeposited in washover lobes behind the island.

In addition to shoreline erosion, storm surges introduce large volumes of salt water onto the marsh surface. The long-term impact of this saltwater inundation is not precisely known. However,

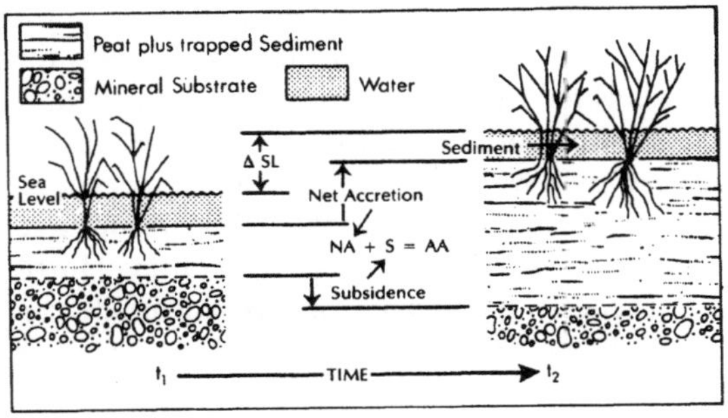

Fig. 22. Factors affecting marsh maintenance. Modified from DeLaune and Smith, 1985.

Gosselink (1984) indicates that hurricanes are major forces on Gulf coastal marshes, and initiate changes that can have significant consequences for years following the storm.

Biological factors

There have been many studies of the biological factors that are critical for sustaining marsh growth; an excellent review of some of these factors can be found in Gosselink (1984). They are extremely complex and cannot be examined in any detail in this discussion. The net result, however, can be evaluated by examining the vertical accumulation rates in the various types of marshlands in coastal Louisiana (Fig. 22). Rates of vertical accretion are known from Cesium 137 profiles and from marker horizons laid down on the surface and tracked through time. Accretion rates vary not only because of plant type, but also in relation to the proximity to streams and the shoreline. Hatton (1981) gives values for various types of marshes and for streamside and inland marshes. The figures from Hatton (1981) are summarized in Table 1.

These figures indicate that, on average, the vertical rate of accretion is about 1.4 cm/yr for streamside settings, while accretion in inland marshes is lower, about 0.75 cm/yr. When these figures are compared to water level changes (effects of subsidence and eustatic sea level rise), it is apparent that the present-day marshes are in extremely delicate balance. In the Eugene Island water level gauge, the rate of water level change is 1.61 cm/yr. Thus the rate of water level rise at Eugene Island is exceeded by only one site in the data presented above. Numerous investigators tend to believe that the rate of eustatic sea level will continue to rise in the future (see section on sea level), and predictions of 1.14 cm/yr by the year 2075 have been made. If this amount is realistic, the relative sea level rise (eustatic sea level and compaction) will attain rates in excess of 2.6 cm/yr at the Eugene Island site, a figure nearly double the average rate of marsh accretion. This projected rise has significant meaning in predicting the future of our wetlands. It also indicates that, even today, the coastal marshes can barely maintain their position relative to rising water levels. Compaction and sea level rise could therefore account for a high percentage of the present rate of marshland loss. Unfortunately, studies of water level changes do not include long-term data sets, and reliable data are not available for a variety of sets. In addition, spatial studies on vertical accumulation rates in coastal wetlands are too sparse to permit definitive statements on accretion rates. It should be pointed out, however, that even the meager data presented are alarming in that they show that, considering only eustatic sea level rise and compaction, these factors could account for an extremely high percentage of the present marshland loss.

Biological degradation. Biological factors that cause degradation of the wetlands are extremely complex, and only in the past few years have ecologists paid much attention to this factor as it affects marsh loss. It is readily accepted that overgrazing of the uplands has had significant impact on degradation of the grasslands and hence has led to major changes in sediment yield. In the coastal marshes, there have been fewer studies of organisms that could influence the rate of marsh decomposition. Gosselink (1984) summarizes some of the more recent studies. Invertebrates in the marshes are exceptionally common, yet little is known about their effects on marsh degradation. Larger consumers, such as snow geese, muskrats, and nutria, are probably responsible for greater marsh destruction than the invertebrates. Smith (1982) reported that, on the Atlantic coast, snow geese can reduce the plant cover by two-thirds in the area where they concentrate and virtually destroy the plants by digging up their roots. O'Neil (1949) indicated that dense concentrations of nutria and/or muskrats can 'eat out' a marsh area. All of these biological factors result in marsh degradation, but quantitative data are lacking on the extent to which these processes affect present-day rates of marshland loss.

Man-induced factors

The rate of marsh loss to open water has been accelerating over the past 50 years and, although the processes leading to this loss are complex and involve natural processes beyond human control, man has increasingly utilized the wetland habitats, and his activities have had some impact on the rate of wetland loss. Quantitatively, however, it is often just as difficult to evaluate these man-induced processes as it is to quantify the natural processes.

Dams and levees
Wetland loss in Louisiana's coastal plain during the past decade can be related in part to man's intervention in the natural system, on both local and regional scales. Modifications to the Mississippi River and its tributaries far removed from Louisiana can have an impact on the coastal environments. In excess of 3 million km^2, covering 31 states and 2 Canadian provinces, lie within the drainage basin of the Mississippi River (Coleman & Wright, 1971). Because of climate and soils in the drainage basin, the Mississippi River has, for long periods of geologic time, been a substantial sediment transporter. With the increasing impact of man in the drainage basin over the past century, the river's suspended sediment regime has been drastically changed. According to Keown et al. (1981), the major modifications to the system have been:

a) Deforestation of the drainage basins for agricultural reasons.
b) Building of the Old River Control Structure to regulate a 30% diversion of the Mississippi River water down the Atchafalaya-Red River systems.
c) Construction of sediment-retention structures (dams) and the placement of channel improvement features on the Missouri River and its tributaries (1953–1967).
d) Construction of sediment-retention structures (dams) and the placement of channel improvement features on the Arkansas River and its tributaries (1963–1970).
e) Construction of streambank protection works (levees) and sediment-retention structures on smaller, higher order streams in the drainage basin.

The summation of these effects has been to decrease the suspended sediment load of the river in recent years. For example, prior to 1963, the sediment load of the Mississippi River, plus the component diverted through the Atchafalaya River, was 434 million tons per year. The present value has declined to 255 million tons per year, a decrease of slightly over 40%. Measurements of bedload indicate a shift in grain size to finer fractions, suggesting the capture of coarser grain sizes by sediment-retention structures on the tributaries.

A decrease in both suspended and bedload sediments in the Mississippi River is highly relevant to the issue of land loss and subsidence in coastal Louisiana. Less sediment in transport by the river means that the increment of sediment added to surfaces in the lower delta during the annual flood will be less. Overbank sedimentation during floods is a process that helps build new land as well as offset the effect of subsidence. The construction of levees along most of the course of the Mississippi has had significant impact on the natural flood overtopping that took place prior to leveeing. Loss of sediment load and leveeing have probably had a significant influence on the rate of wetland loss, but, unfortunately, quantitative data are lacking. Because the sediment load has been cut nearly in half and overbank flooding has not been allowed, marshes that would normally have been nourished no longer receive this nutrient supply, and that thus leads to marsh loss.

Other factors, such as the dredging of the Intracoastal Canal and construction of coastal highways, have had an influence on the aggradation rates of the marsh. Prior to man's heavy use of the wetlands, local drainage and water flow were determined by the distribution of the natural bayous (which were for the most part abandoned delta channels). Highways were initially constructed along the banks or on the high ground (natural levees) of these channels. Later, construction of Highway 90 and other interstate routes began an east-west pattern, strongly influencing the distribution of flood waters. Dredging of the Intracoastal Canal had the same effect; it not only blocked the natural drainage, but in addition allowed salt water to intrude into areas formerly consisting of fresh marshes. Once more, the quantitative impact of these two modifications cannot be evaluated.

Canal dredging

Turner et al. (1982) assessed the impact of dredging canals on total wetland loss in Louisana. In the study area, canal density increased from nearly zero in 1900 to approximately 2.4% of the land area in 1978. Types and sizes of canals constructed range from the Intracoastal Waterway to small access canals into hunting and trapping areas. The majority of canals have been dredged either to provide navigation routes and access to and support of petroleum exploration and exploitation sites or for general coastal navigation purposes. Examples of such canals include the Freshwater Bayou Canal, the Houma Navigation Canal, and the intricate networks of canals that radiate from natural drainage systems and extend into the wetlands in Terrebone and LaFourche parishes.

Early in the history of petroleum exploration, canals were constructed with little regard for the environmental impact. Today, canals are being planned with greater concern for the environment. For example, statistics compiled by the Louisiana Office of Coastal Zone Management indicate a substantial decrease in wetland area directly caused by the average oil and gas activity permitted by the agency during the period 1982–1985. Canals and spoil levees produce an artificial drainage network, which may interrupt the natural drainage density (Turner et al., 1983). This hydrographic interruption can divert surface water and groundwater flow patterns, thereby restricting nutrient and sediment dispersal to some areas while others are affected by increased discharge and possibly saltwater intrusion. Wetlands survive and are maintained by a precariously balanced drainage system. With forced drainage, the soils oxidize and compact, whereas increased water input may cause impoundment and marsh destruction (Turner et al., 1983).

Vegetation distribution in coastal Louisiana is dependent on the salinity tolerance of numerous plant species. Plants that tolerate salty Gulf waters form a narrow band along the coast that widens to the east. Inland of this salt marsh are the brackish water species, which grade inland into totally freshwater marshes (Chabreck, 1982). Introduction of salt water into the brackish and fresh vegetation zones through pipeline and navigation canals has been postulated to cause a slow but progressive deterioration of these wetlands. It should be mentioned, however, that no substantial studies have been carried out in attempting to quantify the amount of salt flux and its influence on the wetlands. In addition, canals enhance the rapid drainage of fresh water from marshes at low tide and the input of saline waters as the tide rises. Ultimately, the non-salt-tolerant species die, leaving the marsh surface exposed to erosion from surface runoff. Before the area can be revegetated, it may decrease 10 cm to 20 cm in elevation (Chabreck, 1982).

Geological environment, substrate lithology, boat traffic, and width of the adjacent spoil bank are major factors controlling the stability of canals through time. In areas with easily eroded substrates and/or a high volume of boat traffic, canal widths increase rapidly through erosion. Heavily traveled canals widen by 2.58 m/yr, while less used canals widen 0.95 m/yr (Johnson & Gosselink, 1982). Considering that canals increase in area through time, and by combining canal area with the area of spoil material, Craig et al. (1980) stated that canals account for 69% of the total wetland loss in Louisiana. It should be pointed out that this figure is derived from statistical correlations and that other factors, especially natural processes, are included in this number. Since quantitative figures are not available for any of the natural processes, and since statistical correlation does not always equate to cause and effect, we feel this figure is excessively high.

Fluid withdrawal

Depressurization of both shallow and deep aquifers, as well as hydrocarbon reservoirs, has resulted in an increase in subsidence over that which could be expected from natural processes. This process is true of normally pressured systems as well as geopressured areas of fluid withdrawal (Trahan, 1982). Jones & Larson (1975) estimated that along the Texas Gulf Coast the economic impact (property damage primarily) of subsidence resulting from groundwater withdrawal totalled over $ 31 million per year during the period 1943–1973. This problem is not unique to the Gulf Coast, but, as summarized by Holzer (1984), over 10,000 km^2 of land in the contiguous states of the U.S. have directly experienced subsidence associated with extraction of groundwater. In Louisiana, subsidence with the loss of pressure in aquifers has been greatest in or near metropolitan areas, where demand for fresh water is high (Keady et al., 1975). Kazmann & Heath (1968) showed that, superimposed on the regional subsidence being experienced by the whole coastal plain, a more local subsidence occurs in the New Orleans area in association with compaction of an underlying aquifer. Groundwater extraction and associated subsidence resulted in a measurable decrease in ground surface elevation within the city of 0.52 m between 1938 and 1964. Similar but somewhat less dramatic problems of subsidence have been experienced in Bâton Rouge, Louisiana (Wintz et al., 1970; Smith & Kazmann, 1978).

Subsidence resulting from fluid withdrawal from a porous subsurface medium can be explained in the following manner. Upon loss of fluid pressure, the reservoir essentially collapses, causing a rearrangement of sedimentary particles and volume reduction. In a sediment column composed of sand bodies encased in clay, as is typical of deltaic deposits, a hydraulic gradient is set up between the clays and sands when pressures in the reservoirs are decreased owing to fluid withdrawal (Poland, 1972). Water then moves from the clays to the sands. Depressuring of the clays results in an increase in stress caused by the weight of the overlying sediments, thus causing compaction and subsidence of the land surface (Kreitler, 1977). The most dramatic effects of this process have been seen with shallow groundwater withdrawal. However, fluid extraction from deeper horizons can contribute to subsidence, although probably to a much lesser extent.

Subsidence over areas of oil and gas fields has been documented from a few fields (Pratt & Johnson, 1926; Gilluly & Grant, 1949), but well-documented accounts of this phenomenon in Louisiana are not readily available. It is likely that some of the shallow producing petroleum fields in Louisiana have undergone subsidence because of hydrocarbon withdrawal, but there has been no quantitative documentation.

Judging from the carefully documented relationships between fluid withdrawal and subsidence, it could be expected that some areas of the low-lying coastal wetlands of Louisiana are affected by this process. Although Geertsma (1973) shows that land subsidence resulting from the production of hydrocarbons seldom is significant, the low relief of the coastal wetlands means that they are susceptible to even small amounts of reduction in surface elevation. Even small depressions caused by fluid withdrawal could allow saltwater intrusion. Because of the numerous factors that control subsidence in Louisiana, it is extremely difficult to estimate the contributions made by oil and gas withdrawal. It is, however, possible that the numerous faults associated with the subsurface reservoirs compartmentalize land subsidence, making it more local than regional. In addition, it should be pointed out that most of the oil and gas fields found in the wetlands are generally associated with relatively deep reservoirs, and the translation of the pressure reductions over such vertical thicknesses is presently not known. Due to the depth of these reservoirs, subsidence related to hydrocarbon extraction is probably not a major factor in overall subsidence rates over relatively short periods of time.

Conclusions

Our present understanding of processes active along Louisiana's coast that have led to loss of the wetlands can be summarized in the following statements:

1. Coastal submergence is directly influenced by two regional and long-term processes: a) Downwarping of the Gulf Coast basin by sedimentary loading and b) global sea level rise. Although geological data conclusively prove regional subsidence exists, experts disagree on rates, primarily because of inadequate data. Little quantitative data are available on such important processes as growth fault movement and salt mobilization and subsurface dissolution. Reasonable estimates of present sea level rise have been made from long-term tide records collected at numerous sites around the globe in regions where subsidence is minimal. These data show that sea level is presently rising.

2. Regional processes that are important to Louisiana's land loss problem include: a) delta switching, which supplies one area with sediment at the expense of other areas in the coastal plain. The stage of the delta cycle in which any one area exists controls to a large degree the rate of land loss. b) Man's intervention in coastal wetlands by building sediment-retention structures in the drainage basin and levees along the river channels severely affects the sediment yield to the coast and overbank nourishment of marshes. Quantification of these impacts is unavailable. c) Normal compaction, dewatering, and oxidation of coastal fine-grained and organic-rich sediments are continually operative in the wetlands and significantly contribute to wetland loss. These processes are at their maximum where the sedimentary sequence is thickest. Since thickness varies considerably across the wetlands, reliable figures are not available for many sites.

3. Important local effects include: a) Manmade modifications to the coastal plain, such as canal dredging and levees, alter the natural flow of water and sediment across the wetlands, which in turn contributes to land loss. Canals, especially large ones such as the Mississippi Gulf Outlet, allow intrusion of salt water into parts of the coastal plain, which changes the plant communities and eventually leads to wetland loss. Quantification of the salt flux and the rate at which this impact operates have not bean ascertained. b) Catastrophic events, such as hurricanes, are extremely detrimental to the shoreline and cause severe coastal erosion. Hurricanes also inundate large areas of the coastal marsh, but the long-term effect of this inundation has not been quantified. c) Biological degradation, such as grazing organisms and decomposers, has not been studied in any great detail, and this effect has not been quantified. A few studies of marsh accretion have been conducted, but not enough sites have been investigated to evaluate the relationship between marsh accretion and relative sea level rise.

In summary, coastal investigators agree that land is lost in Louisiana. However, most processes that are responsible for the land loss cannot be quantified at the present time. Existing data sets are often conflicting and usually reflect multiple interactions. However, after evaluation of the large number of data sets that do exist, and on the basis of our experience, we would rank the causes that are responsible for land loss as follows (in order of decreasing importance):

1. Changes in the depositional site and stage in the delta cycle.
2. Compaction and localized differential subsidence.
3. Sea level changes and their short-term variations.
4. Man's modification of the river system (dams and levees) which results in decreased sediment yield and overbank flooding.
5. Dredging of canals.
6. Biological degradation.
7. Fluid extraction.
8. Short-term catastrophic events (hurricanes) and wave reworking.
9. Regional geosynclinal downwarping.
10. Long-term climate changes.

References

Brooks, M.N., Colquhoun, D.J., Pardi, R.R., Newman, W. &

Abbott, W.J. 1979 Preliminary archaeological and geological evidence for Holocene sea level fluctuations in the lower Cooper River Valley, South Carolina – The Florida Anthropologist, 32: 85–103.

Brown, C.A. 1944 Historical commentary on the distribution of vegetation in Louisiana and some recent observations – Proc. La. Acad. Sci. 8: 35–47.

Chabreck, R.H. 1982 The effect of coastal alteration on marsh plants. In: Boesch, D.F. (ed.) – Proc. Conf. Coastal erosion and wetland modification in Louisiana: causes, consequences, and options – U.S. Dept. Interior, FWS/OBS-82/59: 92–98.

Coleman, J.M. 1976 Deltas: processes of deposition and models for exploration – Continuing Education Publ. (Champaign, Ill.): 102 pp.

Coleman, J.M. & Wright, L.D. 1971 Analysis of major river systems and their deltas: procedures and rationale, with two examples – Coastal Studies Institute Louisiana State University Techn. Rept. 95: 125 pp.

Craig, N.J., Turner, R.E. & Day, J.W., Jr. 1980 Wetland losses and their consequences in coastal Louisiana – Z. Geomorph. 34: 225–241.

DeLaune, R.D. & Smith, C.J. 1985 Release of nutrients and metals following oxidation of fresh water and saline sediment – J. Envir. Qual. 14: 164–168.

Dozier, M.D. 1983 Assessment of changes in the marsh of Southeast Barataria Basin, Louisiana, using historical aerial photographs and a spatial information system – M. Sc. Thesis, Louisiana State Univ., Baton Rouge, La.: 102 pp.

Elliot, T. 1978 Deltas. In: H.G. Reading (ed.): Sedimentary Environments and Facies – Elsevier (New York): 97–142.

Fisher, W.L., L.F. Brown, A.T. Scott & J.H. McGowen 1969 Delta systems in the exploration for oil and gas – Bur. Econ. Geol., Univ. of Texas Austin: 78 pp.

Fisk, H.N. 1944 Geological investigations of the alluvial valley of the lower Mississippi River – U.S. Army Eng. Miss. River Comm. Vicksburg, Miss: 200 pp.

Fisk, H.N. 1952 Geological investigations of the Atchafalaya basin and problems of Mississippi River diversion – U.S. Army Eng. Miss. River Comm. Vicksburg, Miss: 145 pp.

Fisk, H.N. & McFarlan E., Jr. 1955 Quaternary deltaic deposits of the Mississippi River – Geol. Soc. Amer. Spec. Pap. 62: 279–302.

Frazier, D.E. 1967 Recent deltaic deposits of the Mississippi River, their development and chronology – Trans. Gulf Coast Assoc. Geol. Soc. 17: 287–315.

Gagliano, S.M. & Van Beek, J.L. 1970 Geologic and geomorphic aspects of deltaic processes, Mississippi delta system – Center for Wetland Resources, Louisiana State University, Bâton Rouge, La., Rept. 1: 127.

Gagliano, S.M., Meyer-Arendt, K.J. & Wicker, K.M. 1981 Land loss in the Mississippi River Deltaic Plain – Trans. Gulf Coast Assoc. Geol. Soc. 31: 295–306.

Galloway, W.E. 1975 Process framework for describing the morphologic and stratigraphic evaluation of deltaic depositional systems. In: M.L. Broussard (ed.): Deltas, models for exploration – Houston Geol. Soc., Houston, Texas: 87–98.

Geertsma, J. 1973 Land subsidence above compacting oil and gas reservoirs – J. Petrol. Techn.: 734–744.

Gilluly, J. & Grant, U.S. 1949 Subsidence in the Long Beach Harbor Area, California – Geol. Soc. Amer. Bull. 60: 461–530.

Gosselink, J.G. 1984 The ecology of delta marshes of Coastal Louisiana: a community profile – U.S. Fish Wildlife Serv. FWS/OBS-84-09: 134 pp.

Gowanloch, H.N. 1933 Fishes and fishing in Louisiana – Louisiana Dept. Conserv. Bull. 23: 638 pp.

Hathaway, E.S. & Penfound, P.T. 1936 The relation of salinity to plant communities in southeastern Louisiana – Proc. Louisiana Acad. Sci. 3: 68–71.

Hatton, R.S. 1981 Aspects of marsh accretion and geochemistry: Barataria Basin, Louisiana – M.Sc. Thesis, Louisiana State Univ., Bâton Rouge, La.: 116 pp.

Hayes, M.O. 1978 Impact of hurricanes on sedimentation in estuaries, bays, and lagoons. Estuarine Interactions – Academic Press: 323–346.

Hoffman, J.S., Keyes, D. & Titus, J.G. 1983 Projecting future sea level rise: Methodology, estimates to the year 2100, and research needs – U.S. Environmental Protection Agency, EPA 230-09-007: 121 pp.

Holdahl, S.R. & Morrison, N.L. 1974 Regional investigations of vertical crustal movements in the U.S., using precise relievelings and mareograph data. In: R. Green (ed.): Recent crustal movements and associated seismic, and volcanic activity – Tectonophysics 23: 373–390.

Holzer, T.L. 1984 Man-induced land subsidence – Geol. Soc. Amer. Rev. Eng. Geol. 6: 221 pp.

Johnson, W.B. & Gosselink, J.G. 1982 Wetland loss directly associated with canal dredging in the Louisiana coastal zone. In: Boesch, D.F. (ed.): Proc. Conf. Coastal erosion and wetland modification in Louisiana: causes, consequences, and options – U.S. Dept. Interior, FWS/OBS-82/59: 60–72.

Jones, L.L. & J. Larson, 1975 Economic effects of land subsidence due to excessive ground water withdrawal in the Texas Gulf Coast area – Texas Water Resources Inst., Texas A and M Univ. Techn. Rept. 67: 33 pp.

Kahn, J.H. & Roberts, H.H. 1982 Variations in storm response along a microtidal transgressive barrier-island arc – Sediment. Geol. 33: 129–146.

Kazmann, R.G. & Heath, M.M. 1968 Land subsidence related to ground-water offtake in the New Orleans area – Trans. Gulf Coast Assoc. Geol. Soc. 18: 108–113.

Keady, D.M., Lins, T.W. & Russell, E.E. 1975 Status of land subsidence due to ground-water withdrawal along the Mississippi Gulf Coast – Water Resources Res. Inst., Mississippi State Univ.: 25 pp.

Keown, M.P., Dardeau, E.A. & Causey, E.M. 1981 Characterization of the suspended-sediment regime and bedmaterial gradation of the Mississippi River basin – Environmental Laboratory, Potamology Program, USACOE Vicksburg Miss., 2 volumes: 89 pp and 124 pp.

Kolb, C.R. & Van Lopik, J.R. 1958 Geology of the Mississippi deltaic plain-southeastern Louisiana – USACOE, Waterways Exp. Sta. Techn. Rept. 2: 482 pp.

Kraft, J.C. 1971 Sedimentary facies patterns and geologic history of a Holocene marine transgression – Geol. Soc. Am. Bull. 82: 2131–2158.

Kreitler, C.W. 1977 Fault control of subsidence, Houston-Galveston, Texas – *Geoscience and Man,* Louisiana State Univ., Bâton Rouge, La., 18: 7–19.

Lerch, O., Clendenin, W.W. & Stubbs, W.C. 1892 Geology and agriculture of Louisiana – Louisiana Agricul. Experim. Sta., Bâton Rouge, La.: 295 pp.

Louisiana Department of Conservation 1931 The birds of Louisiana: Bulletin 20: 598 pp.

Murray, G.E. 1961 Geology of the Atlantic and Gulf coastal province of North America – Harper and Brothers (New York): 692 pp.

Nummedal, D. 1982 Barrier Islands – Techn. Rep. 82-3, Coastal Research Group, Dept. Geology, Louisiana State University, Bâton Rouge, La.: 79 pp.

Nummedal, D. 1983 Future sea level changes along the Louisiana coast – Shore and Beach 51: 10–15.

Nummedal, D., Penland, S., Gerdes, R., Schramm, W., Kahn, J. & Roberts H. 1980 Geologic response to hurricane impact on low-profile Gulf Coast barriers – Trans. Gulf Coast Assoc. Geol. Soc. 30: 183–195.

O'Neil, T. 1949 The muskrat in the Louisiana coastal marshes – Louisiana Dept. Wildlife & Fisheries, New Orleans, La.: 152 pp.

Penfound, W.T. & Hathaway, E.S. 1938 Plant communities of the southeastern marshes – Proc. Louisiana Acad. Sci. 3: 61–62.

Penland, S. & Boyd, R. 1981 Shoreline changes on the Louisiana barrier coast – IEEE Oceans 81: 209–219.

Poland, J.F. 1972 Subsidence and its control – Amer. Assoc. Petrol. Geol. Mem. 18: 71 pp.

Pratt, W.E. & Johnson, D.W. 1926 Local subsidence of the Goose Creek oil field – J. Geol. 34: 577–590.

Roberts, H.H. 1985 A study of sedimentation and subsidence in the south-central coastal plain of Louisiana – Summary Report to the New Orleans Branch of the U.S. Army Corps of Engineers, New Orleans, La.: 57 pp.

Russell, R.J. 1936 Lower Mississippi River delta: reports on the geology of Plaquemines and St. Bernard Parishes – Louisiana Dept. Conserv. Geol. Bull. 8, 3: 199 pp.

Scruton, P.C. 1960 Delta building and the deltaic sequence. In: Shepard F.P. (ed.): Recent Sediments, N. W. Gulf of Mexico – Am. Assoc. Petrol. Geol., Tulsa, Okla.: 82–102.

Simpson, R.H. & Lawrence, M.B. 1971 Atlantic hurricane frequencies – N.O.A.A. Techn. Mem. NWS SR-58: 10 pp.

Smith, C.G. & Kazmann, R.G. 1978 Subsidence in the Capital Area Ground Water Conservation District – An update – Bâton Rouge La. Capital Area Ground Water Conservation Comm. Bull. 2: 31 pp.

Smith, T.J. 1982 Herbivore-induced changes in salt marsh plant community structure: Wetlands ecology and management – Nat. Inst. Ecol. Internat. Sci. Pub., Jaipur, India: 131–149.

Swanson, R.L. & Thurlow, C.I. 1973 Recent subsidence rates along the Texas and Louisiana coasts as determined from tide measurements – J. Geophys. Res. 78: 2665–2671.

Terzaghi, K. 1943 Theoretical soil mechanics – Wiley (New York): 510 pp.

Trahan, D.B. 1982 Monitoring local subsidence in areas of potential geopressured fluid withdrawal southwestern Louisiana – Trans. Gulf Coast Assoc. Geol. Soc. 32: 231–236.

Turner, R.E., Costanza, R. & Scaife, W. 1982 Canals and wetland erosion rates in coastal Louisiana. In: Boesch D.F. (ed.): Proc. Conf. coastal erosion and wetland modification in Louisiana: causes, consequences and options – U.S. Dept. Interior, FWS/OBS-82/59: 73–84.

Turner, R.E., McKee, K.L., Sikora, W.B., Sikora, J.P., Mendelssohn, I.A., Swenson, E., Neill, C., Leibowitz, S.G. & Pedrazini, F. 1983 The impact and mitigation of man-made canals in coastal Louisiana – Symp. Integration of Ecological Aspects in Coastal Engineering Projects, Rotterdam, The Netherlands: 24–52.

Vail, P.R., Mitchum, R.M., Jr. & Thompson, S. III 1977 Seismic stratigraphy and global changes of sea level Part 4: global cycles of relative changes of sea level. In: Payton C.E. (ed.): Seismic Stratigraphy – Applications to Hydrocarbon Exploration – Amer. Assoc. Petrol. Geol. Mem. 26: 83–97.

Van Beek, J.L. & Meyer-Arendt, K.J. 1982 Louisiana's eroding coastline: recommendations for protection – Office of Coastal Zone Management, Bâton Rouge, La.: 143 pp.

Van Heerden, I. Ll. & Roberts, H.H. 1980 The Atchafalaya delta: rapid progradation along a traditionally retreating coast – Z. Geomorph. N.F. 34: 225–240.

Wells, J.T., Chinburg, S.J. & Coleman, J.M. 1982 Development of the Atchafalaya River delta – Coastal Studies Institute, Louisiana State Univ., prepared for USACOE Waterways Exprim. Sta., Vicksburg, Miss.: 91 pp.

Wintz, W.A., Jr., Kazmann, R.G. & Smith, C.G., Jr. 1970 Subsidence and ground-water offtake in the Bâton Rouge area: Louisiana – Water Resources Res. Inst. Bull. 6: 1–19.

Worzel, J.L. & Burk, C.A. 1979 The margins of the Gulf of Mexico – Amer. Assoc. Petrol. Geol. Mem. 29: 403–419.

Wright, L.D., J.M. Coleman & M.W. Ericson 1974 Analysis of Major River Systems and Their Deltas – Coastal Studies Institute, Techn. Rept. 156, Louisiana State Univ., Bâton Rouge, La.: 114 pp.

Geologie en Mijnbouw 68: 25–34 (1989)

The Hudson Bay Lowland: major geologic features and assets

I. Peter Martini
Department of Land Resource Science, University of Guelph, Guelph, Ont. N1G 2W1, Canada

Received 11 August 1987; accepted in revised form 25 January 1988

Key words: coastal sediments, cold-climate sediments, lowland, peatland, permafrost, water resources

Abstract

The Hudson Bay Lowland is a vast (325,000 km^2), flat (average slope 0.5 m/km) physiographic region of Canada located to the southwest of James Bay and Hudson Bay. It is underlain by Paleozoic and Mesozoic rocks and bounded by Precambrian terrains. Thin Pleistocene till sheets, locally deposited on fluted terrains mantle most of the Lowland, and they are overlain by thin marine and coastal Holocene deposits which have formed during the ongoing regression from an early post-glacial sea, the Tyrrell sea. The present shores of the James Bay and Hudson Bay are but one stage of development of such regressive sequences. More than 90% of the vast emerged Lowland is covered by one of the largest cold wetlands and peatlands of the world. Up to 3–4 m thick peats have developed in the last 5000 years in inland fens and raised bogs.

Fresh water is the major resource of the area, both for hydroelectric power and/or irrigation on a continent wide scale. Other resources not yet fully evaluated, consist of mineral deposits on or near Precambrian inliers, hydrocarbons in the relatively thin Paleozoic sequence, and lignite, kaolin and quartz sand in Mesozoic terrains.

The damage generated by any development in the area, must be carefully weighted against the worldwide importance of this vast peatland on gaseous fluxes and atmospheric balance. Furthermore the coastal zone of the Lowland is a major staging and breeding ground for polar bears, migratory birds and other species. Perhaps assurance of preservation of the still pristine natural Hudson Bay Lowland should be achieved by establishing it as an international heritage park.

Introduction

The Hudson Bay Lowland of central-east Canada has several quasi-unique distinctions: it is one of the largest cold wetlands of this world essentially untouched by man, it developed close to centres of Pleistocene glaciations, and it bounds cold, mesotidal, mediterranean seas which cover the only remnant, sedimentologically active, intracratonic basin of those which covered North America in Paleozoic times (Fig. 1). The objectives of this paper are to illustrate major features of the Lowland and briefly analyze their exploitation potential. Original information gathered over more than a decade of sedimentological and biological multidisciplinary studies of the area is used as well as data from the scanty literature (Haworth et al., 1978; Glooschenko & Martini, 1978; Martini, 1982b; Martini, 1986).

Geology

Precambrian terrains bound the Paleozoic and

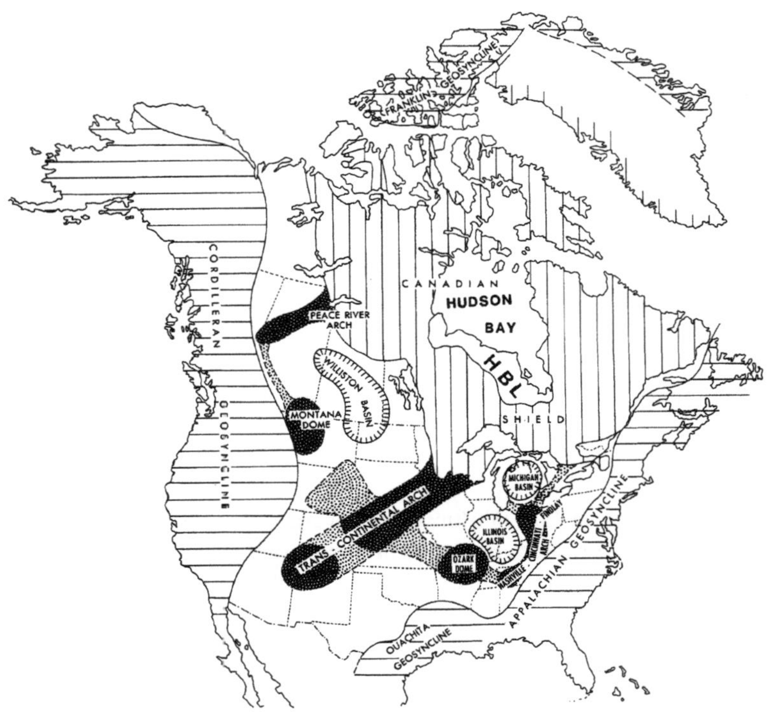

Fig. 1. Location map and Paleozoic basins of North America. HBL = Hudson Bay Lowland. (after Clark & Stearn, 1960).

Mesozoic rocks which form the substratum of the Lowland and most of the seafloor of James Bay and Hudson Bay (Fig. 2). Precambrian inliers of the Sutton Highlands subdivide the Lowland into two zones, a James Bay portion which occupies the geological Moose River Basin and another portion which corresponds to the Hudson Bay Basin (Sanford et al., 1968; Sanford & Norris, 1975; Sanford et al., 1979). These basins have been active since at least early Cambrian times. The origin of the basins is not clear, except that perhaps they are related to Precambrian downwarps associated with the cratonization of former plate collision zones (Stockwell, 1970; Baer, 1970; Donaldson, 1986). An alternative, or complementary hypothesis is that part of the Hudson Bay Basin is directly or indirectly associated with the impact of a large meteorite (asteroid) which may either have generated the crustal depression itself during the Precambrian or may have been influencial in localizing the collision

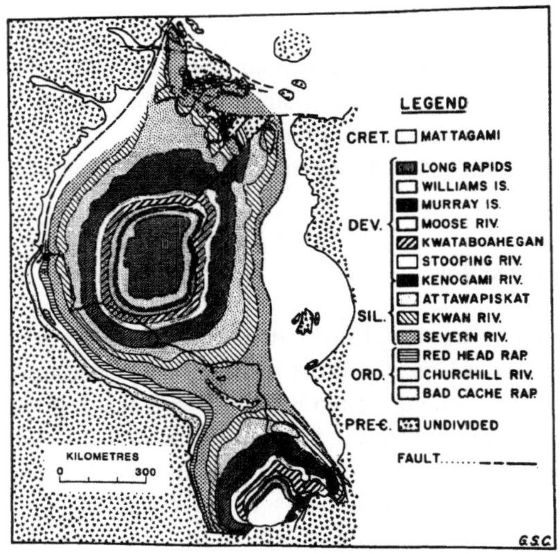

Fig. 2. Geological map of the Hudson Bay area (from Sanford & Norris, 1973).

zones (Beals, 1968). In any case, the Paleozoic sequences record shallow marine environments in both the Hudson Bay and Moose River (James Bay) basins (Sanford et al., 1968; Norris, 1986; Johnson et al., 1986). Well developed Silurian reefs rimmed the persistent Cape Henrietta Maria Arch which subdivided the two basins. Mud (now shaley red beds) was injected from the distal rising Appalachian mountains in Devonian times. Cretaceous sedimentation in the Moose River Basin produced lignite-bearing, fluvial sequences. These Mesozoic terrains contain potentially exploitable deposits of kaolinitic clays (fireclay) and of good quality quartz sand (Telford & Long, 1986). Since Cretaceous times the Lowland has undergone erosion, lately by Pleistocene glaciers (Lee, 1968; Prest, 1970; Shilts, 1982, 1986). The most recent geological events have been the Pleistocene glacial sedimentation followed by an early post-glacial marine submersion (Tyrrell sea) and a still continuing emersion responsible for a thin but widespread offlap sedimentary sequence, and paludification of the Lowland (Lee, 1968; Martini, 1981a, 1982a).

Climate

Similarly to large parts of Alaska, northern Canada, and most of Scandinavia and the U.S.S.R., the Hudson Bay Lowland has a humid microthermal *arctic* climate (Dcf under the Koppen system) (Chapman & Thomas, 1968; Maxwell, 1986). Its northern part, from latitude 58° 50′ N down to latitude 52°, has a mean annual temperature of −3.9° C, the southern part, from 52° N to 50° N, has an average of −1.1° C. Winters are cold with January temperatures varying between mean daily maxima of −15.6° C in the south and −17.8° C in the north, and respective mean daily minima of −26.7° C and −28.9° C. Summers can be rather warm with July average daily maxima of 22.8° C in the south and 20.6° C in the north. Precipitation varies between 66 and 61 cm from the south to the north and the snowfall is respectively 241 and 203 cm. There is no major variation in precipitation, except for a slightly drier, thus more fire-prone area in the northwest. Winds are consistent and strong. In the summer, intense fog patches form, particularly along the Hudson Bay coasts when cold tidal waters of the bay cover the wide warmer tidal flats.

The cold characteristic of the region is in part due to latitude, but mostly to the fact that the flat region is not protected from arctic air masses by any mountainous terrain, and, most of all, to the influx of arctic waters into Hudson Bay and James Bay through the narrow Fury and Hecla Strait in the northwestern corner of Foxe Basin and through the Hudson Strait. These arctic currents move counter-clockwise and have a powerful refrigeration effect on the Lowland (Figs. 3, 4; Dunbar, 1951). By-products of such refrigeration are the presence of an ice-cover for at least six months of the year, and the development of permafrost at very low latitudes (Fig. 4; Markam, 1986). Freeze-up starts in the region in November and break up in May, with ice floes still present in the Bay in July. The ice-cover protects the substratum from any major changes during winter, but, at breakup time, it greatly affects the coastal morphology and sediments and it induces recurring ice jams in the rivers. The northward flowing rivers break first in the south and pile ice, waters and sediments against the still solid ice plugs of the lower reaches and deltas. The permafrost is a Holocene feature. It is still developing as the land emerges from the sea at an isostatic uplift rate of 0.75–1.00 m/century (Hunter, 1970; Webber et al., 1970). While there is no permafrost at the present shores, except in the northernmost part of the Lowland, continuous permafrost develops onland just south of the Hudson Bay, discontinuous permafrost farther south still, and sporadic permafrost in the southernmost parts of the Lowland.

Holocene landscape

Sedimentary features

The Hudson Bay Lowland is characterized by an emergent coastal landscape. Such a landscape is usually portrayed as a series of emerged gravelly beach ridges (Fairbridge & Hillaire-Marcel, 1977; Martini, 1981a). Indeed flights of raised beach

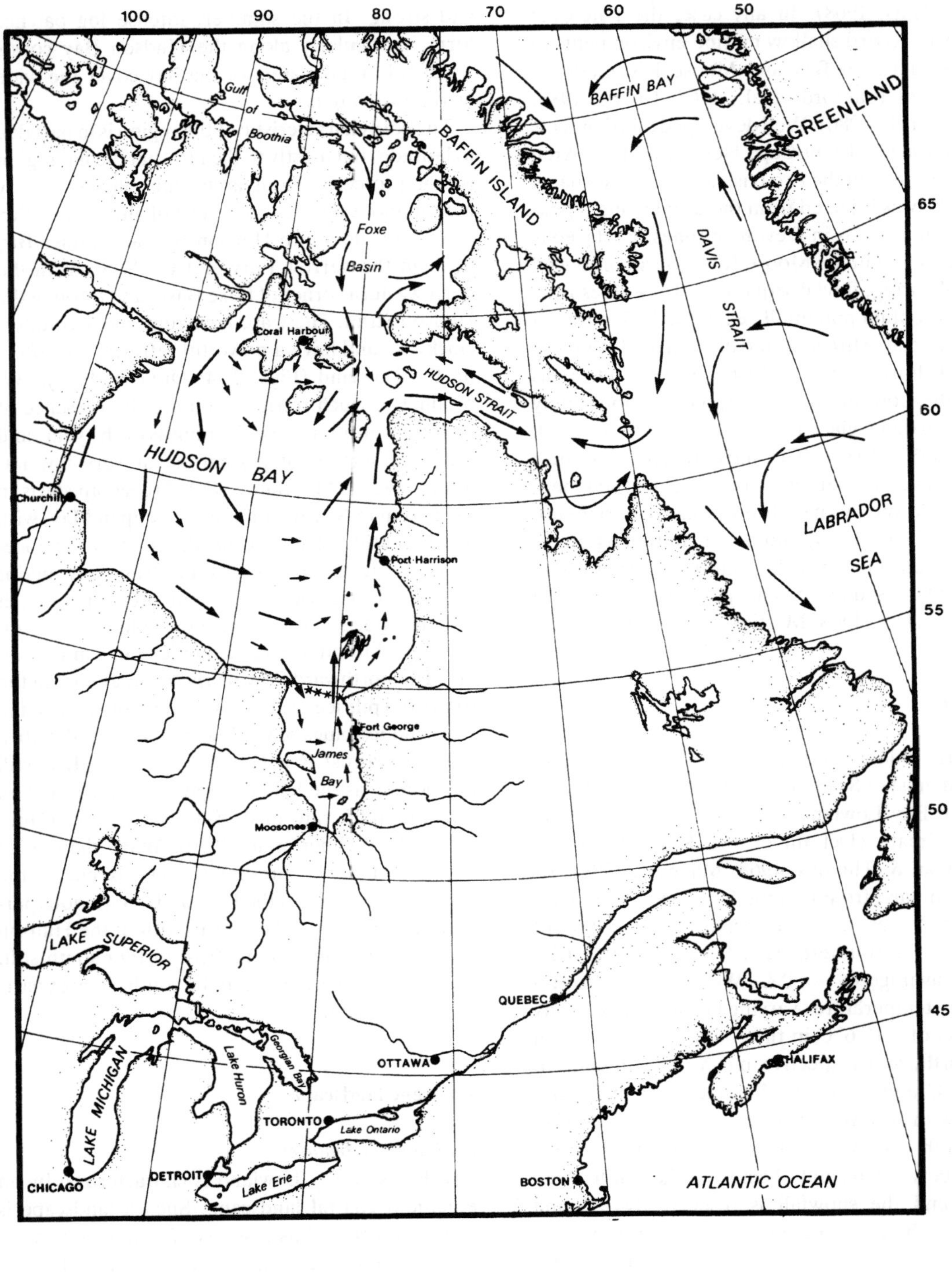

Fig. 3. Marine currents of the Canadian Inland Seas.

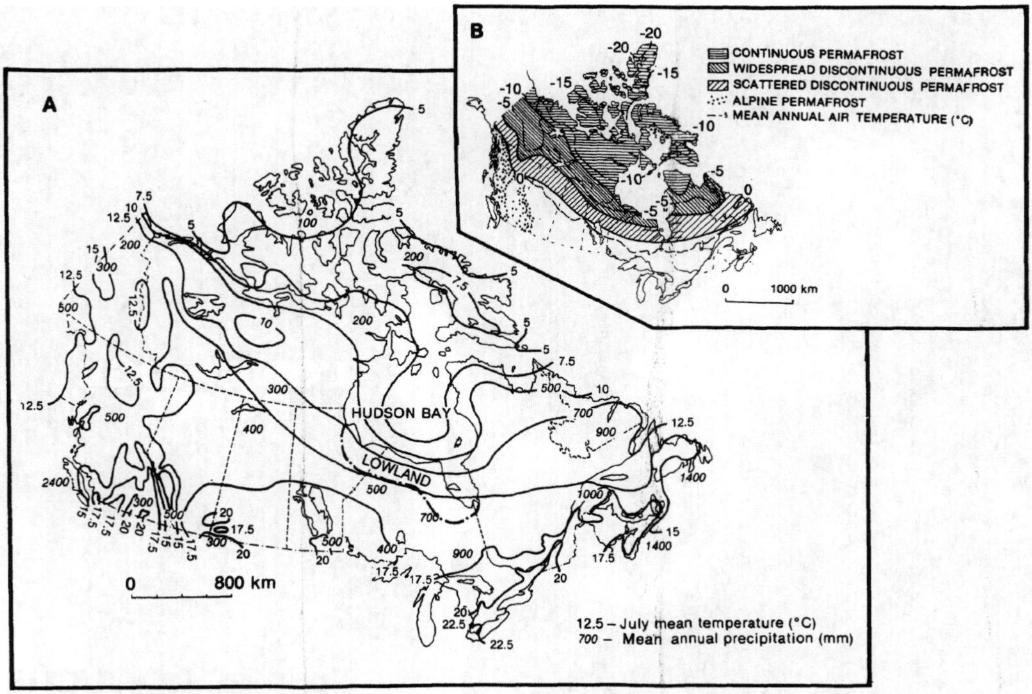

Fig. 4. Cold climatic conditions of central north Canada and the Hudson Bay Lowland. A. July mean temperature and precipitation; B. Permafrost map of Canada.

ridges and strandlines are major characteristics of the region, while a great variety of other features and deposits occur as well, both along the recent and the raised generally flat coasts. A major observation that is generally obscured by the 'beach ridge syndrome' is that those Holocene inland seas were and are mesotidal; thus vast, flat, subtidal and intertidal zones develop and are preserved in the Lowland.

Other features of the area are associated with the complex development of pre-glacial fluvial drainage channels which were later remolded by glaciations, were partially filled by glacial sediments and early post-glacial marine silts (up to 80 m thick), and are now occupied by the lower reaches of the larger rivers (Pelletier, 1968; 1986). Interfluvial regions lack or contain lesser amounts of marine fine deposits, because of lower sedimentation and/or because of erosion during emersion. The effect of differential erosion and fluting of bedrock and glacial deposits are more pronounced in the interfluvial promontory areas. Furthermore, the slightly steeper landscape of this zone develops various types of coastal beach ridges during emersion, and long narrow promontories which protrude as natural jetties across the tidal zone and trap fines (mainly sand and silt) in their updrift sides (Fig. 5; Martini, 1981a).

In these updrift embayments, vast, shallow intertidal flats develop and grade imperceptibly into marshes and inland fens. It is only in these vast intertidal flats and in some parts of the estuarine embayments where the continuous Holocene offlap sedimentary sequence is preserved. From the bottom up, such a sequence consists of: (a) shattered carbonate bedrock (due to vertical crustal isostatic movements and to Pleistocene permafrost conditions); (b) thin diamicts of various types; (c) sublittoral glaciomarine and marine argillaceous silts with a restricted pelecypod fauna and few

Fig. 5. Example of the coastal zone showing narrow transversal promontories and wide tidal flats on their updrift sides.

dropstones, some boulder sized; (d) subtidal and intertidal sandy deposits (up to 4 m thick); (e) well developed marsh sediments (up to 2–3 m thick); (f) organic deposits of freshwater marshes and inland fens characterized by up to several metres (2–3 m) of decomposed graminoid peats, locally containing some woody component. Thicker (up to 3–4 m) peats develop in more mature inland bogs.

The intertidal sediments of the Hudson Bay Lowland contain most of the characteristic features reported in similar deposits of other coasts, such as fining landward and upward sequences, a variety of ripple cross laminations, wavy bedding, flaser bedding, occasional herringbone crosslaminations, some bioturbation and rare body fossils, and crinkly marsh laminations. They differ from other reported intertidal sequences because of cold climate features such as sparse restricted fauna (no intense bioturbation ever occurs in these flats), presence of polymictic, rounded, ice rafted pebbles and occasional boulders locally redistributed by waves and currents, and generally thin marsh laminations bioturbated by root mats of *Puccinellia phryganodes* and slightly larger roots of *Carex sp.*, plants which characterize subarctic and arctic salt marshes (Martini et al., 1980; Martini, 1981b; Scott & Martini, 1982; Martini & Morrison, 1987).

Sediments are carried to the coast by rivers which cannibalize the thin Pleistocene and Holocene sedimentary cover of the Lowland and cut into the carbonate bedrock, and by longshore currents. The intertidal and shallow subtidal sediments of part of the coast are continuously resuspended by waves and ice gouging and are transported and ice rafted toward embayment depocentres. In certain areas the coastal erosion during emergence is complete or quasi-complete, and bare, frost shattered bedrock or thin tills may be exposed in the intertidal zone. Beach ridges develop primarily by storm waves on such a substratum which may be covered by a characteristic, thin, residual deposit called, for want of a better term, 'reconstituted coastal diamicton' (Martini & Protz, 1978). Such a residual deposit is composed of modified Pleistocene till material mixed with fossiliferous marine silts and sand, and with ice rafted fine and coarse angular fragments. The beach ridge deposits are characterized by a relatively well sorted coarse sand and fine gravel. The smaller ridges show plane beds alternating with landward inclined foresets which are formed as the ridges prograde slightly landward. High tidal flats (algal flats) develop behind the ridges and are eventually capped by salt marshes. As the land rises, incipient paludification occurs whereby fens and bogs form in interridge areas of southern regions where only sporadic and discontinuous permafrost occurs, and spruce forests on drier ridges. In northern, continuous permafrost regions of the Lowland, tundra forms near the coast. Farther inland, forest develops covering both the beach ridges and the interridge areas, the latter having been raised, thus sufficiently dried up by ice lenses and peat plateaus. Podzolic soils develop on the drier beach ridges (Protz, 1982).

Wetlands

Along the coast, in varying width depending on the slope of the land, there is a transition from the intertidal marshes inland to meadow marsh, thicket swamps and low shrub marsh. These change landward into coastal fens, and, in the north, into 'edaphic tundra' and peat plateaus.

The coastal marshes are not peat forming environments and are mostly characterized by a *Puccinellia phryganodes* association in the lower part and a *Carex subspathacea* association in the upper part.

More than 90% of the Hudson Bay Lowland is occupied by an unconfined wetland ('muskeg', Radforth, 1973), most of which has developed into a variety of peatlands with a variety of peat types and thicknesses, depending on location and climate.

Most of the Lowland is located within the Boreal Forest Region (taiga) of Canada (Hustich, 1957). The distribution of the wetland types and of the vegetation variation across the Lowland are dictated primarily by the strong north-south change in temperature, by the coastal-inland position, as well as by local topography and substratum which in the north is affected also by permafrost hummocks (Fig. 6; Riley, 1982; Zoltai & Pollett, 1983). The 'midboreal wetland region' B_{Mh} is characterized primarily by fens everywhere, some treed with tamarack (*Larix laricina*), bogs in the northern parts, swamps in parts of the southern interior (where some deciduous trees occur) and near streams, and forests (mainly spruce) on river levees. The 'high boreal wetland region' (B_H) is characterized primarily by fens and bogs and a northward increase in influence of permafrost with development of few peat plateaus and palsas. The 'low subarctic wetland region' (S_L) is characterized by a rapid expansion of peat plateaus associated with the increasing importance of permafrost. The 'high subarctic wetland region' (S_H) has continuous permafrost except at the coast. Lichen-dominated, treed bogs and peat plateaus prevail. In drier areas, that is, on higher raised beach ridges, or on river levees, woodlands occur, primarily of black spruce (*Picea mariana*), white spruce (*Picea glauca* near the coast) with, locally, some balsam fir, trembling aspen (*Populus tremuloides*), balsam poplar (*Populus balsifera*) and white birch (*Betula papirifera*).

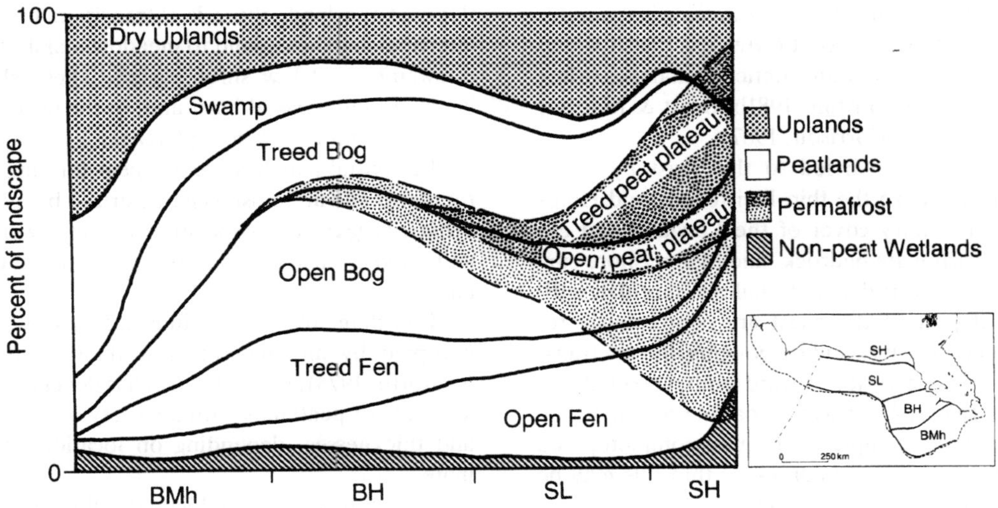

Fig. 6. Wetland types of the Hudson Bay Lowland (after Riley, 1982). SH: High subarctic; SL: Low subarctic; BH: High boreal; BMh: Middle subarctic.

Peat has developed in the Hudson Bay Lowland in the last 5000 years and has reached a thickness of up to 3–4 m in raised bogs, about 100 km inland, and about 2–3 m in inland fens. A quasi-regular increase in the thickness of the peat can be demonstrated from the shore inland, with the net rate of deposition in the southern part of the Lowland being approximately double that of the northern part of the Lowland (Cowell et al., 1983; Martini & Glooschenko, 1985). Where best developed, the peat sequence shows vertical transitions from decomposed, dark, sedge peat (fen) into a sphagnum peat (bog) with occasional woody fragments.

The wetlands are slowly drained by few rivers which experience a nival regime, that is, small discharge under an ice-cover during the winter, large short lived floods at spring breakup, low water conditions during the summer, and in the south, some fall floods associated with rain. It has been calculated that the total river discharge is on the order of 300 km^3/year (Prinsenberg, 1980).

Resources

The Hudson Bay Lowland is essentially untouched by man, except for some damming of rivers for hydroelectric power, and occasional drilling for hydrocarbons (unsuccesful) along the shores. The region has potential a wealth of mineralizations in the Precambrian inliers, and industrial minerals in the Mesozoic sequences, but perhaps the most important economic asset of the land is its enormous supply of freshwater (Johnson et al., 1986; Prinsenberg, 1980). Schemes for diversion of such waters to the Great Lakes of North America and then to the semi-arid midwest of the United States have been proposed, and vary from simple damming and diversion of flow of few major rivers, to the mega-project of damming the mouth of James Bay, transforming it into a gigantic freshwater lake (Milko, 1986). These schemes must be weighted against the potential damages to the present environment, particularly because those shores are used seasonally by polar bears and by some of the largest migratory bird populations of the world which come from South America and the United States up here to breed.

The major assets of the Lowland are related to the immense reserve of fresh water, potential peat resources, large migratory wildlife populations, and, perhaps most of all, the yet not properly studied but obviously important effect that such a vast peatland has on the gaseous flux and atmospheric balance of, for instance, methane, carbon dioxide, sulphur dioxid and nitrogen (Gorham, 1982). Parts of this region are protected under provincial and national wildlife parks. Its importance perhaps should lead to establishing an international heritage park, such that no modifying mega-projects are implemented which may render smaller localized parks ineffective.

Acknowledgement

Funds for this research were provided by the National Science and Engineering Research Council of Canada (Grant A7371) and Environment Canada. Helicopter support was given by ocean and Aquatic Science. Thanks are offered to the colleagues involved in the Hudson Bay Lowland Research Project for their collaboration.

References

Baer, A.J. (ed.). 1970. Basins and geosynslines in the Canadian Shield – Geol. Surv. Can. Pap. 70–40: 265 p

Beals, C.S. 1968. Theories on the origin of Hudson Bay, Part 1: On the possibility of a catastrophic origin for the great arc of eastern Hudson Bay. In: Beals, C.S. (ed.): Science, History, and Hudson Bay, 2 – Dept. Energy, Mines and Resources, Canada: 985–998

Chapman, L.T. & M.K. Thomas. 1968. The climate of northern Ontario – Met. Branch, Dept. Transp. Climat. Stud. Toronto, 6: 58 pp

Clarke, T.H. & C.W. Stearn. 1960. The geological evolution of North America – Ronald Press Co. (New York): 434 pp

Cowell D.W, G.M. Wickware, A.W. Boissoneau, J.K. Jeglum & R.A. Sims. 1983. Hudson Bay Lowland peatland inventory – Proc. Peatland Inventory Methodology Workshop – Land Resource Inst., Agric. Canada, Ottawa: 88–102

Donaldson, J.A. 1986. Precambrian geology. In: I.P. Martini (ed.): Canadian Inland Seas – Elsevier (Amsterdam): 1–15

Dunbar, M.J. 1951. Eastern Arctic waters – Bull. Fish. Res. Board Can. 88: 131 pp

Fairbridge, R.W. & C. Hillaire-Marcel. 1977. An 8,000 yr palaeoclimatic record of the 'Double-Hale' 45-yr solar cycle – Nature 268: 413–416

Glooschenko, W.A. & I.P. Martini. 1978. Hudson Bay Lowlands baseline study. Coastal zone '78. Proc. Symp. Technical. Environmental, Socioeconomic and Regulatory aspects of Coastal Zone Management – ASCE/San Francisco: 663–679

Gorham, E. 1982. Some unsolved problems in peatland ecology – Nat. Can. 109: 533–541

Haworth, S.E., D.W. Cowell & R.A. Sims. 1978. Bibliography of published and unpublished literature on the Hudson Bay Lowland. Great Lakes Forest Research Centre, Rept. O-X-273. Sue St. Marie, Canada: 270 pp

Hunter, G.T. 1970. Postglacial uplift of Forth Albany, James Bay – Can. J. Earth Sci. 7: 574–548

Hustich, I. 1957. On the phytogeography of the subarctic Hudson Bay Lowland – Acta Geogr. 16: 161–198

Johnson, R.D., S.J. Joubin, S.J. Nelson & E. Olsen. 1986. Mineral resources. In: I.P. Martini (ed.): Canadian Inland Seas – Elsevier (Amsterdam): 387–402

Lee, H.A. 1968. Quarternary Geology. In: Beals, C.S. (ed.): Science, History and Hudson Bay – Dept. Energy, Mines and Resources, Ottawa, 2: 503–543

Markam, W.E. 1986. The ice cover. In: I.P. Martini (ed.): Canadian Inland Seas – Elsevier (Amsterdam): 101–116

Martini, I.P. 1981a. Morphology and sediments of the emergent Ontario coast of James Bay, Canada – Geogr. Ann. 63A: 81–94

Martini, I.P. 1981b. Ice effect on erosion and sedimentation on the Ontario shores of James Bay, Canada – Z. Geomorph., N.F. 25: 1–16

Martini, I.P. 1982a. Geomorphological features of the Ontario coast of Hudson Bay – Nat. Can. 109: 415–429

Martini, I.P. (ed.) 1982b. Scientific studies on Hudson and James bays – Nat. Can. 109: 301–1019

Martini, I.P. (ed.) 1986. Canadian Inland Seas – Elsevier (Amsterdam): 494 pp

Martini, I.P. & W.A. Glooschenko. 1985. Cold Climate Peat Formation in Canada, and its relevance to Lower Permian Coal Measures of Australia – Earth Sci. Rev. 22: 107–140

Martini, I.P. & R.I.G. Morrison. 1987. Regional distribution of *Macoma balthica* and *Hydrobia minuta* on the subarctic coasts of Hudson and James Bay, Ontario, Canada – Estuarine, Coastal and Shelf Science 24: 47–68

Martini, I.P., R.I.G. Morrison, W.A. Glooschenko & R. Protz. 1980. Coastal studies in James Bay, Ontario – Geosci. Can. 7: 11–21

Martini, I.P. & R. Protz. 1978. Coastal Geomorphology, Sedimentology and Pedology of southern James Bay, Ontario, Canada – Land Resource Science, Tech. Memo. 78–1, University of Guelph, Ont.: 207 pp

Maxwell, J.B. 1986. A climate overview of the Canadian Inland Seas. In: I.P. Martini (ed.): Canadian Inland Seas – Elsevier (Amsterdam): 79–99

Milko, R. 1986. Potential ecological effects on the proposed GRAND Canal diversion project on Hudson and James bays – Arctic 39: 316–326

Norris, A.W. 1986. Review of Hudson Platform Paleozoic Stratigraphy and Biostratigraphy. In: I.P. Martini (ed.): Canadian Inland Seas – Elsevier (Amsterdam): 17–42

Pelletier, B.R. 1968. Submarine physiography, bottom sediments, and models of sediment transport in Hudson Bay. In: P.J. Hood (ed.): Earth Sci. Symp. Hudson Bay – Geol. Surv. Can. Pap. 68–53: 100–135

Pelletier, B.R. 1986. Seafloor morphology and sediments. In: Martini, I.P. (ed).: Canadian Inland Seas – Elsevier (Amsterdam): 143–162

Prest, V.K. 1970. Quaternary Geology of Canada. In: J.W. Douglas (ed.): Geology and Economic Minerals of Canada – Dept. of Energy, Mines and Resources, Geol. Surv. Can., Economic Geol. Ser. 1: 676–764

Prinsenberg, S.J. 1980. Man-made changes in the freshwater input rates of Hudson and James Bays – Can. J. Fish. Aq. Sci. 37: 736–742

Protz, R. 1982. Development of podzolic soils in the Hudson and James Bay Lowlands, Ontario – Nat. Can. 109: 501–510

Radforth, N.W. 1973. The muskeg of the Hudson Bay Lowland. In: B. Kay (ed.): Physical environment of the Hudson Bay Lowland. University of Guelph Symposium – University of Guelph, Canada: 83–113

Riley, J.L. 1982. Hudson Bay Lowland floristic inventory, wetlands catalogue and conservation strategy – Nat. Can. 109: 543–555

Sanford, B.V., A.C. Grant, J.A. Wade & M.S. Barss. 1979. Geology of eastern Canada and adjacent areas – Geol. Surv. Can., Map 1401A, 1: 2,000,000

Sanford, B.V., A.W. Norris & H.H. Bostock. 1968. Geology of the Hudson Bay Lowlands (Operation Winisk) – Geol. Surv. Can., Pap. 67–60: 1–45

Sanford, B.V. & A.W. Norris. 1973. The Hudson Platform. In: R.G. McCrossan (ed.): The future petroleum provinces of Canada – their geology and potential – Canad. Soc. Petrol. Geol., Mem. 1: 387–409

Sanford, B.V. & A.W. Norris. 1975. Devonian stratigraphy of the Hudson Platform – Geol. Surv. Can., Mem, 379: 372 pp

Scott, D.B. & I.P. Martini. 1982. Marsh foraminifera zonations in western James and Hudson bays – Nat. Can. 109: 399–414

Shilts, W.W. 1982. Quaternary evolution of the Hudson-James Bay Region – Nat. Can. 109: 309–332

Shilts, W.W. 1986. Glaciation on the Hudson Bay Region. In: I.P. Martini (ed.): Canadian Inland Seas – Elsevier (Amsterdam): 55–78

Stockwell, C.H. 1970. Geology of the Canadian Shield: Introduction. In: R.J.W. Douglas (ed.): Geology and Economic Minerals of Canada – Geol. Surv. Can., Econ. Geol. Rept. 1: 44–54

Telford, P.G. & D.E.F. Long 1986. Mesozoic geology of the Hudson Platform. In: I.P. Martini (ed.): Canadian Inland Seas – Elsevier (Amsterdam): 43–54

Webber, P.J., J.W. Richardson & J.T. Andrews. 1970. Post glacial uplift and substrate age at Cape Henrietta Maria, southeastern Hudson Bay, Canada – Can. J. Earth Sci. 7: 317–325

Zoltai, S.C. & F.C Polett. 1983. Wetlands in Canada: their classification, distribution and use. In: A.J.P. Gore (ed.): Mires: A. Swamp, Bog, Fen, and Moor. Regional Studies – Elsevier (Amsterdam): 245–268

Geologie en Mijnbouw 68: 35–48 (1989)

Cyclic morphologic changes of the ebb-tidal delta, Texel Inlet, The Netherlands

Sha Li Ping
Comparative Sedimentology Division, Institute of Earth Sciences, University of Utrecht, Budapestlaan 4, 3508 TA Utrecht, The Netherlands

Received 11 November 1987; accepted in revised form 25 March 1988

Key words: Cyclic, ebb-tidal deltas, Frisian Islands, morphodynamics, tidal inlets, Wadden Sea

Abstract

Cyclic morphological changes occur in the ebb-tidal delta system of Texel Inlet (The Netherlands). This geomorphological cycle lasts about 70 years. The cycle starts with the development of a main ebb channel in the southern half of the inlet. A large ebb delta shoal forms north of this ebb channel. The shoal grows upwards into the inter- to supra-tidal zone and moves eastwards under the influence of wind and waves. The flood channel north of the shoal is forced to rotate clockwise, and it approaches the shoreline of Texel. The marginal ebb channel in the southern part of the inlet develops due to the tidal currents deflected to the south by the eastward migrating shoal and slowly rotates clockwise, forced by the small flood marginal channel that adjoins the mainland coast to the east. The cycle is completed by shoal attachment to the southern tip of Texel Island, which causes the northern marginal channel of the inlet to be buried. The eastward migration rate of the shoals is about 60–70 m per year, which involves a sediment transport rate of order of 580 to 0.64 × 10^6 m^3/year.

Introduction

The geomorphology of ebb-tidal deltas is controlled by tidal currents, waves and wind, and their interaction (Oertel, 1972, 1975; Hubbard, 1975; Hubbard et al., 1977, 1979; Nummedal et al., 1977; Nummedal & Penland, 1981; Hayes, 1980). Conversely, the geomorphology of ebb deltas can influence the dynamic processes, especially the flow pattern. In many cases, currents have determined the orientation and the distribution of shoals. In other cases, shoals clearly function as shields that determine the pattern of inlet water flow (Oertel, 1977). The interrelationships between the morphology of ebb deltas and the dynamic conditions commonly result in their cyclic development (Oertel, 1977; FitzGerald, 1984). Geomorphological records clearly show a similar cyclic pattern in Texel Inlet (Fig. 1) and other tidal inlets along the Wadden Islands (Joustra, 1971; Luck, 1976; Rijkswaterstaat, 1). Nevertheless, the interrelationships between the geomorphology and the dynamic processes are not clearly understood. The dynamic processes in the ebb-tidal delta of Texel Inlet are particularly complex, caused by the interaction of strong currents, waves and wind. The cyclic geomorphological pattern in the ebb delta of Texel Inlet is discussed in this paper with attention centred on the dominant processes involved.

The evidence presented here is based mainly on the analysis of historical data. The tidal range is calculated from records of the tidal station at Den Helder, wind data are compiled from records of 10 years (1965–1975) from the southern shore of the

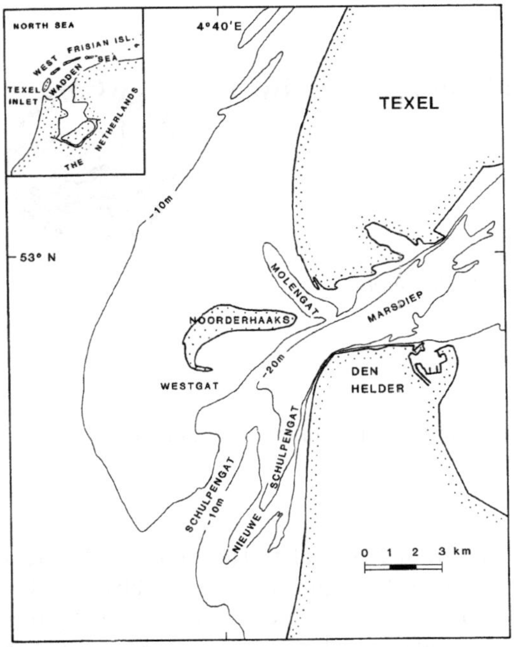

Fig. 1. Location map of the ebb-tidal delta of Texel Inlet.

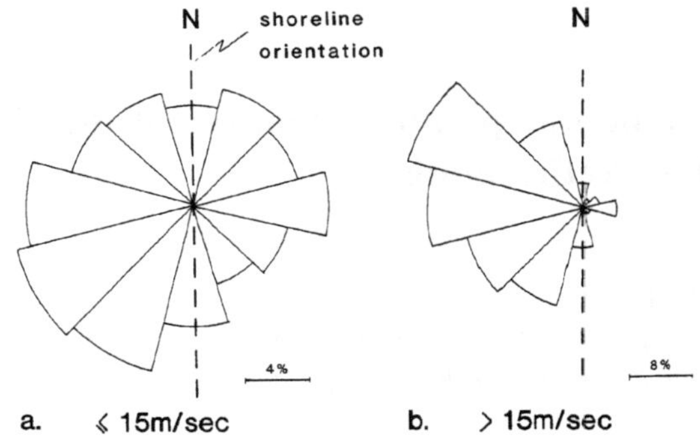

Fig. 2. a. Directional distribution of wind (<15 m/s, 97% of total observations). b. Directional distribution of wind (>15 m/s, 3% of total observations) (data from the Ministry of Transport and Public Works).

inlet. Mean wave height is calculated from observations by the Ministry of Transport and Public Works and the wave energy flux is calculated by the method of Nummedal & Stephen (1978). Old hydrographic charts and echo-sounding maps from the Ministry of Transport and Public Works provide an extensive data base to reconstruct the morphologic evolution of the ebb delta and the inlet.

Physical setting

Texel Inlet (Fig. 1) is the most westerly tidal inlet of the West Frisian Islands. It faces west, is about 2.5 km wide and the maximum depth is about 50 m. The well-developed ebb-tidal delta extends ca. 10 km seawards. The tidal basin behind the inlet (the most western part of the Wadden Sea) covers 713 km^2, 122 km^2 of which (17%) are intertidal flats. The inlet is almost entirely tide-influenced, as the input of fresh water is negligible. Supply of clay and sand to the Wadden Sea occurs from the North Sea through the inlets (Postma, 1982; Sha, 1987; Wiersma & Van Alphen, 1988). At present there is hardly any direct input of terrestrial sediments, especially sand-size particles, into this part of the Wadden Sea.

The inlet has a semidiurnal tide with a diurnal inequality. The mean tidal range is 1.38 m and the mean tidal prism is about 10^9 m^3 (Reus, 1980). The yearly average mean wind velocity is 7 m/s (Eisma, 1980). The prevailing wind is from the WSW; dominant winds (≥7 Beaufort) come mainly from the

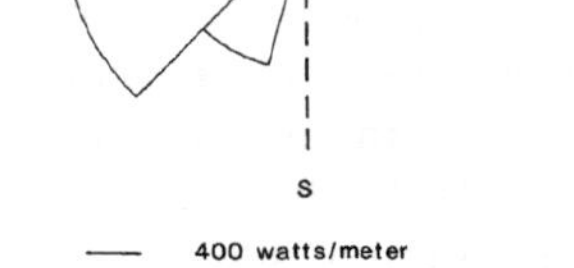

Fig. 3. Wave energy flux distribution (data from the Ministry of Transport and Public Works). Wave data were compiled from data based on regular measurements in 1975.

WNW (Fig. 2a & b). The mean wave height offshore is 1.16 m. Most of the mean annual wave energy flux is directed normally to the shoreline (Fig. 3). The resultant longshore component of wave power is relatively small, so that no significant longshore sediment transport is generated by waves. Southward sand transport may occur during storms as storm waves are mainly from the WNW. Tidal currents do transport sand northwards in the offshore area along the Dutch coast, but no reliable quantitative estimate has been made of the amounts involved.

Sand transport pattern

A short summary of the sand transport pattern in the ebb delta of Texel Inlet (Fig. 4; Sha, 1987) is

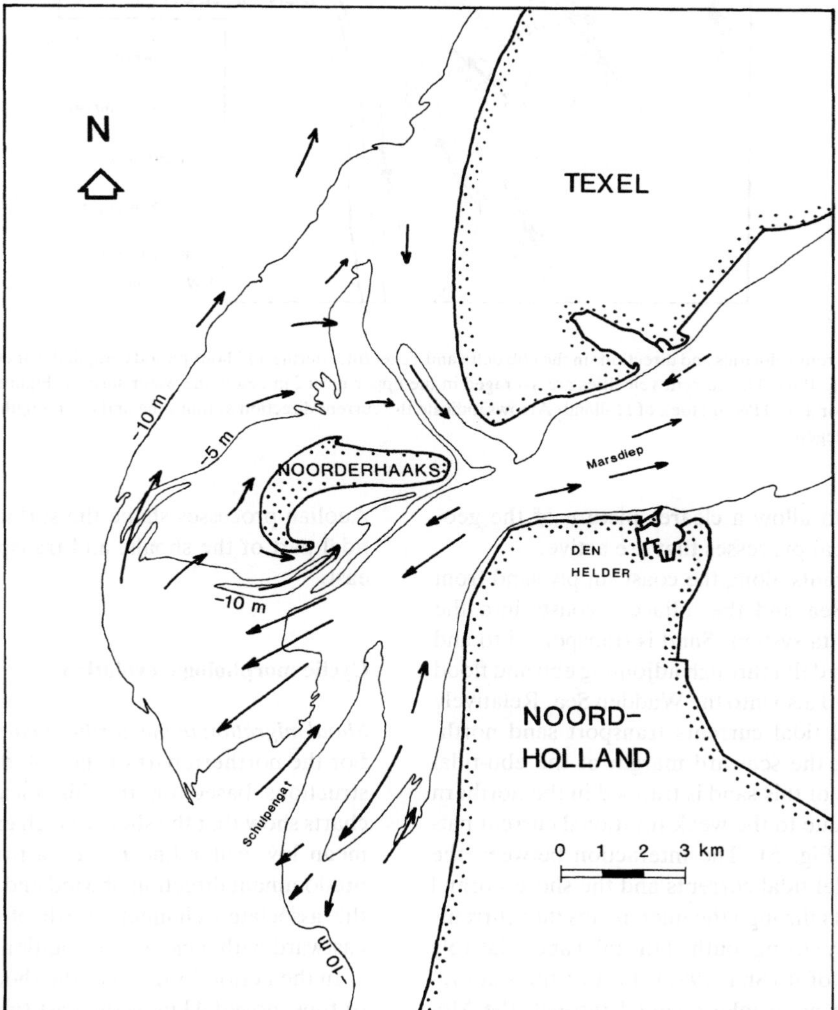

Fig. 4. The sand transport model proposed (Sha, 1987). It is drawn on the basis of bedforms, sedimentary structures, geomorphology and current data. Arrows indicate the net sand transport direction.

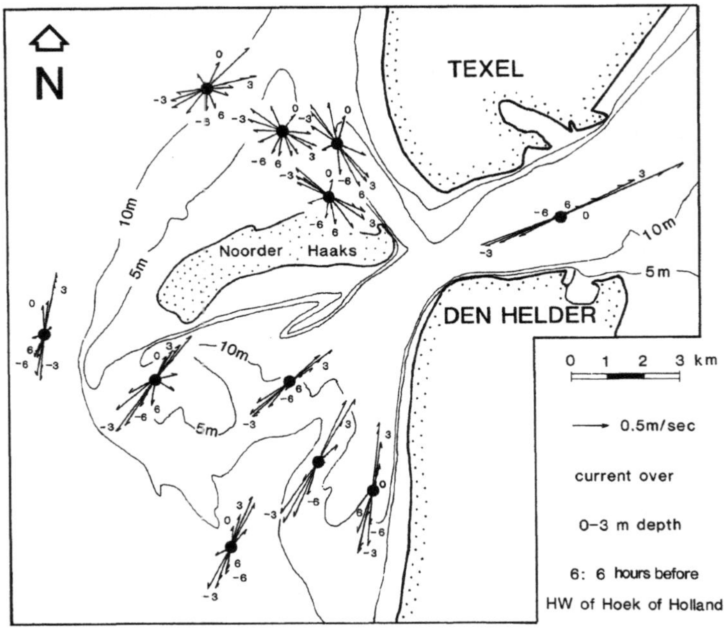

Fig. 5. Tidal current velocities and directions in the ebb delta and Texel Inlet during a 12 hour period compiled from the Stream Atlas of the Netherlands, 1963. The current velocities are averaged in the upper most 3 m below the water surface. Figures indicate the time (hours) before or after HW of Hoek of Holland. Arrows indicate the current direction at that time and their lengths represent current velocities (see legend).

given here to allow a clearer picture of the geomorphological processes that are active.

Tidal currents along the coast supply sand from the North Sea and the adjacent coasts into the ebb-tidal delta system. Sand is transported to and from the ebb delta through adjoining ebb and flood channels, and also into the Wadden Sea. Relatively strong flood tidal currents transport sand northwards along the seaward margin of the ebb-tidal delta. Some of this sand is trapped in the northern delta shoal due to the weak rotational current pattern there (Fig. 5). The interaction between the shore-parallel tidal currents and the shore-normal tidal currents through the inlet means that currents are relatively strong south of the inlet and relatively weak north of it (Sha, 1987). Part of the sand returns to the main inlet channel through the Molengat channel. Westerly waves modify the shoal morphology. They erode the seaward margin of the shoal and deposit sand on its landward margin. Aeolian processes shape the surface of the supratidal part of the shoals, and transport sand to the east.

Cyclic morphologic evolution

Shoal migration in the northern part of the ebb delta
For the northern part of the ebb delta, the reconstructions based on the historical hydrographic charts show that the shoals which extend above the mean low water line move eastward, i.e. in the predominant direction of wind and waves, and that the associated channels north of them also shift eastward with a clockwise rotation.

In the period 1851–1908, the shoal 'Onrust' (Fig. 6) thus moved 4 km to the east (about 67 m/year). This shoal first appeared near LWL in 1838. Between 1908–1916, the shoal attached to Texel Island and now forms the sand spit at its southern tip.

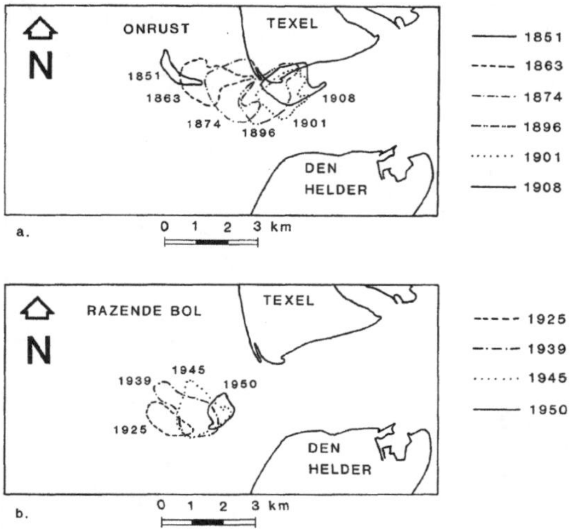

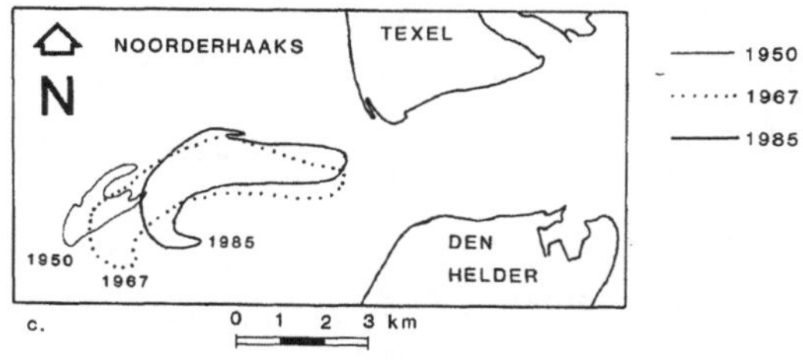

Fig. 6. Migration of the inter- to supra-tidal shoals in the ebb delta of Texel Inlet. a. The shoal 'Onrust' from 1851 to 1908. b. The shoal 'Razende Bol' from 1925 to 1950. c. The shoal 'Noorderhaaks' from 1950 to 1981.

The Noordergat channel north of the 'Onrust' was already situated in front of, and parallel to the shoreline of Texel in 1796 (Fig. 8a). It approached the southern tip of the island and was filled by the 'Onrust' shoal when it became attached to the island.

The present Noorderhaaks shoal (Fig. 7) is a combination of two shoals, 'Razende Bol' and 'Noorderhaaks' (Fig. 6b & 6c). The Razende Bol (Fig. 6b) has been permanently above the mean low water line since 1925. It migrated 1.5 km eastward from 1925 to 1950 (60 m/year). Between 1956 and 1960, the shoal 'Noorderhaaks' (Fig. 6c) west of the Razende Bol extended eastward and the two

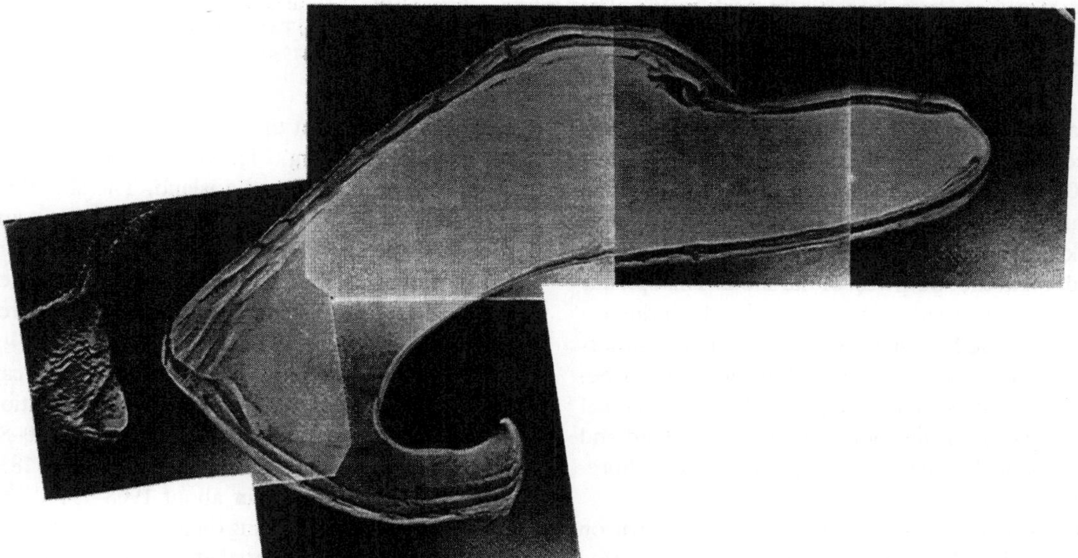

Fig. 7. Areal Photo showing the present shape of Noorderhaaks (from Ministry of Transport and Public Works). Refer to Fig. 1 for location and scale of Noorderhaaks.

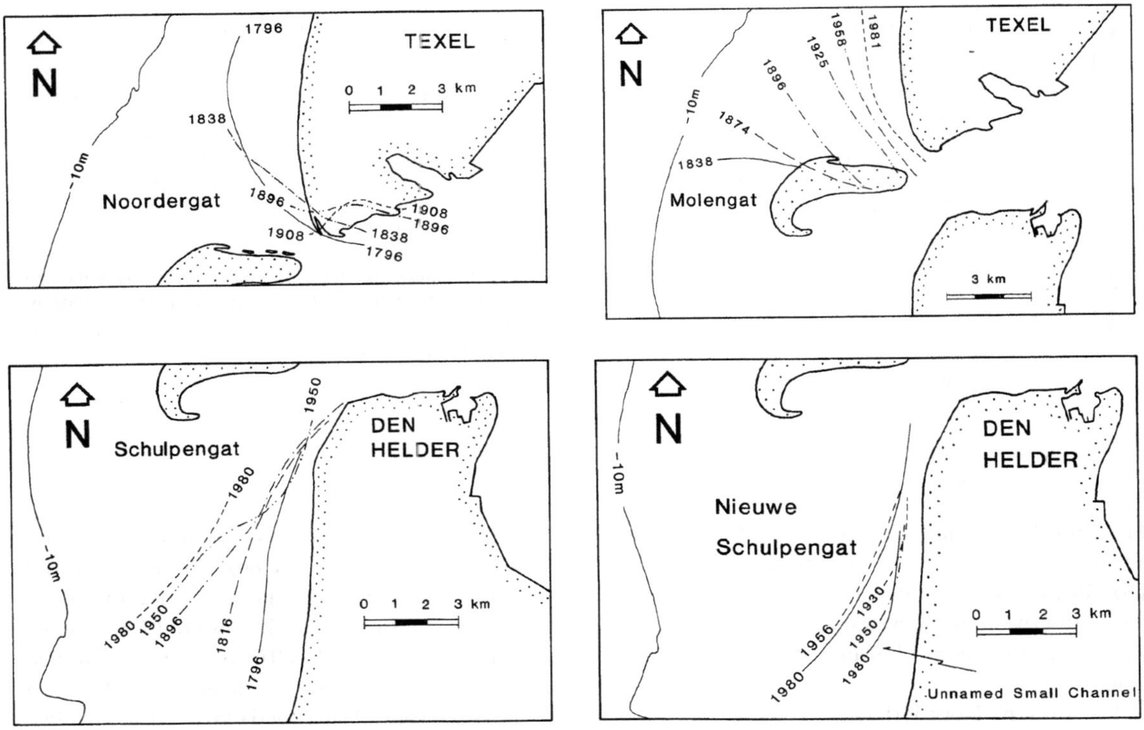

Fig. 8. Migration of some channels in the ebb delta of Texel Inlet in time. a. (upper left). The channel axis of the Noordergat channel 1796–1908. b. (upper right). The channel axis of the Molengat channel 1838–1981. c. (lower left). The channel axis of the Schulpengat channel 1796–1980. d. (lower right). The channel axis of the Nieuwe Schulpengat channel 1930–1980.

shoals coalesced. The eastern margin of this coalesced shoal has hardly moved further east since 1950, but the western margin of the shoal underwent rapid erosion and retreated eastward about 1.7 km from 1950 to 1980 (34 m/year). There is a sand spit at the southwest of the shoal. This spit has gradually recurved to the east under the influence of waves. The Molengat channel north of Noorderhaaks (Fig. 8b) has rotated about 90 degrees between 1838–1981. At present the Molengat channel nearly parallels the shoreline of Texel Island and the axis of the channel is approaching the shoreline.

These two morphological cycles are shown on cross section A′–Q′ (Fig. 9b & 9c). In the period from 1796 to 1851 the Noordergat channel approached the shoreline, resulting in shoreline retreat there. As 'Onrust' moved into the inlet after 1851, the shoreline at that point prograded due to the channel recurving (Fig. 8a, line 1908) and the shoal attaching itself to the island. The shoreline retreated again after 1896 as the Molengat approached the shoreline (Fig. 9c). If the Noorderhaaks moves further into the inlet and becomes attached to Texel Island in the future, shore progradation will occur again at the same place.

For the period since 1796, two major cycles can be distinguished in the appearance and migration of the tidal shoals above MLW. It took about 70–80 years from the emergence of 'Onrust' in about 1838 to its shore-attachment in about 1908–1916. The second cycle has been going on for the last 60 years, since the Razende Bol shoal emerged permanently above MLW in 1925. It is suggested that this cycle terminates as the shoal becomes attached to Texel Island.

Main ebb channel development

The channel development along the shoreline south of the inlet has a complex pattern. Generally, the channels developed parallel to the shoreline, and rotated slowly clockwise, while, at the same time, a new, small channel developed to the east. This can be seen in the Schulpengat (Fig. 8c & 8d).

The development of the main ebb channel in the ebb-tidal delta is related to the cyclic shoal migration. The Westgat (the main channel to the west) reached its maximum extension in 1838 (Fig. 10a). Subsequently, as the shoal gradually moved into the inlet, the current strengths to the west and southwest increased. In this period (1863–1896), the main channel (Westgat-Marsdiep) deepened further (Fig. 9d) and the tidal shield between the main ebb channel and the Schulpengat channel was breached in 1896 (Fig. 10b). After the shoal 'Onrust' became attached to the shoreline, the Westgat channel reached a maximum depth of 18 m in 1925 (Fig. 9d). Through onshore migration of the second shoal (Razende Bol) by 1950 (Fig. 10c), the tidal currents were deflected to the south, since the shoal was located in front of the inlet mouth. The southward extension of the spit (Fig. 10c & Fig. 1) at the eastern end of the shoal shows the interaction between the ebb currents and the eastward moving shoal. This caused the shoaling of the Westgat (1925–1981) (Fig. 9d) and the development of the 'Nieuwe Schulpengat' (Fig. 9e & Fig. 1). Correspondingly, the tidal discharge decreased in the Westgat and increased in the Nieuwe Schulpengat.

Ebb-tidal delta volumetric change, sand budget and sand transport rate

The total sand budget of the ebb delta calculated using the method of Dean & Walton (1975) is $490 \times 10^6\,m^3$. Due to the migration and growth of the shoal complex in the north, there is a cyclic change in the distribution of sand in the ebb-tidal delta. When the bar reached its maximum size for the two examples known, the part above the mean water line was about 6.5 km long and over 1 km wide and contained at least $32.5 \times 10^6\,m^3$ sand (about 7% of the total sand budget). During onshore migration, it may have decreased in size, but it still brought a large amount of sand to the shoreline. For example, in about 1908, the shoal 'Onrust' attached to the shore which thus gained sand of a minimum of $9.2 \times 10^6\,m^3$ (about 2% of the total sand budget).

The estimated eastward sediment transport rate by shoal migration was on average at least $0.61 \times 10^6\,m^3$/year. This is about 0.13% of the total sand budget in the delta.

Discussion

Mechanism of the Cyclic Morphologic Process

A comparison of the length of the main ebb channel (measured on the −10 m line) in different years and the tidal prism at different inlets along the West Frisian Islands shows a strong correlation between the two (Fig. 11). Although the channel length for a certain tidal prism at an inlet is variable through time, the upper and lower limits of the channel length in the history of the inlets correlate almost linearly with the tidal prism through that inlet. This means that when the channel length is below the upper limit, the channel is in a developing stage, and when it approaches the upper limit, the main ebb channel loses its efficiency, starts to silt up and then is progressively abandoned.

The formation of the shoal is related to the development of the main ebb channel. Conversely, the shoal migration influences the channel development.

As the main ebb channel develops, it extends seawards and interrupts the shore-parallel tidal currents (Sha, 1987). This leads to a large dynamic 'shadow' area north of the main channel and as a result, there a large delta shoal develops and the water depth decreases significantly over a large area. Onshore waves dominate the shallow bottom, form swash bars and push them onshore. When the shoal rises above the low water line, aeolian transport becomes important. It dominates as the shoal becomes supra-tidal. Thus, both the westerly waves and winds result in the eastward migration of the 'Onrust' and 'Noorderhaaks'. The importance of aeolian transport on the Noorderhaaks was observed during field work in 1986 (Fig.

a.

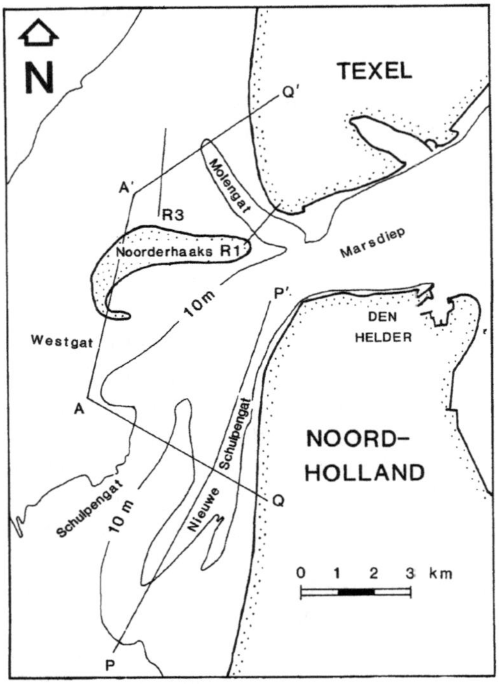

Fig. 9. Geomorphological evolution along several cross-sections in the ebb delta of Texel Inlet.

a. Location map (left).

b & c. Cross-section A′–Q′ from 1796–1896 and from 1896–1981 (below).

b.

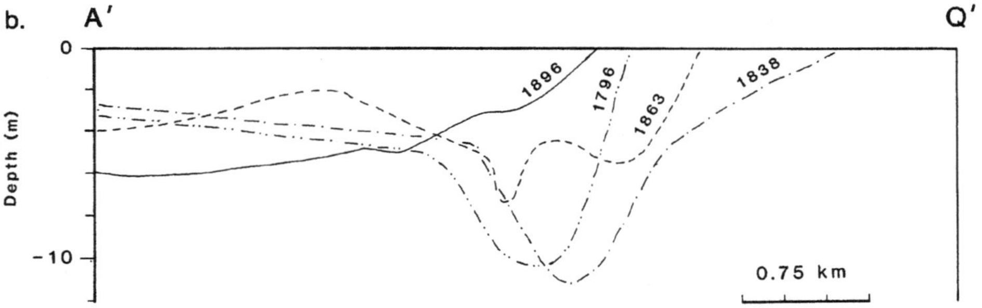

c.

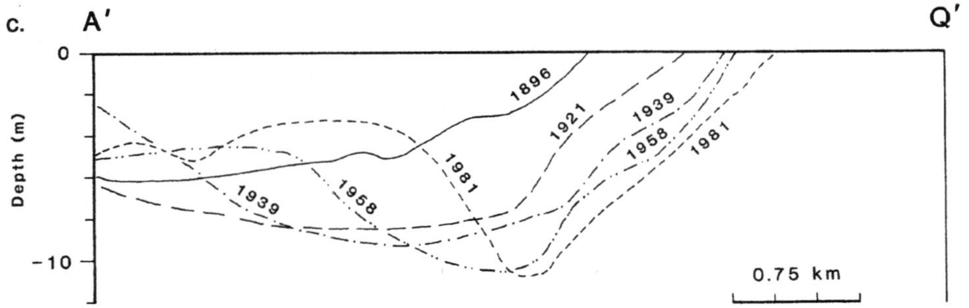

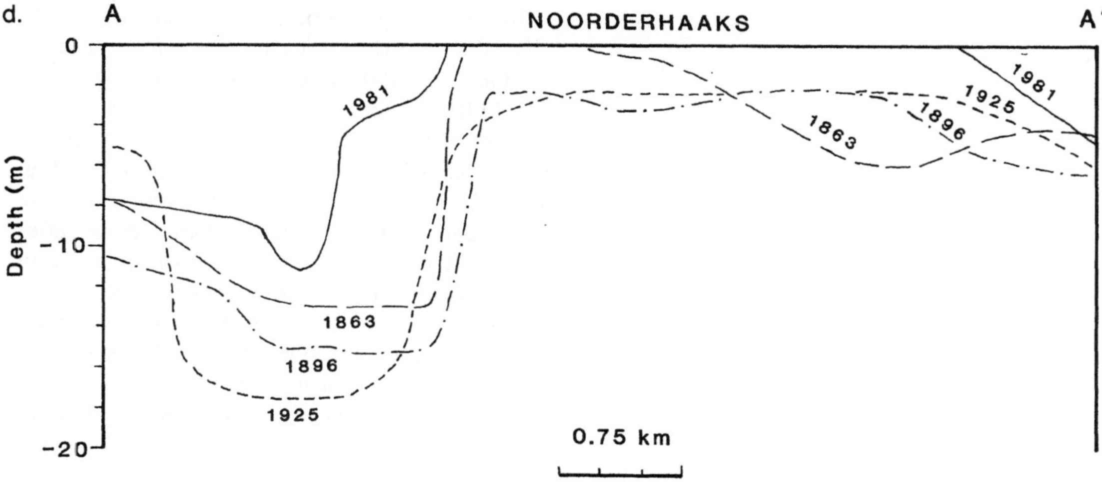

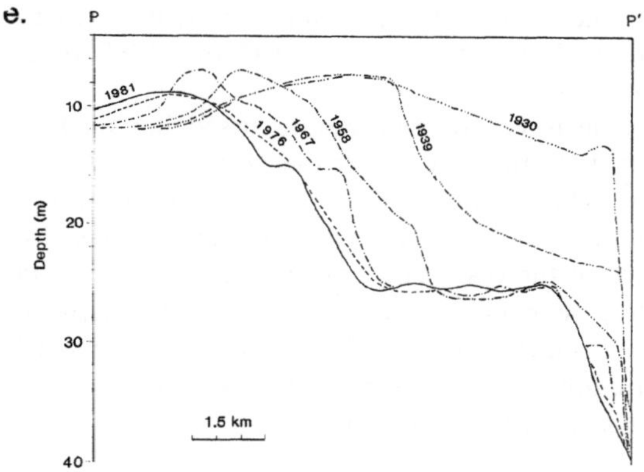

d. Cross-section A–A′ from 1863–1981 (above).

e. Cross-section P–P′ from 1930–1981 (left).

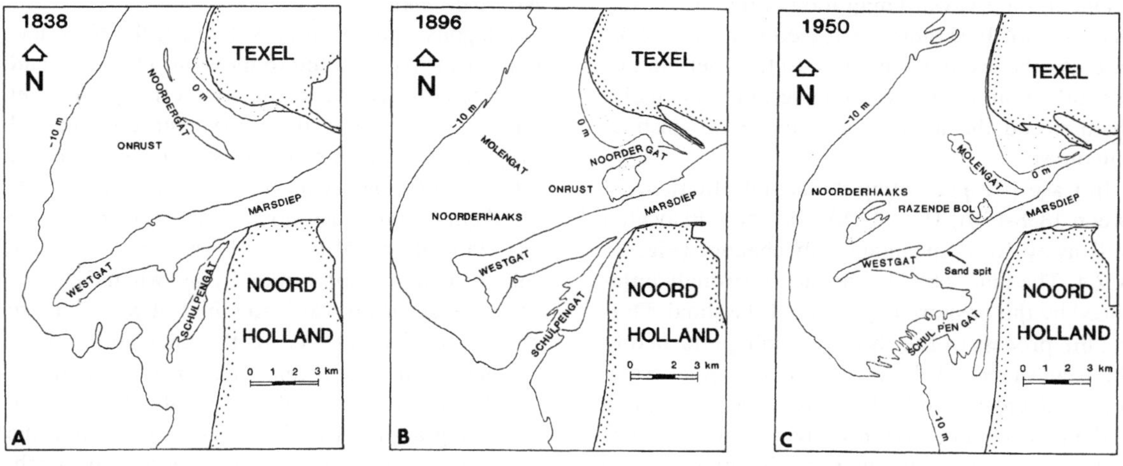

Fig. 10. The bathymetric maps of the ebb delta in (a) 1838, (b) 1896 and (c) 1950. Note the sand spit and channel-shoal development.

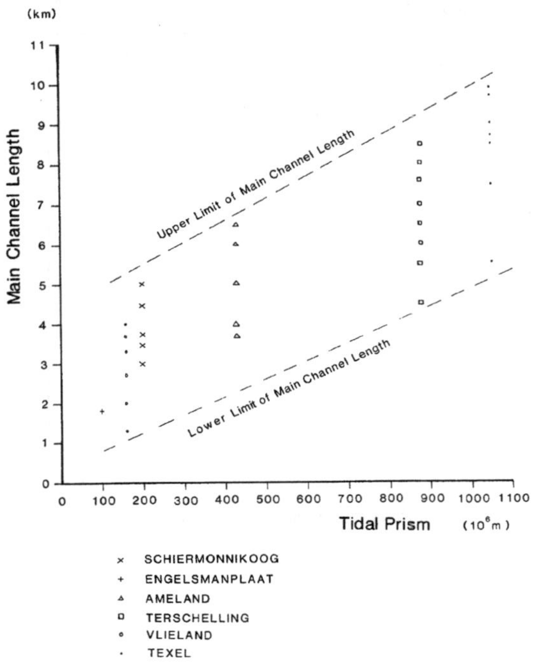

Fig. 11. Relation of the main ebb channel length in different years (1796–1981) and the tidal prism in various inlets along the West Frisian Islands. Tidal prism through individual inlets is assumed to be stable.

12). The runnel along the beach of the shoal was completely filled during a single storm.

Due to the eastward migration of the shoals, the channels north of them are forced to rotate clockwise. As the shoal moves close to the inlet mouth, depending on its location relative to the main ebb channel and the inlet, there can be two distinct outcomes:

In Case 1 (representing the actual situation between 1838-1916) (Fig. 13A), the shoal is located well away from the main ebb channel (Fig. 13 A.2.). The main ebb channel is not strongly influenced by the onshore migration of the shoal. During the process of the shoal becoming attached to the island (Fig. 13 A.3.), the inlet channel cross-section decreases and tidal currents become strong in the main channel. At this stage, a new cycle may start due to the development of the main channel (Fig. 13 A.4.).

In Case 2 (representing the situation since 1925) (Fig. 13B), the shoal is located more in front of the inlet mouth and the onshore migration (Fig. 10c & Fig. 13 B.2, B.3 & B.4) of the shoal deflects the ebb currents to the south. The spit extended southwestward from the east margin of the Razende Bol (Fig. 10c). This led to silting of the Westgat and the development of the Nieuwe Schulpengat after 1925. After a further attachment of the shoal to Texel Island, the main ebb channel may breach again through the area to the west (cf. FitzGerald, 1982), where waves eroded the sand during the first cycle. A new cycle then starts (Fig. 13 B.5 & B.6).

The end of a cycle and beginning of a new one are likely to be triggered by storm waves, e.g. the attachment of 'Onrust' to the Texel Island was completed by a storm (Ministry of Transport and Public Works, Department of Hoorn, 1987, pers. comm.).

One aspect, which is not discussed due to a lack of data, is the influence of the closure of the IJsselmeer-Afsluitdijk in 1932 on the ebb-tidal delta. It is known that the tidal range increased after the closure of the dike, and possibly there has been an increase of the tidal prism. This may have extended the main channel seaward or switched the channel orientation. The closure of the dike may influence consequently the cycle duration.

Practical implication: the future of the Noorderhaaks

Analogous to the history of Onrust, the Noorderhaaks shoal is expected to move into the inlet and to become attached to the southern tip of Texel Island. At the same time, the main ebb channel may open up again to the west. However, recent data do not show a continued and significant eastward migration of the shoal. This may be related to the stabilization of the southern shoreline of the inlet by construction works. If, and when the shoal moves eastward towards the inlet, it will decrease the cross sectional area of the inlet and thereby increase tidal current velocities. Stronger tidal currents erode the east side of the shoal. Eastward shoal migration thus is balanced by erosion of the east side of the shoal by tidal currents. Slow northward migration, however, does occur. Possibly, a

Fig. 12. Photo of eolian dunes on Noorderhaaks (May 12, 1986). Note that the wind dunes migrated towards a wet beach runnel. Hand shovel/spatula for scale.

further northerly movement of the Noorderhaaks will lessen the influence of the ebb currents coming through the inlet on the shoal and finally will lead to attachment of the shoal to the land during a storm event.

Geological record

The observed cyclic process of the ebb delta of the inlet system has important implications for the understanding of the geological record. For example, inlet sequences produced by successive shoreward moving shoals will show that paleo-channels are successively buried by shoals, in an overall progradational barrier/spit sequence (Fig. 14). In ebb deltas, the preservation of sand bodies is the end-product of several cyclic processes. Due to the frequent migration of the channel-shoal system, the preservation of a single channel sequence may be rare, instead a sequence of channel floor sheet deposits, formed by the lateral shifting of channels, especially of marginal channels, is more likely to be preserved. In systems with a scale and dynamics comparable to those of Texel Inlet, these surfaces will normally be formed at depths of about 10–20 m (depending on the inlet size).

Conclusion

The ebb-tidal delta and inlet system of Texel Inlet, shows a cyclic morphologic development. This process involves a maximum extension of the main ebb channel, formation of a large delta shoal to the north, onshore migration of the shoal, and flood channel rotation north of the inlet. The cycle ends with the attachment of the shoal to Texel Island. The development and silting up of the main ebb channel are related to the tidal prism of the inlet (Fig. 11) and are influenced by onshore migration

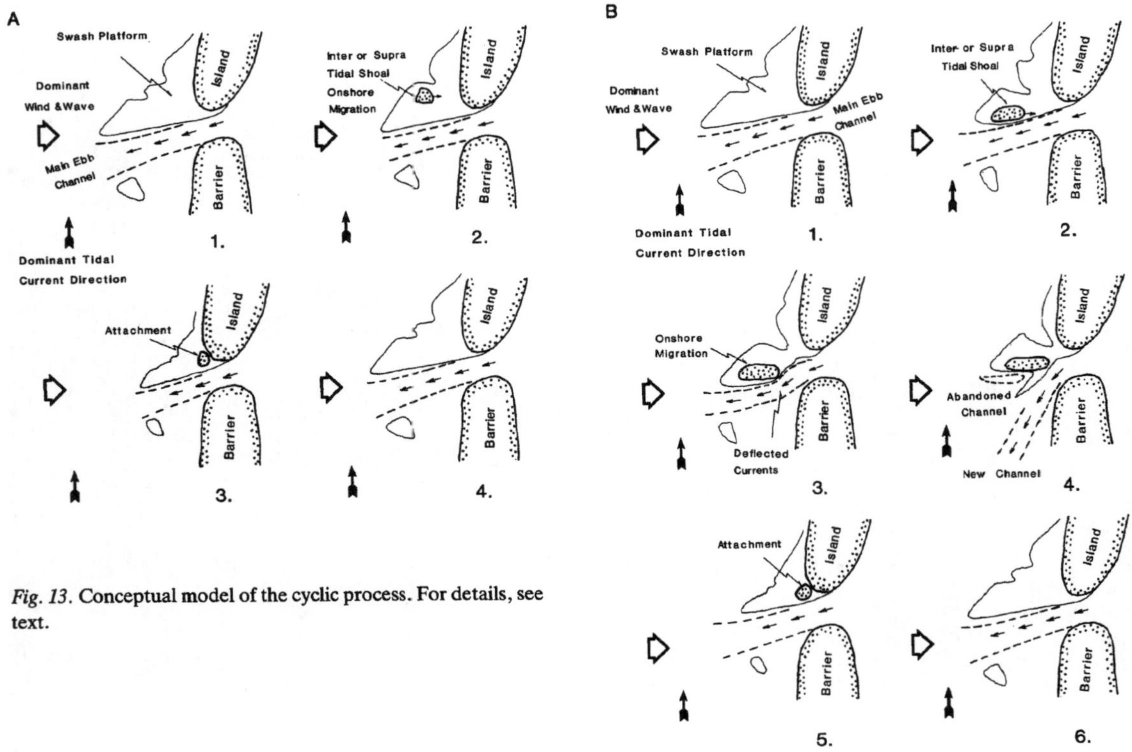

Fig. 13. Conceptual model of the cyclic process. For details, see text.

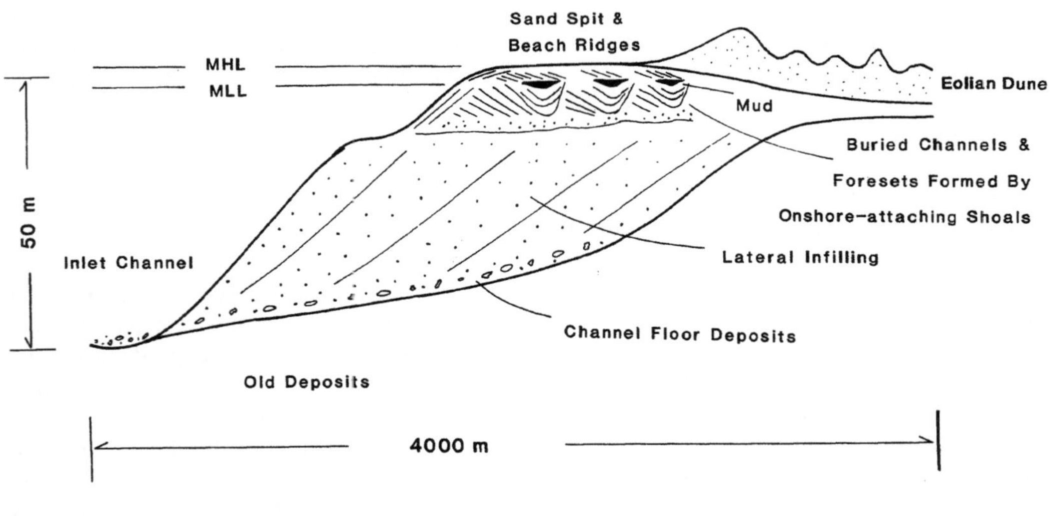

Fig. 14. A schematic model showing the inlet sequence produced by inlet migration and successive onshore-attaching shoals.

and enlargement of the shoal-complex.

The morphologic development of the inlet-delta system clearly is a feed back response system of dynamic processes and morphologic development. In fact, it is in this way that the inlet maintains dynamic equilibrium.

Eastward migration of Noorderhaaks is presently slowed down by erosion of ebb currents from the inlet at the eastern margin of the shoal. Eastward migration may possibly resume after the shoal has shifted further north and it may ultimately become attached to Texel Island, probably during a major storm.

Acknowledgements

I acknowledge the support of the Netherlands Institute for Sea Research and of the Ministry of Transport and Public Works for this research project. I thank Dr P.L. de Boer, Dr D. Eisma. Dr S.D. Nio, Dr Tj.C.E. Van Weering and Dr J. Wiersma, for their very active support and help during the research work. Special thanks are due to Dr P.L. de Boer, in particular for his very careful work on my crude manuscript. Very helpful checking and suggestions from Dr D. Eisma, Tj. C.E. Van Weering, J. Wiersma, and many others, are appreciated. Dr J.S. Pethick has corrected the language of the manuscript. The study department of the Ministry of Transport and Public Works in Hoorn supplied many valuable data, which is gratefully acknowledged. Finally, my thanks are for the help at sea by the captains and crew of the Mss. 'Navicula' and 'Aurelia' and the technicians of the Netherlands Institute for Sea Research.

References

Dean, R.G. & Walton, Jr.T.L. 1975 Sediment transport processes in the vicinity of inlets with special reference to sand trapping. In: Cronin, L.E. (ed.): Estuarine Research 2. Academic Press (New York, San Francisco, London): 129–150.

Eisma, D. 1980 Natural forces. In: K.S. Dijkema, H.-E. Reineck & W.J. Wolff (eds): Geomorphology of the Wadden Sea area – A.A. Balkema (Rotterdam): 20–31.

FitzGerald, D.M. 1982 Sediment bypassing at mixed energy tidal inlets – Proc. 18th Coast. Eng. Conf. ASCE (Cape Town, South Africa): 1094–1118.

FitzGerald, D.M. 1984 Interactions between the ebb-tidal delta and landward shoreline: Price Inlet, South Carolina – J. Sediment. Petrol. 54: 4.

Hayes, M.O. 1980 General morphology and sediment patterns in tidal inlets – Sediment. Geol. 26: 139–156.

Hubbard, D.K. 1975 Morphology and hydrodynamics of the Merrimack River ebb-tidal delta. In: Cronin, L.E. (ed.): Estuarine Research 2. Academic Press (New York, San Francisco, London): 253–266.

Hubbard, D.K., Barwis, J.H. & Nummedal, D. 1977 Sediment transport in four South Carolina inlets – Proc. 'Coastal Sediments '77', Amer. Soc. Civil Eng. New York: 582–601.

Hubbard, D.K., Oertel, G.F. & Nummedal, D. 1979 The role of waves and tidal currents in the development of tidal-inlet sedimentary structures and sand body geometry: examples from North Carolina, South Carolina and Georgia – J. Sediment. Petrol. 49: 1073–1092.

Joustra, D.Sj. 1971 Geulbeweging in de Buitendelta's van de Waddenzee – Rijkswaterstaat Intern. Rapt. W.W.K. 71-14. (unpubl.): 27 pp.

Luck, G. 1976 Inlet changes of the East Frisian Islands – Proc. 15th Conf. Coast. Eng. 2, ASCE (New York): 1938–1957.

Nummedal, D., Oertel, G.F., Hubbard, D.K. & A.C. Hine 1977 Tidal inlet variability: Cape Hatteras to Cape Canaveral – 'Coastal Sediments '77', ASCE (New York): 543–562.

Nummedal, D. & S. Penland 1981 Sediment dispersal in Norderneyer Seegat, West Germany. In: S.D. Nio, R.T.E. Schuttenhelm & Tj. C.E. van Weering (eds): Holocene Marine Sedimentation in the North Sea Basin. Intern. Ass. Sediment. Spec. Publ. 5. Blackwell (Oxford, London, Edinburg, Boston, Melbourne): 187–210.

Nummedal, D. & Stephen, M.F. 1978 Wave climate and littoral sediment transport, Northeast Gulf of Alaska – J. Sediment. Petrol. 48, 2: 359–371.

Oertel, G.F. 1972 Sediment transport on estuary entrance shoals and the formation of swash platforms – J. Sediment. Petrol. 42: 858–863.

Oertel, G.F. 1975 Ebb-tidal deltas of Georgia Estuaries. In: Cronin, L.E. (ed.): Estuarine Research 2. Academic Press (New York, San Francisco, London): 267–276.

Oertel, G.F. 1977 Geomorphic cycles in ebb deltas and related patterns of shore erosion and accretion – J. Sediment. Petrol. 47: 1121–1131.

Postma, H. 1982 Hydrography of the Wadden Sea – A.A. Balkema. (Rotterdam): 75 pp.

Reus, J.H. 1980 Ontwikkeling Zeegat van Texel – Rijkswaterstaat Intern. Notitie wwkz-80. H248 (unpubl.): 21 pp.

Rijkswaterstaat, 1: A picture film (8 mm) on the evolution of inlets along the West Wadden Sea. Ministry of Transport and Public Works of The Netherlands.

Sha, L.P. 1987 Sand transport patterns in the ebb-tidal delta off Marsdiep Inlet, Wadden Sea, The Netherlands. (Abstract) – KNGMG Symp. Coastal Lowlands Geology and Geotechnology, The Hague, The Netherlands, 1987 – Abstracts and

Programme 58.
Wiersma, J. & Van Alphen, J.S.L.J. 1988 The morphology of the Dutch shoreface between Hook of Holland and Den Helder. In: P.L. De Boer, A. Van Gelder & S.D. Nio (eds): Tide-influenced sedimentary environments and facies – Reidel (Dordrecht): 101–111.

Geologie en Mijnbouw 68: 49–72 (1989)

Intraplate stresses and the stratigraphic evolution of the North Sea Central Graben

Henk Kooi[1], Sierd Cloetingh[1] & Gijs Remmelts[2]
[1] *Vening Meinesz Laboratory, Institute of Earth Sciences, University of Utrecht, P.O. Box 80.021, 3508 TA Utrecht, The Netherlands; Present address: Department of Sedimentary Geology, Institute of Earth Sciences, Free University, P.O. Box 7161, 1007 MC Amsterdam, The Netherlands;*
[2] *Geological Survey of The Netherlands, P.O. Box 157, 2000 AD Haarlem, The Netherlands*

Received 21 Januari 1988; accepted 28 April 1988

Key words: stratigraphic modelling, intraplate stresses, tectonic evolution of northwestern Europe, quantitative subsidence analysis, paleo-stress fields, Dutch Central Graben, paleobathymetry, sea level changes, Vail curves, basin tectonics

Abstract

We present results of stratigraphic modelling and quantitative analysis of subsidence data for the southern part of the North Sea Basin. Tectonic subsidence curves are given for fifteen wells in the northernmost segment of the Dutch North Sea and the southern part of the Dutch Central Graben. These curves have been supplemented with tectonic subsidence curves for eight wells from the Broad Fourteens and West Netherlands Basins. Subsidence analysis and thermo-mechanical modelling show that Late Jurassic and Early Cretaceous multiple stretching phases with a finite duration are required to explain the observed stratigraphic record. Our analysis demonstrates the important role of intraplate stresses in the evolution of these basins. The paleo-stress curve inferred from the stratigraphic modelling shows a trend with a change from tensional and neutral stresses during Mesozoic times to a stress regime of more overall compressional character during Cenozoic times. Superimposed on this long-term trend are short-term stress fluctuations. This paleo-stress curve and the associated stratigraphic record of the Dutch North Sea Basin sheds light on the record of paleo-stress measurements in the Northwestern European platform and is consistent with data on the kinematic evolution of the Tethys belt. These findings demonstrate the key-importance of tectonics and stress-induced vertical motions – related to rifting events in the northern Atlantic region and the interaction of the Eurasian and African plates – in controlling the stratigraphic evolution of the North Sea Basin.

Introduction

During the last few years considerable progress has been made in quantative modelling of sedimentary basin development (see for a review Beaumont & Tankard 1987). These studies have demonstrated the important role of thermo-mechanical properties of the lithosphere in models of basin evolution (e.g. Watts et al. 1982). Furthermore, they have quantified the contributions of a variety of geodynamic processes to the vertical motions of the lithosphere at sedimentary basins. These processes include thermal contraction induced by cooling of the lithosphere amplified by the loading of sediments that accumulate in these basins (Sleep 1971), isostatic response to crustal attenuation by stretching (McKenzie 1978), and flexural bending in response to vertical loading (Beaumont 1978). The increase in flexural rigidity associated with long-term cooling of the lithosphere upon rifting has been shown to explain adequately the long-term widening of rifted basins and the associated long-term (on time scales of several tens of Ma) sea level record (Watts 1982).

A new key element in basin modelling is the incorporation of intraplate stresses in stratigraphic modelling (Cloetingh 1988). Cloetingh et al. (1985) and Cloetingh (1986) have demonstrated that fluctuations in intraplate stress fields provide a tectonic explanation for short-term (1–5 Ma) fluctuations in apparent sea level. These fluctuations in intraplate stress fields are superimposed on the tectonic subsidence induced by the driving mechanisms mentioned above (see Fig. 1). Vertical motions of the lithosphere at basin flanks (or apparent sea level changes) are induced at a rate and magnitude consistent with analysis of the seismic stratigraphic record (Vail et al. 1977, Haq et al. 1987). Figure 2 schematically illustrates the effects of changes in intraplate stress on the stratigraphy at the edge of a sedimentary basin calculated for an elastic lithosphere. When horizontal compression occurs, the peripheral bulge flanking the basin is magnified, resulting in uplift of the basin flanks and seaward migration of the shore line. An offlap develops and an apparent fall in sea level results, possibly exposing the sediments, thus causing the development of an unconformity. Simultaneously, the basin centre deepens, resulting in a steeper basin slope. For a horizontal tensional intraplate stress field, the flanks of the basin subside. This results in a landward migration of the shore line and an apparent rise in sea level so that renewed deposition with a corresponding facies change is possible. In this case the centre of the basin shoals, and the basin slope is reduced. The synthetic stratigraphy at the basin edge is schematically shown for the following three situations: long-term widening of the basin with cooling in the absence of an intraplate stress field (Fig. 2a); the same case with a superimposed transition to 500 bar compression (Fig. 2b); and the case of a stress change to 500 bar tension (Fig. 2c).

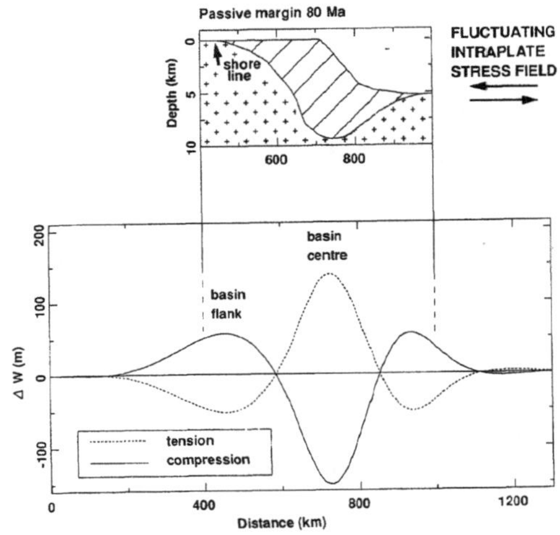

Fig. 1. Flexural deflections at a sedimentary basin induced by changes in the intraplate stress field. Sign convention: uplift is positive, subsidence is negative. Above: an 80 Ma old passive margin initiated by stretching. The wedge of sediments flexurally loads an elastic plate. The thickness of this plate varies horizontally due to lateral changes in the temperature structure of the lithosphere. Below: the vertical deflections induced by a change to 1 kbar compression (solid curve). The flank of the margin is uplifted and the basin centre subsides. A change to 1 kbar tension (dashed curve) induces uplift of the basin centre and subsidence of the basin flank. The shape and magnitude of these stress-induced deflections evolve through time not only because of the increasing load, but also due to changes in the thermal structure of the lithosphere (after Cloetingh et al. 1985).

These short-term cycles in apparent sea level have been traditionally interpreted in terms of a glacioeustatic control. However, with the exception of the Oligocene events, there is no evidence in the geological and geochemical records for significant Mesozoic and Cenozoic glacial events prior to Middle Miocene times (Frakes 1979). Glacio-eustatics, therefore, cannot explain those major parts of the apparent sea level record where glacial phases are thought to have been insignificant (Pitman & Golovchenko 1983). Furthermore, changes in intraplate stress fields associated with tectonic reorganizations in the lithosphere also explain the existence of a strong correlation (Bally 1982) between the timing of plate reorganizations and rapid low-

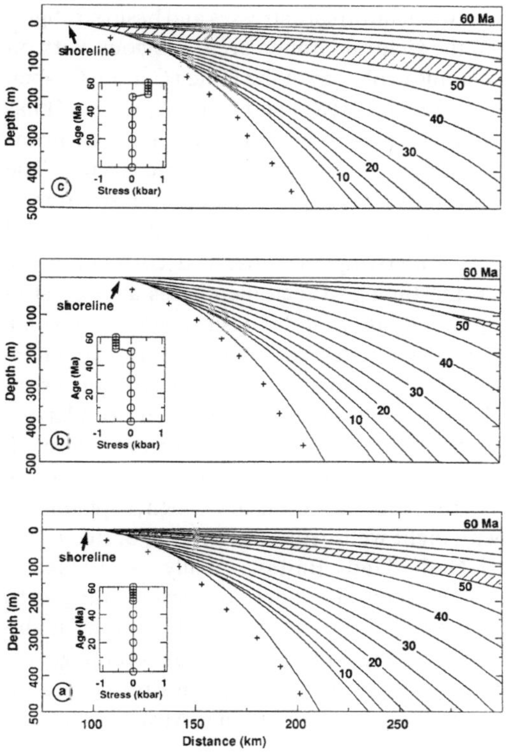

Fig. 2. Synthetic margin stratigraphy for a 60 Ma old basin, which is initiated by lithospheric stretching followed by thermal subsidence and flexural infilling of the resulting isostatic depression. Shading indicates the position of a sedimentary package bounded by isochrons of 50 Ma and 52 Ma after basin formation. a) Stratigraphy with zero-intraplate stresses. b) Effect of a stress change to 500 bar compression at 50 Ma. Stress-induced uplift of the peripheral bulge induces narrowing of the basin and a phase of rapid offlap, followed by a long-term phase of gradual onlap due to thermal subsidence. c) Effect of a stress change to 500 bar tension at 50 Ma. Stress-induced downwarping of the peripheral bulge causes widening of the basin and a phase of rapid basement onlap.

erings in sea level shown in the Vail et al. (1977) curves.

Simultaneously, considerable progress has been made recently in the study of the lithospheric stress field itself. Detailed analysis of earthquake focal mechanisms, in-situ stress measurements and analysis of break-out orientation logs taken in wells drilled for commercial purposes have demonstrated the existence of consistently oriented stress provinces in the lithosphere (Zoback 1985, Klein & Barr 1986). At the same time, numerical modelling (Wortel & Cloetingh 1981, 1983; Cloetingh & Wortel 1985, 1986) has yielded better understanding of the causes of the observed variations in stress level and stress directions in the various lithospheric plates. These studies have demonstrated a causal relationship between processes at plate boundaries and deformation in the plate's interiors (e.g. Johnson & Bally 1986). Similarly, substantial progress has been made in quantifying the stress levels associated with compressional and extensional deformation. Folding of oceanic lithosphere in the Bay of Bengal, for example, has been shown to require compressional stress levels of the order of several kbars (McAdoo & Sandwell 1985), a stress level that is in excellent agreement with independent estimates of the stress level in the northeastern Indian Ocean (Cloetingh & Wortel 1985). The formation of sedimentary basins by lithospheric stretching requires tensional stress levels also on the level of a few kbars (Cloetingh & Nieuwland 1984, Houseman & England 1986).

Given current active research in sea level fluctuations, the possibility that the effect of intraplate stresses is significant has been explored for various regions. Schlanger (1986) suggested that intraplate stress variations provide a possible explanation for the enigmatic high frequency (order of a few Ma) variations in Cretaceous apparent sea levels. Hallam (in press) discussed the implications of such variations for the Jurassic sea levels of northwestern Europe. Meulenkamp & Hilgen (1986) explored a causal relation between Neogene stratigraphy in the eastern Mediterranean and intraplate stress variations. De Vries-Klein (1987) examined the stratigraphic record of the Paleozoic Appalachian foreland basin and concluded that local variation and local tectonic styles and associated changes in stress field might have masked the possible expressions of global fluctuations in sea levels.

Recently, we have concentrated on the North Sea area to explore the potential use of the sea level record as a new source of information on paleostresses (Cloetingh 1986, Lambeck et al. 1987, Cloetingh et al. 1987). This choice was primarily motivated by the documented occurrence of important temporal changes in tectonic regime, a feature that is inherent to the location of the North Sea area relative to the North Atlantic rift system and the Alpine fold belt (P.A. Ziegler 1982). Furthermore, the North Sea area has played a crucial role in the construction of the Exxon 'global' sea level charts (see Vail et al. 1977, their figure 5). However, the traditional interpretation of the sea level record in terms of glacio-eustatics has been the subject of increasingly intensive debate. As has been realized by several authors (Parkinson & Summerhayes 1985, Miall 1986, Summerhayes 1986, Hallam, in press) several features displayed in the Vail third-order cycles may actually reflect North Sea Basin tectonics. Therefore, the North Sea area is of key importance to examine the relative contributions of glacio-eustatics and tectonics to the observed sea level fluctuations.

The tectonic history of the North Sea area can be divided into several stages (P.A. Ziegler 1975, 1978, 1982; W.H. Ziegler 1975). The North Sea Permo-Triassic basin was formed after the Variscan geosynclinal stage. Sedimentation took place in the Northern and Southern Permian Basin separated by the Ringkobing-Fyn and Mid North Sea High. Rifting occurred during Early Triassic to Early Cretaceous, abated subsequently and ceased altogether during the Late Paleocene. Rifting prevailed contemporaneously with the opening of the Atlantic, with the strongest rifting pulses occurring during Late Jurassic and Early Cretaceous. During Late Cretaceous and early Tertiary fault block-movement – and associated differential subsidence – were reactivated in response to the Laramide compressional phases, resulting in some very prominent basin inversions. Subsequently, a wide Cenozoic basin developed which is nearly unaffected by faulting. As demonstrated by P.A. Ziegler (1982) there is a correlation between the timing of most of the short-term lowerings in sea level and the major rifting and compressional episodes of Mesozoic and Tertiary ages.

In the present paper we discuss results of a quantitative analysis of subsidence data and modelling of basin stratigraphy for the southern part of the North Sea Basin. In doing so we focus on the relation between intraplate stresses and basin stratigraphy in the Central Graben area. Previous work on the tectonic evolution of the Dutch Central Graben by P.A. Ziegler (1987) and Van Wijhe (1987) has shown a correlation between the timing of inversion tectonics and Alpine orogenic events. The present study aims to provide a quantitative framework to improve understanding of the mechanical aspects underlying the observed correlation between the timing of tectonic phases and the stratigraphic evolution of the basins. We begin with a brief review of observational evidence for consistently oriented regional stress fields in Northwestern Europe. Subsequently, we present results of a quantitative analysis of subsidence data from fifteen wells along two transects through the southern part of the North Sea Central Graben, supplemented with eight wells from the Broad Fourteens and West Netherlands Basins. Finally, we present a thermomechanical model for basin stratigraphy along one of these transects. This model is discussed in the light of pertinent data on the geodynamic evolution of Northwestern Europe and the interaction of the Eurasian and African plates.

Regional stress field in Northwestern Europe

The existence of a present-day compressive regional stress field in the upper crust of the Alpine foreland of Northwestern Europe has been inferred from a wide range of observational techniques. The stress pattern as determined by in-situ stress techniques and analysis of break-out orientation logs from oil wells is largely consistent with stress directions inferred from analysis of focal mechanisms of earthquakes (Klein & Barr 1986). The general trend of the horizontal component of maximum compressive stress σ_1 is SE-NW (about 140°), a trend that is remarkably uniform over large distances. These features provide strong evidence for far-field stress propagation away from the Alpine collision from (Klein & Barr 1986, see Fig. 3). Local deviations from the overall stress pattern are observed mainly along zones of active strain release such as the Rhine Graben (Illies et al. 1981).

Paleo-stress patterns for the Alpine foreland

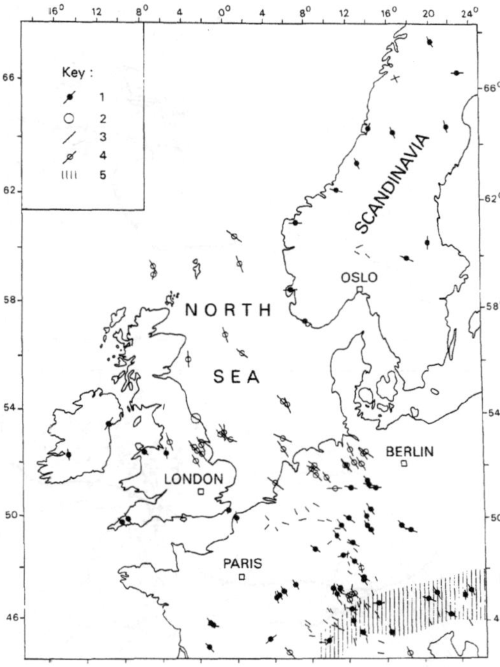

Fig. 3. Compilation of observed maximum horizontal present-day stress directions in the Alpine foreland. 1 = the direction of maximum horizontal stress from in-situ measurements, 2 = a horizontal stress equal in all directions found from in-situ measurements, 3 = the direction of maximum horizontal stress inferred from earthquake focal mechanism studies, 4 = the direction of maximum horizontal stress based on break-out analysis, 5 = Alpine fold belt. The data indicate stress propagation away from the Alpine fold belt in the platform region (after Klein & Barr 1986).

have been deduced from measurements on microstructures in sedimentary rocks such as stylolites, tensional joints and from movements on small faults (Letouzey 1986, Bergerat 1987). These authors demonstrated the relatively uniform character of the paleo-stress patterns in both time and space in the Cenozoic (Fig. 4). The regional stress field during Cenozoic times seems to be primarily controlled by European-African plate convergence and continental collision. Although the regional stress field is of overall compressive character, it has been subject to important temporal variations. Changes in magnitude (Letouzey 1986) and rotation of principal stress axes in adjustment to migration of the European-African rotation pole are observed (Bergerat 1987, Letouzey 1986). Inversion of parts of subsiding graben systems during the Late Cretaceous and early Tertiary and the timing of these events also provide strong evidence for transmission of compressive stresses from the Alpine collision front throughout the Alpine foreland over distances of the order of 1000 km (P.A. Ziegler 1987).

Subsidence analysis of well data

The sedimentary record provides key information on vertical motions of the lithosphere and, consequently, on the tectonic subsidence underlying the formation and evolution of a basin. As the tectonic subsidence of a basin is amplified by sediment loading, a correction for this effect has to be applied before the driving tectonic subsidence can be extracted from the stratigraphy. This procedure is commonly called backstripping (Steckler & Watts 1978) or geohistory analysis (Van Hinte 1978).

We have calculated (water loaded) tectonic subsidence through time using methods discussed by Seckler & Watts (1978), Sclater & Christie (1980) and Bond & Kominz (1984). In this study we have adopted the Harland et al. (1982) time scale. Lithological effects, in particular compaction and extreme densities of evaporites, have been corrected for. Each stratigraphic unit between two chronostratigraphic horizons has been assigned a sand, silt, shale, carbonate, anhydrite and halite percentage with each lithology responding according to its own compaction scheme. Minimum and maximum limits of lithological effects have been tested which showed that associated uncertainties in tectonic subsidence are of the order of up to several tens of metres. We have used the stratigraphic nomenclature of The Netherlands (NAM & RGD 1980) for the interpretation of the stratigraphy. Paleo-water depth data have been used for three wells together with information on long-term eustatic sea level changes. For the other wells we adopted an arbitrarily constant paleo-water depth of 100 metres. We have assumed Airy isostasy in the analysis of the subsidence data. By doing so we have ignored the

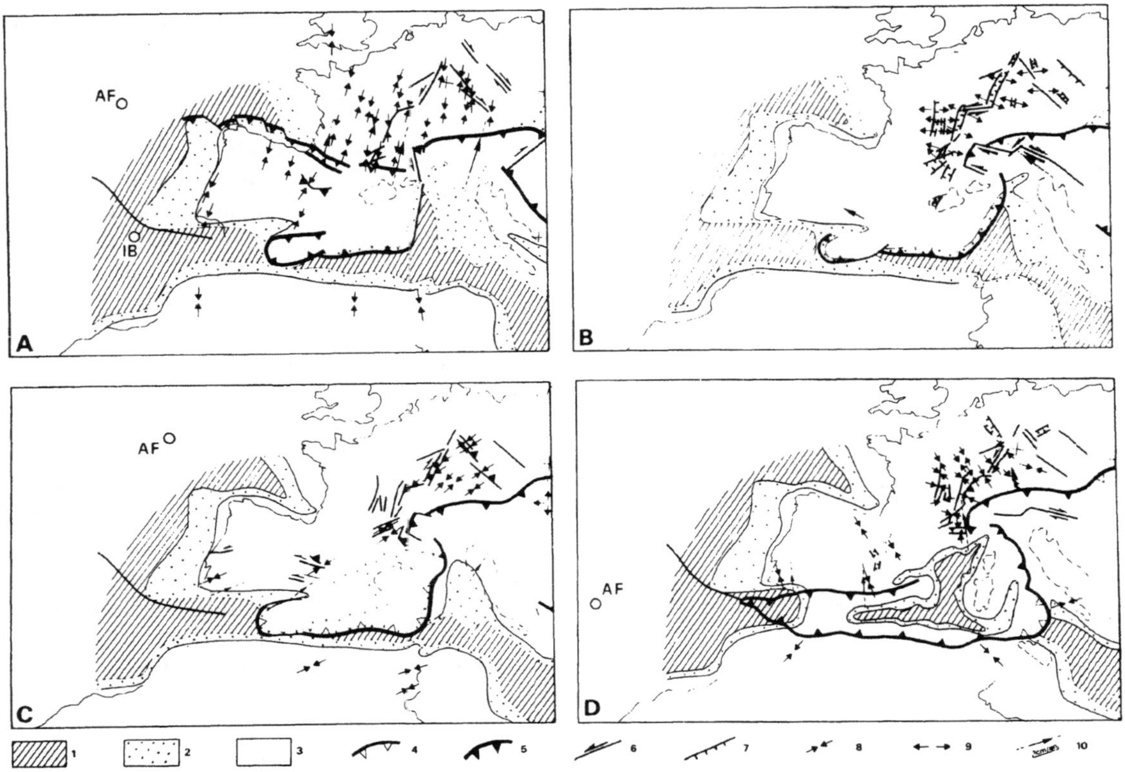

Fig. 4. Paleo-stress directions in the Alpine foreland. a) Late Eocene (40 Ma). b) Oligocene (35–30 Ma). c) Early Miocene (22–20 Ma). d) Late Miocene to post-Miocene (7–4 Ma). 1 = oceanic crust, 2 = thinned continental crust, 3 = continental crust, 4 = subduction zone, 5 = overthrust belt, 6 = strike-slip fault, 7 = normal fault, 8 = azimuth of the main maximum stress σ_1, 9 = azimuth of the main minimum stress σ_3, 10 = relative vector of motion Africa/Eurasia in centimetres/year (after Bergerat 1987).

effect of a finite strength of the lithosphere on its response to sediment loading. This assumption, however, does not significantly affect the shape of the inferred subsidence curves (Watts et al. 1982). This also implies that flexural effects of intraplate stresses have been retained in these curves. Similarly, we have ignored the reduced basement cooling due to the blanketing effect of sediments (Turcotte & Ahern 1977, Lucazeau & Le Douaran 1985). To incorporate this effect in the analysis would require detailed knowledge of the thermal structure of the lithosphere throughout the basin evolution. Furthermore, due to the long time scale on which lithospheric cooling operates (order of tens of Ma), a correction for the effect of blanketing (order of magnitude of up to 100 metres) will not significantly alter the slope of the tectonic subsidence curves (Lucazeau & Le Douaran 1985).

The locations of the wells used in the present analysis are given in Fig. 5 and Table 1. The wells have been selected for three regions; the northernmost part of the Dutch North Sea, the Dutch Central Graben area (54° N) and the Broad Fourteens/West Netherlands Basins, respectively. Because commercial bore holes are normally placed on prospective structures some deviations from the regional subsidence pattern can be expected. These features will be treated further on together with estimates of the magnitude of uplift phases, associated with an erosional expression in the stratigraphy, obtained from regional seismic lines.

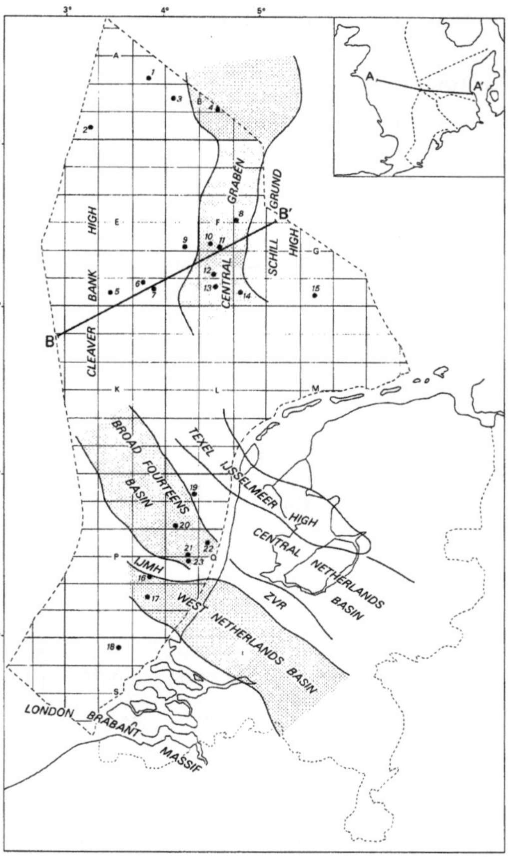

Fig. 5. Simplified tectonic map of the Dutch Central North Sea. Wells used in the present study are indicated by dots. Numbering of the wells refers to convention used in Table 1. Otherwise, wells are named after the blocks in which they are situated. The location of the stratigraphic cross-section AA' through the northernmost part of the Dutch Central Graben is shown in the inset. Line marked by BB' indicates the position of a NOPEC seismic line. ZVR and IJMH denote Zandvoort Ridge and IJmuiden High respectively.

Northernmost part of the Dutch North Sea

Tectonic subsidence curves for four wells in the northernmost part of the Dutch sector of the North Sea are displayed in Fig. 6. Horizontal dashed lines in the tectonic subsidence curves indicate a truncation in subsidence and correspond to stratigraphic unconformities. This representation is clearly a simplification as an erosional hiatus is usually associated with complex motions of the basement.

Evidence is lacking for net Middle Triassic to Late Jurassic subsidence in the area of the A-wells on the western flank of the Central Graben (Mid North Sea High). This is in agreement with observed deep truncation caused by Middle Jurassic upwarping of the Central North Sea rift dome and uplift of the rift flanks during the Late Triassic rifting phase. Evidence for subsidence in the latest Jurassic can be found in the Graben area (B14-1) at an earlier stage (Callovian) than in the flank region, which suggests that either renewed subsidence first occurred in the graben, or that Callovian sediments have been removed from the flanks during the Oxfordian. The latter possibility is associated with a short-term fall in relative sea level, which can be explained by relaxation of tensional stresses. This mechanism is strongly supported by the presence of a short-wavelength Oxfordian-Portlandian crustal bulge on both sides of the Central

Table 1. Summary of boreholes used in this study.

Nr.	Well	Latitude (°*N*)	Longitude (°*E*)	Penetration Depth (m)
1	A12-1	55°24′00.1″	03°48′33.9″	3404
2	A16-1	55°07′28.1″	03°15′11.9″	2678
3	B13-1	55°18′01.2″	04°04′30.2″	2837
4	B14-1	55°12′16.9″	04°34′35.5″	2525
5	E17-1	54°05′38.8″	03°25′26.2″	3727
6	E18-1	54°07′40.9″	03°53′10.3″	2463
7	E18-2	54°09′05.0″	03°46′38.2″	4445
8	F9-2	54°30′56.8″	04°43′32.5″	2382
9	F10-1	54°21′57.4″	04°13′59.1″	3433
10	F11-1	54°22′04.0″	04°35′39.0″	2920
11	F11-2	54°24′54.6″	04°27′38.8″	2663
12	F14-1	54°11′39.5″	04°30′32.0″	2977
13	F17-1	54°08′43.6″	04°31′41.9″	3871
14	F18-1	54°05′54.0″	04°44′32.0″	3718
15	G17-1	54°04′29.9″	05°30′41.0″	3955
16	P12-1	52°23′18.9″	03°52′38.4″	3554
17	P15-1	52°15′36.8″	03°51′32.8″	3224
18	S2-1	51°57′10.1″	03°33′28.7″	1766
19	Q1-2	52°53′04.6″	04°18′52.4″	3072
20	Q4-1	52°41′14.6″	04°07′20.1″	3105
21	Q7-1	52°30′16.5″	04°13′16.7″	2573
22	Q8-2	52°35′42.8″	04°23′33.6″	2535
23	Q10-2	52°29′57.6″	04°13′27.9″	2337

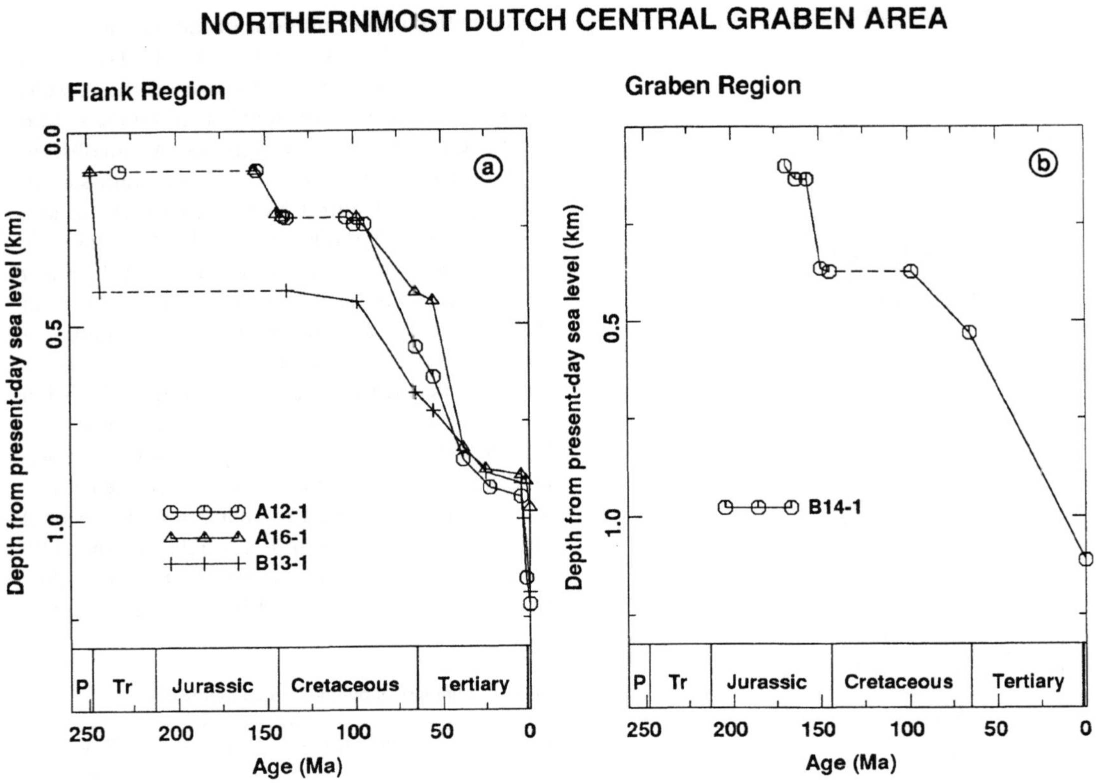

Fig. 6. Water loaded tectonic subsidence curves for four wells in the northermost part of the Dutch Central Graben. Curves have been constructed ignoring long-term changes in sea level and paleobathymetry. For location of the wells see Fig. 5 and Table 1. Wells located in the Central Graben and wells located on the flanks have been separated. Note the overall convex upward subsidence pattern of these curves.

Graben (P.A. Ziegler 1982). This phase of differential subsidence is followed by a phase of rapid subsidence corresponding to the deposition of the Kimmeridge Clay. Kimmeridgian subsidence also influenced the flanks of the graben. The B13-1 well is located on a small salt piercement and the associated salt tectonics might explain its deviating subsidence pattern. Subsidence is fairly continuous from Early Cretaceous on. For the Graben region the Late Cretacous subsidence rate appears to be relatively low. Inspection of a regional seismic NOPEC line, however, has shown that roughly half of the Late Cretaceous subsidence in the area of the B14-1 well has been compensated by salt movements, which explains the relatively low subsidence rate for this particular period inferred from the B14-1 well. Decelleration of subsidence is observed during Oligocene and Miocene times, which in turn is followed by extremely rapid subsidence during Pliocene-Quaternary times. As demonstrated by the subsidence curves for the wells in the flank region, this recent phase of subsidence is not characterized by spatially uniform subsidence rates over the area.

The absence of thick sequences of Jurassic and Early Cretaceous syn-rift sediments produces the observed overall convex upward subsidence pattern. This suggests that Jurassic and Early Cretaceous syn-rift subsidence was largely prohibited by active subcrustal heating and tectonic activity and that strong subsidence due to crustal attenuation only began after this active heating abated. On the other hand, a striking characteristic of all four wells in the northernmost part of the Dutch North Sea area is the nearly linear (post-Early Cretaceous) subsidence until mid-Tertiary times, followed by strong decelleration during Oligocene and Miocene times. This subsidence pattern deviates from the predictions of a simple thermal model for the evolution of cooling lithosphere. Although influence of active thermal processes can not be excluded, these observations create the impression that other factors have an equally important influence on the inferred subsidence characteristics.

We have investigated the effects of paleobathymetry (e.g. Gradstein et al. 1985, in press) and long-term eustatic sea level changes on the inferred tectonic subsidence. The underlying cause of long-term sea level changes (as shown in Fig. 7) is the temporal variation in thermal structure of oceanic lithosphere, which in turn is dependent on changes in the age distribution of the ocean floor. Therefore, a major control on first order sea level cycles (on the order of several tens of Ma) is the formation of new ocean basins and the destruction of preexisting ocean basins (Angevine et al. 1988). Estimates for paleobathymetry for the Cenozoic based on analysis of several wells in the Viking Graben and the Central Graben (Gradstein 1988, pers. comm.) show consistent characteristics over distances of several hundred kilometres. Apart from differences in absolute values of the inferred amplitudes, the results are largely consistent with independent paleobathymetry data for the Danish and Dutch sectors and for the central and northern part of the North Sea Basin (Barton & Wood 1984, Wood 1981). Based on these data we have constructed paleo-water depth curves for the Cenozoic, which were combined with Mesozoic data from P.A. Ziegler (1982), Frandsen et al. (1987) and Jensen & Buchardt (1987). In general, paleo-water depth estimates are subject to great uncertainties. Therefore, we have used upper limits of estimated paleo-water depths to test extreme effects of this component on the tectonic subsidence.

Fig. 8 shows the results for the analysis of the tectonic subsidence of the B14-1 well employing these paleo-water depth data and the long-term sea level curves of Kominz (1984) given in Fig. 7. Correcting for long-term sea level changes (Fig. 8a) results in net Early Cretaceous uplift, reduced Late Cretaceous subsidence and higher Cenozoic tectonic subsidence rates. Application of this correction, therefore, enhances the convex upward tectonic subsidence pattern. Adding a correction for changes in paleo-water depth (Fig. 8b) further changes the inferred subsidence pattern. Increasing water depths during Early Cretaceous yield a strong increase in tectonic subsidence. Continuing deep water conditions toward the Paleocene do not further affect the Late Cretaceous tectonic subsidence rates. More than a kilometre of sediments can be taken up in the process of net shallowing toward the present, which strongly reduces the amount of Cenozoic tectonic subsidence necessary to explain the observed sediment thickness. For comparison a synthetic tectonic subsidence curve for instantaneous depth dependent stretching (Royden & Keen 1980) at 130 Ma is shown in Fig. 8b. We adopted a crustal stretching factor $\delta = 1.17$. The subcrustal attenuation factor β controls the relative amount of initial syn-rift subsidence and thermal subsidence. We have used $\beta = 1.35$ for the construction of the synthetic tectonic subsidence curve shown in Fig. 8b. Incorporation of paleo-water depth information thus results in a concave upward subsidence pattern for low-magnitude long-term sea level

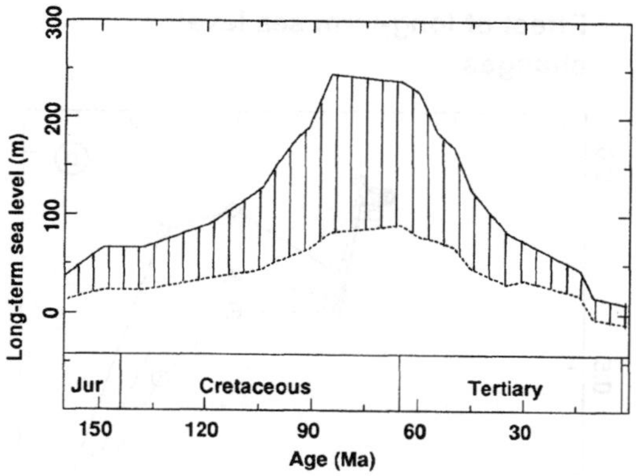

Fig. 7. Long-term sea level changes used in the analysis of two wells in the northernmost part of the Dutch Central Graben and one well in the Dutch part of the Central Graben (54° N). Continuous an dashed lines indicate maximum and minimum sea level, respectively (after Kominz 1984).

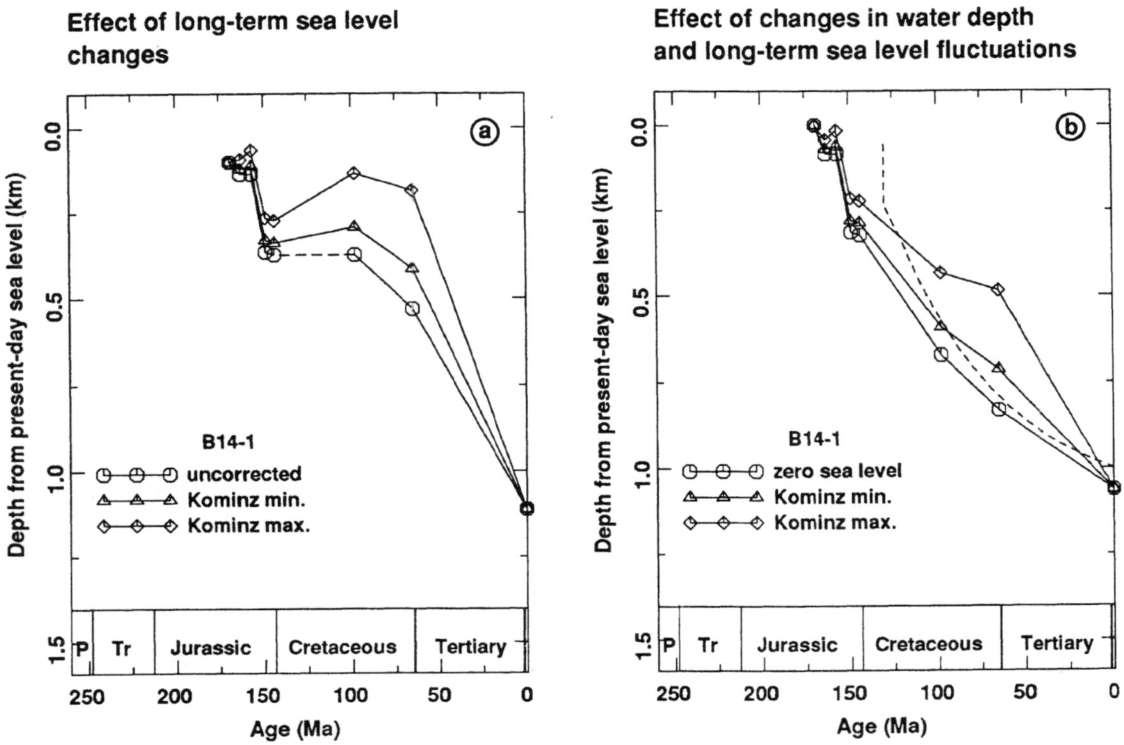

Fig. 8. Tectonic subsidence for the B14-1 well corrected for paleo-water depth and for long-term sea level changes (Kominz 1984) as indicated in Fig. 7. a) Corrected for long-term sea level changes only. The lowermost curve denotes the uncorrected tectonic subsidence. b) Corrected for both long-term sea level changes and variations in paleo-water depth. The dashed curve denotes a synthetic tectonic subsidence curve for instantaneous stretching at 130 Ma with crustal stretching by a factor of $\delta = 1.17$ and subcrustal attenuation by a factor of $\beta = 1.35$. Note the improved fit to this curve when both corrections are applied, especially for low magnitudes of the long-term sea level changes.

changes. However, the paleo-water depths used do not seem to produce the concavity predicted by a simple cooling plate model. This feature is probably caused by a lack of resolution in the Early Cretaceous part of the subsidence curve. Water depths might have been very shallow or zero near the end of the Early Cretaceous (approximately mid-Aptian), which is supported by the presence of the Early Cretaceous unconformity. Therefore, rapid subsidence might have started later than suggested by the subsidence curves of Fig. 8.

We also applied corrections for paleobathymetry and long-term changes in sea level for the A12-1 well, which contains more detailed information on Cenozoic subsidence (Fig. 9). The applied corrections for this well also give rise to dramatic modifications in the pattern of calculated tectonic subsidence. Tertiary subsidence rates are high during Paleocene and Eocene followed by uplift during Oligocene and Miocene times. This suggests that considerable differences between paleo-water depths in the northern and southern segment of this part of the North Sea Basin have occurred, especially during the Paleocene. This might be related to documented Paleocene rifting activity in the Arctic-North Atlantic domain (P.A. Ziegler 1982). Lower estimates for early Tertiary paleo-water depths give rise to a decrease in early Tertiary tectonic subsidence and at the same time reduces or nullifies middle to late Tertiary uplift (Fig. 9b).

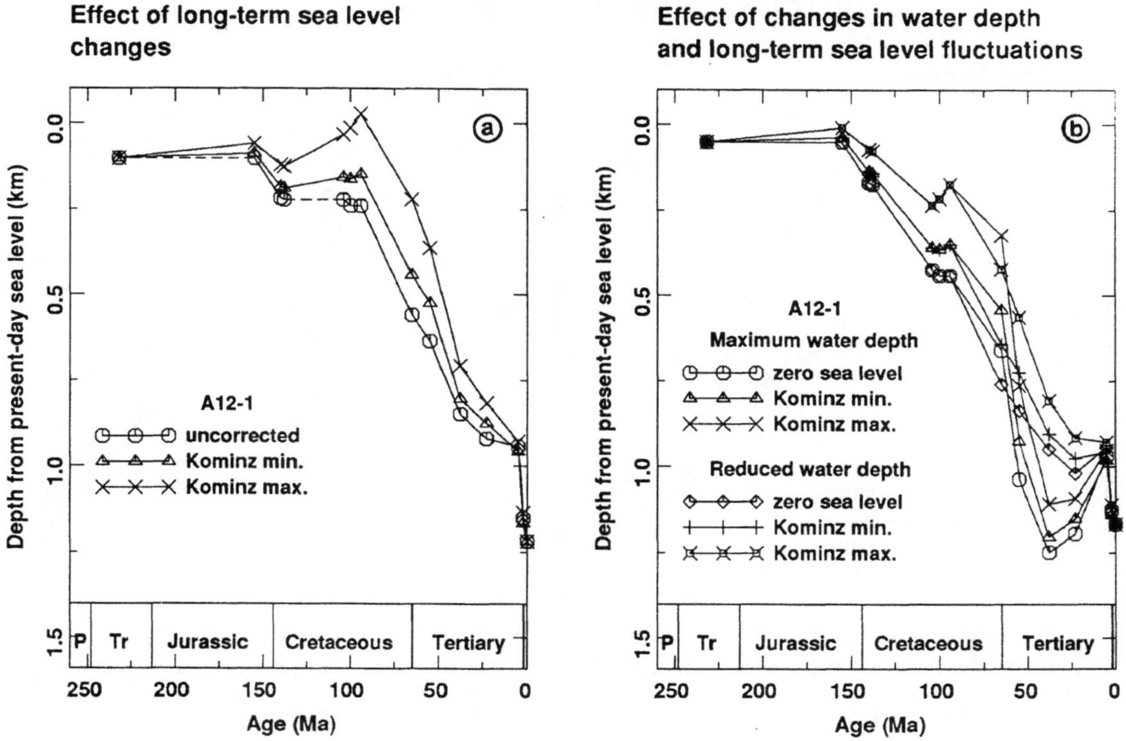

Fig. 9. Tectonic subsidence for the A12-1 well corrected for paleo-water depth variations and long-term sea level changes (Kominz 1984) as indicated in Fig. 7. a) Corrected for long-term sea level changes only. The lowermost curve denotes the uncorrected tectonic subsidence. b) Corrected for both long-term sea level changes and variations in paleo-water depth. A maximum and a reduced paleo-water depth curve have been used.

It should be kept in mind that the paleo-water depth changes used in the foregoing analysis are by their nature subject to large uncertainties. This analysis demonstrates that the interpretation of North Sea Basin stratigraphy in terms of long-term sea level changes and tectonic subsidence depends strongly on accurate information on paleo-water depths.

Central Graben area (54° North)

Stratigraphic data from eleven wells in the Dutch sector of the North Sea Basin around 54° North have been analyzed (Fig. 5). In conjunction with these well data we used a seismic NOPEC line (position indicated by BB' in Fig. 5), which traverses the Dutch Central Graben region and shows the profound effect of salt tectonics and the Laramide graben inversion in this part of the Dutch North Sea. Curves displaying the (water loaded) tectonic subsidence are given in Fig. 10.

Rapid Early and Late Jurassic subsidence is characteristic for the wells in the graben region. This phase of subsidence is the response to the Late Triassic and Middle Jurassic rifting phases. The latter phase corresponds in most wells to a period of no net subsidence or a strong decrease in subsidence rate. This feature might be associated with the uplift of the Central North Sea rift dome (P.A. Ziegler 1982). Subsequent extremely high subsidence rates during Callovian and the Oxfordian are in accordance with the observed opening of the southern part of the Central Graben during these

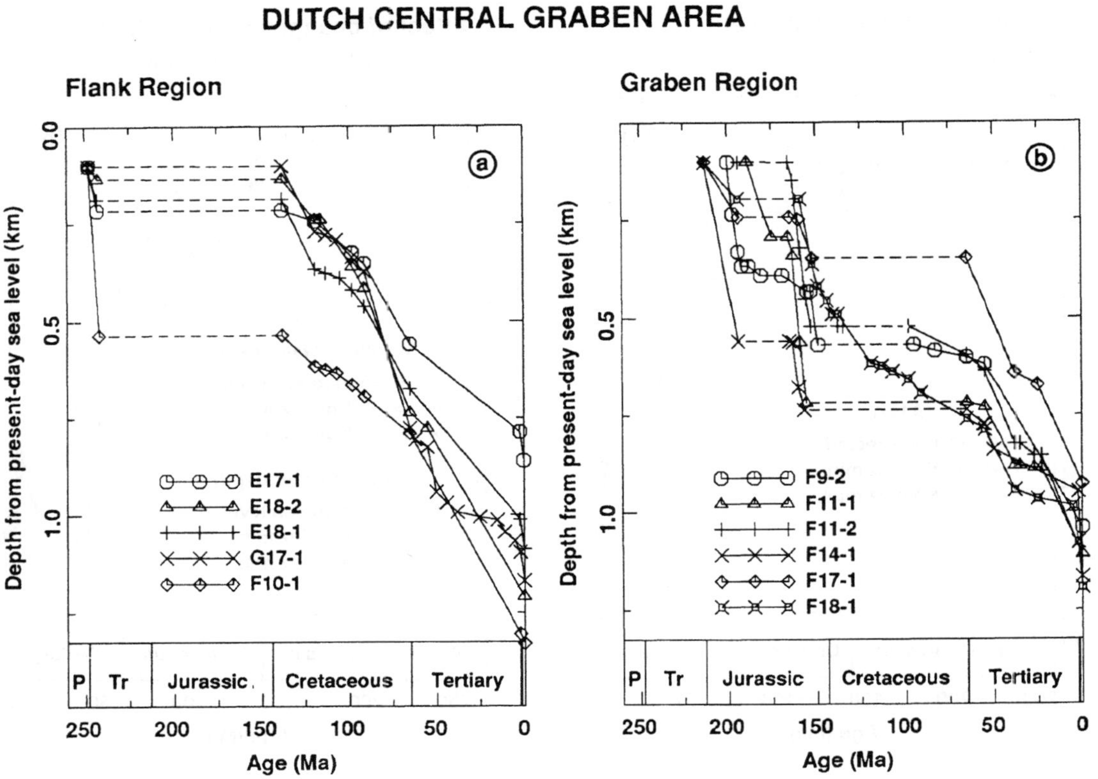

Fig. 10. Water loaded tectonic subsidence curves for eleven wells in the Dutch part of the Central Graben (54° N). For location of the wells see Fig. 5 and Table 1. Figure conventions as in Fig. 6. For the curves shown in Fig. 10 effects of long-term sea level changes and paleobathymetry have been ignored. Note the distinct differences in subsidence characteristics of these regions, indicating differential subsidence between Graben areas and flanks. Also note the overall linear to convex upward subsidence pattern displayed by these curves.

times (Herngreen & Wong 1988, this issue pp 73–105). The Cretaceous hiatuses might in part be associated with tectonic activity during the earliest Cretaceous and an increase in tectonic activity during Aptian times. To a large extent these hiatuses, however, can be attributed to deep erosion of Cretaceous and locally even Jurassic deposits caused by Senonian inversion tectonics (Heybroek 1975, P.A. Ziegler 1987). The F11-1 and F14-1 wells have been drilled on the inversion axis. The F18-1 and F9-2 wells are situated on the eastern and the F11-2 well on the western side of the inversion. The slow Late Cretaceous subsidence rates inferred from analysis of these wells might in part be due to a condensed sequence that formed in response to salt tectonics. High subsidence rates do seem to occur during early Tertiary and diminish towards the Present (well F14-1 and F18-1). These wells also show an extremely high Quaternary subsidence rate, a feature that can also be inferred for the other wells from inspection of Quaternary isopach maps (Bjorslev Nielsen et al. 1986).

The results for the wells in the flank region show remarkably coherent results (see Fig. 10). Jurassic pronounced differential subsidence of the graben and flank regions is evident. High subsidence rates on the flanks of the graben are characteristic for Cretaceous times. The timing of the initiation of this phase of Cretaceous subsidence suggests a key role of the earliest Cretaceous rifting phase in the

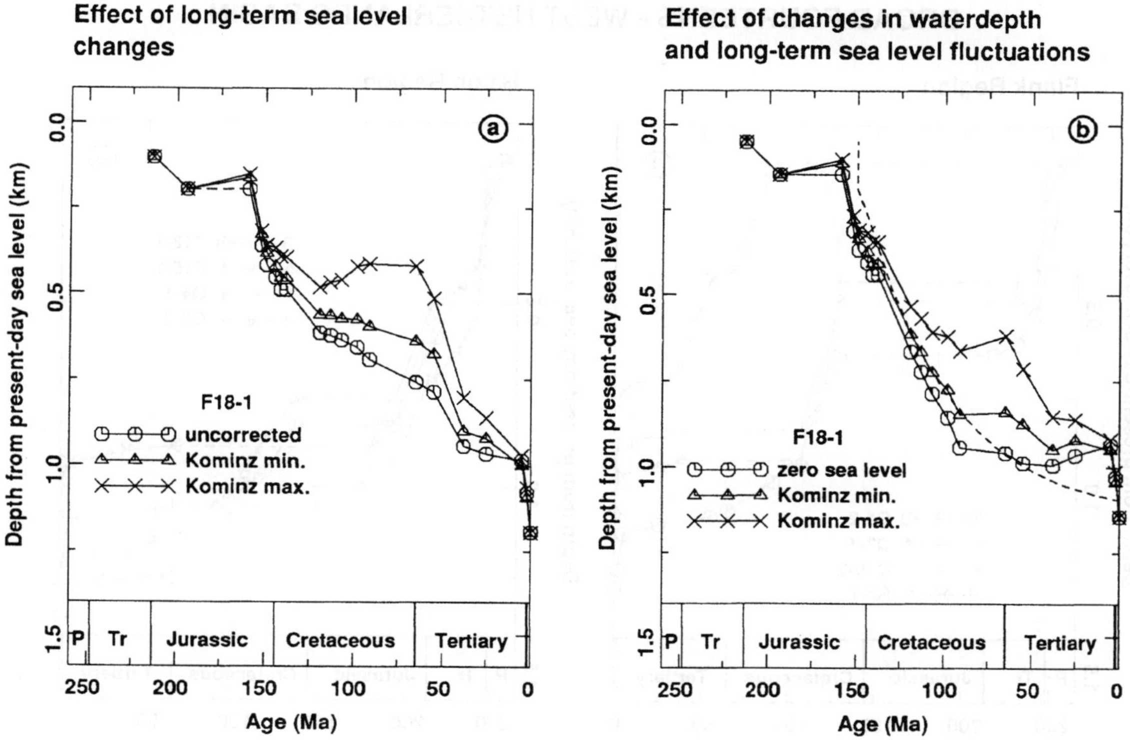

Fig. 11. Tectonic subsidence for the F18-1 well corrected for paleo-water depth fluctuations and long-term sea level changes (Kominz 1984) as indicated in Fig. 7. a) Corrected for sea level changes only. The lowermost curve denotes the uncorrected tectonic subsidence. b) Corrected for both long-term sea level changes and variations in paleo-water depth. The dashed curve denotes a synthetic tectonic subsidence curve for instantaneous stretching at 150 Ma with crustal stretching by a factor of $\delta = 1.2$ and subcrustal attenuation by a factor of $\beta = 1.4$. Correction for both low magnitude long-term sea level changes and variations in paleobathymetry yields a subsidence pattern which closely resembles predictions from thermal models.

subsidence history of the area.

We have incorporated paleo-water depth data and the long-term sea level curves from Kominz (1984) in the subsidence analysis of the F18-1 well (Fig. 7). Again low-magnitudes for long-term sea levels, combined with conservative estimates of paleo-water depths yield a relatively good fit to a synthetic tectonic subsidence curve for instantaneous crustal stretching at 150 Ma with $\delta = 1.2$ and subcrustal attenuation with $\beta = 1.4$ (Fig. 11). Strong deviations from the thermally induced subsidence can be observed for the Pliocene and the Quaternary.

The analysis of the well data for the Dutch Central Graben area (54° N) has shown the great impact of the Late Jurassic and probably Early Cretaceous rifting phases on the subsidence history of this region. Inspection of the subsidence curves given in Figs. 10 and 11 shows that the wells on the flanks and in the Graben each have distinctly characteristic and different subsidence histories.

Broad Fourteens/West Netherlands Basins

The location of eight wells located along a NNE-SSW trending line crossing the Broad Fourteens Basin and the western part of the West Netherlands Basin is given in Fig. 5. The tectonic subsidence curves reconstructed from these well data are dis-

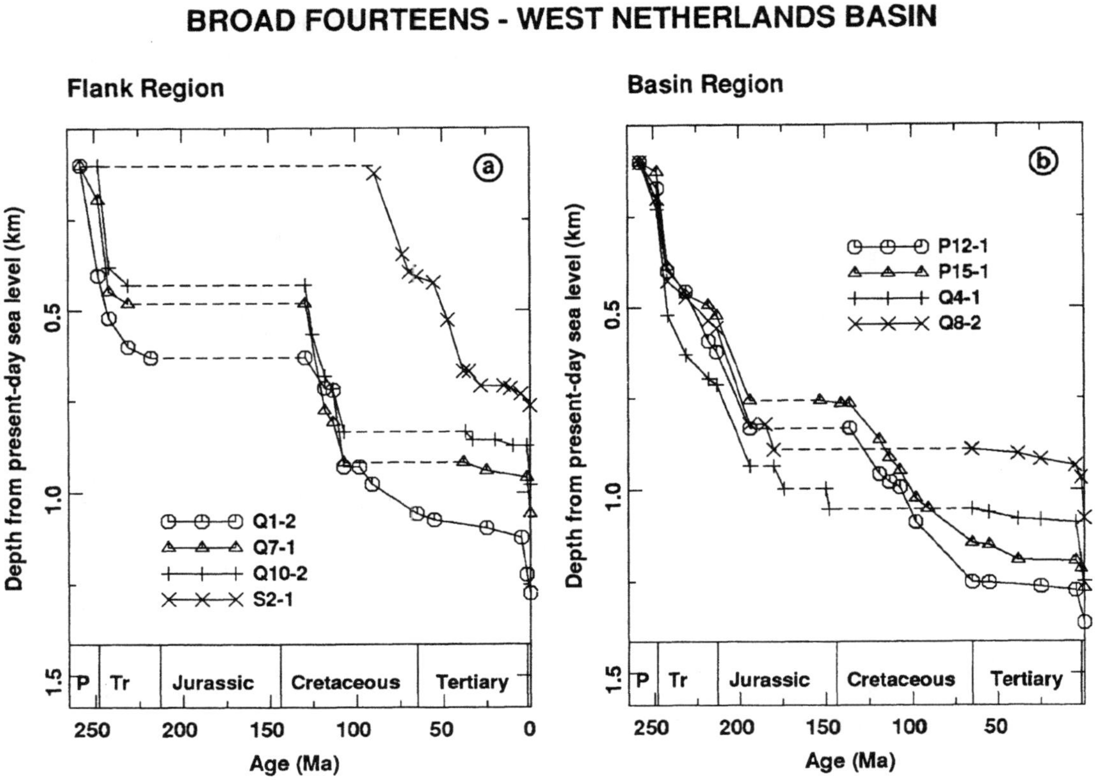

Fig. 12. Water loaded tectonic subsidence curves for eight wells in the Broad Fourteens and West Netherlands Basins. For location of the wells see Fig. 5 and Table 1. Wells located in the deepest parts of the basins are displayed separately from the wells located on the flanks. Figure conventions as in Fig. 6. For the curves shown in this figure effects of long-term sea level changes and paleobathymetry have been ignored. Note the overall (stepwise) convex downward subsidence pattern.

played in Fig. 12. Similarly to the procedure outlined in the previous sections we have differentiated between well data from the flanks and the basin interiors. Most of the subsidence curves demonstrate tectonic subsidence since the beginning of the Late Permian, which corresponds to the initiation of the Zechstein basin. The progressively northward increase in the thicknesses of Zechstein deposits is reflected in the subsidence curves. This period is followed by very high subsidence rates during Early Triassic to Late Jurassic. As described in the previous two sections, a similar phase of rapid Early Triassic subsidence has been observed for the flank region of the Central Graben (B13-1, F10-1). These high subsidence rates attest to the rift-stage character of the southern North Sea Basin at these times. Subsidence characteristics are apparently uniform in all wells during Permo-Triassic times and probably also during the earliest Jurassic. An important erosional phase has removed much of the Jurassic, and earliest Cretaceous subsidence record from basin interiors and has even truncated Triassic deposits from the adjacent highs. Interruptions of these periods of no net subsidence are observed in the Q4-1 and Q8-2 wells (located in the Broad Fourteens Basin) with evidence of minor, but fast, Aalenian and Kimmeridgian subsidence.

Following the Early Cretaceous rifting phase, renewed subsidence has occurred in the West Netherlands Basin (P-wells). At the same time Late Cretaceous and early Tertiary phases of basin inversion have truncated much of the Early Cretaceous deposits in the Broad Fourteens Basin. In the wells not subjected to inversion (Q1-2, P12-1, P15-1), a pattern of concave upward tectonic subsidence is observed, which is in striking contrast with the linear Cretaceous pattern of tectonic subsidence in the Central Graben area around 54° N and the convex upward subsidence pattern for the same period for the northernmost part of the Dutch Central Graben described in the previous sections. However, as demonstrated earlier, this feature is masked by sedimentation not keeping pace with subsidence in the more northerly areas. Late Jurassic and Early Cretaceous rift/wrench tectonics seem to have been important in the region of the Broad Fourteens/West Netherlands Basin (Van Wijhe 1987).

Tertiary subsidence rates are low for all wells with exception of the well located on the flank of the London-Brabant Massif (S2-1), which Late Cretaceous to Oligocene subsidence pattern deviates strongly from the other wells. This period of tectonic stability abruptly ends during Pliocene-Quaternary times with a phase of renewed rapid subsidence which can be observed for all wells.

The timing of the main subsidence phases for the Triassic, Jurassic and earliest Cretaceous as inferred from the wells from the Broad Fourteens/West Netherlands Basins is in accordance with the findings for the Central Graben area described in the two previous sections. On the other hand, conspicuous differences exist from one region to another for Cretaceous and Tertiary subsidence. The overall subsidence characteristics shown in the foregoing analysis of well data provide useful constraints for the stratigraphic modelling of North Sea Basin stratigraphy presented in the following section.

Stratigraphic modelling

In this section, we discuss models for the stratigraphy of the northernmost transect described in the quantitative subsidence analysis of the well data (see Fig. 5). Stratigraphic modelling for the southern part of the Central Graben, the Broad Fourteens and West Netherlands Basins will be published in future work.

Our modelling approach is based on the stretching assumption (McKenzie 1978, Royden & Keen 1980, Steckler 1981). Depth-dependent stretching has been incorporated by employing a crustal (δ) and subcrustal (β) stretching factor. The former defines the assymptotic depth to which the basin subsides, the latter independently delimits the amount of thermal subsidence. The thermal calculations have been carried out using a finite-difference approach (Verwer 1977). This enables us to incorporate finite stretching and multiple stretching phases into the modelling. Jarvis & McKenzie (1980) have shown that instantaneous stretching is adequate only for stretching events with a duration of less than 20 Ma. In general, for significant lateral heat flow during extension, deviations will occur from the predictions of the instantaneous stretching model (Cochran 1983). Analyses of gravity data by Barton & Wood (1984) point to the presence of relatively weak lithosphere underlying the North Sea Basin. These authors obtained a best fit to the observed gravity field with an elastic thickness of only 5 kilometres. Thorne (1986) obtained estimates for the effective elastic thickness varying between 8 and 25 kilometres for several periods during the Cretaceous and Cenozoic. Such low estimates for flexural rigidities are not unusual for basins located on continental lithosphere. Detailed analysis of the northern Canadian Sverdrup Basin has yielded estimates of the equivalent elastic thickness of less than 30 km (Stephenson et al. 1987). In contrast, estimates of equivalent elastic thicknesses of old oceanic lithosphere are characteristically of the order of 40–50 km (McAdoo et al. 1985). In our modelling of North Sea subsidence we have adopted a value of 10 km for the effective elastic thickness of the lithosphere. Compaction is calculated according to an exponential porosity-depth relation (Sclater & Christie 1980)

$$\varphi(z) = \varphi_0 e^{-cz} \qquad (1)$$

where φ_0 and c denote the surface porosity and the decay constant respectively.

A structural cross-section published by P.A. Zie-

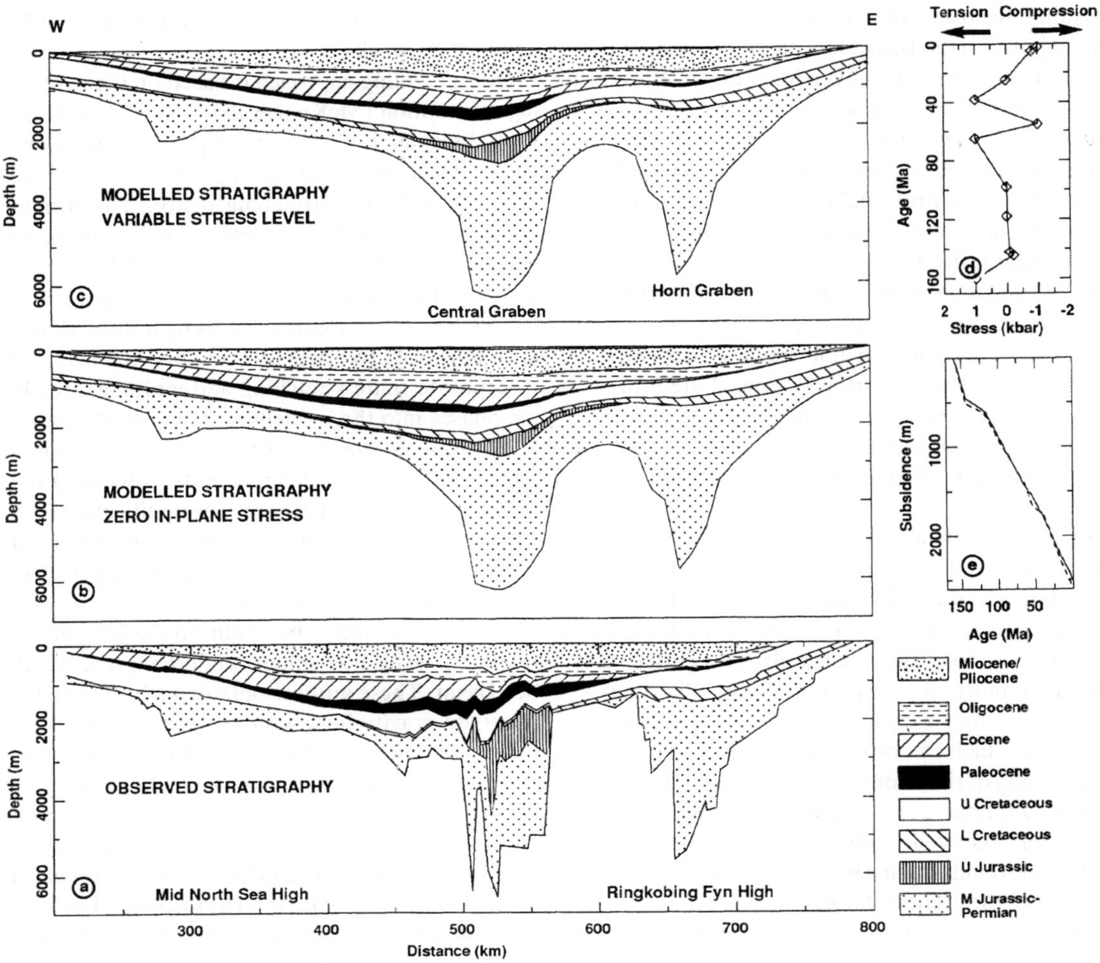

Fig. 13. Stratigraphic modelling for the transect through the northernmost part of the Dutch Central Graben. Location shown in Fig. 5 as AA'. a) Documented stratigraphy (after P.A. Ziegler 1982). The pre-Quaternary stratigraphy has been restored by removing the thick Quaternary deposits from the original section by P.A. Ziegler (1982). Note the great thickness of Tertiary deposits relative to the amount of Cretaceous sediments. b) Best fit to observed stratigraphy obtained in the absence of intraplate stress changes. The basin created by the subsidence due to thermal contraction and crustal thinning is flexurally filled with sediments. c) Modelled stratigraphy for changes in intraplate stress level displayed in Fig. 13d. Note the absence of Paleocene sediments over the horst separating the two graben systems and a phase of Eocene offlap. d) The intraplate stress pattern inferred from the stratigraphic model displayed in Fig. 13c. e) Basement subsidence amplified by sediment and water loading in the graben plotted as a function of age at the 540 km location in Figs. 13b, c. Solid curve denotes the subsidence in the absence of intraplate stress changes. Dashed curve indicates the subsidence for a variable stress level.

gler (1982) with a length of 600 kilometres from Denmark towards the British coast is shown in Fig. 13a. This section cuts through the Ringkobing-Fyn and Mid North Sea High and traverses the Danish Central Graben and the Horn Graben. The Cenozoic basin displays both onlap and offlap phases. The cross-section does not show differential compaction over the basement highs and lows, which is predicted when using porosity-depth relations from Bond & Kominz (1984). This might be an indication of either an overestimate of the amount of porosity reduction due to compaction, or overpressuring of the graben sediments. For reasons of convenience we adopted the former possibility and employed a surface porosity φ_0 of 25% and a decay constant c of 0.1 km^{-1}. Strictly speaking, modelling of the pre-Cretaceous basin would require a model of graben formation and fault tectonics. In the present study we did not incorporate the effects of faulting in our thermo-mechanical models. As the present work focuses on the Cretaceous and Tertiary basin, this seems to be a reasonable simplification. Forthcoming, more detailed, numerical modelling of the Mesozoic stratigraphy will explore the effect of faulting on basin stratigraphy.

As shown in the analysis of the well-data, negligence of corrections for paleobathymetry would result in concave downwards tectonic subsidence curves for this region. Hence, ignoring these variations in paleobathymetry in forward modelling would require either periods of active stretching near the end of Cretaceous times or delayed subsidence due to active subcrustal heating during the Cretaceous. However, incorporation of changes in paleo-water depth in the analysis of the well-data produces a tectonic subsidence pattern which is more in accordance with a period of stretching activity during Late Jurassic and probably also Early Cretaceous. We have assumed this period of stretching to extend from Callovian (163 Ma) to Aptian (119 Ma) and ignored the thermal subsidence due to Mid-Jurassic and Triassic periods of extension. The latter assumption results in an overestimate of the amount of crustal thinning (Hellinger et al. in press). As Late Jurassic sediments are mainly confined to the Central Graben, we assumed the Jurassic period of stretching (163–144 Ma) to have affected primarily the crust, without inducing much thermal subsidence. The Cretaceous period of stretching (144–119 Ma) has been taken as the period of more extensive lithospheric thinning.

The Quaternary basin forms a problem on its own. Up to 1000 metres of Quaternary sediments are present in the North Sea area (Bjorslev Nielsen et al. 1986), a thickness that is comparable to the thickness of Miocene-Pliocene deposits. The associated acceleration in subsidence can not be accounted for by filling of pre-existing bathymetric lows and must therefore be associated with an acceleration in tectonic subsidence. Such short-term phases of rapid subsidence are not limited to the North Sea Basin and are a feature commonly observed in pull-apart basins, such as the Neogene Californian basins with documented subsidence rates of the order of a few km/Ma (Christie-Blick & Biddle 1985, Pitman & Andrews 1985). The present analysis concentrates on the pre-Quaternary basin evolution, as better constraints on the origin of the Quaternary basin are required to discriminate between several potential models for Quaternary North Sea Basin subsidence.

We have simulated the presence of the Central and Horn Graben by incorporating pre-existing isostatic depressions in the models. This approach probably implies an overestimate of the flexural component of the basement response but allows us to account for compaction of the sediments that fill in the graben. Inspection of Fig. 13a shows that Early Cretaceous deposits are absent or very thin on the Mid North Sea High and the Rynkobing Fyn High. The environment of deposition for much of the Early Cretaceous deposits, especially for the Danish sector, indicates relatively deep-marine conditions as exemplified by mass flow deposits and mid-Early Cretaceous pelagic carbonates (Frandsen et al. 1987, Jensen & Buchardt 1987). At the same time, much of the Rynkobing Fyn High and parts of the Mid North Sea High were subaerially exposed during Berriasian to Barremian times (P.A. Ziegler 1982). The latter indicates a lack of tectonic subsidence. As mentioned previously, subcrustal attenuation can be used to independently delimit the amount of lithospheric heating. We employed this mechanism to constrain the amount of Early Cretaceous syn-rift sediments and adopted a maximum paleo-water depth of 70 metres at early Aptian. On the other hand, restriction of the thickness of post-rift Aptian, Albian and Late Cretaceous sediments in the modelling is a consequence of an increase in paleo-water depth. We employed a maximum paleo-water depth of 470 metres. This increase in paleo-water depth compensates for both the rapid thermal subsidence following the rifting phase and for a long-term rise in sea level. For the latter we adopted the minimum curve from Kominz (1984) (see Fig. 7). Subsequently, for the modelling of the Tertiary basin we incorporated gradual shallowing towards Quaternary times. These water depths conform reasonably well with the paleo-water depths used in the subsidence analysis of the well-data.

Fig. 13b shows our preferred model derived in the absence of intraplate stress changes. The total crustal stretching used in the modelling has a maximum (δ) of 1.2 in the Central Graben and decreases rapidly towards the basin edges (Table 2). Integration of these stretching factors yields a mean crustal extension of 9% for this cross-section, which corresponds to roughly 60 kilometres. This value is probably an overestimation of the total post-Mid-Jurassic extension due to the ongoing thermal subsidence of previous stretching events (Hellinger et al. in press) and possible contibutions from physico-chemical processes affecting the density of the crust. For most of the section pre-Late Jurassic sediments are not very thick due to deep Mid-Jurassic erosion, which is associated with the uplift of the Central North Sea rift dome (P.A. Ziegler 1982). Therefore, the stretching parameters used in the modelling also account for at least a large part of the subsidence due to previous Triassic to Mid-Jurassic extension.

Table 2. Stretching factors.

Position (km)	β	δ
200	1.25	1.05
240	1.25	1.06
280	1.35	1.07
310	1.41	1.08
340	1.44	1.09
350	1.47	1.10
370	1.47	1.11
380	1.47	1.12
390	1.49	1.13
400	1.60	1.13
410	1.70	1.14
450	1.71	1.15
470	1.73	1.16
510	1.73	1.16
520	1.73	1.17
530	1.63	1.20
540	1.63	1.20
550	1.53	1.16
560	1.43	1.12
610	1.43	1.10
680	1.44	1.09
730	1.41	1.08
750	1.35	1.07
760	1.30	1.06
780	1.25	1.05
790	1.18	1.04
800	1.05	1.02

A substantially better fit to the observed stratigraphy can be obtained by incorporating fluctuating intraplate stress levels in the analysis (Fig. 13c). The stress-induced differential vertical movements strongly enhance the quality of the modelling of details of the stratigraphy. The intraplate stress pattern inferred from the stratigraphic modelling is characterized by Late Jurassic to Teriary tensional and neutral stress levels followed by compressive stresses during Tertiary times (Fig. 13d). Superimposed on this long-term trend are fluctuations in stress level of a shorter duration. Synthetic subsidence curves for the base Upper Jurassic (Fig. 13e) demonstrate the characteristic stress-induced short-term deviations from the long-term subsidence predicted by thermal models of basin evolution. The paleo-stress levels vary from 1 kbar tension in Callovian, Early Paleocene and Early Oligocene to 1 kbar compression at Early Eocene and Quaternary. These are stress levels throughout the basin profile, as a uniform plate thickness has been used in the modelling. These changes in intraplate stress levels in the modelled cross-section can be induced by either a rotation of the stress field, or by changes in the magnitude of the principal stresses. The local effects of the horst and the grabens are evident. Late Jurassic relaxation of tension has removed part of the Late Jurassic sediments from the flanks of the Central Graben and has produced an erosional unconformity. Similarly, a change from a tensional to a compressional stress regime during the Paleocene induces downwarping of the grabens and uplifting of the intervening horst. The subsequent release of compression in the Eocene has the opposite effect, resulting in a relatively thin package of Eocene deposits in the Central Graben. The Tertiary stress

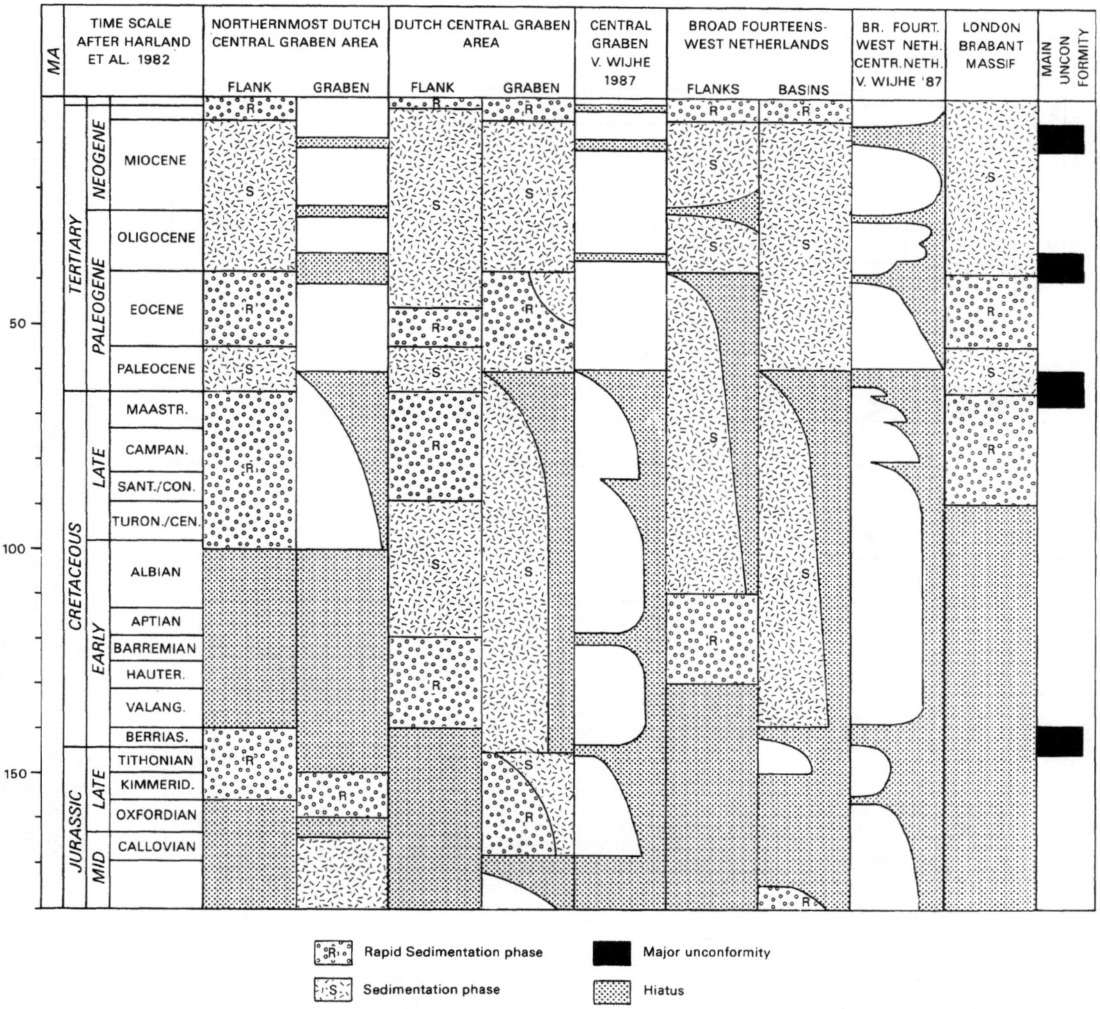

Fig. 14. Stratigraphic correlations for the well data used in this study supplemented with stratigraphic compilations made by Van Wijhe (1987). Hiatuses have been correlated to infer the main unconformities for the Dutch North Sea Basin. Similarly, main phases of rapid sedimentation and normal/slow rate sedimentation have been indicated. The column on the right gives the results of the correlation.

pattern inferred from the stratigraphic modelling is relatively well-constrained. The Cretaceous paleostress pattern is subject to larger uncertainties, a feature that is inherent to less stratigraphic resolution for this particular time slice. The numerical values of the stresses are dependent on rheological properties of the lithosphere under the North Sea Basin and give therefore the order of the stress magnitudes.

Discussion and conclusions

Figure 14 summarizes the timing of sedimentation phases and the periods characterized by hiatuses and unconformities in the Dutch part of the North Sea. Comparison of the overall stratigraphy inferred from this study supplemented with the outcome of earlier work by Letsch & Sissingh (1983) and Van Wijhe (1987) demonstrates the occur-

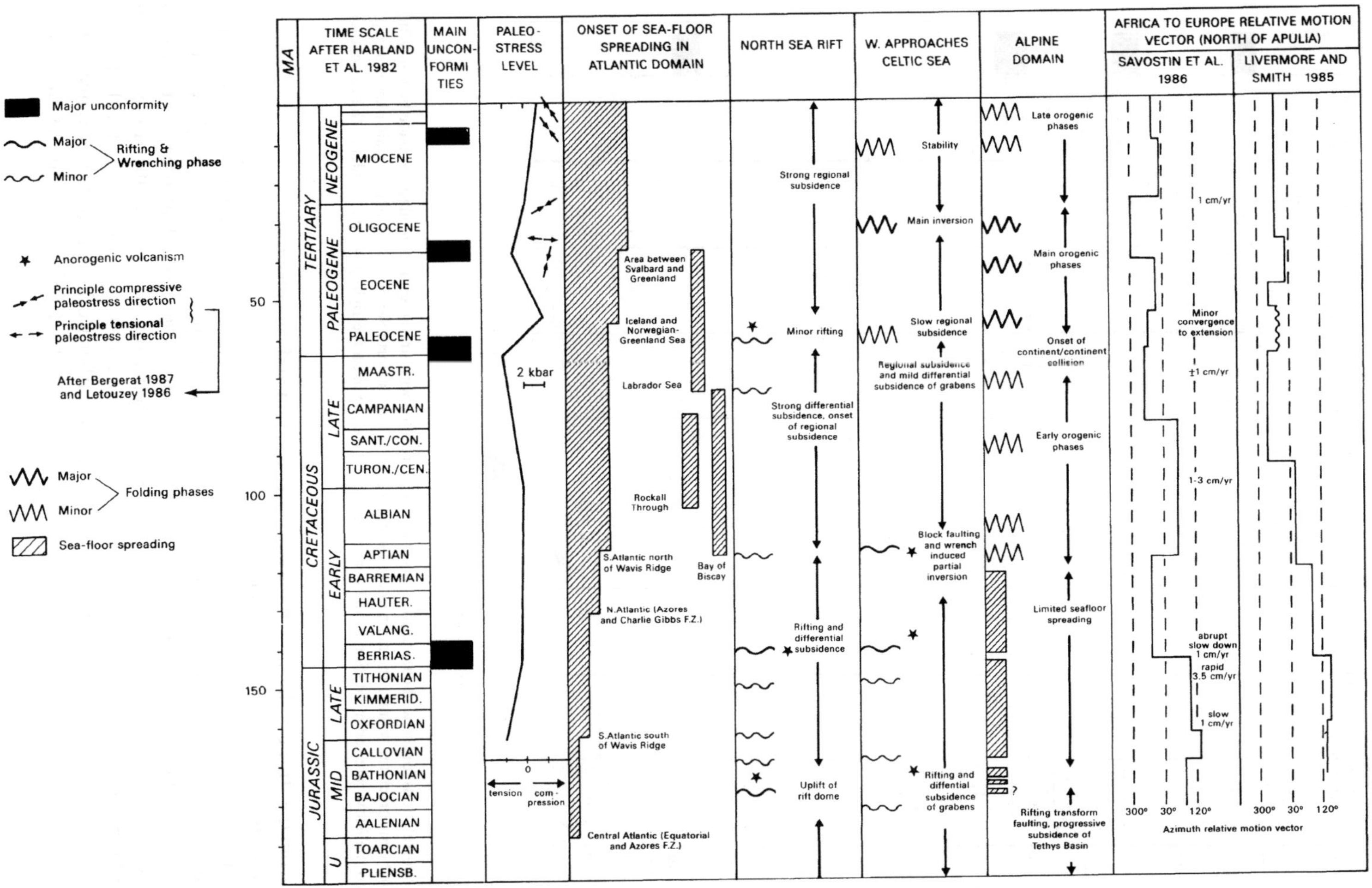

Fig. 15. Synthetic paleo-stress curve as inferred from the forward modelling of the North Sea Basin stratigraphy. Also shown in this column are paleo-stress orientation data from Bergerat (1987). The column on the far left shows stratigraphic correlations for the Dutch North Sea Basin (Fig. 14). In the central part of the figure information is displayed on the timing of onset of sea-floor spreading in the Atlantic domain and timing of tectonic events in the North Sea rift, the Western Approaches/Celtic Sea and in the Alpine domain (after P.A. Ziegler 1982). The columns on the right show Africa relative to Europe plate motion data after Savostin et al. (1986) and Livermore & Smith (1985).

rence of 4 widespread unconformities occurring both in the Central Graben and the West Netherlands/Broad Fourteens Basins. One of these unconformities – at the Jurassic/Cretaceous boundary – coincides with the timing of a stretching phase in the North Sea Basin. The other three major unconformities which occur at the Cretaceous/Tertiary boundary, during mid-Oligocene times and at Late Miocene times, respectively cannot be directly associated with stretching phases in the North Sea area.

The timing of the unconformities is related to changes in the Africa to Europe relative motion vector documented by Savostin et al. (1986) and Livermore & Smith (1985). The incorporation of intraplate stresses in stratigraphic modelling, however, enables us to explore the dynamics underlying some of the observed correlations between plate kinematics and the apparent sea level record. Fig. 15 gives a comparison of the timing of the main unconformities with P.A. Ziegler's (1982) timing of tectonic events in Northwestern Europe. Also shown is the paleo-stress curve resulting from the present modelling of the stratigraphy. The trend of the curve with a change from overall tension and neutral stresses during Mesozoic times to a stress regime of more overall compressional character is consistent with the documented (P.A. Ziegler 1982) change from a Mesozoic regime of rift/wrench tectonics, associated with the terminal stages of the breakup of Pangea, to a tectonic regime dominated by Alpine collision phases. Superimposed on this long-term trend are stress fluctuations of a shorter period. The paleo-stress curve of Fig. 15 displays a strong phase of compression during Early Eocene times corresponding in timing with the occurrence of strong folding phases in the Alpine domain. For Late Eocene to Early Oligocene we predict a stress regime of more tensional character, concomittant with the timing of initiation of rifting in the European platform, and an associated observed tensional paleo-stress field (Letouzey 1986, Bergerat 1987). The predicted overall increase in the level of the post-Early-Oligocene compression is consistent with the observed (Letouzey 1986, Bergerat 1987) rotation of the paleo-stress field in Northwestern Europe from NE-SW oriented Late Oligocene/Early Miocene compression to the present NW-SE orientation of the largest compressive stress, a direction which is more perpendicular to the strike of the Central Graben basins (Klein & Barr 1986, see also Fig. 15). The paleo-stress curve has been derived under the assumption of an elastic model for the mechanical properties of the lithosphere in the North Sea Basin. Incorporation of a more realistic rheology with a depth-dependent finite strength in the modelling will result in lower values for the inferred stress levels. The values given for the stress levels should, therefore, be considered to provide upper limits. Our work strongly suggests that the record of short-term relative sea level fluctuations inferred from the stratigraphic record of the North Sea Basin is to a large extent dominated by the effects of large-scale intraplate stresses.

The present study has demonstrated that the incorporation of fluctuations in intraplate stress levels in quantitative stratigraphic modelling of the North Sea Basins provides a powerful tool to enhance our insight in observed correlations of tectonic events in Northwestern Europe and the stratigraphic evolution of the North Sea area. We have shown that a paleo-stress curve inferred from the seismic stratigraphic record is consistent with independent data sets on the tectonic history of Northwestern Europe. The outcome of this tectonic modelling sheds light on the relative role of tectonics and eustasy in the relative sea level record of Northwestern Europe, and hence, provides a new angle to resolve the regional versus global character of Vail's short-term sea level changes. Our findings suggest that tectonics might dominate the apparent sea level record, even during periods with a non-ice free world.

Acknowledgements

Partial support for this work was provided by the Netherlands Organization for Scientific Research NWO and by NATO grant 0148/87. P.A. Ziegler, M.J.R. Wortel, W. Zagwijn, A. Lokhorst, J. Thorne, T. McGee and C. Leyzers-Vis are thanked for helpful discussions and comments. Merlin Exploration Services is acknowledged for generous permission to make use of seismic reflection data from NOPEC lines. F. Gradstein and Y. Poslawski (Geologic BV) kindly put bore hole data at our disposal. Th. Wong, P.A. Ziegler and D. Van Wijhe furnished preprints of papers.

References

Angevine, C.L., S.R. Linneman & P.L. Heller 1988 Supercontinent breakup: effect on eustatic sea level and the oceanic heat flux. In: N.J. Vlaar, G. Nolet, M.J.R. Wortel & S.A.P.L. Cloetingh (eds): Mathematical Geophysics – Reidel (Dordrecht): 389–399

Bally, A.W. 1982 Musings over sedimentary basin evolution – Phil. Trans. Roy. Soc. Lond. Ser. A 305: 325–338

Barton, P. & R. Wood 1984 Tectonic evolution of the North Sea basin: crustal stretching and subsidence – Geophys. J.R. Astron. Soc. 79: 987–1022

Beaumont, C. 1978 The evolution of sedimentary basins on a viscoelastic lithosphere: Theory and examples – Geophys. J.R. Astron. Soc. 55: 471–498

Beaumont, C. & A.T. Tankard (eds) 1987 Sedimentary basins and basin-forming mechanisms – Can. Soc. Petrol. Geol. Mem. 12: 1–527

Bergerat, F. 1987 Stress fields in the European platform at the time of Africa-Eurasia collision – Tectonics 6: 99–132

Bjorslev Nielsen, O., S. Sorensen, J. Thiede & O. Skarbo 1986 Cenozoic differential subsidence of the North Sea – Am. Assoc. Petrol. Geol. Bull. 70: 276–298

Bond, G.C. & M.A. Kominz 1984 Construction of tectonic subsidence curves for the early Paleozoic miogeocline, southern Canadian Rocky Mountains: Implication for subsidence mechanisms, age of breakup, and crustal thinning – Geol. Soc. Am. Bull. 95: 155–173

Cloetingh, S. 1986 Intraplate stresses: a new tectonic mechanism for fluctuations of relative sea level – Geology 14: 617–620

Cloetingh, S. 1988 Intraplate stresses: a new element in basin analysis. In: K. Kleinspehn & C. Paola (eds): New perspectives in basin analysis – Springer (New York): 305–330

Cloetingh, S., K. Lambeck & H. McQueen 1987 Apparent sea-level fluctuations and a palaeostress field for the North Sea region. In: J. Brooks & K. Glennie (eds): Petroleum Geology of North West Europe – Graham & Trotman (London): 49–57

Cloetingh, S., H. McQueen & K. Lambeck 1985 On a tectonic mechanism for regional sealevel variations – Earth Planet. Sci. Lett. 75: 157–166

Cloetingh, S. & F. Nieuwland 1984 On the mechanics of lithospheric stretching and doming: a finite element analysis – Geol. Mijnbouw 63: 315–322

Cloetingh, S. & R. Wortel 1985 Regional stress field of the Indian plate – Geophys. Res. Lett. 12: 77–80

Cloetingh, S. & R. Wortel 1986 Stress in the Indo-Australian Plate. In: Johnson, B. & A.W. Bally (eds): Intraplate deformation: characteristics, processes and causes – Tectonophysics 132: 49–67

Christie-Blick, N. & K.T. Biddle 1985 Deformation and basin formation along strike-slip faults. In: N. Christie-Blick & K.T. Biddle (eds): Strike-slip deformation, basin formation and sedimentation – Soc. Econ. Paleont. Spec. Pub. 42:1–34

Cochran, J.R. 1983 Effects of finite rifting times on the development of sedimentary basins – Earth Planet. Sci. Lett. 66: 289–302

De Vries-Klein, G. 1987 Current aspects of basin analysis – Sediment. Geol. 50: 95–118

Frakes, L.A. 1979 Climates throughout Geologic Time – Elsevier (Amsterdam): 310 pp

Frandsen, N., O.V. Vejbaek, J.J. Moller & O. Michelsen 1987 A dynamic geological model of the Danisch Central Trough during the Jurassic-Early Cretaceous. In: J. Brooks & K.W. Glennie (eds): Petroleum Geology of North West Europe – Graham & Trotman Ltd. (London): 453–468

Gradstein, F.M., F.P. Agterberg, J.C. Brouwer & W. Schwarzacher 1985 Quantitative stratigraphy – Unesco & Reidel (Dordrecht): 598 pp

Gradstein, F.M., M.A. Kaminsky & W.A. Berggren, in press: Cenozoic foraminiferal biostratigraphy of the Central North Sea. In: F. Roegl & F.M. Gradstein (eds): Proc. II Workshop Agglutinated Benthic Foraminifera (Vienna)

Hallam, A., in press: A reevaluation of Jurassic eustasy in the light of new data and the revised Exxon curve – Soc. Econ. Paleont. Spec. Pub. 42

Harland, W.B., A.V. Cox, P.G. Llewellyn, C.A.G. Pickton, A.G. Smith & R. Walters 1982 A geologic time scale – Cambridge University Press (London): 131 pp

Haq, B., J. Hardenbol & P.R. Vail 1987 Chronology of fluctuating sea level since the Triassic (250 million years to present) – Science 235: 1156–1167

Hellinger, S.J., J.G. Sclater & J. Giltner, in press: Mid-Jurassic through Mid-Cretaceous extension in the Central Graben of the North Sea: Part 1, estimates from subsidence

Herngreen, G.F.W. & Th.E. Wong 1988 Revision of the Late Jurassic stratigraphy of the Dutch Central North Sea Graben. In: W.J.M. van der Linden, S.A.P.L. Cloetingh, J.P.K. Kaasschieter, J. Vandenberghe, W.J.E. van de Graaff & J.A.M. van der Gun (eds): Proc. KNGMG Symp. Coastal Lowlands: Geology and Geotechnology, The Hague, 1987 – Kluwer Acad. Publ. (Dordrecht): 73–105 (this issue)

Heybroek, P. 1975 On the structure of the Dutch part of the Central North Sea Graben. In: A.W. Woodland (ed.): Petroleum and the continental shelf of north-west Europe – Appl. Sci. Publ. (Ltd): 339–349

Houseman, G. & P. England 1986 A dynamical model of lithosphere extension and sedimentary basin formation – J. Geophys. Res. 91: 719–729

Illies, J.H., H. Baumann & B. Hoffers 1981 Stress pattern and strain release in the Alpine foreland – Tectonophysics 71: 157–172

Jarvis, G.T. & D.P. McKenzie 1980 Sedimentary basin formation with finite extension rates – Earth Planet. Sci. Lett. 48: 42–52

Jensen, T.F. & B. Buchardt 1987 Sedimentology and geochemistry of the organic carbon-rich Lower Cretaceous Sola Formation (Barremian-Albian), Danish North Sea. In: J. Brooks & K.W. Glennie (eds): Petroleum Geology of North West Europe – Graham & Trotman Ltd. (London): 431–440

Johnson, B. & B. Bally (eds) 1986 Intraplate deformation: characteristics, processes and causes – Tectonophysics 132: 1–278

Klein, R.J. & M.V. Barr 1986 Regional state of stress in western Europe. In: O. Stephensson (ed.): Rock stress and rock stress measurements – Centek (Lulea, Sweden): 33–44 pp

Kominz, M.A. 1984 Oceanic ridge volumes and sea level change – and error analysis – Am. Assoc. Petrol. Geol. Mem. 36: 109–127

Lambeck, K., S. Cloetingh & H. McQueen 1987 Intraplate stresses and apparent changes in sealevel: the basins of northwestern Europe. In: C. Beaumont & A.T. Tankard (eds): Sedimentary basins and basin-forming mechanisms – Can. Soc. Petrol. Geol. Mem. 12: 259–268

Letouzey, J. 1986 Cenozoic paleo-stress pattern in the Alpine foreland and structural interpretation in a platform basin – Tectonophysics 132: 215–231

Letsch, W.J. & W. Sissingh 1983 Tertiary stratigraphy of the Netherlands – Geol. Mijnbouw 62: 305–318

Livermore, R.A. & A.G. Smith 1985 Some boundary conditions for the evolution of the Mediterranean region. In: D.J. Stanley & F.C. Wezel (eds): Geological evolution of the Mediterranean Basin – Springer (New York): 83–98

Lucazeau, F. & S. Le Douaran 1985 The blanketing effect of sediments in basins formed by extension: a numerical model. Application to the Gulf of Lion and Viking graben – Earth Planet. Sci. Lett. 74: 92–102

McAdoo, D.C., C.F. Martin & S. Poulouse 1985 Seasat observations of flexure: evidence for strong lithosphere – Tectonophysics 116: 209–222

McAdoo, D.C. & D.T. Sandwell 1985 Folding of oceanic lithosphere – J. Geophys. Res. 90: 8563–8569

McKenzie, D.P. 1978 Some remarks on the development of sedimentary basins – Earth Planet. Sci. Lett. 40: 25–32

Meulenkamp, J.E. & F.J. Hilgen 1986 Event stratigraphy, basin evolution and tectonics of the Hellenic and Calabro-Sicilian arcs – Developm. Geotect. 21: 327–350

Miall, A.D. 1986 Eustatic sea level changes interpreted from seismic stratigraphy: A critique of the methodology with particular reference to the North Sea Jurassic record – Am. Assoc. Petrol. Geol. Bull. 70: 131–137

Ned. Aardolie Maatsch. & Rijks Geol. Dienst 1980 Stratigraphic nomenclature of the Netherlands – Verhandelingen K.M.G.M.G.: 32 pp

Parkinson, D.N. & C.P. Summerhayes 1985 Synchronous global sequence boundaries – Am. Assoc. Petrol. Geol. Bull. 69: 685–687

Pitman III, W.C. & J.A. Andrews 1985 Subsidence and thermal history of small pull-apart basins. In: Christie-Blick, N. & K.T. Biddle (eds): Strike-slip deformation, basin formation and sedimentation – Soc. Econ. Paleont. Spec. Pub. 42: 45–49

Pitman III, W.C. & X. Golovchenko 1983 The effect of sealevel change on the shelf edge and slope of passive margins – Soc. Econ. Paleont. Miner. Spec. Pub. 33: 41–58

Royden, L. & C.E. Keen 1980 Rifting process and thermal evolution of the continental margin of eastern Canada determined from subsidence curves – Earth Planet. Sci. Lett. 51: 343–361

Savostin, L.A., J.C. Sibuet, L.P. Zonenshain, X. Le Pichon & M.J. Roulet 1986 Kinematic evolution of the Tethys belt from the Atlantic Ocean to the Pamirs since the Triassic – Tectonophysics 123: 1–35

Schlanger, S.O. 1986 High-frequency sea level oscilations in Cretaceous time: An emerging geophysical problem – Am. Geophys. Un. Geodyn. Series 15: 61–74

Sclater, J.G. & P.A.F. Christie 1980 Continental stretching: an explanation of the post-mid-Cretaceous subsidence of the central North Sea basin – J. Geophys. Res. 85: 3711–3739

Sleep, N.H. 1971 Thermal effects of the formation of Atlantic continental margins by continental break up – Geophys. J. Roy. Astron. Soc. 24: 325–335

Steckler, M.S. 1981 The thermal and mechanical evolution of Atlantic-type continental margins – (Ph.D. thesis), Columbia University: 261 pp

Steckler, M.S. & A.B. Watts 1978 Subsidence of the Atlantic-type continental margin off New York – Earth. Plan. Sci. Lett. 41: 1–13

Stephenson, R.A., A.F. Embry, S.M. Nakiboglu & M.A. Hastaoglu 1987 Rift-initiated Permian to Early Cretaceous subsidence of the Sverdrup Basin. In: C. Beaumont & A.T. Tankard (eds): Sedimentary basins and basin-forming mechanisms – Can. Soc. Petrol. Geol. Mem. 12: 213–231

Summerhayes, C.P. 1986 Sealevel curves based on seismic stratigraphy: their chronostratigraphic significance – Palaeogeogr., Palaeoclimatol., Palaeoecol. 57: 27–42

Thorne, J.A. 1986 A quantitative analysis of North Sea subsidence – Pettijohn Symp. New Perspectives in Basin Analysis, University of Minnesota (Minneapolis) – Program and Abstracts: 31

Turcotte, D.L. & J.L. Ahern 1977 On the thermal and subsidence history of sedimentary basins – J. Geophys. Res. 82: 3762–3766

Vail, P.R., P.M. Mitchum Jr. & S. Thompson III 1977 Global cycles of relative changes of sea level – Am. Assoc. Petrol. Geol. Mem. 26: 83–97

Van Hinte, J.E. 1978 Geohistory analysis- application of micropaleontology in exploration geology – Am. Assoc. Petrol. Geol. Bull. 62: 201–222

Van Wijhe, D.H. 1987 Structural evolution of inverted basins in the Dutch offshore – Tectonophysics 137: 171–219

Verwer, J.G. 1977 A class of stabilized three-step Runge-Kutta methods for the numerical integration of parabolic equations – J. Comp. Appl. Math. 3: 155–166

Watts, A.B. 1982 Tectonic subsidence, flexure and global

changes of sea level – Nature 297: 469–474

Watts, A.B., G.D. Karner & M.S. Steckler 1982 Lithospheric flexure and the evolution of sedimentary basins – Phil. Trans. Roy. Soc. Lond. Ser. A 305: 249–281

Wood, R.J. 1981 The subsidence history of Conoco well 15/30-1, Central North Sea – Earth Planet. Sci. Lett. 54: 306–312

Wortel, R. & S. Cloetingh 1981 On the origin of the Cocos-Nazca spreading center – Geology 9: 425–430

Wortel, R. & S. Cloetingh 1983 A mechanism for fragmentation of oceanic plates – Am. Assoc. Petr. Geol. Mem. 34: 793–801

Ziegler, P.A. 1975 North Sea basin history in the tectonic framework of North-Western Europe. In: A.W. Woodland (ed.): Petroleum and the continental shelf of north-west Europe – Appl. Sci. Publ. (Ltd): 131–150

Ziegler, P.A. 1978 North-Western Europe: tectonics and basin development – Geol. Mijnbouw 57: 589–626

Ziegler, P.A. 1982 Geological atlas of Western Europe – Shell Internationale Petroleum Maatschappij/Elsevier, Amsterdam – 130 pp

Ziegler, P.A. 1987 Late Cretaceous and Cenozoic intra-plate compressional deformations in the Alpine foreland – a geodynamical model – Tectonophysics 137: 389–420

Ziegler, W.H. 1975 Outline of the geological history of the North Sea. In: A.W. Woodland (ed.): Petroleum and the continental shelf of north-west Europe – Appl. Sci. Publ. (Ltd): 165–190

Zoback, M.D. 1985 Wellbore breakout and in situ-stress – J. Geophys. Res. 90: 5523–5530

Geologie en Mijnbouw 68: 73–105 (1989)

Revision of the 'Late Jurassic' stratigraphy of the Dutch Central North Sea Graben

G.F.W. Herngreen & Th.E. Wong
Geological Survey of The Netherlands,
P.O. Box 157, 2000 AD Haarlem, The Netherlands

Received 1 October 1987; accepted in revised form 25 November 1987

Key words: Dutch Central North Sea Graben, Late Jurassic stratigraphy, depositional environment, geological history, sea-level changes, micropaleontology, palynology

Abstract

The 'Late Jurasic' stratigraphy of the Dutch Central North Sea Graben is revised. The sediments, ranging in age from Callovian to Ryazanian, are grouped in two lithological units: the mainly non-marine Central Graben Group (with Lower Graben Sand, Middle Graben Shale, Upper Graben Sand, Puzzle Hole, and Delfland formations) and the distinct marine Scruff Group (with Kimmeridge Clay, Scruff Greensand, and Clay Deep formations). The latter two formations are new and are introduced formally.

Basic palynological and micropaleontological data are included to support age assignments of various formations. Several log correlations, range charts, distribution maps, facies maps and seismic sections are given to illustrate the stratigraphic framework. Finally, a synopsis of the geological history is presented, with special attention to sea-level changes and ensuing coastal developments, illustrating the relationships with the Danish and Norwegian sectors.

Introduction

The Central North Sea Graben can be regarded as the southern extension of the Central Graben (Fig. 1). According to Heybroek (1975), the graben appears to be a complicated block-faulted feature with maximum subsidence at its centre, flanked by intermediate blocks which step down from the bordering highs in the west, east and south (Fig. 1). Modern structural interpretations can be found in Clark-Lowes et al. (1987) and Van Wijhe (1987). The graben margins are usually coincident with linear, north-south oriented Zechstein salt piercements, but large salt domes and piercements also occur near the centre of the basin (Fig. 2). Since rifting was dominant in the Mid to Late Jurassic, the geological development of the area can be divided in pre-rifting, rifting, and post-rifting phases (see the chapter on the geological history).

Since the early Seventies, the stratigraphy of the Dutch Central North Sea Graben has consistently received attention from the Geological Survey of The Netherlands (RGD). This attention was not only justified by correlation problems between the sedimentary sequence in the northern and southern parts of the basin. The different and often conflicting opinions held by consultants and oil companies (see, for example, Heybroek 1975), also underscored the need to review the Central Graben stratigraphy.

In 1980 the Nederlandse Aardolie Maatschappij and the Rijks Geologische Dienst (NAM & RGD, 1980) published a monograph entitled Stratigraphic nomenclature of The Netherlands. Herngreen &

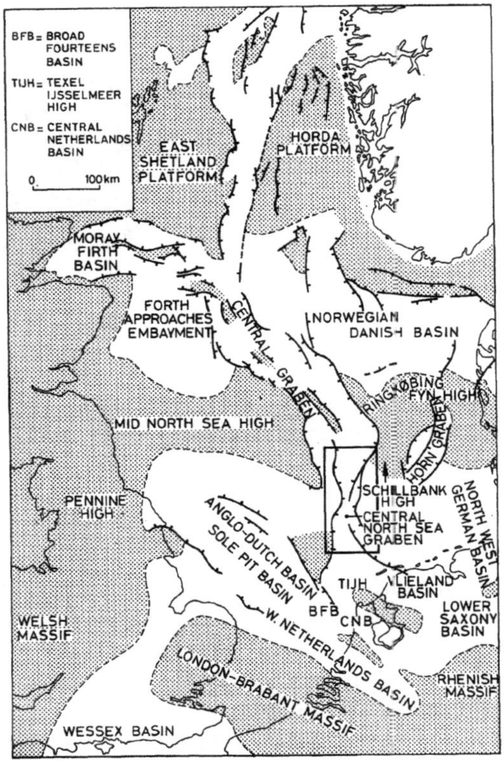

Fig. 1. Structural framework of the North Sea area (modified after Rawson & Riley; 1982).

De Boer (1985) showed that some of the data given by NAM & RGD for the 'Upper Jurassic' required re-interpretation in the light of new palynological evidence. The main conclusions from their study may be summarized as follows:

1. The Scruff (Green) Sand overlying the Main Kimmeridge Clay is Portlandian to Early Berriasian (Ryazanian). This lithostratigraphic unit, widely known as F18 Sand, Main Sand, etc., was not mentioned in NAM & RGD (1980).
2. The age of the Kimmeridge Clay deposits in the F3-3 reference well is not younger than Late Kimmeridgian. No evidence was found for a Portlandian and/or Berriasian (Ryazanian) age of the Kimmeridge Clay Formation in the boreholes mentioned by NAM & RGD (1980).
3. The organically rich clay above the Scruff Sand, which was provisionally indicated as Scruff Shale, is Berriasian (Ryazanian).
4. The Upper Kimmeridge Clay Member in the F11-2 reference well is of roughly Early Kimmeridgian age and not Kimmeridgian-Berriasian (Ryazanian) as pointed out in NAM & RGD (1980, p. 38) or indicated as Late Kimmeridgian in text-figure 9 of the same monograph.
5. Based on age dating and the depositional environment the Lower Kimmeridge Clay Member (NAM & RGD 1980) is considered to be equivalent to the restricted marine Middle Graben Shale Formation.
6. The Puzzle Hole Formation in reference well F11-2 is Middle to Late Oxfordian and not Early Kimmeridgian. The rapidly alternating sands and shales with minor coal seams in the southern part of the Central North Sea Graben show close resemblance to the Delfland strata.

After the presentation of the first results in September 1984 at the Symposium on Jurassic Stratigraphy in Erlangen, the RGD continued to carry out palynological and micropaleontological studies on Late Jurassic and Early Cretaceous sediments of the Dutch Central North Sea Graben. In dinoflagellate and sporomorph analyses emphasis was put on (side-wall) cores; the foraminiferal and ostracod work was mainly carried out on cuttings. During the last two years several of the questions which arose after the compilation of the revised rock stratigraphic diagram in Herngreen & De Boer (1985) were treated in more detail. These concern particularly:

1. the relationship between the Upper Graben Sand and the Puzzle Hole formations;
2. the age and development of strata equated with the Puzzle Hole sediments in the area south of F11;
3. a more accurate dating of the Scruff (Green) Sand and of some sand bodies in the Kimmeridge Clay and Middle Graben Shale formations; and
4. the distribution of the Scruff Shale, at present formally designated Clay Deep Formation.

It is now considered justified to give a formal description of some lithostratigraphic units used informally in Herngreen & De Boer (1985). This opportunity is also taken to present additional information on other formations described by NAM & RGD (1980) and to define a regrouping of some units.

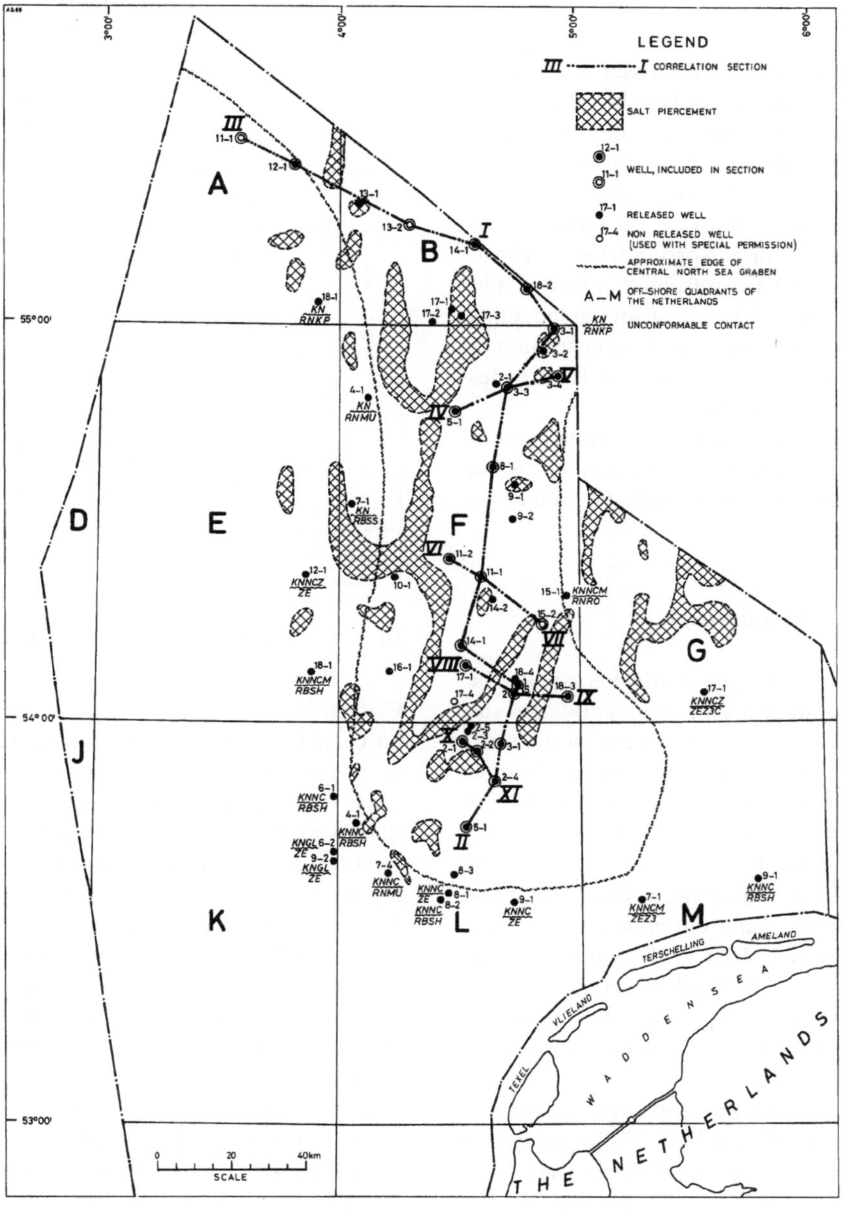

Fig. 2. General outline of the Dutch Central North Sea Graben, showing position of relevant drill-holes, cross-sections (Figs 18–22 and Encl. 1), salt domes and piercements.

Stratigraphy

All released wells of the area under study were included in the present report (see Fig. 2 for location of these wells). In a few cases, additional relevant information has been added for which permission for publication was obtained from the respective oil companies. Several log correlations through the Dutch Central North Sea Graben have been constructed and are presented in Encl. 1 and Figs 18–22. Finally, the lithostratigraphic units were tied to recent seismic data. A brief account of these results and a description of the seismic facies are given in the next chapter.

We will first discuss a number of problems concerning the dual interpretation of Kimmeridgian and Portlandian and the Jurassic-Cretaceous boundary.

ENGLAND		STANDARD AMMONITE ZONES	FRANCE	SOVIET UNION	
PORTLANDIAN	Late	lamplugi	PORTLANDIAN	VOLGIAN	Late
		preplicomphalus			
		primitivus			
	Early	'oppressus'			Middle
		anguiformis			
		kerberus			
		okusensis			
		glaucolithus			
		albani			
KIMMERIDGIAN	Late	fittoni			
		rotunda			
		pallasioides			
		pectinatus			Early
		hudlestoni			
		wheatleyensis			
		scitulus			
		elegans			
	Early	autissiodorensis	KIMMERIDGIAN	KIMMERIDGIAN	
		eudoxus			
		mutabilis			
		cymodoce			
		baylei			
OXFORDIAN		pseudocordata to cordatum	OXFORDIAN	OXFORDIAN	

Fig. 3. **Stratigraphy of the Kimmeridgian and Portlandian.**

Kimmeridgian and Portlandian

There is still considerable confusion about the meaning of the stage names Kimmeridgian and Portlandian (Fig. 3). Since the middle of the nineteenth century both terms have been used either with a 'British' meaning (*sensu anglico*) or in a continental sense (*sensu gallico*). In Britain, the Kimmeridgian spans the interval from the *baylei* Zone up to and including the *fittoni* Zone. The Portlandian extends from the *albani* Zone at the base to the *lamplughi* Zone at the top. In France, however, the base of the Portlandian, as established by d'Orbigny (1842–1851), is taken at the base of the *Gravesia* beds which are equated with the *elegans* Zone.

According to the recommendations and resolutions of the First and Second Jura Colloquia held in Luxembourg in 1962 and 1967 (Maubeuge 1964, 1970), the Kimmeridge stage has been defined by the following ammonoidal zones for both northern (Boreal) and southern (Mediterranean) Europe:

top : Zone à *Aulacostephanus autissiodorensis* (Boreal) and Zone à *Hybonoticeras beckeri* (Mediterranean)
base : Zone à *Pictonia baylei* (Boreal) and Zone à *Sutneria platynota* (Mediterranean)

Unfortunately, no decision was reached about the base and the name of the terminal Jurassic stage. If we were to use the term Portlandian in the original sense (in Britain it is customary to equate Portlandian with the Portland Beds and higher Jurassic strata), i.e., starting with the *albani* Zone and ending with the *lamplughi* Zone, a new stage would have to be introduced for the period between the *elegans* and *fittoni* zones. This confusing situation may explain the tendency in the northwestern European literature (e.g. Rawson & Riley 1982; Zeiss 1983) to use the subdivision into Oxfordian, Kimmeridgian (continental sense and Luxembourg) and Volgian. This classification has been added in the extreme right column in Fig. 3.

According to Cope (1985), the Soviet Union would now rather use the Tithonian than Volgian as their standard. However, Dr. V.A. Zakharov recently assured us that this statement is not correct (V.A. Zakharov pers. comm. 1987).

Here, the conventional North European or Boreal subdivision of the Late Jurassic into Oxfordian, Early and Late Kimmeridgian, and Portlandian will be used. This division is followed for reasons outlined in Herngreen & De Boer, 1985: to facilitate comparison with NAM & RGD 1980 and to tie in with the classical subdivision that is widely used by the northwest European stratigraphers.

Jurassic – Cretaceous boundary

The Portlandian standard ammonite zonation has been established by Wimbledon (in Cope et al. 1980) and is based on earlier schemes by Casey (1973) and Wimbledon & Cope (1978). Casey also proposed a subdividion of the Ryazanian stage for Britain. According to Wimbledon (1985), the *oppressus* Zone fauna is anomalous and he inferred a gap between the *primitivus* and *anguiformis* zones.

Zeiss (1983) discussed the ammonite sequences for the latest Jurassic-earliest Cretaceous; the zonation for the Berriasian in the Tethyan/Mediterranean realm is essentially based on Le Hégarat (1973). It has repeatedly been pointed out (e.g. Casey 1973; Hancock 1972; Rawson et al. 1978; Wimbledon 1985) that the Mediterranean Berriasian overlaps the terminal Jurassic stage. In other words, the Jurassic/Cretaceous boundary is drawn at a somewhat higher level in the Boreal realm than in the Tethyan/Mediterranean realm (see Fig. 4). Zeiss (1983) correlated the Tithonian/Berriasian boundary with the transition *oppressus/primitivus* zones. Cope (1985) placed this boundary at the topmost part of the *primitivus* Zone.

We support the usage of Portlandian as the name for the latest Jurassic stage. Moreover, we prefer to avoid a mixed Boreal (Portlandian) and Tethyan/Mediterranean (Berriasian) terminology for intervals spanning the Jurassic/Cretaceous boundary. For this reason, we use Ryazanian as the earliest Cretaceous stage.

Lithostratigraphy

General:
In revising the 'Late Jurassic' sediments we have followed two directives:
1) All non-marine sediments are included in the Central Graben Group, whilst the marine sediments are grouped in the Scruff Group.
2) The original classification of NAM & RGD (1980) has been followed as much as possible. This implies that existing names for lithological units will be used in emended forms.

The lithological units will be discussed in stratigraphic order; see Fig. 5 for a general scheme. For details on the biozonation, the reader is referred to Encl. 2.

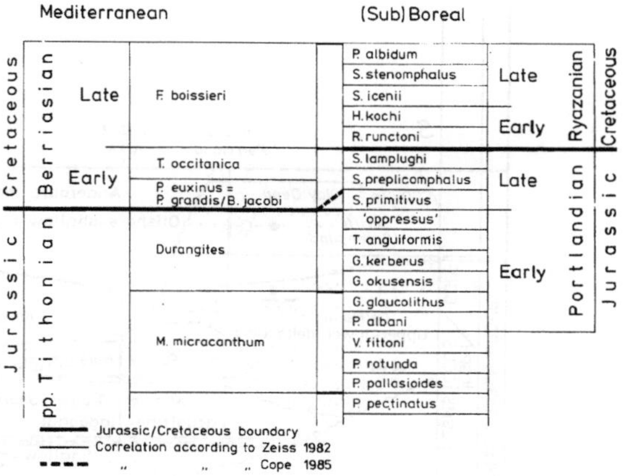

Fig. 4. Stratigraphic position of the Jurassic-Cretaceous boundary in the Boreal and Mediterranean areas.

Central Graben Group (CG): NAM & RGD (1980), emended
General – Traditionally, this Group comprises the Lower Graben Sand Formation, the Middle Graben Shale Formation and the Upper Graben Sand Formation. It is proposed here to add the Delfland Formation and the Puzzle Hole Formation to this Group.

Name – Name derived from the Central North Sea Graben (NAM & RGD, 1980).

Reference section – Well F3-3, 2547–3652 m, coord.: N 54° 50′ 45.5″, E 4° 42′ 29.3″ (Encl. 1). For the Delfland and Puzzle Hole successions, see the reference sections for these formations.

Definition – (emended after NAM & RGD, 1980). Group of formations deposited in a predominantly paralic environment and consisting of shales, sandstones, and coal seams. In the Dutch Central North

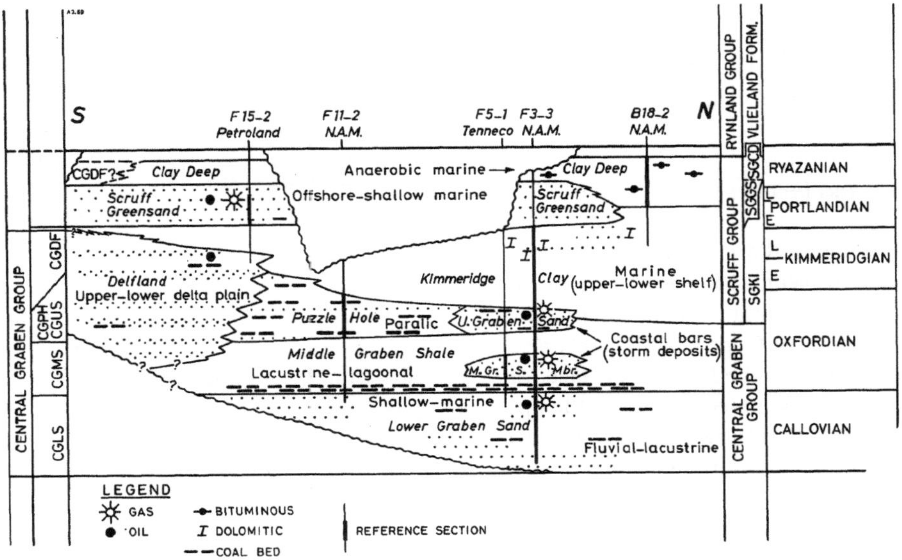

Fig. 5. **Rock stratigraphic diagram of the Central Graben and Scruff Groups, showing depositional environments and the occurrence of oil and gas in the various reservoir rocks.**

Sea Graben this succession unconformably overlies the Altena Group (generally the Aalburg Shale Formation and locally the Werkendam Shale Formation). In the Roer Valley Graben, the West Netherlands Basin, and the main part of the Broad Fourteens Basin, the Central Graben Group is situated between the marine sediments of the Altena and Rijnland Groups.

However, in the Central Netherlands and Vlieland Basins the Group overlaps sediments assigned to the Niedersachsen Group. In the Dutch Central North Sea Graben the Group interfingers with marine shales of the Kimmeridge Clay Formation of the Scruff Group.

Age – Callovian to Ryazanian.

Subdivision – five formations are distinguished, as follows:

Delfland Formation (CGDF)
Puzzle Hole Formation (CGPH)
Upper Graben Sand Formation (CGUS)
Middle Graben Shale Formation (CGMS)
Lower Graben Sand Formation (CGLS)

Lower Graben Sand Formation (CGLS). NAM & RGD (1980), emended.

Name – Named after the Central North Sea Graben (NAM & RGD, 1980).

Reference section – Well F3-3, 3090–3652 m; co-ord.: N 54° 50′ 45.5″, E 4° 42′ 29.3″ (NAM & RGD, 1980).

Definition – Section of greyish-brown, very fine to fine, well-sorted sandstones, occurring in beds generally less than 10 m thick, with intercalations of thin greyish-brown silty to sandy shales. The formation is generally carbonaceous with some distinct coal layers. The individual sandbodies display a rather restricted lateral extension. Especially in higher parts of this unit the gamma ray log pattern of these beds shows a generally coarsening upward sequence. The formation, which is confined to the Central Graben, ranges in thickness from 40 m to 562 m, with maxima attained in the SW (fault bounded) corner of block F3 (Fig. 6). The formation rests unconformably on strata of the Werkendam Shale Formation, the Aalburg Shale Forma-

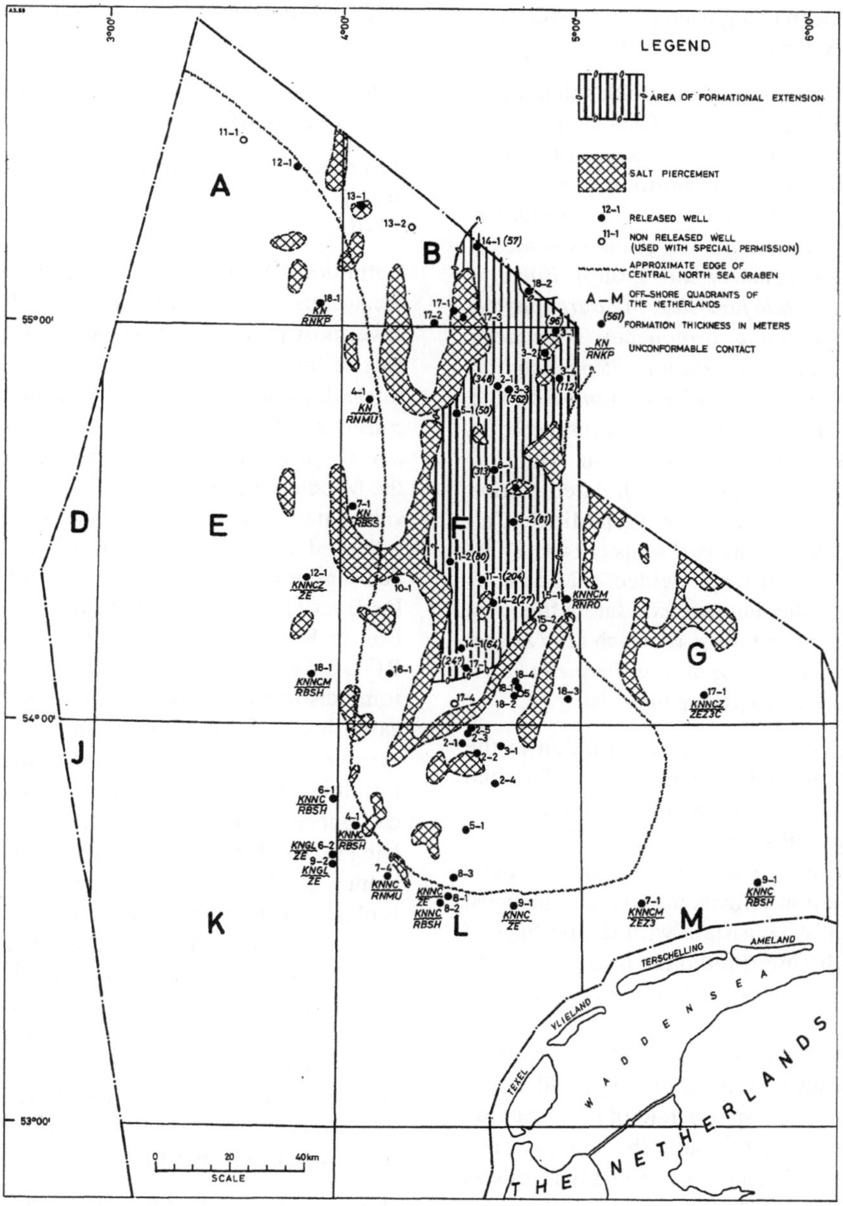

Fig. 6. Geographical distribution of Lower Graben Sand Formation.

tion, or the Upper Germanic Triassic Group. Occasionally, the underlying Werkendam Shale Formation may be developed in a coarse-grained oolitic facies, which makes the boundary between the two formations difficult to pick on the logs. The Lower Graben Sand Formation is conformably overlain by the Middle Graben Shale Formation. In the Danish part of the Central Graben the formation has also been recognized (Koch, 1983; Jensen et al., 1986 & Frandsen et al., 1987).

Age – Callovian, presumably not older than Middle Callovian. A Middle to Late Callovian age can be established on the basis of dinoflagellate associations. Significant species permitting a more precise dating are: *Acanthaulax senta, Ctenidodinium* spp. (a.o. *C. continuum, C. gochtii-kettonense* group, *C. sellwoodii-stauromatos* group), *Energlynia acollaris, Lithodinia jurassica, Meiourogonyaulax borealis, Pareodinia prolongata, Rigaudella aemula, Scriniodinum galeritum, Stephanelytron* spp., *Systematophora* spp., *Wanaea fimbriata,* and *W. thysanota.* There is a consistent presence of a characteristic sporomorph association with *i.a. Contignisporites* sp. 1. In other well-dated marine deposits, this species is not older than Middle Callovian. Samples from only two wells (F14-1: 2109–2115 m and F17-3: 2040 m) yielded a Middle or Middle to Late Callovian ostracod fauna the most characteristic representatives of which are *Fastigatocythere interrupta interrupta, Fuhrbergiella horrida horrida,* and *Lophocythere bipartita.*

Sedimentary history – The great variation in the thickness of the formation points to a differential subsidence with larger sediment accumulations in the more rapidly subsiding areas (e.g. F3-3).

Since no marine organisms occur in the lower part of the formation a fluvial origin seems the most likely here. This is consistent with the findings of Koch (1983), who interpreted the basal part of the J-2 unit to have been deposited in a braided river plain.

In the course of time deltaic conditions with occasional marine incursions started to prevail, as can be deduced from the sparse microfauna, rich dinoflagellate assemblages, and the log patterns. Coarsening upward sequences probably represent various channel-mouth bars, whereas clays formed as delta plain deposits.

There is great similarity with the Danish area where, according to Jensen et al. (1986), the formation was deposited 'in a marginal marine environment with strong fluvial influence, probably a delta with lagoons and interdistributary bays'.

Middle Graben Shale Formation (CGMS). NAM & RGD (1980), emended.

Name – Named after the Central North Sea Graben (NAM & RGD, 1980).

Reference section – Well F3-3, 2670–3090 m; co-ord.: N 54° 50′ 45.5″, E 4° 42′ 29.3″ (NAM & RGD, 1980).

Definition – Section of grey, locally very silty carbonaceous shales with some thin coal seams in the lowermost part of the formation. In the northern part of the F-quadrant (e.g. in F2, F3, and F5) a single thick sandstone bed may be intercalated. It is considered to have such a divergent development that we propose it to be a separate member called the Middle Graben Sand Member. The formation varies in thickness and its geographical distribution is limited to the northern part of the Central North Sea Graben (Fig. 7). According to Herngreen & De Boer (1985), the formation also comprises the Lower Kimmeridge Clay Member of NAM & RGD (1980). In the lower part of the formation some distinct coal seams occur. They are laterally extensive and form important lithostratigraphic markers. One such coal seam at the base of the formation reflects the contrast between the basal carbonaceous part of the formation and the underlying sands of the Lower Graben Sand Formation. Hence, the boundary between the two formations is placed at the base of this coal seam. Toward the north the formation is overlain by the Kimmeridge Clay Formation. In southern direction the formation is (in part) equivalent to the Puzzle Hole Formation. In the Danish part of the Central Graben the J4-unit as described by Koch (1983) displays a similar lithology, and therefore Jensen et al. (1986) and Frandsen et al. (1987) correctly assign the name Middle Graben Shale Formation to this unit.

Age – Early-Middle Oxfordian. The age determination based on dinoflagellates yields a latest Callovian-Middle Oxfordian age as indicated by *Acanthaulax senta, Leptodinium subtile, Polystephanephorus paracalathus, Scriniodinium crystallinum, Systematophora valensii, Systematophora* spp., *Wanaea fimbriata* and *W. thysanota.* Among the sporomorphs the frequent presence of the *Contig-*

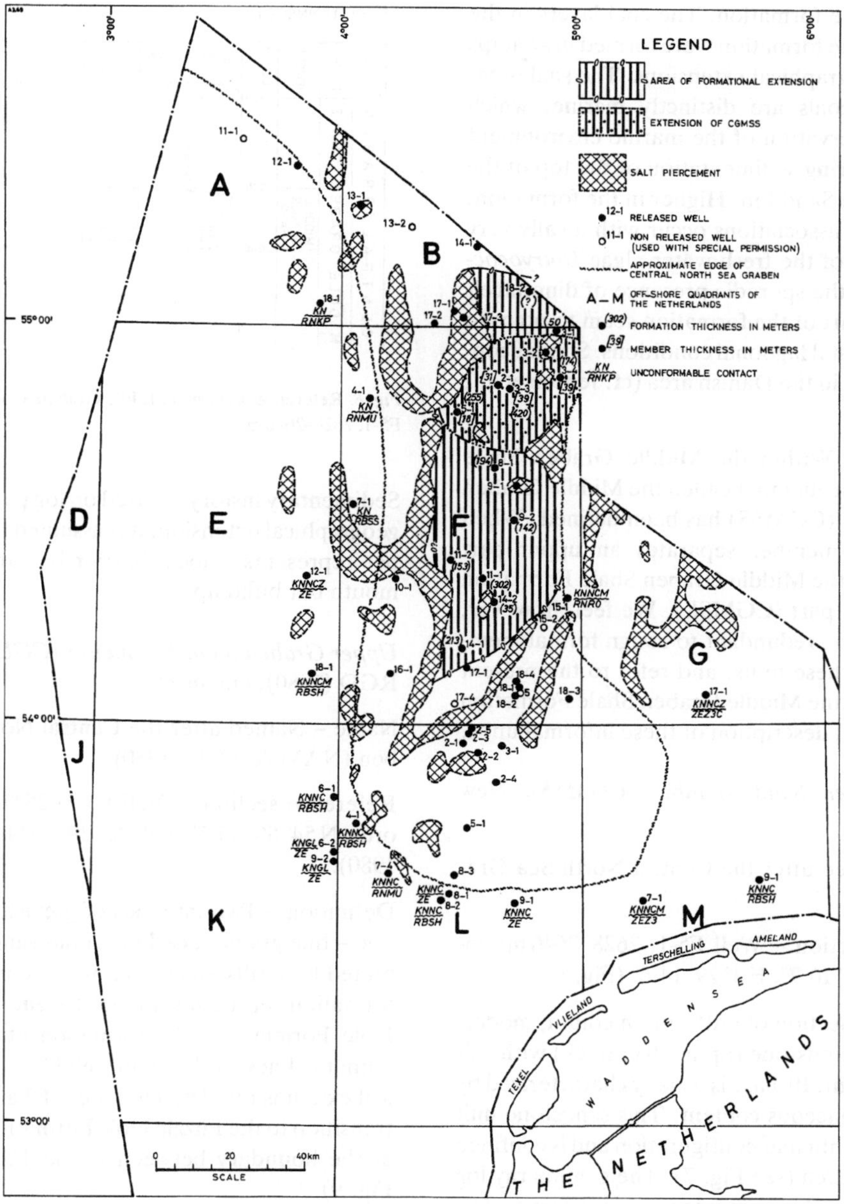

Fig. 7. Geographical distribution of Middle Graben Shale Formation.

nisporites sp. 1–3 complex can be noted. The formation is characterized by a typical ostracod assemblage (consisting of *Galliaecytheridea 'praedissimilis', G. 'prae-postsinuata', G. 'incurvata'*) which represents the Early-Middle Oxfordian.

Sedimentary history – Herngreen & De Boer (1985) concluded to a paralic depositional environment for the formation. The coal layers in the lower part of the formation were formed in swamps with large geographical extensions. The shales between these coals are distinctly marine, which points to preservation of the marine environment established during sedimentation of the top of the Lower Graben Sand Fm. Higher in the formation, only terrestric associations occur with locally very high numbers of the fresh water algae *Botryococcus*. This and the sporadic presence of dinoflagellates in that part of the formation seem to point to lacustrine-coastal lagoonal conditions. Similar conditions existed in the Danish area (cf. Jensen et al., 1986).

Subdivision – Within the Middle Graben Shale Formation, one member called the Middle Graben Sand Member (CGMSS) has been defined. When present, this member separates an upper part (CGMSU) of the Middle Graben Shale Formation from its lower part (CGMSL). We feel, however, that it would be redundant to assign formal member status to these units, and refer to the present description of the Middle Graben Shale Formation for the general description of these informal units.

Middle Graben Sand Member (CGMSS). New member.

Name – Named after the Central North Sea Graben.

Reference section – Well F5-1, 2628–2646 m; coord.: N 54° 47′ 10.7″, E 4° 29′ 14.6″ (Fig. 8).

Definition – Section of buff, fine-medium, moderately sorted sandstone (up to 20 m thick) with calcareous cement. Its base is usually characterized by a high carbonaceous content. This sandstone unit has a rather lenticular configuration and is confined to the F3-F5 area (see Fig. 7). The gamma ray log shows a generally coarsening upward trend. The top and base of the member are characterized by distinct log breaks defining the contacts with the under- and overlying non-calcareous shales of the Middle Graben Shale Formation.

Age – Early-Middle Oxfordian; based on its position within the Middle Graben Shale Formation.

F5-1 (TENNECO)

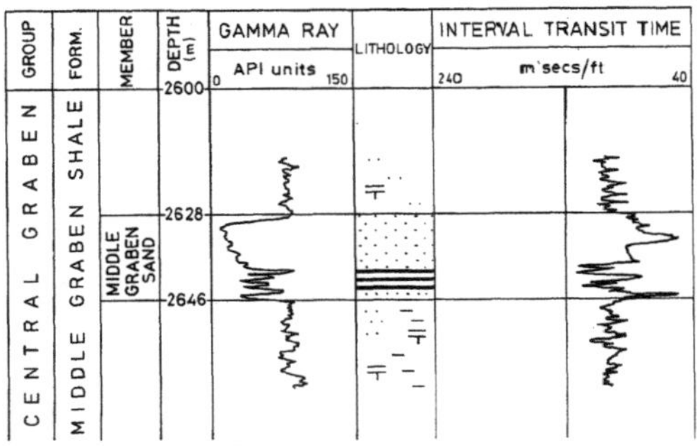

Fig. 8. Reference section of Middle Graben Sand Member; well F5-1, 2628–2646 m.

Sedimentary history – Based on log patterns and its geographical extension, it is assumed that the number represents a local barrier bar and/or deltaic mouth bar build up.

Upper Graben Sand Formation (CGUS). NAM & RGD (1980), emended.

Name – Named after the Central North Sea Graben (NAM & RGD, 1980).

Reference section – Well F3-3, 2547–2670 m; coord.: N 54° 50′ 45.5″, E 4° 42′ 29.3″ (NAM & RGD, 1980).

Definition – Two intervals of greyish-brown, very fine – fine grained, carbonaceous sandstones, separated by a silty shale section. In our concept the formation represents a basal tongue of the Puzzle Hole Formation. The formation attains a maximum thickness of 123 m in well F3-3. Its geographical extent is restricted to blocks F3 and F5 and the transition to the Puzzle Hole Formation takes place at the boundary between blocks F5 and F8 (see Fig. 9).

Generally, the formation conformably overlies the Middle Graben Shale Formation and pinches out in the Kimmeridge Clay Formation toward the north. Consequently, the upper limit of the formation is also marked by the contact of its sandy beds with the shales of the Kimmeridge Clay Formation.

Age – Middle-Late Oxfordian. In the type section

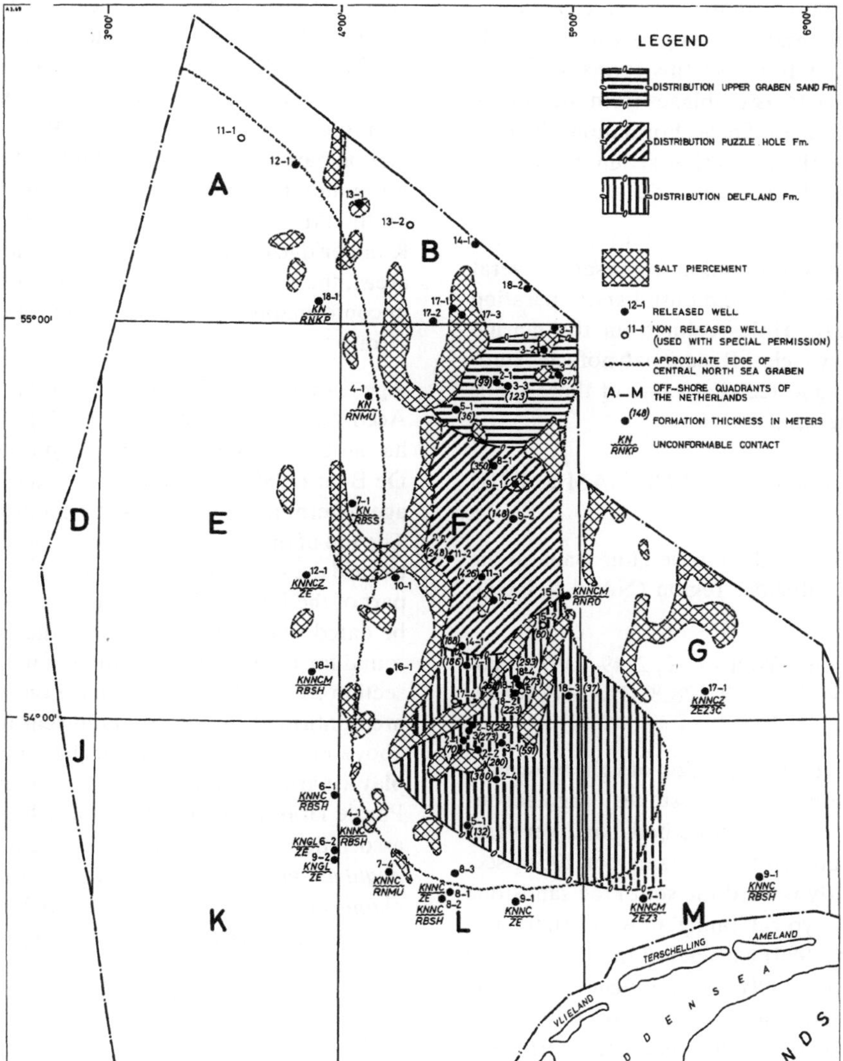

Fig. 9. Geographical distribution of Upper Graben Sand, Puzzle Hole and Delfland formations.

this age is based solely on superposition. A definite late Middle-Late Oxfordian age could be established from dinoflagellates in well F3-4 (in the cored interval from 2755.4–2777.5 m). Among the diagnostic forms are *Ellipsoidictyum* sp., *Glossodinium dimorphum,* and *Leptodinium arcuatum.* The top occurrence of *Systematophora valensii* has been recorded in this formation. Among the sporomorphs the first representatives of the *Concavissimisporites-Impardecispora* complex can be noted. The few ostracods found in this formation show more resemblance to assemblages from the underlying Middle Graben Shale Formation than to those of the overlying marine sediments of the Kimmeridge Clay Formation.

Sedimentary history – The sand-bodies of the Upper Graben Sand Formation represent several coastal bars and/or deltaic mouth bars, separating the paralic Puzzle Hole realm from the marine Kimmeridge Clay facies. Recent palynofacies studies have even suggested that some of these sands were storm generated.

Puzzle Hole Formation (CGPH). NAM & RGD (1980), new combination.

Name – Named after the Puzzle Hole Bank in the Dutch northern offshore region (NAM & RGD, 1980).

Reference section – Well F11-2, 2149–2397 m; coord.: N 54° 24′ 54.6″, E 4° 27′ 38.8″ (NAM & RGD, 1980).

Definition – Regular alternation of light brownish-grey silty carbonaceous claystones, argillaceous siltstones with thin sandstone beds, and very frequent coal seams (10 to 20 seams per 100 m section). Particularly toward the south the sandstone beds display a typical fining upward character. NAM & RGD (1980) considered the Puzzle Hole Formation to be a silty-sandy tongue within the Kimmeridge Formation, as part of the Scruff-Group. Because of its predominantly terrestric-paralic character we have placed the Puzzle Hole Formation in the Central Graben Group. Geographically, the formation occupies an intermediate position between the Delfland Formation, into which it grades toward the south, and the Upper Graben Sand Formation in the north (Fig. 9). The latter is thought to be a northern tongue of the Puzzle Hole Formation. The transition to the Delfland Formation is rather gradual and takes place in the southern part of the F quadrant. The Puzzle Hole Formation displays a typical 'nervous' pattern on both gamma ray and sonic logs due to the rapid alternation of thin sandstones, siltstones, claystones, and coal seams, by which it differs from the Delfland and Upper Graben Sand Formations. These formations have thicker sand beds with fewer intercalated coal seams. The Puzzle Hole Formation partially overlies the Middle Graben Shale Formation but to the south it rests unconformably on the Altena Group. Due to the removal of the Kimmeridge Clay Formation by erosion over large areas, the Puzzle Hole Formation is often covered unconformably by basal sediments of the Rijnland Group.

Age – Late Oxfordian-Early/Late Kimmeridgian. According to NAM & RGD (1980) the formation has an Early Kimmeridgian age, but Herngreen & De Boer (1985) have already pointed out that this age determination was based on a misconception. The top of the formation seems to become younger toward the south. By the use of ostracods, the basal part of the Puzzle Hole Formation in well F8-1 can be dated as Late Oxfordian and the same age was found for the top of the formation in the reference section F11-2 (*Pseudocordatum* Zone). These ages are confirmed by the palynological findings. The above-mentioned intervals in F8-1 and F11-2 display a marine character (an uncommon facies in the Puzzle Hole Formation) with a rich dinoflagellate association in F11-2. Characteristic forms are *Gonyaulacysta jurassica* (abundant), *Ctenidodinium chondrum, Hystrichosphaerina orbifera, Occisucysta monoheuriska,* and *Systematophora.* Among the sporomorphs, *Trilites minutus* and representatives of the *Contignisporites* complex occur.

Sedimentary history – The sediments of the Puzzle Hole Formation were deposited in a paralic (lower delta plain) environment. In this facies the coals were formed in numerous swamps.

Delfland Formation (CGDF). NAM & RGD (1980), status novum and extended.

Name – Named after the polder authority of Delfland in the province of Zuid-Holland (NAM & RGD, 1980).

Reference section – Well Nieuwerkerk-1, 968–1942 m; coord.: N 51° 57′ 00.2″, E 4° 37′ 32.4″ (NAM & RGD, 1980).

Additional reference section – Well F18-1, 2417–2690 m, coord.: N 54° 5′ 54″, E 4° 44′ 32″ (Encl. 1).

Additional reference section – Well K15-1, 1559–2270 m, coord.: N 53° 13′ 28.2″, E 3° 53′ 47.6″ (NAM & RGD 1980).

Definition – Sequence of rapidly alternating sands, siltstones, and shales, with some coal beds and with common lignitic matter. The siltstones and shales are generally multi-coloured (grey, green, beige, and pink) and become increasingly reddish-mottled toward the south. The gamma ray log pattern of the sandbeds shows a change from predominantly coarsening upward in the north (F17) to predominantly fining upward in the south (L & K quadrants). In the original concept of NAM & RGD (1980) the rank of Group had been assigned to this formation. Its traditional subdivision into Lower Delfland, Fourteens Clay and Upper Delfland formations was mainly based on the fact that they considered the Fourteens Clay to be a major marine intercalation between the Lower and Upper Delfland formations. Herngreen & De Boer (1985) pointed out that the Fourteens Clay in the type well K15-1 distinctly showed terrestrial assemblages and only a restricted marine influence at the top of the formation. This does not justify the threefold subdivision proposed by NAM & RGD (1980). Moreover, the subdivision seems to be a local phenomenon. We favour the concept of an undivided Delfland Formation with the Fourteens Clay (CGDFF) as a member of this formation (see NAM & RGD (1980) for more information on type section, etc.). We realize that in the Delfland Formation other lithostratigraphic units with member rank may be recognized in other basins, but it would take us beyond the scope of the present paper to deal in detail with this formation outside the Central North Sea Graben.

Age – Oxfordian – Ryazanian. A Kimmeridge age can be attributed to the Formation in the F18-L2-L3 area. In F17-4, however, an Oxfordian age, and probably even a Middle Oxfordian age, can be established on the basis of palynological data. Ostracod assemblages are similar to those usually found in the Middle Graben Shale, and thus confirm this age determination. This implies that the Delfland Formation in the southern part of the Central North Sea Graben is distinctly older than the original type section K15-1 (of NAM & RGD) in the Broad Fourteens Basin. Sporomorphs from the basal part of the formation (F17-4) are similar to the Oxfordian assemblages from the Middle Graben Shale Formation and the Upper Graben Sand Formation. The Late Kimmeridge associations are very well developed in F18-2 and F18-5; they show first occurrences of the *Trilobosporites bernissartensis* group, *Pilosisporites* spp., and *Kraeuselisporites* sp., and last appearance of *Lycopodiacidites irregularis*.

Sedimentary history – The sediments of the Delfland Formation mainly represent upper to lower delta plain deposits. Occasional high frequencies of *Botryococcus* point to local lacustrine conditions.

Scruff Group (SG): NAM & RGD (1980), emended.

General – According to the original definition of NAM & RGD (1980) the Scruff Group consists of the Kimmeridge Clay Formation and the Puzzle Hole Formation. We propose in this report that the Puzzle Hole Formation be transferred to the Central Graben Group. In addition, we introduce two new formations, which are included in the Scruff Group: the Scruff Greensand Formation and the Clay Deep Formation.

Name – Named after the offshore Upper Scruff Bank, which lies adjacent to the reference well F3-3 (NAM & RGD, 1980).

Reference section – Well F3-3, 1628 to 2547 m; coord.: N 54° 50′ 45.5″, E 4° 42′ 29.3″ (NAM & RGD, 1980; emended).

Additional reference section – Well F3-1, 2265 to 2810 m; coord.: N 54° 59′ 40″, E 4° 54′ 18″ (Encl. 1).

Definition (emended after NAM & RGD, 1980) – Group of formations deposited in a predominantly marine environment. Most of this group rests conformably on several formations of the Central Graben Group. The sediments of the Rijnland Group usually unconformably overlie the Scruff Group.

Age – Late Oxfordian – Ryazanian.

Subdivision – Three formations are distinguished as follows:

Kimmeridge Clay Formation (SGKI)
Scruff Greensand Formation (SGGS)
Clay Deep Formation (SGCD)

Kimmeridge Clay Formation (SGKI). NAM & RGD (1980), emended.

Name – Name derived from the British stratigraphic nomenclature where it is applied to similar clayey deposits which are wide-spread in the general North Sea region (NAM & RGD, 1980).

Reference section – Well F3-3, 1780–2547 m; coord.: N 54° 50′ 45.5″, E 4° 42′ 29.3″ (NAM & RGD, 1980; emended).

Definition – 'In the northern part of the Central North Sea Graben, the Formation is a sequence of medium to dark olive-grey, generally silty shales with numerous thin dolomite streaks (expressed on wire-line logs with a characteristic peaked appearance). Fossil fragments are common in lenses; lignite particles occur frequently. Towards the south, the carbonate streaks and the olive hue gradually disappear. Furthermore, the shales become increasingly silty to sandy' (from NAM & RGD, 1980). In the northern part of the area a twofold division can be made of the formation. The upper part is characterized by the presence of numerous dolomitic beds, whereas the lower part is predominantly clayey. In seismic sections from this area (e.g. Fig. 14) a slight intraformational unconformity coincides with the boundary between these lithologies. The formation rests conformably on the sediments of the Central Graben Group. Generally, the formation is conformably overlain by the Scruff Greensand Formation. North of F3, where the latter formation is not present, the Clay Deep Formation locally rests conformably on the Kimmeridge Clay Formation. The twofold lithological division, as described by us, can also be noted in the Danish part of the Central Graben. There, the Upper and Lower units are named respectively, Farsund and Lola Formation (Jensen et al., 1986). In our new concept the formation corresponds with the Main Kimmeridge Clay as described by Herngreen & De Boer (1985). The same authors also discarded the subdivision into the Upper and Lower Kimmeridge Member of NAM & RGD (1980). Figure 10 shows the geographical distribution of the formation. We agree with the argumentation put forward by Doré et al. (1985), who dealt with the problems in connection with correlation of the Kimmeridge Clay Formation and the Humber Group from the type areas in England with several North Sea basins. In this respect we had two options: 1) to extend the Humber Group into the Dutch area and reject the Scruff Group; or 2) to introduce a new name for the Kimmeridge Clay Formation in the Dutch area (as Jensen et al. 1986 have done in the Danish area).

Both solutions would have been stratigraphically correct, but at this moment we prefer to follow NAM & RGD (1980) and use well-established formation names rather than adopt a new nomenclature.

Age – Late Oxfordian-Berriasian (NAM & RGD, 1980). In the type section F3-3 this age does not seem to be appropriate, since the age of the top of the Kimmeridge Clay in that well is near the Kimmeridgian-Portlandian boundary (see Herngreen & De Boer, 1985). An earliest Portlandian and latest Kimmeridgian age could be established in F3-1 from side-wall samples from the sandy interval 2427–2510 m. The age determinations based on micropaleontology and palynology are in good agreement. The Kimmeridge Clay Formation is generally characterized by a rich dinoflagellate association. Some species with stratigraphic significance are *Cannosphaeropsis thula, Ctenidodinium chondrum, C. culmulum, C. panneum, Egmontodinium polyplacophorum, Epiplosphaera bireticulata, Gochteodinia mutabilis, Gonyaulacysta jurassica, 'Gonyaulacysta' longicornis, Histiophora ornata, Leptodinium arcuatum/eumorphum, L. subtile, Muderongia* sp.A in Davey 1982, *Occisucysta balia, 'Oligosphaeridium pulcherrimum'*-acme, *Scriniocassis dictyota, Scriniodinium crystallinum, S. inritibilum, S. luridum,* and *Stephanelytron* spp. In the holomarine facies there are strongly selected sporomorph assemblages, dominated by bisaccates

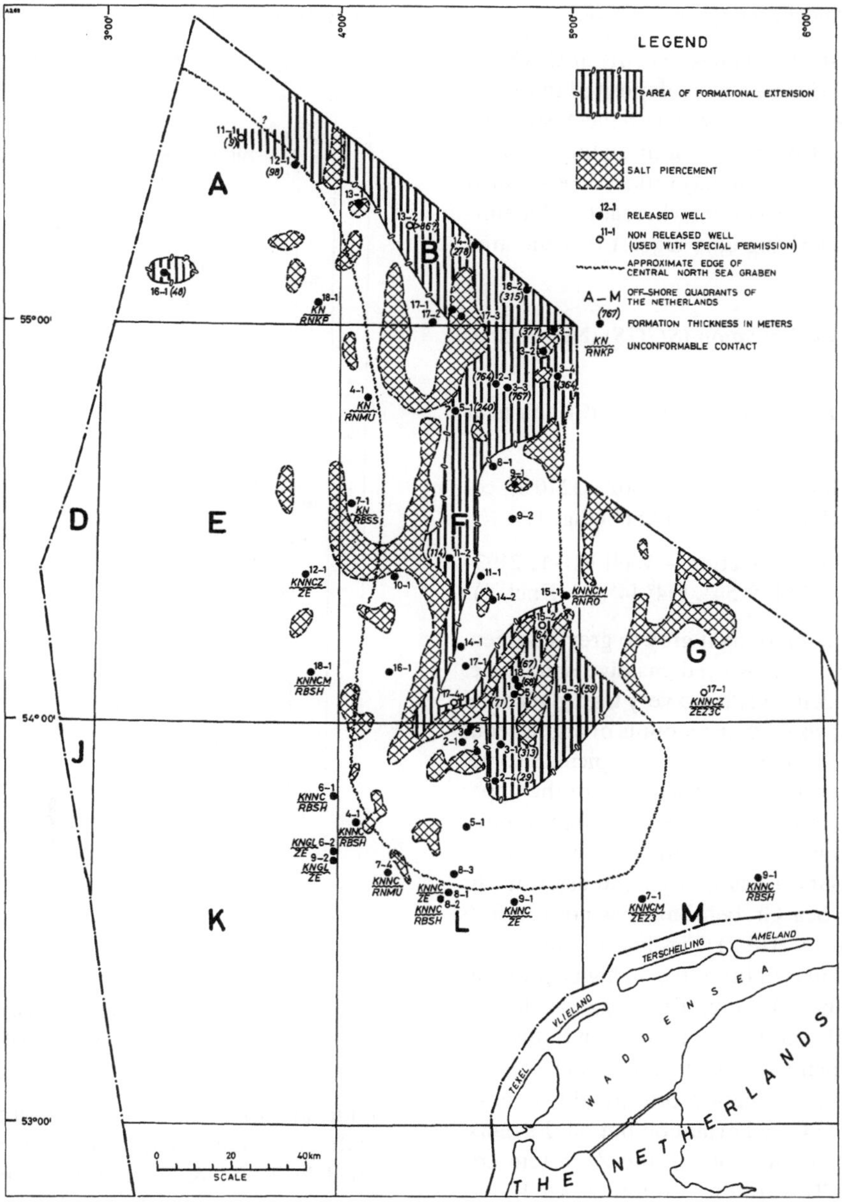

Fig. 10. Geographical distribution of Kimmeridge Clay Formation.

and *Calliasporites*. Micropaleontologically, two marine associations can be distinguished within the Kimmeridge Clay Formation. The upper part of the formation bears the ostracods *Galliaecytheridea compressa* and *G. spinosa* and the foraminifer *Saracenaria triquetra,* whereas the lower part is characterized by the ostracods *Galliaecytheridea dissimilis, G. gracilis, G. dorsetensis, Macrodenti-*

na cicatricosa, and *Epipleura eleonorae.*

Sedimentary history – The sediments of the Kimmeridge Clay Formation were deposited in an upper to lower shelf environment with increasing terrestrial influence toward the south. In the northern area the lower clayey unit may reflect deposition in a tranquil, deep environment whereas the dolomitic beds in the overlying unit suggest that the area became shallower.

Scruff Greensand Formation (SGGS). New formation.

Name – Named after the Upper Scruff Bank near Well F3-3.

Reference section – Well F15-2, 3021–3276 m; coord.: N 54° 14′ 18.54″, E 4° 50′ 44.86″ (Fig. 11).

Additional reference section – Well F18-1, 2193–2350 m; coord.: N 54° 5′ 54″, E 4° 44′ 32″ (Encl. 1).

Definition – Section of light grey to greyish-green, very fine – fine, well-sorted massive sandstone. The formation generally has a very high glauconite content and shows varying amounts of argillaceous matter. Large amounts of sponge spiculae sometimes form a conspicuous constituent of the upper part of the formation. The general configuration of the formation varies from distinct massive to sheet-like. The formation has a large extension in the Central North Sea Graben but does not reach its northern part (e.g. B18-2) (Fig. 12). The formation onlaps on the flanks of the adjacent highs, probably overstepping them locally. This suggests that we are dealing with a complex of numerous individual sandbodies which were stacked and coalesced. In the type area the formation conformably overlies the Kimmeridge Clay Formation but on the flanks it may rest unconformably on the Zechstein erosional surface. Toward the south (e.g. L5-2), biostratigraphic and seismic data suggest an unconformable contact between the Scruff Greensand Formation and the underlying Delfland Formation. There seems to be a conformable contact with the overlying part of the Clay Deep Formation (e.g. L3-1 and F3 block). When the Clay Deep Formation is absent the sediments of the Rijnland

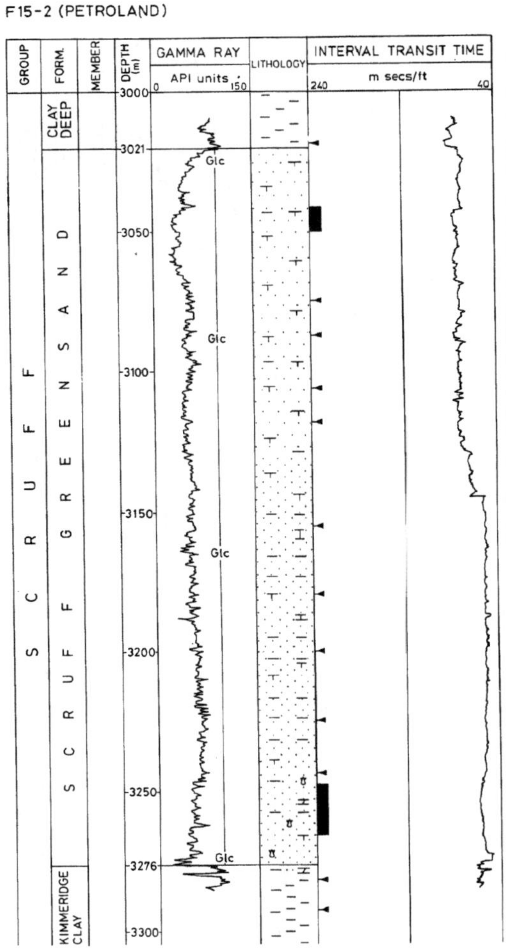

Fig. 11. Reference section of Scruff Greensand Formation; well F15-2, 3021–3276 m (by courtesy of Petroland B.V.).

Group are found to rest unconformably on the Scruff Greensand. The formation was not mentioned in NAM & RGD (1980) and since then several names such as the Scruff Sand (Herngreen & De Boer, 1985), F18 Sand, Main Sand, and Vlieland Sand have been used to denote this unit.

Age – Portlandian-Early Ryazanian. This age is based on palynology and superposition, since no microfauna has been encountered yet in the formation. The reference section in F15-2 is character-

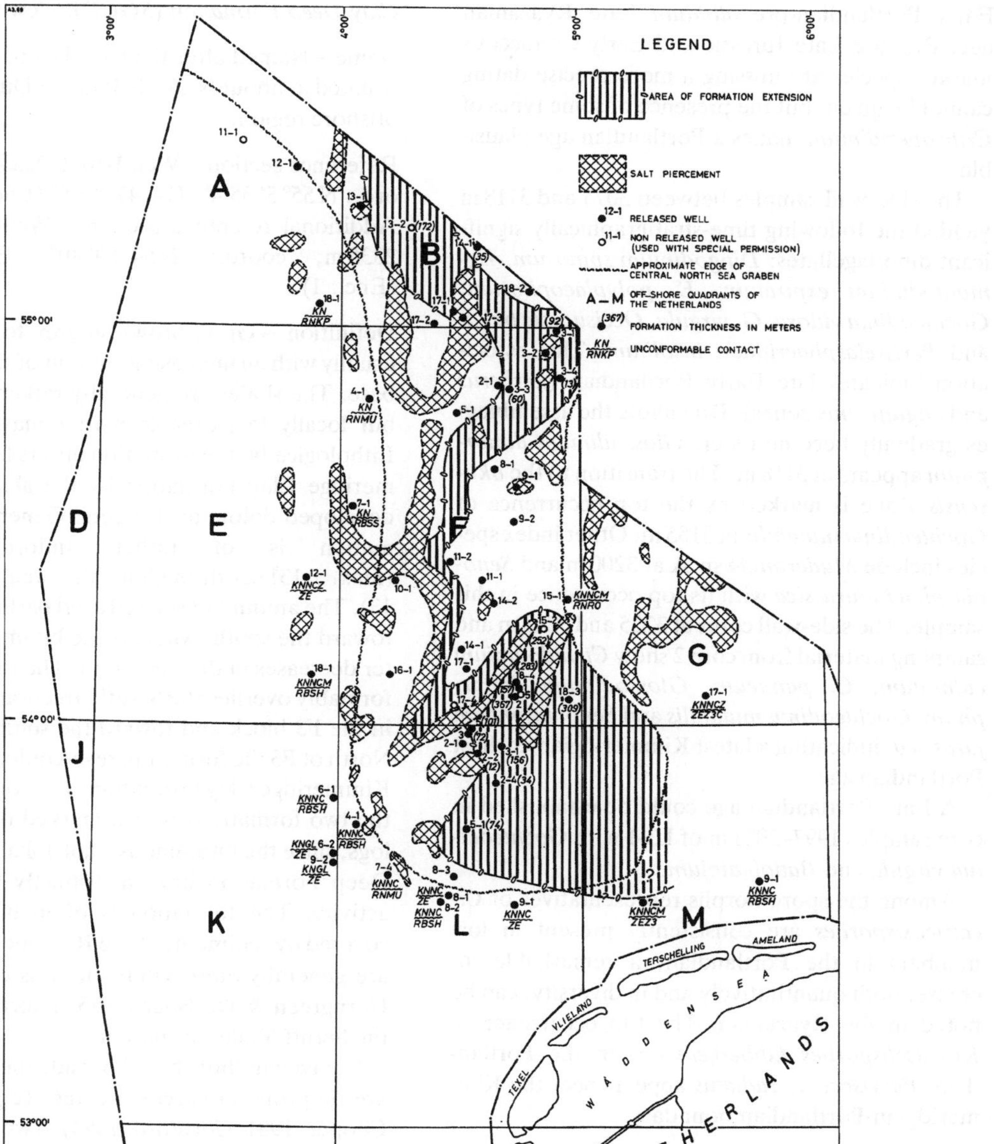

Fig. 12. Geographical distribution of Scruff Greensand Formation.

ized by a rich dinoflagellate association which made the following age determination possible. Two cores (no. 1 at 3042–3050.87 m and no. 2 at 3248–3266.5 m) and eight side-wall samples between 3075 and 3243 m were examined.

The samples from core 1 are poor in fossils; they contain for example *Cannosphaeropsis thula* and *Gochteodinia villosa,* indicating an indifferent late

Early Portlandian-pre *albidum* Late Ryazanian age. Because Late Jurassic and Early Cretaceous marker species are missing a more precise dating cannot be given, but the presence of some types of *Cribroperidinium* makes a Portlandian age plausible.

The side-wall samples between 3075 and 3118 m yielded the following time-stratigraphically significant dinoflagellates: *Dingodinium spinosum, Egmontodinium expiratum, E. polyplacophorum, Gochteodinia villosa, G. virgula, Occisucysta balia,* and *Perisseiasphaeridium insolitum.* This association indicates late Early Portlandian (*kerberus* and *anguiformis* zones). Downhole the assemblages gradually become older. *Glossodinium dimorphum* appears at 3118 m. The transition to the *okusensis* Zone is marked by the top occurrence of *Gochteodinia mutabilis* at 3155 m. Other index species include *Muderongia* sp.A at 3200 m and *Senoniasphaera jurassica* with its top occurrence in this sample. The side-wall cores at 3225 and 3243 m and sampling material from core 2 show *Ctenidodinium culmulum, C. panneum, Glossodinium dimorphum, Gochteodinia mutabilis* and *Senoniasphaera jurassica,* indicating a latest Kimmeridgian-earliest Portlandian age.

A Late Portlandian age could be established in core samples 1997–2021 m of L3-1 with *Gochteodinia virgula* and *Batioladinium.*

Among the sporomorphs representatives of *Cicatricosisporites* are consistently present in low numbers in the Portlandian; a remarkable increase, both quantitatively and in diversity, can be noted in the Ryazanian. The top occurrence of *Kraeuselisporites tubbergensis* is in the Portlandian; *Parvisaccites radiatus* appears near the Kimmeridgian-Portlandian boundary.

Sedimentary history – The Scruff Greensand Formation was deposited in a shallow marine environment as a complex of barrier sands which in the course of time were partly reworked and coalesced into transgressive sheet sands.

Clay Deep Formation (SGCD). New formation.

Name – Named after the Clay Deep, a depression situated at about N 55°, E 4° in the Dutch northern offshore region.

Reference section – Well B18-2, 2225–2357 m; coord.: N 55° 5′ 35.4″, E 4° 47′ 48.6″ (Encl. 1).
Additional reference section – Well F3-1, 2265–2335 m; coord.: N 54° 59′ 40″, E 4° 54′ 54.18″ (Encl. 1).

Definition – Grey, brownish-grey to black shales usually with an increasing amount of silt toward the base. The shales are generally rather bituminous, but locally less organic matter may be present. Lithologically, the formation differs from the Kimmeridge Clay Formation by the absence of well developed dolomite stringers. Generally, the formation is of rather uniform thickness (max. ± 131 m) throughout the area studied (Fig. 13). The amount of silt and sand particles increases toward the south, whereas the bituminous character decreases in that direction. The formation conformably overlies the Scruff Greensand Formation in the F3 block and toward the south (e.g. L3-1). North of F3 the formation rests conformably on the Kimmeridge Clay Formation. The contact between the two formations is well marked on gamma ray logs, since the bituminous ('hot') shales of the Clay Deep Formation have a distinctly higher radioactivity. The formation is often unconformably covered by sediments of the Rijnland Group which are generally more sandy and less carbonaceous. Herngreen & De Boer (1985) informally called it the Scruff Shale Formation.

Equivalent 'hot shales' outside the Dutch sector are oil prone in varying degrees (e.g. Barnard & Cooper, 1981; Cornford, 1984). Two examples are the Tau 'hot shale' Member of the Borglum Formation of the Fiske sub-basin and the Mandal Formation in the adjacent Norwegian sector of the Central Graben (cf. Cornford, 1984). Jensen et al. (1986) describe a similar unit (informally named 'Hot unit') from the Danish Central Trough, which they consider to be patchily developed as a 'hot' organic rich facies of the Kimmeridge Clay.

Age – Late Early Portlandian – Ryazanian, pre-

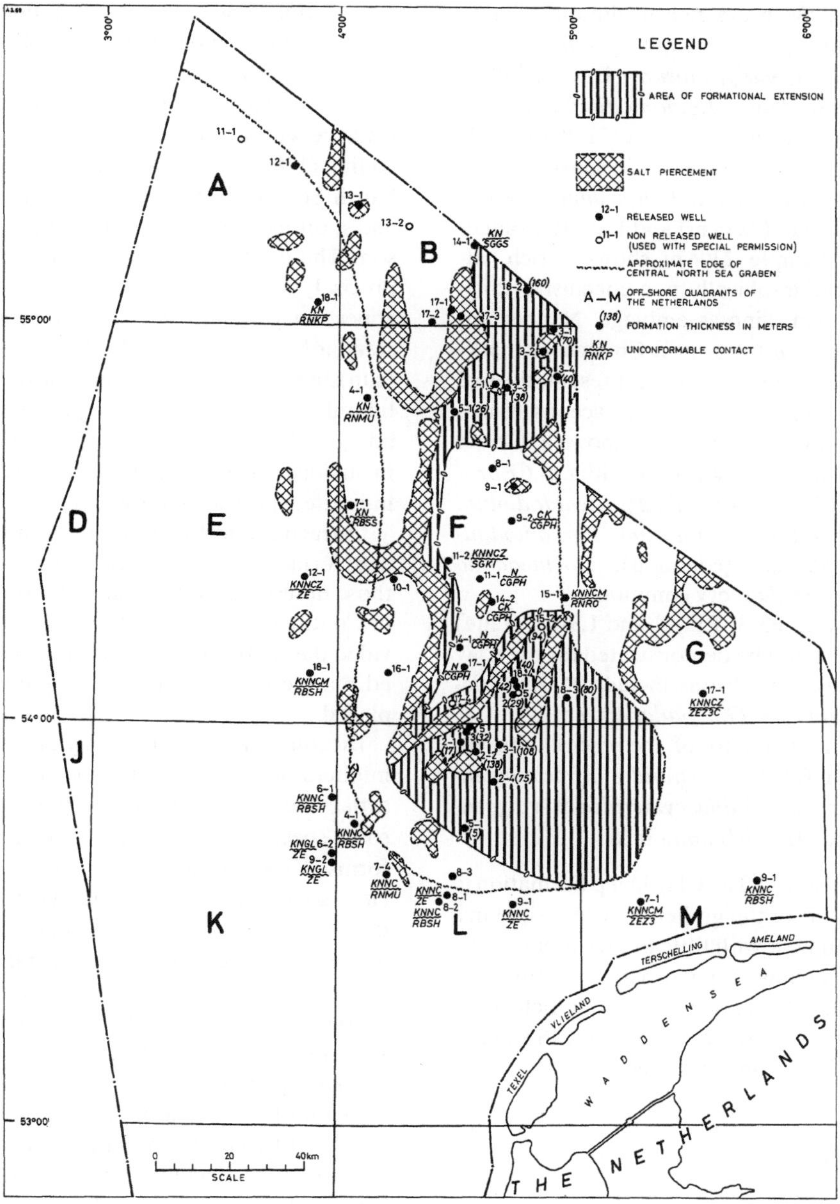

Fig. 13. Geographical distribution of Clay Deep Formation.

albidum Zone. From the reference section B18-2, core 2242–2260 m and four side-wall samples between 2279 and 2343 m were examined. The eight core samples contain for example *Batioladinium radiculatum, Gochteodinia villosa* subsp. *multifurcata, Gonyaulacysta* sp. A/B plexus (Davey, 1979; 1982) and *Oligosphaeridium diluculum* indicating a pre-*albidum* Late Ryazanian age. Side-wall samples 2279 and 2288 m are rich in dinoflagellates. Time-significant species include *Batioladinium radiculatum, B.* cf. *varigranosum* (Davey, 1982), *Cannosphaeropsis thula, Egmontodinium expiratum/torynum, Gonyaulacysta* sp. A (2288 m) and B (2279 m), and *Occisucysta* sp. A in Davey (1979; recently described as *Tehamadinium daveyi*) representing a Ryazanian (2288 m) and Late Ryazanian (2279 m) age. Sample 2304 m shows a rich, although timestratigraphically non-conclusive Portlandian/Ryazanian, dino-assemblage. Neither uppermost Jurassic nor lowermost Cretaceous marker types were recognized; some representatives of *Batioladinium* suggest Jurassic/Cretaceous transitional strata. Next sample 2343 m shows a distinct late Early Portlandian association with *i.a. Cribroperidinium* sp. A in Davey (1982), *Ctenidodinium panneum, Dingodinium spinosum, Egmontodinium polyplacophorum* (common), *Gochteodinia villosa* and *G. virgula* (very common).

Elsewhere, e.g. F3-3, F15-2, and L3-1 a (Late) Ryazanian age could be demonstrated for the Clay Deep Fm. Diagnostic forms include *B. radiculatum, E. torynum,* and *Gonyaulacysta* sp. B.

Characteristic elements of the microfauna are the foraminifer *Haplophragmoides infracalloviensis* and the ostracods *Galliaecytheridea* cf. *volgaensis, G. teres* and *Mandelstamia sexti.*

Sedimentary history – The Clay Deep Formation in the northern part of the graben was deposited in a marine environment with little or no water circulation, causing euxinic (anaerobic) conditions favourable for the formation of organic-rich, bituminous shales. Toward the south (e.g. F18 and L3) open marine conditions prevailed.

Biostratigraphy

A total account of the biostratigraphy in this review article would take too much space. Details on the micropaleontology and palynology will be dealt with in a future publication. For the sake of brevity, the reader is referred to Encl. 2.

The palynomorph zonation schemes are based exclusively on cored material and side-wall samples. Dinoflagellate zones are defined by top-occurrence (first appearance downhole) of the nominate species. In general this zonation can be correlated with information from the British Isles, for instance with a scheme established by Woollam & Riding (1983). In the southern part of the Central North Sea Graben fewer marine intervals occur, and sporomorphs are used for dating and comparison. The limits of the sporomorph zones are less precisely known than those of the dinoflagellate zones. In general, they cannot be drawn at the standard ammonite levels, but they represent a (sub)stage accuracy. Nevertheless, some major boundaries in the sporomorph scheme could be adequately dated with dinoflagellates in marine incursions. Several zonal markers represent new, in-house, species and/or genera.

Ostracods and foraminifera of the Central Graben formations show little correspondence with those described from equally old strata in England or Continental Europe. Marine intercallations provided the framework in which faunas from restricted marine or non-marine environments could be placed.

Biozonation is based mainly on ostracods; foraminifera play only a minor role in the area. Nodosariae are present in marine intervals, as well as some species of *Epistomina* and a few arenaceous forms. Paralic facies may yield *Ammodiscus* species, sometimes in great numbers. Assemblages dominated by the ostracod genus *Cypridea,* that are of common occurrence in the transitional Jurassic-Cretaceous fresh-to-brackish water deposits in all parts of the world are not present. The genus *Cypridea,* is only represented by an early form *C. valdensis praecursor.*

Within the Central Graben area, relatively minor local variations in depositional environments gave rise to distinct microfaunal communities. As a result, some of the assemblages recognized in this study have coincident ranges.

The ranges of assemblages presented in Encl. 2 are based on material from drill cuttings; due to caving they are less accurate than would have been the case if cores had been used. Since this means that total ranges are not always precisely known, the bars have been left open-ended.

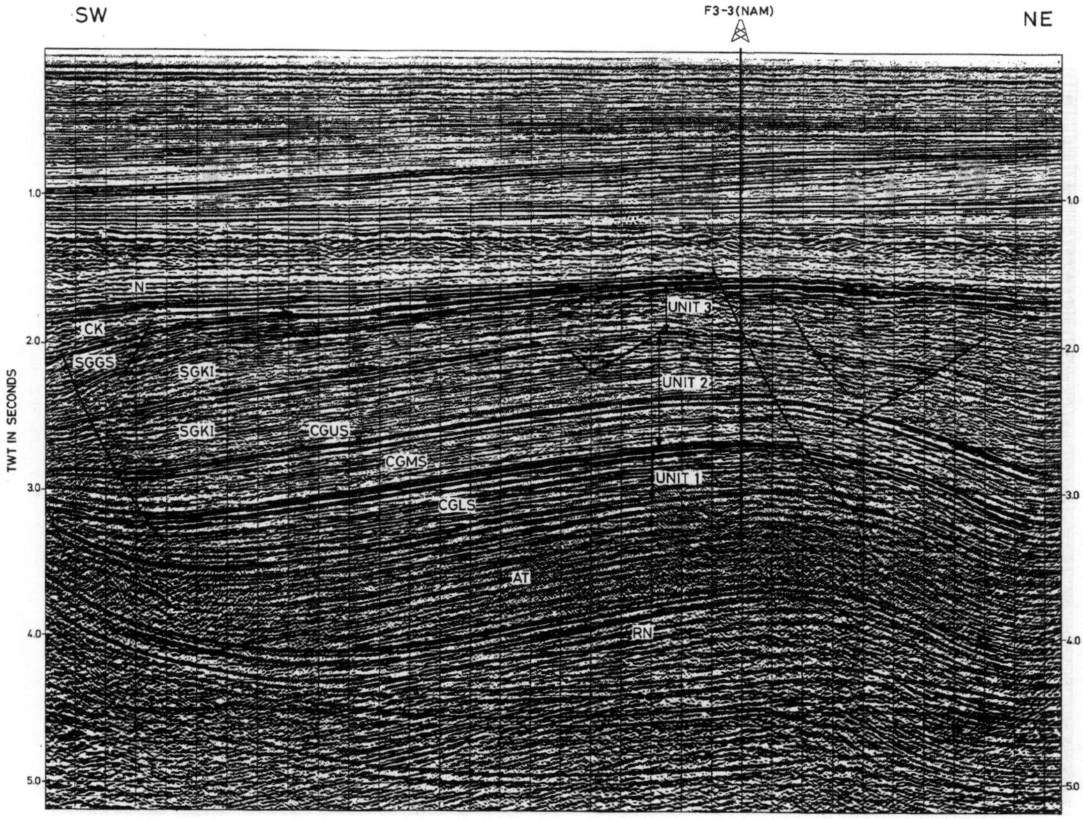

Fig. 14. Seismic section, Dutch Central North Sea Graben, showing sediment succession through well F3-3 and character of seismic facies units 1–3. RN = Upper Germanic Trias, AT = Altena Group (Early-Late Jurassic), CGLS = Lower Graben Sand Fm., CGMS = Middle Graben Shale Fm., CGUS = Upper Graben Sand Fm., SGKI = Kimmeridge Clay Fm., SGGS = Scruff Greensand Fm., CK = Chalk Group (Late Cretaceous), N = Tertiary (by courtesy of Mobil Producing Netherlands Inc.).

Seismic facies description

The Callovian – Ryazanian sediments of the Dutch Central North Sea Graben can be considered to be one seismic sequence with only minor internal seismic unconformities. The base of the complete sequence is formed by the Callovian unconformity. The top may be formed either by the conformable contact with the Valanginian, or by one of the unconformities corresponding to the bases of the Rijnland Group, the Chalk, or the Tertiary.

As Vail et al. (1977) showed, reflections within a seismic sequence parallel stratal surfaces and have the same chronostratigraphic significance as stratal surfaces. Consequently, all primary reflections recognized within the sequence represent chronostratigraphic surfaces and do not necessarily correspond to the lithostratigraphic boundaries as defined in this article. Despite this problem, an attempt has been made to link our lithostratigraphic horizons with seismic data from the studied area. Furthermore, we were able to follow characteristic seismic events over significant distances. Notwithstanding the presence of local lateral changes and diachronous lithostratigraphic boundaries, a general description of seismic facies units can be presented. The latter consist of one or more lithostratigraphic units. According to Mitchum et al.

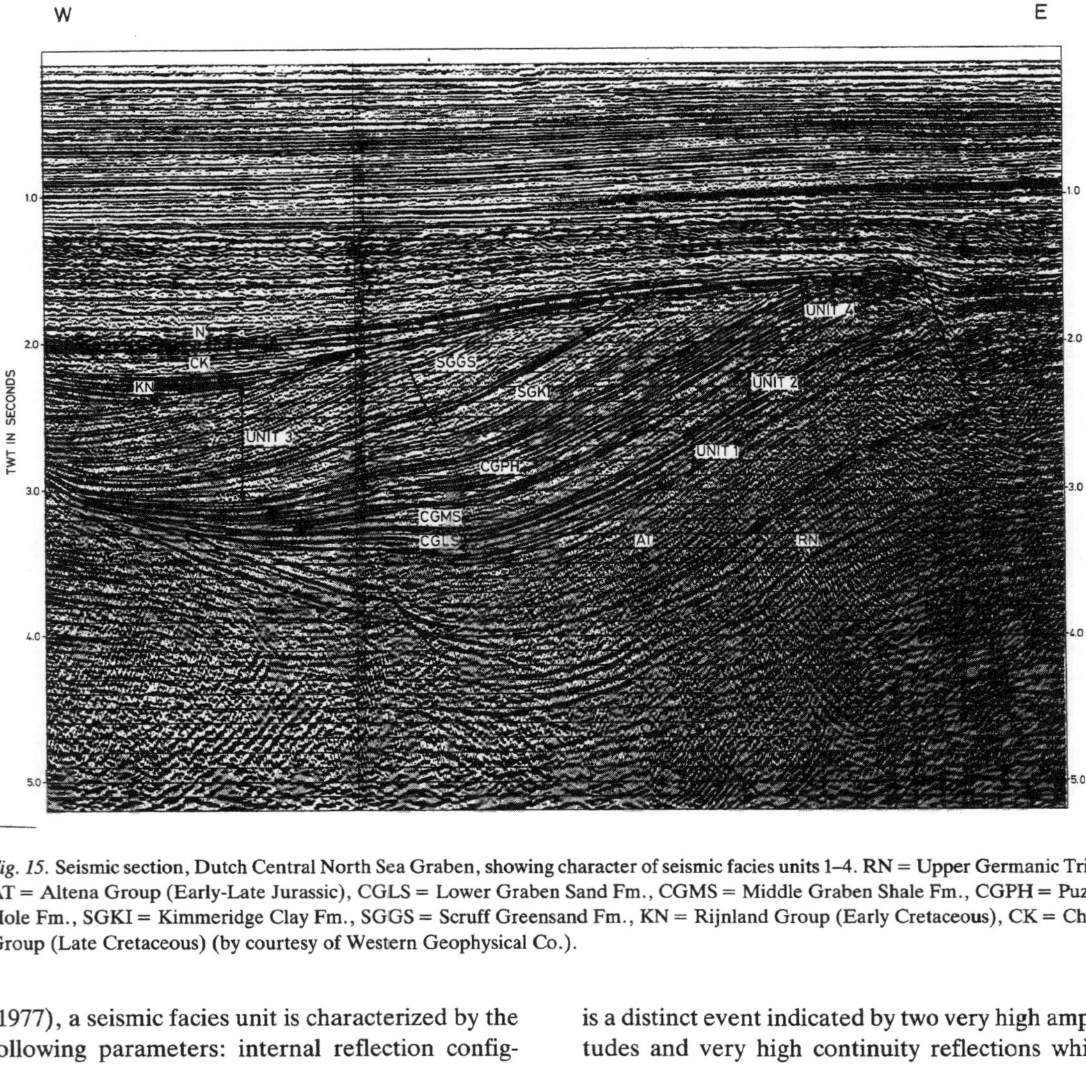

Fig. 15. Seismic section, Dutch Central North Sea Graben, showing character of seismic facies units 1–4. RN = Upper Germanic Trias, AT = Altena Group (Early-Late Jurassic), CGLS = Lower Graben Sand Fm., CGMS = Middle Graben Shale Fm., CGPH = Puzzle Hole Fm., SGKI = Kimmeridge Clay Fm., SGGS = Scruff Greensand Fm., KN = Rijnland Group (Early Cretaceous), CK = Chalk Group (Late Cretaceous) (by courtesy of Western Geophysical Co.).

(1977), a seismic facies unit is characterized by the following parameters: internal reflection configuration, seismic amplitude, frequency, reflector continuity, and interval velocity. Hence, these parameters should define a facies unit which is distinctive from adjacent units.

The following seismic facies units within the Callovian-Ryazanian sequence can be recognized (see Figs 14 & 15):

Unit 1 – Corresponding to the Lower Graben Sand Fm. This unit is characterized by high-amplitude parallel reflections with relatively low frequency and medium continuity. At its top, however, there is a distinct event indicated by two very high amplitudes and very high continuity reflections which can be followed over large distances.

Unit 2 – Corresponding to the sequence consisting of the Middle Graben Shale Fm., Upper Graben Sand Fm., and the lower part of the Kimmeridge Clay Fm. Occasionally a slight intra-Kimmeridge Clay unconformity can be identified which separates this unit from the upper part of the Kimmeridge Clay Fm. That upper part displays a different seismic facies and it onlaps the lower part of the Kimmeridge Clay. The unit is characterized by mostly low amplitude, high frequency parallel reflectations which are extremely continuous. Due to extremely low amplitudes, the unit locally is seismically transparent. Of the two relatively higher amplitude reflectors one can be easily tied to the bases of the Upper Graben Sand, the other one to the Kimmeridge Clay.

Unit 3 – Corresponding to the upper part of the Kimmeridge Clay and the Scruff Greensand Fm. At the top of this unit the Clay Deep Fm. may be included, but this formation is usually too thin to be seismically recognizable. The unit has low amplitude, medium to high continuity, and high frequency reflectors which can be either parallel or divergent.

Unit 4 – Corresponding to the Puzzle Hole and Delfland formations. This unit is very characteristic and easily recognizable because of its many closely spaced, high amplitude, and relatively low frequency reflectors. These reflectors have either low continuity when they form a subparallel configuration or medium continuity when they are parallel.

Geological history

Prior to the rifting phase, predominantly continental sediments were deposited during the Carboniferous, Permian, and Triassic. Evaporites were formed in the Zechstein. Marine sedimentation started in the latest Triassic with the onset of the large transgressive phase known as the Rhaetian transgression which continued into the Early Jurassic. With a minor unconformity the sediments of the Altena Group overly the Upper Germanic Trias Group. This unconformity is referred to as the Early Cimmerian event. During the Early Dogger another tectonic pulse known as the Mid Cimmerian phase truncated the Early Jurassic section (Ziegler, 1977).

With the opening of the Central North Sea Graben during the Callovian, sediments started to be deposited on the eroded surfaces of the Aalburg Shale and Werkendam Formations. Based on the description of the Bathonian – Callovian (and questionable Bajocian) deltaic-coastal section from the Lulu-1 well in the Danish Central Trough (Frandsen, 1986), we assume that the opening may have occurred earlier in the north than in the Dutch sector further south. This configuration seems consistent with the Bajocian – Bathonian paleogeographical map of Ziegler (1982b). This map does not indicate sedimentation for the area of the Central North Sea Dome. Deposition of the Lower Graben Sand Formation started in approximately mid-Callovian times and initially took place under fluvial and lacustrine conditions. In the course of time these conditions became deltaic and marginally marine. Differential subsidence of the area during this period was responsible for the widely differing thicknesses of the formation. The Callovian transgressive trend, which correlates well with the documented sea level changes of Hallam (1978, 1981, 1984) and Vail & Todd (1981), declines at the beginning of the Oxfordian. At that time extensive vegetated swamps came into existence, as shown by the carbonaceous shales and coal seams of the Middle Graben Shale Formation. The majority of the Middle Graben Shale deposits reflect stable conditions in a lagoonal setting.

A clearly regressive development can be noted during the Middle Oxfordian, as indicated by the progradation of the Puzzle Hole Formation and the Upper Graben Sand Formation. This corresponds with a shift of the coastline from south to north under influence of the regressive trend. There seems to be a close relationship between this Mid-Oxfordian regression and the age of the Zuidwal volcanism (144 ± 1 M.a. cf. Dixon et al., 1981). This is in accordance with the findings of Cloetingh et al. (1987), which underline the congruence between sea level fluctuations and the paleostress field in the North Sea region. According to their concept, 'periods of gradual increase in sea level are associated with times of more gradual build-up of tension and stretching, while the lowering of sea level is associated with discrete rifting episodes'.

The next transgressive sequence started in the Late Oxfordian with the onset of the marine sedimentation of the Kimmeridge Clay Formation. The marine influence introduced by this second transgression can also be observed in the contemporaneous Upper Graben Sand Formation of F3-4.

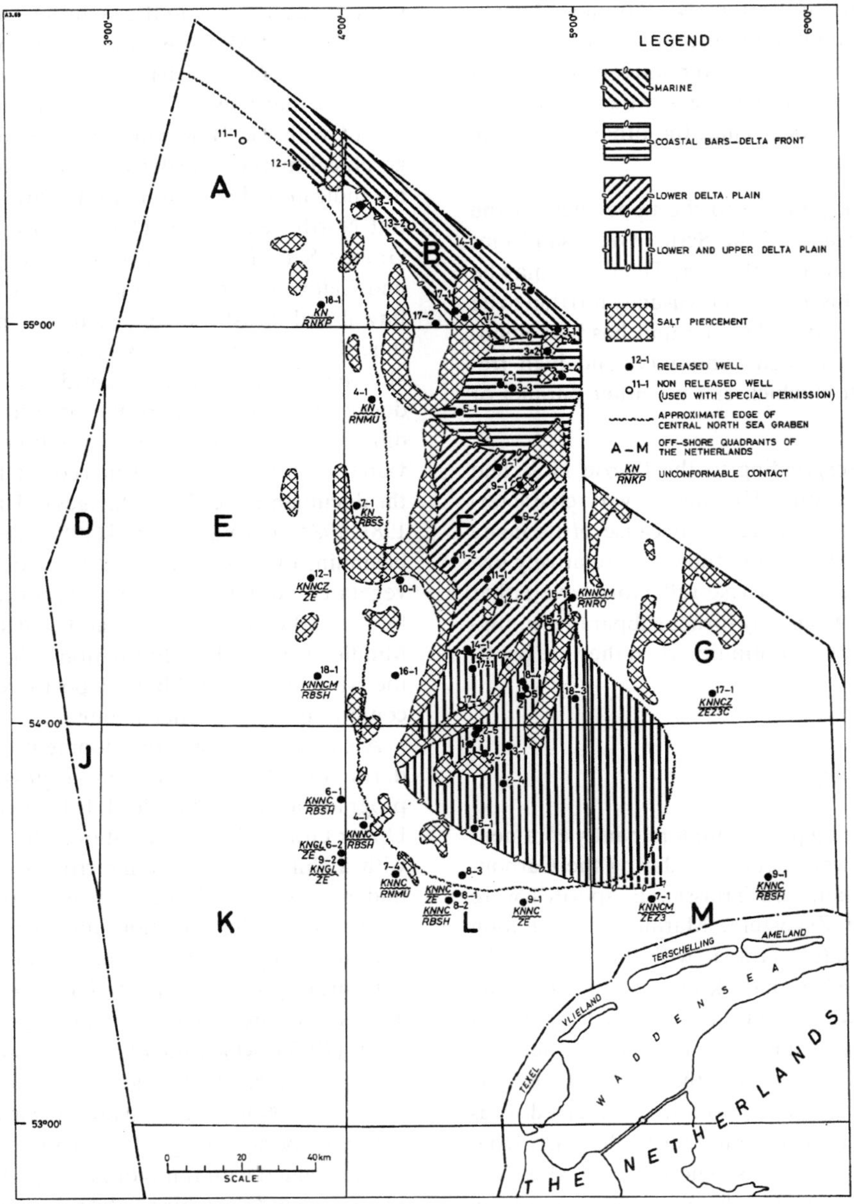

Fig. 16. Facies map for Middle-Late Oxfordian times.

The north-south direction of the transgression is indicated by the progressively younger age of the base of the Kimmeridg Clay Formation noted when correlating F3-3 via F11-2 with the F18 and L2 area.

A paleogeographic map of the area at this time (Fig. 16) shows from south to north: the lower and upper delta plains with deposition of the Delfland Formation (e.g. L2-F17), the paralic environment

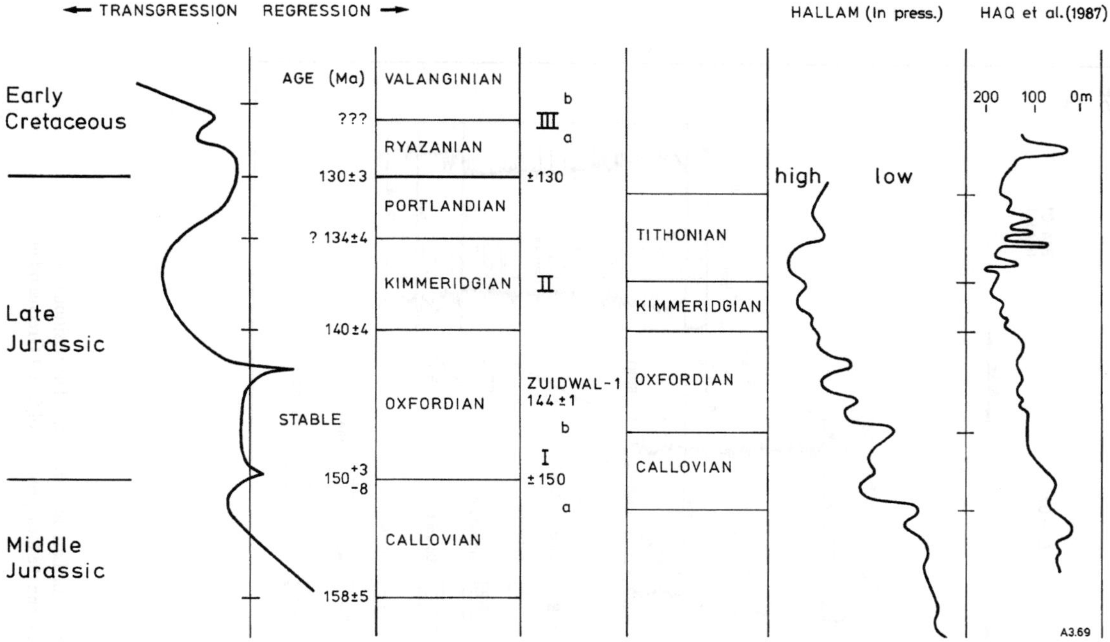

Fig. 17. Late Jurassic-Earliest Cretaceous sea-level curve for the Dutch Central North Sea Graben, with similar curves determined by Hallam (In press) and Haq et al. (1987). Absolute ages after Kennedy & Odin (1982); dating of the Zuidwal Volcanism according to Dixon et al. (1981).

with increasing marine influences reflected by sediments of the Puzzle Hole Formation (F11-2, F11-1 to F8-1), the coastal barrier complex and storm deposits represented by the Upper Graben Sand Formation (F5-1 and F3-3) and the shallow marine environment with sedimentation of the Kimmeridge Clay Formation (B14-1 and B18-2).

Our biostratigraphic data suggest that in the southern part of the Central North Sea Graben (e.g. F18-1, F18-2, L2-2, and L2-4) the Delfland sediments are not older than Kimmeridgian. On these grounds it may be postulated that opening of the graben in this area only started in the Kimmeridgian. Moreover, the marine influence of the second transgression could be observed here around the Kimmeridgian-Portlandian boundary. An important interruption of the otherwise tranquil marine sedimentation of the Kimmeridge Clay Formation is formed by the apparently high-energy conditions in which the Scruff Greensand Formation was deposited. The latter formation represents a complex of barrier sands which in time became partially reworked and coalesced into transgressive sheet sands overstepping the graben margins and covering adjacent areas. Recently, Spencer et al. (1986) discussed the complex genesis of similar Late Jurassic sands in the Central Graben of Norway. These sands may have been storm generated, as Bailey et al. (1981) suggested and accumulated in topographic depressions on downfaulted graben margins. Spencer et al. (1986) postulated that halokinetic movements may also have played an important role. In the northernmost part of the Central North Sea Graben (e.g. B18-2) no Scruff Greensand is developed and pelitic sedimentation continued into the Late Ryazanian. The depositional environment locally became rather restricted and the ensuing anoxic conditions favoured the formation of the bituminous 'hot shales' of the Clay Deep Formation. Rawson & Riley (1982) attributed the

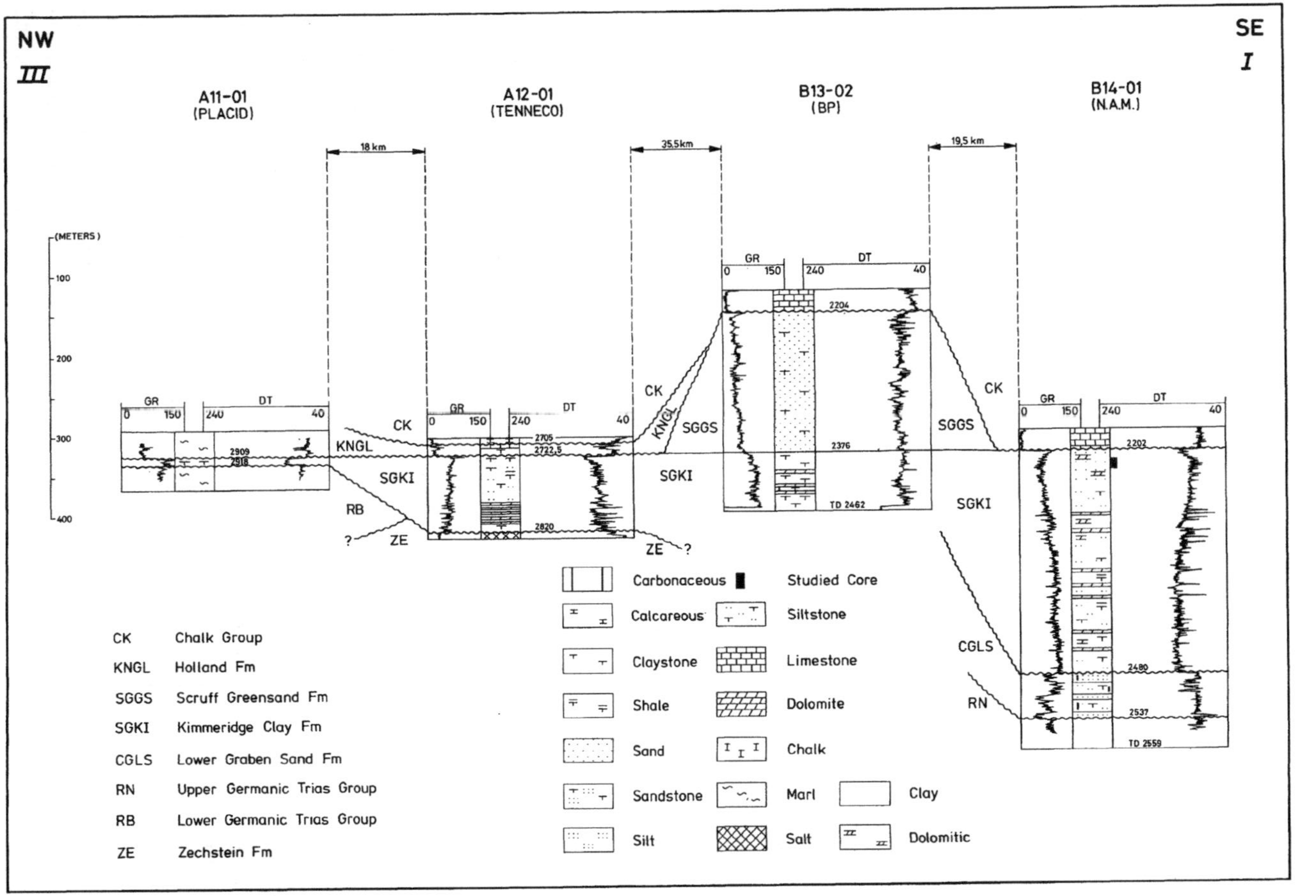

Fig. 18. Section III-I. This northernmost cross-section shows the limited development of the Late Jurassic strata. The position of the section is marked in Fig. 2. Note the thinning of the Kimmeridge Clay in NW direction and the local presence of Scruff Greensand in B13-2.

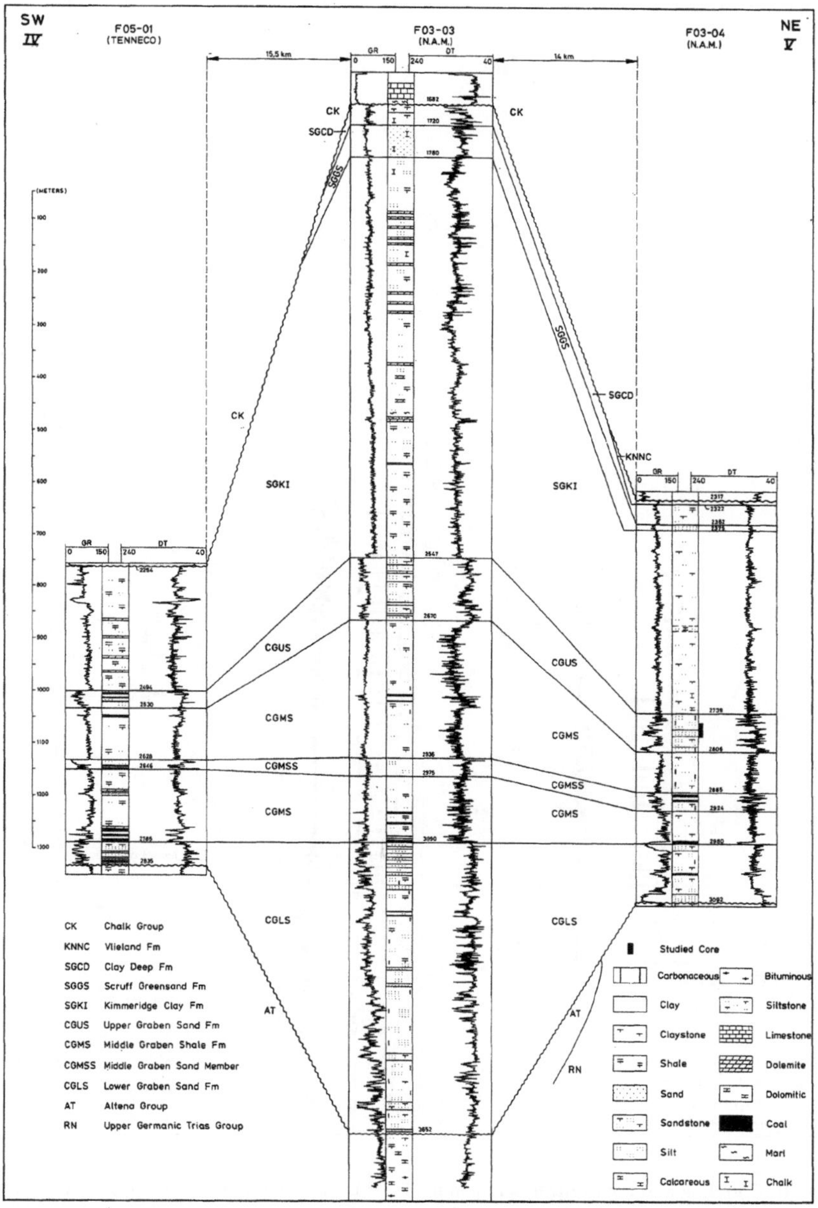

Fig. 19. Section IV-V. Correlation of Late Jurassic strata in the northern part of the F-quadrant. The position of the section is marked in Fig. 2. In F3-3 the various formations reach a remarkable thickness. The Middle Graben Sand Member and the Upper Graben Sand Formation are consistently present in this area.

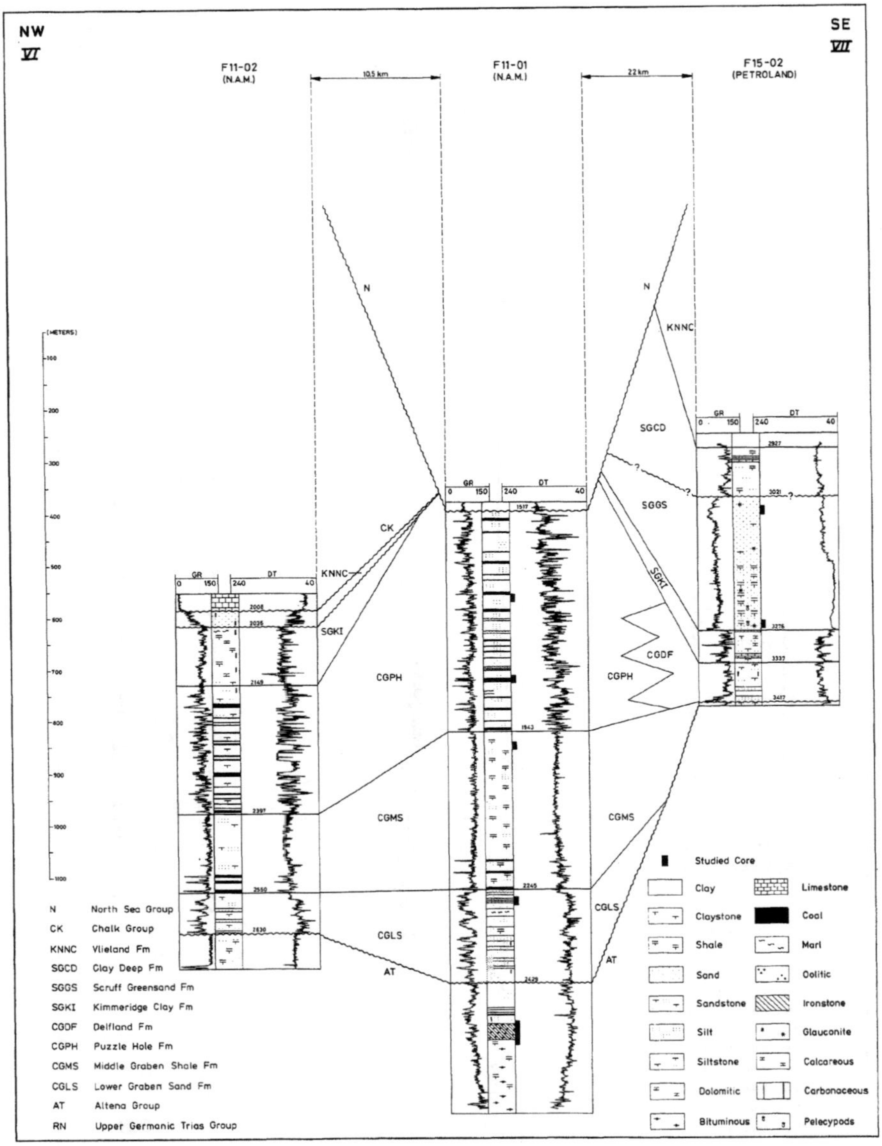

Fig. 20. Section VI-VII. Correlation of Late Jurassic strata in the F11-F15 area. The position of the section is marked in Fig. 2. The Central Graben Group is rather completely developed. Note the typical log pattern of the Puzzle Hole Formation (due to the numerous coal beds) and the presence of the well-marked coal beds at the base of the Middle Graben Shale Fm.

anoxic shales, which were formed under conditions of stagnant water circulation, to 'tectonically controlled basin enclosure'. The Clay Deep Formation represents one of the southernmost occurrences of the Kimmeridge Clay 'hot shale' facies of Barnard & Cooper (1981) in the North Sea area. Due to both shallower burial depths and decreasing organic matter, the formation becomes less mature in

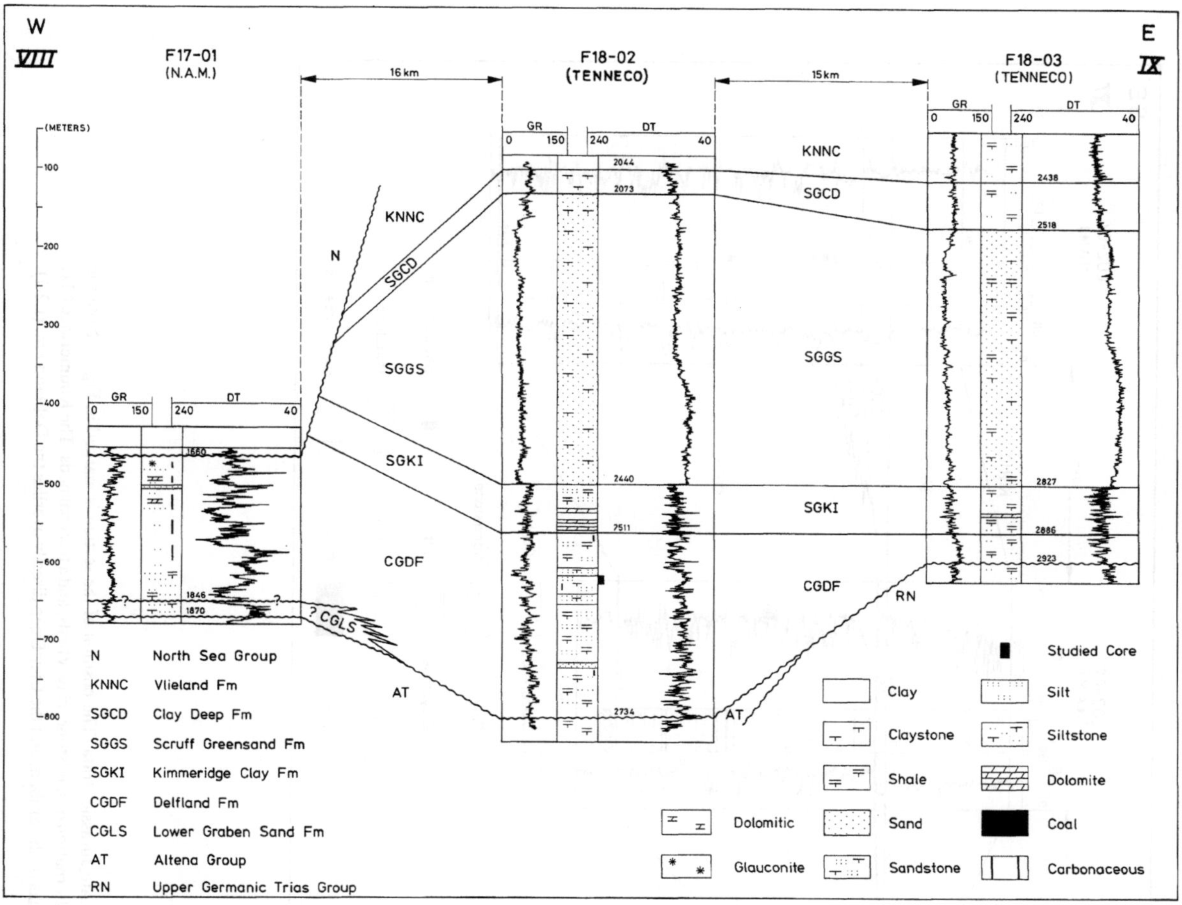

Fig. 21. Section VIII-IX. Correlation of Late Jurassic strata in the F17-F18 area. The position of the section is marked in Fig. 2. The Puzzle Hole Fm. of the F11 and F14 area has changed laterally into the Delfland Fm. Note the conspicuous thickness of up to several hundreds of meters of the Scruff Greensand Fm. The Clay Deep Fm. is hardly bituminous in this area and may be recognized by its slightly higher gamma ray readings.

the southern part of the Central North Sea Graben. We consider the deposition of the Clay Deep Formation on the Scruff Greensand Formation to be a transgressive development (phase 3a, Fig. 17).

During the latest Ryazanian and continuing into the Early Cretaceous, a rapid transgression of the Vlieland Formation occurred. Several authors have pointed to the regionally abrupt lithological change and the large hiatus around the Jurassic-Cretaceous boundary. This level is generally called Late Cimmerian unconformity or base Cretaceous unconformity (e.g. Heybroek, 1975; NAM & RGD, 1980; Ziegler, 1975; Ziegler, 1978). Anderson et al. (in: Michelsen 1982) picture a regional NW European 'Late Cimmerian' tectonic phase, accompanied by a significant sea level drop but they interpreted this as a general submarine unconformity with continuing sedimentation. According to Ziegler (1982a, b), it is a eustatic lowering of the sea level which manifests itself regionally as a regression. We think that there are distinct hiatuses at the margin of the Central North Sea Basin but

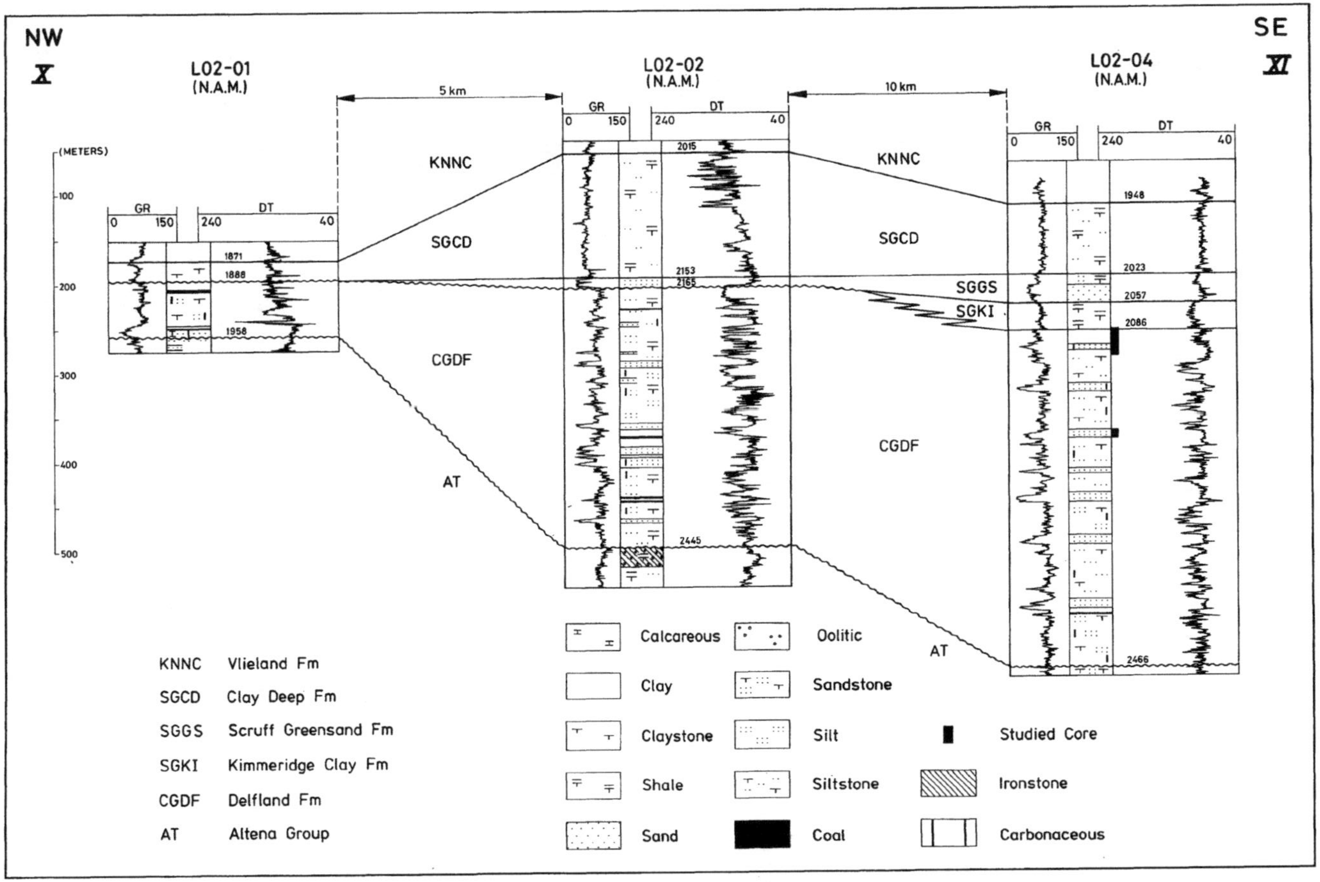

Fig. 22. Section X-XI. Southernmost cross-section of Late Jurassic strata. The position of the section is marked in Fig. 2. Note the well-developed Delfland Fm., characterized by a rapid alternation of sandstone, clayey beds, and a few coal beds. The Kimmeridge Clay and Scruff Greensand wedge out rapidly toward the flanks of the graben, and the Clay Deep directly overlies the Delfland Fm. (L2-1).

that overall the sedimentation was continuous. There is a conspicuous facies change from the anaerobic conditions of the Kimmeridge Clay to the well-oxygenated bottom conditions of the Vlieland Formation. This is in line with the findings of Rawson & Riley (1982), who see the Kimmeridge Clay – Valhall Formation boundary as an isochronous facies change rather than a regional unconformity. Very recently, Vejbaek (1986) reached the same conclusion, i.e., that a major unconformity at the base of the Valanginian in the northern North Sea, as claimed by Vail & Todd (1981), cannot be recognized as such in the Danish Central Trough.

In the Late Cretaceous, differential subsidence of the Graben ended and the 'Chalk sea' transgressed across the entire North Sea region. The major inversion at the end of the Cretaceous and the beginning of the Tertiairy, caused by the Subhercynian and Laramide compressional events, occurred in a number of pulses. The general uplift resulted in removal of the Late and Early Cretaceous and part of the Late Jurassic sequence. This succession is truncated at increasingly deeper levels toward the axis of inversion. Additionally, halokinesis also played an important role in disrupting the pattern of basin subsidence and sedimentation. Finally, in the Late Paleocene or Early Eocene, inversion movements ceased and the area was covered by a thick sequence of Tertiary marine clastics.

Acknowledgements

We express our gratitude to the Director of the Geological Survey of The Netherlands (Rijks Geologische Dienst) for his permission to publish this paper. We also acknowledge the contributions of A.N. Plomp, B.M. Schroot and L.J. Witte. We are indebted to Dr. H.A. van Adrichem Boogaert and R. Smit for their valuable suggestions and critical reading of the manuscript, and to Dr. W. Sissingh (formerly Nederlandse Aardolie Maatschappij) for productive discussions. The value of this paper was greatly increased by the availability of rock samples and/or proprietary basic data which were put at our disposal by the British Petroleum Exploratie Mij. Nederland, Mobil Producing Netherlands Inc., Petroland B.V., Placid International Oil Co., Nederlandse Aardolie Maatschappij, and Western Geophysical Co. We sincerely thank the managements of these companies for their cooperation and permission to publish the results and/or the data. The drawings were prepared by J. Houkes, H. Witteveen, and A. Koers. Finally, appreciation is expressed to Miss C.M. Beuming for typing the manuscript.

References

Bailey, C.C., Price, I & A.M. Spencer 1981 The Ula Oil Field, Block 7/12, Norway. In: Norwegian Symposium on Exploration, Bergen – Norsk Petrol. For. Article 18: 1–26.

Barnard, P.C. & Cooper, B.S. 1981 Oils and source rocks of the North Sea area. In: Illing, L.V. & G.D. Hobson (eds): Petroleum geology of the continental shelf of Northwest Europe – Inst. of Petroleum, London: 169–175.

Casey, R. 1973 The ammonite succession at the Jurassic – Cretaceous boundary in eastern England. In: Casey, R. & P.F. Rawson (eds): The Boreal Lower Cretaceous – Geol. J., Spec. Iss. 5: 193–266.

Clark-Lowes, D.D., Kuzemko, N.C.J. & D.A. Scott (1987) Structure and petroleum prospectivity of the Dutch Central Graben and neighbouring platform areas. In: J. Brooks & K. Glennie (eds): Proc. 3rd Conf. on Petroleum Geology of NW Europe, London – Graham & Trotman, London, 1: 337–356.

Cloetingh, S., Lambeck, K. & McQueen, H. (1987) Apparent sea level fluctuations and a paleostress field for the North Sea region. In: J. Brooks & K. Glennie (eds): Proc. 3rd Conf. on Petroleum Geology of NW Europe, London, Graham & Trotman, London, 1: 49–58.

Cope, J.C.W. 1985. Report of the Terminal Jurassic Stage Working Group – Proc. Int. Symp. Jurassic Stratigraphy, Erlangen, September 1–8, 1984, I: 111–116.

Cope, J.C.W., Duff, K.L., Parsons, C.F., Torrens, H.S., Wimbledon, W.A. & J.K. Wright 1980 A correlation of Jurassic rocks in the British Isles. Part 2: Middle and Upper Jurassic – Geol. Soc. Lond., Spec. Rept 15: 109 pp.

Cornford, C. 1984 Source rocks and hydrocarbons of the North Sea. In: Glennie, K.W. (ed.): Introduction to the Petroleum Geology of the North Sea – Blackwell: 171–204.

Davey, R.J. 1979 The stratigraphic distribution of dinocysts in the Portlandian (Latest Jurassic) to Barremain (Early Cretaceous) of Northwest Europe – AASP Cont. Ser. 5B: 49–81.

Davey, R.J. 1982 Dinocyst stratigraphy of the latest Jurassic to Early Cretaceous of the Haldager No. 1 borehole, Denmark – Geol. Surv. Denmark, Ser. B., 6: 1–56.

Dixon, J.E., Fitton, J.G. & R.T.C. Frost 1981 The tectonic significance of post-Carboniferous igneous activity in the North Sea Basin. In: Illing, L.V. & G.D. Hobson (eds):

Petroleum geology of the continental shelf of Northwest Europe – Inst. of Petroleum, London: 121–137.

Doré, A.G., Vollset, J. & G.P. Hamar 1985 Correlation of the offshore sequence referred to the Kimmeridge Clay Formation – relevance to the Norwegian sector. In: Thomas, B.M. et al. (eds): Petroleum geochemistry in exploration of the Norwegian Shelf – Norwegian Petroleum Society – Graham & Trotman: 27–37.

Frandsen, N. 1986 Middle Jurassic deltaic and coastal deposits in the Lulu-1 well of the Danish Central Trough – Dan. Geol. Unders. Ser. A. 9: 1–23.

Frandsen, N., Vejbaek, O.V., Møller J.J. & O. Michelsen (1987) A dynamic geological model for the Danish Central Trough during the Jurassic – Early Cretaceous. In: J. Brooks & K. Glennie (eds): Proc. 3rd Conf. on Petroleum Geology of NW Europe, London – Graham & Trotman, London, 1: 453–468.

Hallam, A. 1978 Eustatic cycles in the Jurassic – Palaeogeogr., Palaeoclim., Palaeoecol. 23 (1/2): 1–32.

Hallam, A. 1981 A revised sea-level curve for the early Jurassic – J. Geol. Soc. Lond. 138 (6): 735–743.

Hallam, A. 1984 Pre-Quaternary sea-level changes – Ann. Rev. Earth Planet. Sci. 12: 205–243.

Hallam, A (in press) A reevaluation of Jurassic eustasy in the light of new data and the revised Exxon curve. In: Wilgus, C.K. (ed.): Sea level changes – an integrated approach – S.E.P.M. Spec. Publ. 42.

Hancock, J.M. 1972 (ed.) Crétacé de l'Angleterre, Pays de Galles, Ecosse – Lexique Strat. Int. 1: 162 pp.

Haq, B.U., Hardenbol, J. & P.R. Vail 1987 Chronology of fluctuating sea levels since the Triassic – Science 235: 1156–1167.

Herngreen, G.F.W. & K.F. De Boer 1985 Palynology of the 'Upper Jurassic' Central Graben, Scruff and Delfland groups in the Dutch part of the North Sea Continental Shelf – Proc. Int. Symp. Jurassic Stratigraphy, Erlangen, September 1–8, 1984 – Geological Survey of Denmark, Copenhagen: III: 695–713.

Heybroek, P. 1975 On the structure of the Dutch part of the Central North Sea Graben. In: Woodland, A.W. (ed.): Petroleum and the continental shelf of North-West Europe – Applied Science Publishers Ltd, Barking, Essex: 1: 339–351.

Jensen, T.F., Holm, L., Frandsen, N. & O. Michelsen 1986 Jurassic – Lower Cretaceous lithostratigraphic nomenclature for the Danish Central Trough – Dan. Geol. Unders., Ser. A. 12: 65 pp.

Kennedy, W.J. & G.S. Odin 1982 The Jurassic and Cretaceous time scale. In: Odin, G.S. (ed.): Numerical dating in stratigraphy – Wiley: I: 557–592.

Koch, J.O. 1983 Sedimentology of Middle and Upper Jurassic sandstone reservoirs of Denmark – Geol. Mijnbouw 62: 115–129.

Le Hégarat, G. 1973 Le Berriasien du Sud-Est de la France – Doc. Lab. Géol., Fac. Sci., Lyon, 43: 1–576.

Maubeuge, P.L. 1964 Résolutions du Colloque. In: Maubeuge, P.L. (ed.): Colloque du Jurassique à Luxembourg 1962 – Publ. Inst. Grand-Ducal. Sect. Sci. nat., Phys. Math., Luxembourg: 77–80.

Maubeuge, P.L. 1970 Résolutions du deuxième Colloque International du Jurassique. In: Colloque du Jurassique à Luxembourg 1967 – Musée d'Histoire Nat., G.-D. Luxembourg: 38 pp.

Michelsen, O. (ed.) 1982 Geology of the Danish Central Graben – Dan. Geol. Unders. Ser. B, 8: 133 pp.

Mitchum, R.M., Vail, P.R. & J.B. Sangree 1977 Seismic stratigraphy and global changes of sea level, part 6: Stratigraphic interpretation of seismic reflection patterns in depositional sequences – AAPG Mem. 26: 117–133.

Nederlandse Aardolie Maatschappij B.V. & Rijks Geologische Dienst 1980 Stratigraphic nomenclature of The Netherlands – Kon. Med. Geol. Mijnb. Gen., Verh. 32: 77 pp.

Orbigny, A.d' 1842–1851 Paléontologie française – Terrains jurassiques, I, Céphalopodes, Paris: 642 pp.

Rawson, P.F., Curry, D., Dilley, F.C., Hancock, J.M., Kennedy, W.J., Neale, J.W., Wood, C.J. & B.C. Worssam 1978 A correlation of Cretaceous rocks in the British Isles – Geol. Soc. Lond. Spec. Rept 9: 70 pp.

Rawson, P.F. & L.A. Riley 1982 Latest Jurassic – Early Cretaceous events and the 'Late Cimmerian unconformity' in North Sea area – AAPG Bull. 66: 2628–2648.

Spencer, A.M., Home, P.C. & V. Wiik 1986 Habitat of hydrocarbons in the Jurassic Ula Trend, Central Graben, Norway. In: Spencer, A.M. et al. (eds): Habitat of hydrocarbons on the Norwegian Continental Shelf – Norwegian Petroleum Society – Graham & Trotman: 111–127.

Vail, P.R. & R.G. Todd 1981 Northern North Sea Jurassic unconformities, chronostratigraphy and sea-level changes from seismic stratigraphy. In: Illing, L.V. & G.D. Hobson (eds): Petroleum geology of the continental shelf of Northwest Europe – Inst. of Petroleum, London: 216–235.

Vail, P.R., Todd, R.G. & J.B. Sangree 1977 Seismic stratigraphy and global changes of sea-level, part 5: Chronostratigraphic significance of seismic reflections – AAPG Mem. 26: 99–116.

Van Wijhe, D.H. 1987 Sructural evolution of inverted basins in the Dutch Offshore – Tectonophysics 137: 171–219.

Vejbaek, O.V. 1986 Seismic stratigraphy and tectonic evolution of the Lower Cretaceous in the Danish Central Trough – Dan. Geol. Unders., Ser. A, 11: 46 pp.

Wimbledon, W.A. 1985 The Portlandian, the terminal Jurassic stage in the Boreal realm. In: Michelsen, O. & A. Zeiss (eds): Proc. Int. Symp. Jurassic Stratigraphy – Geological Survey of Denmark, Copenhagen: II: 533–549.

Wimbledon, W.A. & J.C.W. Cope 1978 The ammonite faunas of the English Portland Beds and the zones of the Portlandian Stage – J. Geol. Soc., Lond., 135 (2): 183–190.

Woollam, R. & J.B. Riding 1983 Dinoflagellate cyst zonation of the English Jurassic – Inst. Geol. Sciences, Rept 83 (2): 1–40.

Zeiss, A. 1983 Zur Frage der Aequivalenz der Stufen Tithon/Berrias/Wolga/Portland in Eurasien und Amerika. Ein Beitrag zur Klärung der weltweiten Korrelation der Jura-/Kreide-grenzschichten im marinen Bereich – Zitteliana 10: 427–438.

Ziegler, P.A. 1977 Geology and hydrocarbon provinces of the North Sea – Geojournal 1: 7–32.
Ziegler, P.A. 1978 North-Western Europe: tectonics and basin development – Geol. Mijnbouw 57: 589–626.
Ziegler, P.A. 1982a Faulting and graben formation in western and central Europe – Phil. Trans. R. Soc. London A, 305: 113–143.
Ziegler, P.A. 1982b Geological Atlas of Western and Central Europe – Shell Int. Petrol. Maatsch. B.V., distributed by Elsevier: 130 pp.
Ziegler, W.H. 1975 Outline of the geological history of the North Sea. In: Woodland, A.W. (ed.): Petroleum and the Continental Shelf of North-West Europe – Applied Science Publishers Ltd, Barking, Essex: 1: 165–190.

Errata added in prove
Fig. 6. Thickness of CGLS in F11-1 should read 184 instead of 204.
Fig. 9. Thickness of CGPH in F14-2 is 54.
Fig. 11. Delete side-wall core at ± 3180 m.
Fig. 12. Thickness of SGGS in B14-1 should read 0 instead of 35.

Geologie en Mijnbouw 68: 107–120 (1989)

KEYNOTE ADDRESS

The Netherlands during the Tertiary and the Quaternary: A case history of Coastal Lowland evolution

W.H. Zagwijn
Geological Survey of The Netherlands,
P.O. Box 157, 2000 AD Haarlem, The Netherlands

Received 23 January 1988; accepted 29 February 1988

Key words: Basin studies, paleogeography, intra-plate tectonics, Coastal Lowlands, Tertiary, Quaternary, The Netherlands, North Sea Basin

Abstract

The Netherlands and the adjoining southern region of the North Sea form part of a subsiding area with a complicated tectonic and sedimentary history. This area was either a shallow sea or a coastal lowland. After a compressional stage at the onset of the Tertiary, tensional forces dominated from the Oligocene onward and induced the formation of an intraplate rift system. The relationship between this system and sediment supply by rivers originating in the hinterland is discussed.

In the Quaternary, depocentres shifted considerably. Here a role was played by changes in sea level due to build up of inland ice and repeated climatic changes, leading to increased sediment discharge. In the later part of the Quaternary, inland ice itself invaded the basin and reshaped the landscape.

Introduction

The area occupied by The Netherlands is of particular interest for the study of coastal lowlands, because it forms part of an intraplate basin with an intricate tectonic history complicated by the presence since Oligocene times of a SE-NW-running rift-fault system. The evolution of the coastal plains during the Miocene and the Holocene (treated in some detail in the original Keynote lecture) has been published elsewhere (Zagwijn & Hager 1987; Zagwijn 1986) and will not be dealt with here.

A general discussion of the tectonic framework is followed by an analysis of the paleogeographic evolution in the Tertiary (which differed between the earliest Tertiary and the Oligocene) and the Quaternary, under changing influences of climate, sediment supply, and position of depocentres.

Tectonic framework

The tectonic framework (Fig. 1) has two main elements: one, the Hercynian Ardenno-Rhenish Massif, which has been subject to uplift since the Miocene and even more so since the Middle Pleistocene (Fuchs et al. 1983); the other, the Graben systems (Upper Rhine Graben, Leine Graben, Lower Rhine Embayment, Rur Graben, and Central Graben), which became active at various times during the Eocene and Oligocene. Toward the northwest this Graben system grades into the intraplate basin of the southern part of the North Sea. The northwestern branch of the rift system, which underlies the southern half of the present Netherlands and shows a NW-SE-trending fault system, is of special interest in the present context. Toward the southeast this fault system is linked to the Lower Rhine Embayment, which was originally part of the Hercynian Massif, and has been subject to considerable downwarping since the Early Oligocene. This occurred later than the formation of the southern branch of the rift system, the Upper Rhine Graben. The northeastern branch of the triple system, the Leine Graben, was active in Oligocene times but is now inactive (Ziegler 1982).

The structural map showing the base of the Miocene (Fig. 2, adapted from Van Doorn et al. 1985) is illustrative of some of the essential features of the rift system underlying the region (Fig. 18). The main structural lows are the deep Central Graben and the Zuiderzee Basin, which extends southeastward into the Venlo Graben. The latter only became a shallow graben very late in the Tertiary, having previously been a relative high with much the same character as the Peel blocks, which still belong to a structural high (Van Rooijen et al. 1984). In the northeastern part of The Netherlands a different tectonic style prevails, one related to salt structures and associated faults.

The structural map of the base of the Quaternary (Fig. 3) shows some additional basins at least partly representing rejuvenations of much older Mesozoic structures, such as the West Netherlands Basin, the Broadfourteens Basin, and the Vlieland Basin (Zagwijn & Doppert 1978).

Paleogeographic evolution during the Tertiary

The Tertiary and Quaternary sedimentary fill in the Lower Rhine Graben system and the southern part of the North Sea Basin is shallow marine (littoral and epineritic) and continental. The continental deposits more specifically belong to coastal plain and deltaic environments. Generally speaking, the relative subsidence and the rate of sedimentation were in equilibrium, but changes in sea level and sudden changes in sedimentation rates produced some extensive hiatuses. However, subsidence never became so rapid that it produced sea depths of considerably more than 100 metres. Thus, the isopach maps presented here reflect mainly differences in down warping.

During the earliest part of the Tertiary, subsidence patterns were very different from those of Oligocene and later times (Letsch & Sissingh 1983; Keizer & Letsch 1963). This is exemplified by the map showing the thickness of the Lower Eocene deposits (Fig. 5). Two basins can be recognized, one situated in the SW part of The Netherlands (called the Voorne Trough) and the other further north. These two basins were separated by a high, the Mid-Netherlands ridge, which had been formed by Late Cretaceous to Early Tertiary inversion of older basins (Heybroek 1974; Ziegler 1982).

During the Oligocene this situation changed completely. First, a large shallow basin was formed that extended over the ancient Mid-Netherlands high and also protruded toward the southeast into the Lower Rhine Embayment, the formation of which had just started (Fig. 6). Faulting was still of little importance. Of the older basin structures, only a remnant of the Voorne Trough persisted for some time. Except in former coastal areas, the marine sediments were fine grained (Boom Clay) and remarkably uniform thoughout the basin (Van den Bosch & Hager 1984).

During the deposition of Upper Oligocene sediments, faulting was already evident (Fig. 7). In particular, the Central Graben of The Netherlands and its southeastern continuation into the Lower Rhine Embayment stands out clearly. In the innermost part of the Embayment, close to the Hercynian Massif, coastal lowlands developed where in places peat and organic lake deposits could accumulate. These deposits are now found as lignites and sapropelic oil-bearing coals. The marine sediments are more variable than they were before, sands dominating closer to the coast and clays further away from it. Remnants of pre-existing basins, such as the Voorne Trough, are no longer active.

The area in which Upper Oligocene marine deposits are preserved is conspicuously smaller than the area covered by Lower Oligocene sediments.

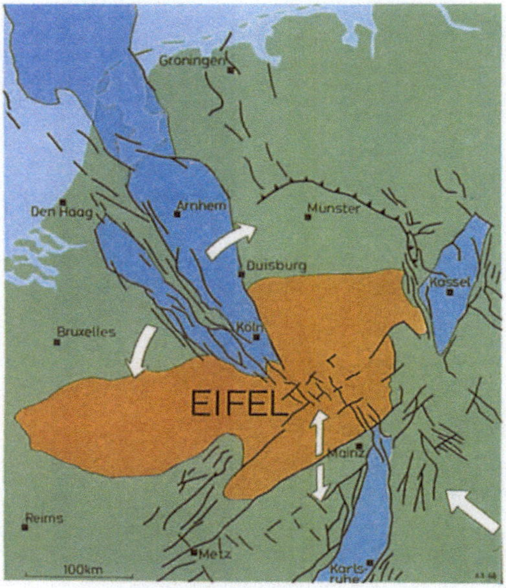

Fig. 1. General tectonic framework of the Western European Cenozoicum rift system (after Fuchs et al. 1983).

Fig. 2. Structural map and depth contours of the base of the Miocene for The Netherlands (after Van Doorn et al. 1984).

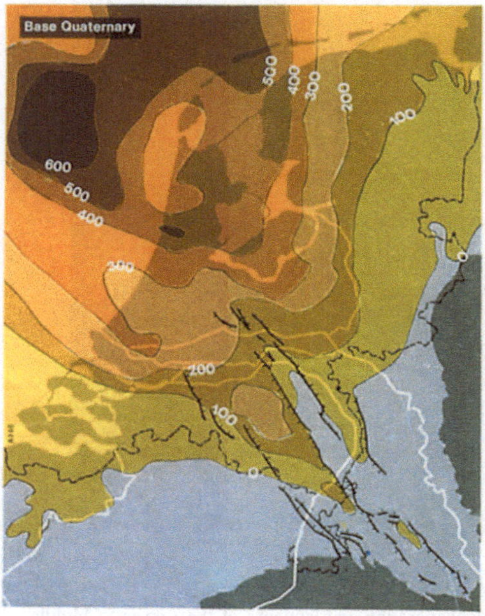

Fig. 3. Structural map and depth contours (below present sea level) of the base of the Quaternary for The Netherlands (after Zagwijn & Doppert 1978).

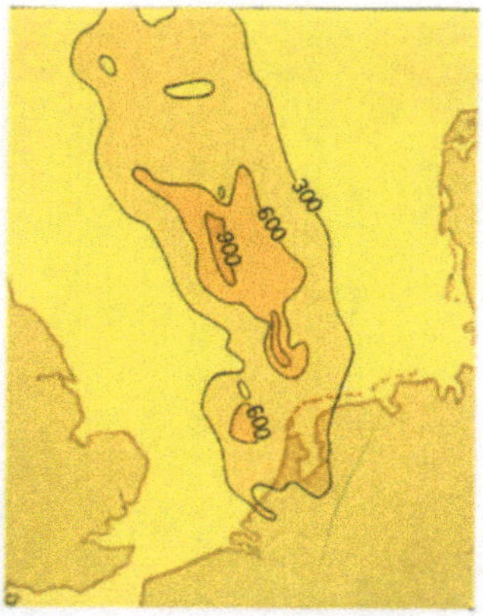

Fig. 4. Depth contours of the base of the Quaternary where i lies deeper than 300 m below present sea level in the North Se: Basin (modified from McCave et al. 1977).

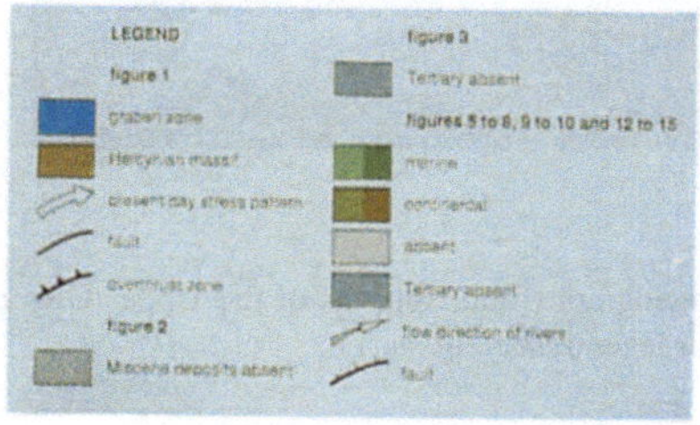

Fig. 4a. Legend for figures 1 to 3, 5 to 10 and 12 to 15

Fig. 5. Paleogeography and isopachs of the Lower Eocene deposits (after Letsch & Sissingh 1983). (For legend, see Fig. 4a.)

Fig.6. Paleogeography and isopachs of the Lower Oligocene deposits. (For legend, see Fig. 4a.)

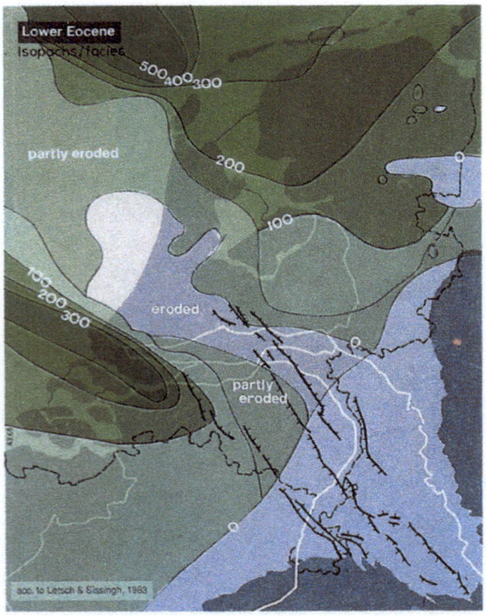

Fig. 7. Paleogeography and isopachs of the Upper Oligocene deposits. (For legend, see Fig. 4a.)

Fig. 8. Paleogeography and isopachs of the Lower to Middle Miocene deposits. (For legend, see Fig. 4a.)

Fig. 9. Paleogeography and isopachs of the Middle to Upper Miocene deposits. (For legend, see Fig. 4a.)

Fig. 10. Paleogeography and isopachs of the Lower Pliocene deposits. (For legend, see Fig. 4a.)

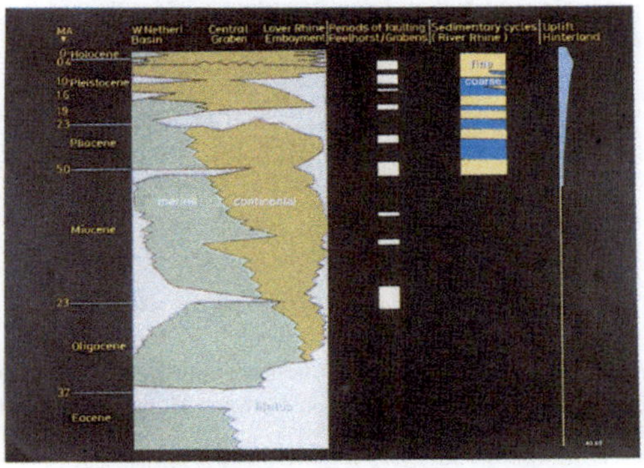

Fig. 11. Generalized distribution of Tertiary and Quaternary deposits in relation to time in the Graben zone between the Lower Rhine Embayment and the western part of The Netherlands, with emphasis on facies distribution and unconformities.

Fig. 12. Paleogeography and isopachs of the Lower Pleistocene (Praetiglian and Early Tiglian, dated between 2.3 and 1.8 MA). (For legend, see Fig. 4a; the coastline during the Praetiglian regression is shown by the red line.)

Fig. 13. Paleogeography and isopachs of the Lower Pleistocene (Late Tiglian, 1.8–1.6 MA). (For legend, see Fig. 4a.)

Fig. 14. Paleogeography and isopachs of the Lower Pleistocene (Eburonian to Menapian, 1.6–1.0 MA). (For legend, see Fig. 4a.)

Fig. 15. Paleogeography and isopachs of the Lower to Middle Pleistocene (Bavelian and early Cromerian, 0.9 to 0.45 MA). (For legend, see Fig. 4a.)

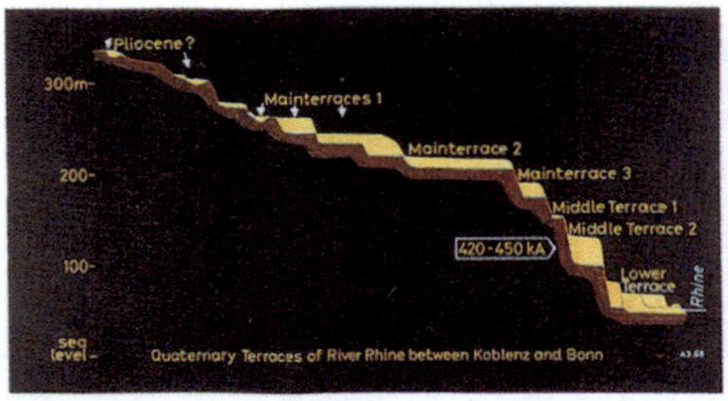

Fig. 16. Quaternary terrace sequence of the river Rhine in the uplift area of the Rhenish Massif between Bonn and Koblenz.

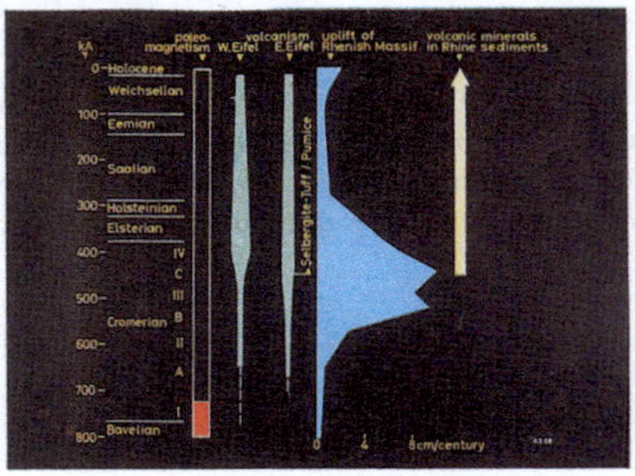

Fig. 17. Table showing the relation between the uplift of the Rhenish Plateau, Eifel volcanism, and the Quaternary stratigraphy of The Netherlands.

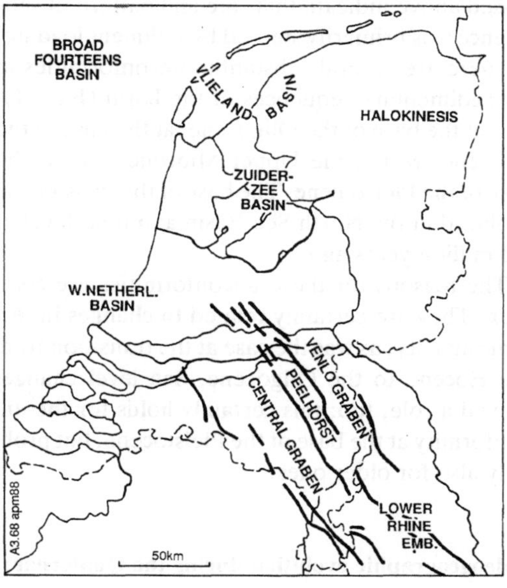

Fig. 18. Nomenclature of tectonic units in The Netherlands and adjacent areas.

This may be partially due to later erosion, but to a large extent the pattern seems to be original and to have been created by non-deposition (synsedimentary erosion) and to the formation of condensed sections outside the area of stronger subsidence. At the transition between the Early and Late Oligocene the sea level dropped sharply (Vail et al. 1977), and it is conceivable that outside the area of strongest subsidence the sea became so shallow that bottom currents and wave action could easily remove any sediment.

In the Miocene the area occupied by coastal lowlands increased considerably (Figs. 8 and 9). Particularly during the Middle Miocene and the early part of the Late Miocene, peat formation was widespread. The position of the coastline, and thus the area of the coastal lowlands, varied considerably. However, in the inner part of the Lower Rhine Embayment peat accumulation continued for a very long period, at least 6 million years, and was not interrupted by either marine or fluvial sedimentation (Hager 1986). The result was a continuous accumulation of peat, originally to a thickness of about 300 m and now compressed to a 100 m thick seam of brown coal. This main seam and its associated seams together form the largest single brown coal occurrence in the world, the total geological reserve amounting to 55 billion tons of coal. Several special conditions contributed to this peculiar peat accumulation: the coastal lowland environment, the sea level fluctuation, a warm and moist climate, and very slow tectonic subsidence with little faulting during that particular part of the Miocene. For further details, reference is made to Zagwijn & Hager (1987).

Two major cycles of transgression and regression occurred during the Miocene. Around the Oligo-Miocene boundary a general break in sedimentation can be observed in many places, probably with the exception of the deepest parts of the Central and Rur Grabens. Subsequently the extent of marine sedimentation increased, and this was probably related to a marine transgression which penetrated the Lower Rhine Embayment. Later, in Early Miocene to Middle Miocene times, a regression took place and the coastline shifted to the west. At times, peat accumulation in the Central Graben area even extended west of the present city of Eindhoven. (Fig. 8). During renewed transgression the coastline again shifted far to the east, over a distance of more than 80 km during the later part of the Middle Miocene. When regression started once again at the end of that period, it continued throughout the remainder of Miocene and even Pliocene times (Fig. 9).

The evidence from the Miocene makes it even clearer that depocentres lay in marine nearshore environments in front of deltas which were the source of sediment supply (Zagwijn & Doppert 1978). Besides the Central Graben area, a new centre of subsidence stands out, the Zuiderzee Basin. This area strongly subsided during the later part of the Miocene (Fig. 9), a period which is characterized in the Central Graben itself by increased fault activity.

Finally, it is evident that the area of marine sedimentation became considerably larger during the Miocene compared with the situation which prevailed in the Late Oligocene. However, further west, off the present west coast of The Netherlands and in the part of the basin far away from the coast, the Miocene sequence is either absent or strongly condensed. The primary reason for this particular phenomenon seems to be that the sediment brought from the hinterland by a precursor of the Rhine was trapped in the two subsiding areas before it reached the more central areas of the southern part of the North Sea Basin. Only the finest-grained material is found in these areas, and coarser sandy sediments only occur closer to the ancient coasts. A more detailed study of the sediments shows that there were more minor sedimentary cycles than those related to the two major transgression-regression movements. Furthermore, some minor breaks in the sedimentation can be recognized, and these breaks were at least partially related to differential tectonic movements of fault blocks.

During the Pliocene (Fig. 10) the development in the southeastern region continued as in the preceding period: growth of the delta of the ancient river Rhine and, related to this growth, a shift of the depocentre in the Central Graben system toward the northwest. In the northeast the paleogeographic situation had changed, because by now the sea regressed from the North-German lowlands and an extensive delta developed at the mouth of an ancient river system that drained the northern lowlands and the Baltic region. Seaward of this delta there is a depocentre which had shifted northward from the Zuiderzee Basin toward the Vlieland Basin since the Late Miocene.

In sum, the following can be said about the tectonic and paleogeographic evolution since the Oligocene. From then on tensional movements within this part of the European plate led to the formation of a SE-NW-trending rift fault system in the southern part of The Netherlands. Depocentres and areas of maximum subsidence lay in the Central Graben; later on also further north in the Zuiderzee Basin, seaward of ancient deltas that were formed by precursor river systems of the Rhine.

In the last part of the Tertiary a delta of another river system developed in the north. As will soon be seen, the latter system became very important in the Early Pleistocene, that is, in the period between 2.3 and 0.7 million years ago. The data discussed above indicate that rate and pattern of subsidence was primarily steered by sediment loading.

There are several substantial unconformities in the sedimentary sequences of the basin (Fig. 11), i.e., at the base of the Oligocene, at the base of the Miocene, within the Upper Miocene, and at the base of the Pleistocene. The base of the Pleistocene is placed in the North Sea Basin at a time-level of 2.3 million years ago.

The reasons for these unconformities are complex. They are certainly related to changes in tectonic activity, as was the case at the transition from the Eocene to the Oligocene. Sea level changes played a role, too; this certainly holds for the unconformity at the base of the Pleistocene, but probably also for older ones.

Paleogeographic evolution during the Quaternary

During the Quaternary, profound changes took place in the southern part of the North Sea Basin. Whereas during Oligocene and Miocene times the main centres of deposition were situated on the present mainland, the Quaternary depocentre shifted toward the present North Sea (Fig. 4). Thicknesses of more than 900 m have been reported from the Central North Sea, but even further south this series attains considerable thickness, ranging from over 400 metres in the western part of The Netherlands to almost twice that, locally offshore. The second important feature is the increase in the size of the two deltas, one in the south (Rhine and other rivers) and the other in the north (North German-Baltic system) during the Early Pleistocene. Starting about 1.7 Ma ago, this expansion led to the formation of a single large delta which was even much larger than the present area of The Netherlands. In fact, the size of this delta system was similar to that of the largest deltas of the world.

Another interesting feature is that sedimentation rates increased during the Quaternary to about ten times the rate prevailing in the Neogene (Zagwijn & Doppert 1978). The reasons for this are manifold, one being the increasing uplift of the hinterland, but the main reason must have been the climatic changes during the Quaternary with re-

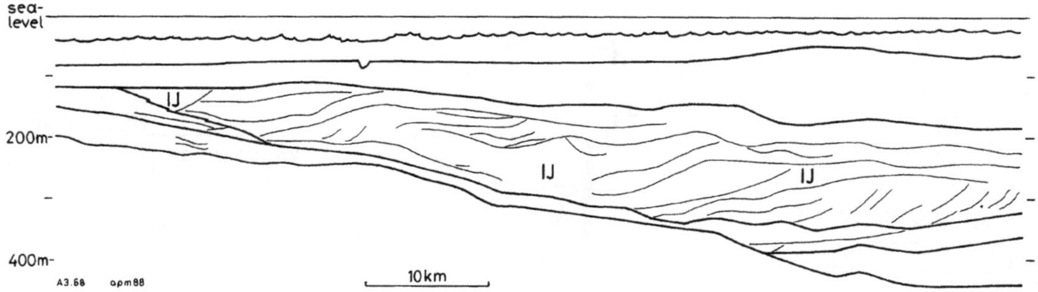

Fig. 19. West-East section situated off the shore of the western part of The Netherlands, based on seismic data and showing Lower Pleistocene delta progradation (IJ = IJmuiden Ground Formation). Source: Cameron et al. 1984.

peated cold intervals. Those were of such intensity that the entire area of Western Europe was north of the tree-line and therefore was not only deforested but also became part of the permafrost region. This pattern is known to have prevailed after the first cold phase of this kind, i.e. the Praetiglien, which lasted from about 2.3 to 2.1 million years ago. Under these deforested and permafrost conditions, erosion of the hinterland and sediment supply into the basin increased enormously.

Delta progradation accelerated considerably after the onset of the Pleistocene, as shown for instance (Fig. 19) by offshore seismic sections along the western part of The Netherlands (Cameron et al. 1986; Laban et al. 1984). The first progradational sequence of this kind, which belongs to the IJmuiden Ground Formation, has recently been dated to represent the first cold phase of the Pleistocene, i.e. the Praetiglian.

The discussion of the further history of basin development during the Quaternary must be prefaced by some remarks concerning the significance of the change in climate during this period. Pollen-analytic investigations in particular have provided a detailed picture of this climatic evolution, which was characterized by frequent shifts from very cold glacial conditions to warm-temperate interglacial conditions and vice versa. Comparison of our results with the well-known climatic record published by Shackleton & Opdyke 1976 that are based on oxygen isotope measurements in deep-sea sediments, shows a high degree of correspondence (Fig. 20). In both cases – dated by paleomagnetism – distinctly cold conditions occurred first around 2.3 million years ago and since that time there were several glacial-interglacial cycles. For the last million years a cyclicity of 100,000 years has predominated, but other periods have also been observed. Although discussion of the causes of these cycles may not have led to agreement on all details, the general opinion is nevertheless that the main cause is astronomical and of the Milankovitch type. This means that they are the result of variations in the movements of the earth in relation to the sun, corresponding with variations in the amount of solar radiation received at earth's surface (Berger & Pestiaux 1984; Berger 1985).

Another factor related to the glacial-interglacial cycles is the expansion and decay of large ice-caps, particularly in the northern hemisphere, and the corresponding pattern of falling and rising sea levels. As early as 2.3 million years ago a very large ice-cap covered North America (Easterbrook 1982) and this development must have been accompanied by a considerable drop of the sea level, estimated at 80 to 100 metres. At the same time, during the Praetiglian, a regression occurred in the southern part of the North Sea Basin (Fig. 12, coastline indicated by a red line), followed by a transgression in the next warm stage, the Tiglian (Zagwijn 1975). This regression is related not only to the already-mentioned strong deltaic progradation but also to a hiatus in the basal sediments of the Pleistocene (Fig. 21), representing the oldest of three major Pleistocene unconformities. All of these phenomena can be considered the result of

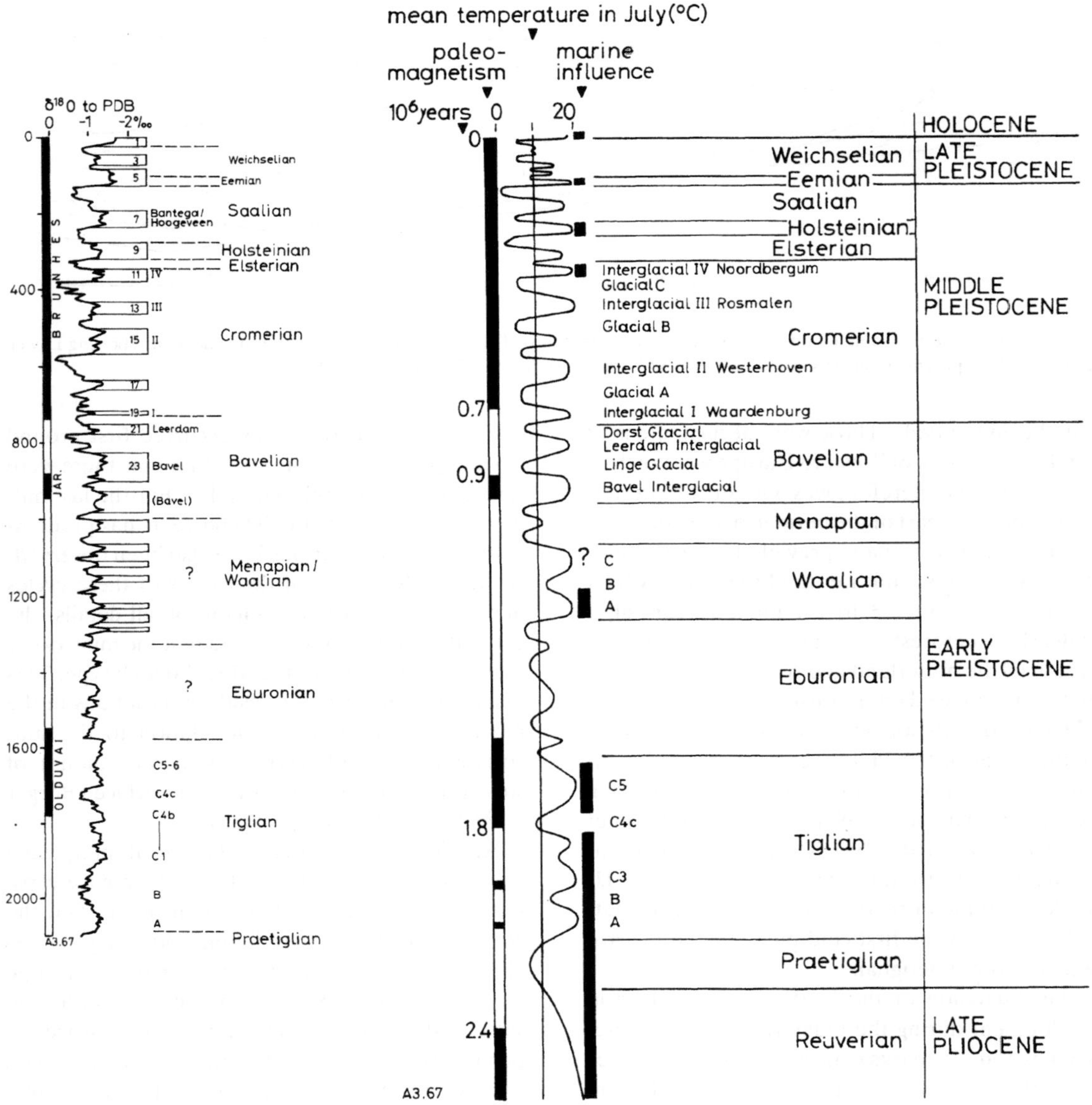

Fig. 20. Comparison of the oxygen-isotope curve of core V28–239 (according to Shackleton & Opdyke 1976) with the Quaternary climatic stratigraphy for The Netherlands (right-hand side). Correlation of the isotope stages (1–21) with climatic stages is tentative.

the first build-up of large areas of inland-ice over the northern continents and the associated drop in sea level. It should be noted that there was a marked change in the pattern of subsidence during that same period. Centres of deposition shifted to the west, i.e., to the West Netherlands Basin, and to the north, to the Vlieland Basin (Fig. 12).

After the Tiglian transgression conditions changed again, because in the Late Tiglian, i.e., in the period between 1.8 and 1.6 million years ago, delta progradation became rapid and the two deltas merged to form one large delta protruding far into

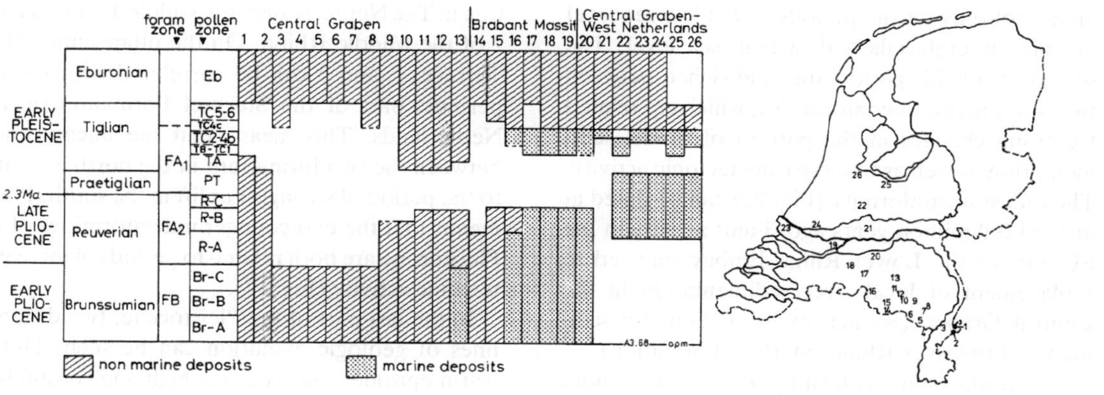

Fig. 21. The basal Pleistocene unconformity in the Central Graben region and adjacent areas. Inset: location of sections 1–26.

the southern part of the present North Sea (Fig. 13). In this period fault activity in the Central Graben area was low and an area of stronger subsidence only occurred in the northern region.

Later, during the Early Pleistocene between about 1.6 and 1.0 million years ago, similar conditions prevailed, but fault activity in the Central Graben region resumed. During this period depocentres lay in the extreme southeastern part of the Central Graben and in a northwestern area whose extension into the North Sea Basin is not yet known (Fig. 14).

The process of deltaic progradation and narrowing of the southern part of the North Sea Basin continued until the transition to the Middle Pleistocene, when the deltaic plain of the two river systems extended far northward and the coastline was probably in the vicinity of the present Dogger Bank (Fig. 15). During this interval, between 900,000 and 450,000 years ago, regression reached a maxi-

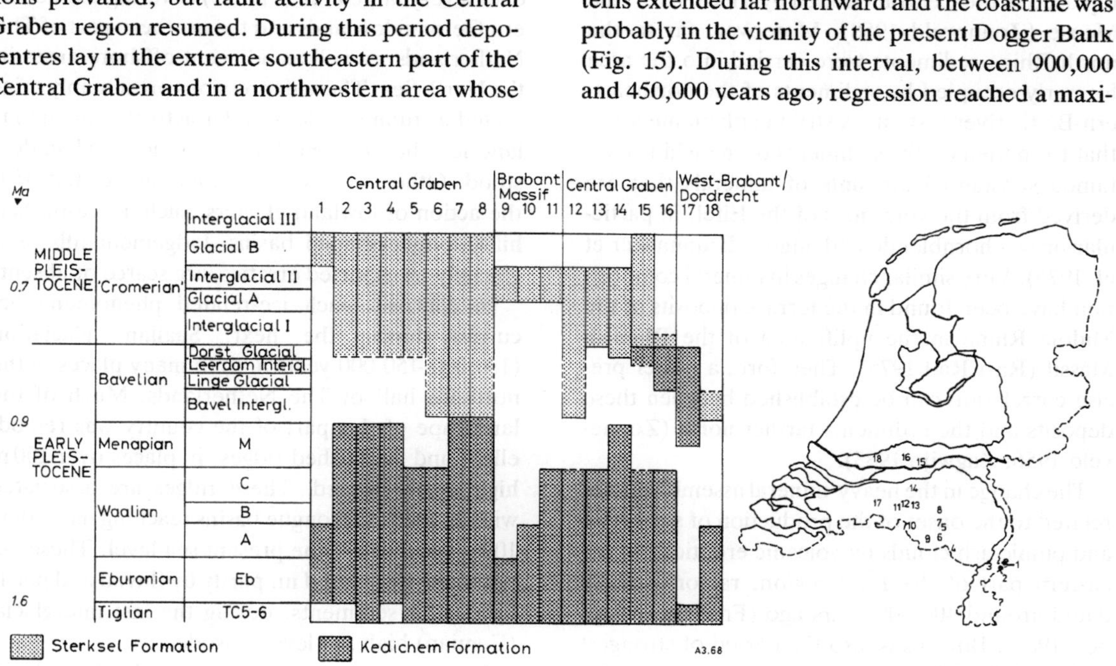

Fig. 22. The unconformity at the top of the Lower Pleistocene in the Central Graben region and adjacent areas. Inset: location of sections 1–18.

mum, not only during periods with a low sea level but also interglacials with a high sea level. The sediments of this period are sandwiched between two other major unconformities, which are related to strong changes in the pattern of fluvial sedimentation as well as to increasing tectonic activity. The oldest unconformity (Fig. 22) can be dated at around 0.9 million years ago. Fault activity in the SE part of the Lower Rhine Embayment led to replacement of local river sedimentation in the Central Graben (Kedichem Formation) by sediments of the river Rhine (Sterksel Formation).

The youngest unconformity, dating from about 450,000 years ago, is related to even greater changes in pattern. The influx of northeastern rivers into the basin had ceased, probably because the drainage system had been impaired by the invasion of Scandinavian inland-ice. After a period of strong fault activity during deposition of the Sterksel Formation, the Rhine no longer took a northwestern course through the Central Graben but flowed northward from Bonn in its present valley, being replaced in the Central Graben by a local river, the Meuse (Zonneveld 1958). Moreover, far to the north Rhine sediments were carried into the area formerly occupied by sediments of the northeastern-Baltic river system. A striking phenomenon is that from then on the sediments of the Rhine contained substantial amounts of minerals that are derived from the volcanoes of the Eifel, in particular brown hornblende and augite (Brunnacker et al. 1976). Very similar changes in mineral composition have been found in the terrace deposits of the Middle Rhine in the uplift area of the Rhenish Massif (Razi Rad 1975). Therefore, a rather precise correlation can be established between these deposits and the sediments farther north (Zonneveld 1956; Zagwijn 1985).

The change in the heavy-mineral assemblage was related to the onset of the production of selbergite and pumic-ash clouds by volcanic eruptions in the eastern part of the Eifel region, radiometrically dated around 400,000 years ago (Frechen & Lippolt 1965). But this is also the period of strongest uplift of the Rhenish Plateau (Figs. 16 and 17; Fuchs et al. 1983). All of these data made it possible to correlate the oldest sediments of the Urk Formation in The Netherlands with Middle Terraces 1 and 2 of the Middle Rhine. On the other hand, Main Terraces 2 and 3 can be correlated with the uppermost part of the Sterksel Formation of The Netherlands. This means that the unconformity between the two formations in the north is related to the period of strongest uplift in the south. It also means that the two younger unconformities of the Pleistocene are both related to periods of increased tectonic activity (Fig. 17).

In the last part of the Pleistocene, two different lines of geologic evolution can be seen. During warm episodes sea level was high and coastal lowlands developed in the present coastal areas of The Netherlands during each interglacial. During cold periods, Scandinavian inland-ice invaded the lowlands of Poland and northern Germany several times and reached the northern part of The Netherlands at least twice. The first of these invasions was in the Elsterian, around 350,000 years ago. During this stage very deep erosion valleys, locally reaching depths of more than 300 metres, were formed over a very large area (Fig. 23) covering the northern German lowlands and the northern part of The Netherlands as well as adjacent offshore areas in the North Sea. These depressions presumably originated as tunnel valleys under actively moving inland-ice, but the mechanism is not well understood. Other features commonly associated with the action of lowland glaciers, such as ice-pushed hills, tongue-shaped basins, lodgement tills, and glacially transported erratics, are scarce or absent.

In contrast, such ice-related phenomena occurred during the next, Saalian, glaciation (180,000–150,000 years ago) in many places in the northern half of The Netherlands. Much of the landscape of this part of the country was remodelled, and ice-pushed ridges, in places over 100 m high, were formed. These ridges are associated with deep glacial tongue basins reaching more than 100 metres below the present sea level. These depressions were filled in, partly by fluvial and partly by marine sediments, during the last interglacial (Eemian) high sea level episode.

For much of the time during the last glacial (115,000 to 10,000 years ago) The Netherlands lay in the permafrost zone. The greater part of the

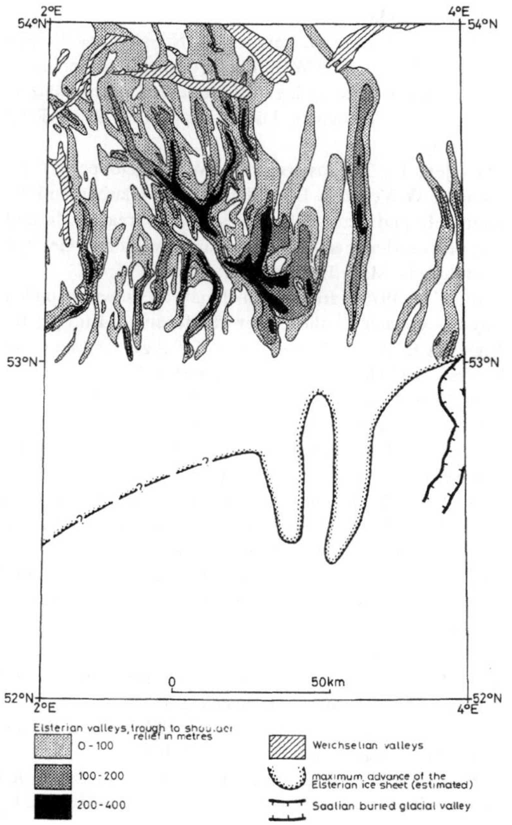

Fig. 23. Deep erosion valleys of the Elsterian glaciation in an offshore area of the southern part of the North Sea. Based on seismic survey. Source: Cameron et al. 1986.

bottom of the southern part of the North Sea was exposed. This time, however, inland-ice did not reach as far south and halted near Hamburg.

Sediments of local and aeolian origin that were formed under a permafrost regime are abundantly present. Thus, much of the old Saalian relief was levelled by erosion and the depressions were filled in by sediment.

For the time being, the story of the geological evolution of The Netherlands ends with the development of the present coastal plains during the last 8,000 years. Many factors have played a role in this evolution, among the most important were the effect of the eustatic rise of the sea level and its interaction with the pattern of the land surface as shaped during the last glacial. Closely related factors include the pattern of the drainage system, its discharge and sediment transport. But other factors have been of importance as well, such as the tidal range, the availability of sediment to build up the coastal plain during rising sea level rise, and the climate. Man's influence increasingly affected the process of coastal lowland building (for the first time in geological history). First in an indirect way, by removing trees and thus influencing the run-off of rivers, and later by draining wet-lands, digging peat, building dikes to control flooding, and, finally, by reclamation of flooded land. For a discussion of this last episode in the geological evolution of the coastal lowlands of The Netherlands, the reader is referred to Zagwijn (1986).

Acknowledgements

The author is indebted to Mr. A.P. Marselje for his excellent drawings, to Mrs. M.E.I. Jouini for the typing of the manuscript, and to Mrs. I. Seeger for reading the English text. Permission to publish the results was given by the Director of the Geological Survey of The Netherlands.

References

Berger, A. 1985 The astronomical theory of paleoclimates – World Climate Programme Newsletter, January 1985: 5 pp

Berger, A. & P. Pestiaux 1984 Modelling the astronomical theory of paleoclimates in the time and frequency domain – Invited paper EC Climatology Symp., Sophia Antipolis, October 1984: 20 pp

Brunnacker, K., W. Boenigk, A. Koci & W. Tillmans 1976 Die Matuyama/Brunhes Grenze am Rhein und an der Donau – Neues Jahrb. Geol. Paläont. Abh. 151: 358–378

Cameron, T.D.J., C. Laban, C.S. Mesdag & R.T.E. Schüttenhelm 1986 Indefatigable, Sheet 53° N-02° E. Quaternary Geology – Ordnance Survey, Southampton

Cameron, T.D.J., C. Laban & R.T.E. Schüttenhelm 1984 Flemish Bight, Sheet 52° N-02° E. Quaternary Geology – Ordnance Survey, Southampton

Easterbrook, D.J. 1982 Revision of North American Pleistocene chronology based on paleomagnetic, fission-track and amino acid dating – XI INQUA Congress Moscow, Abstr. II: 68

Frechen, J. & H.J. Lippolt 1965 Kalium-Argon-Daten zum Alter des Laacher Vulkanismus der Rheinterrassen und der

Eiszeiten – Eiszeitalter Gegenw. 30: 157–172

Fuchs, K., K. von Gehlen & H. Mälzer (eds) 1983 Plateau uplift: The Rhenish Shield – a case history, Berlin: 411 pp

Hager, H. 1986 Peat accumulation and syngenetic clastic sedimentation in the Tertiary of the Lower Rhine basin (F.R. Germany) – Mém. Soc. géol. France, N.S. 149: 51–56

Heybroek, P. 1974 Explanation to the tectonic maps of The Netherlands – Geol. Mijnbouw 53: 43–50

Keizer, J. & W.J. Letsch 1963 Geology of the Tertiary in The Netherlands – Kon. Ned. Geol. Mijnb. Gen. Verh. 21: 147–172

Laban, C., T.D.J. Cameron & R.T.E. Schüttenhelm 1984 Geologie van het Kwartair in de zuidelijke bocht van de Noordzee – Meded. Werkgr. Tert. Kwart. Geol. 21: 139–154

Letsch, W.J. & W. Sissingh 1983 Tertiary stratigraphy of The Netherlands – Geol. Mijnbouw 62: 305–318

McCave, I.N., V.N.D. Caston & N.G.T. Fannin 1977 The Quarternary of the North Sea. In: F.W. Shotton (ed.): British Quaternary Studies – Oxford, 187–204

Razi Rad, M. 1975 Schwermineraluntersuchungen zur Quartär-Stratigraphie am Mittelrhein – Thesis Univ. Köln: 164 pp

Shackleton, N.J.S. & N.D. Opdyke 1976 Oxygen-isotope and paleomagnetic stratisgraphy of Pacific Core V28-239 Late Pliocene to Latest Pleistocene – Geol. Soc. Am. Mem. 145: 449–464

Vail, P.R., R.M. Mitchum, R.G. Todd, J.M. Widmier, S. Thompson, J.B. Sangree, J.N. Bubb & W.G. Hatlelid 1977 Seismic stratigraphy and global change of sea level – Am. Ass. Petr. Geol. Mem. 26: 49–212

Van den Bosch M. & H. Hager 1984 Lithostratigraphic correlation of Rupelian deposits (Oligocene) in the Boom area (Belgium), the Winterswijk area (The Netherlands) and the Lower Rhine District (F.R.G.) – Meded. Werkgr. Tert. Kwart. Geol. 21: 123–138

Van Doorn, Th.H. M., C.J. Leyzers Vis, N. Salomons, W. van Dalfsen, H. Speelman & W. Zijl 1985 Aardwarmtewinning en grootschalige warmte-opslag in tertiaire en kwartaire afzettingen – Rijks Geologische Dienst Rapport nr. 85 KAR 02 EX: 108 pp

Van Rooijen, P., J. Klostermann, J.W.Chr. Doppert, C.K. Rescher, J.W. Verbeek, B.C. Sliggers & P. Glasbergen 1984 Stratigraphy and tectonics in the Peel-Venlo area as indicated by Tertiary sediments in the Broekhuizervorst and Geldern T_1 boreholes – Meded. Rijks Geol. Dienst 38-1: 1–27

Zagwijn, W.H. 1975 Variations in climate as shown by pollen analysis, especially in the Lower Pleistocene of Europe. In: Wright, A.E. & F. Moseley (eds): Ice Ages: Ancient and Modern – Seel House Press (Liverpool): 137–152

Zagwijn, W.H. 1985 An outline of the Quaternary stratigraphy of the Netherlands – Geol. Mijnbouw 64: 17–24

Zagwijn, W.H. 1986 Nederland in het Holoceen – Haarlem/'s-Gravenhage: 46 pp

Zagwijn, W.H. & J.W.Chr. Doppert 1978 Upper Cenozoic of the Southern North Sea Basin: Palaeoclimatic and Palaeogeographic evolution – Geol. Mijnbouw 57: 577–588

Zagwijn, W.H. & H. Hager 1987 Correlations of continental and marine Neogene deposits in the south-eastern Netherlands and the Lower-Rhine District – Meded. Werkgr. Tert. Kwart. Geol. 24: 59–78

Ziegler, P.A. 1982 Geological Atlas of Western and Central Europe – Shell/Elsevier ('s-Gravenhage): 130 pp

Zonneveld, J.I.S. 1956 Das Quartär der südöstlichen Niederlande – Geol. Mijnbouw, N.S. 18: 379–385

Zonneveld, J.I.S. 1958 Litho-stratigrafische eenheden in het Nederlandse Pleistoceen – Meded. Geol. Stichting, N.S. 12: 31–64

Geologie en Mijnbouw 68: 121–129 (1989)

Geological and geotechnical conditions of the Beaufort Sea coastal zone, Arctic Canada

P.J. Kurfurst & S.R. Dallimore
Geological Survey of Canada, 601 Booth Street, Ottawa, Ontario, Canada, K1A OE8

Received 28 August 1987; accepted in revised form 29 February 1988

Key words: Beaufort Sea coastal zone, geotechnical, ground ice, permafrost, thaw settlement

Abstract

The coastal zone of the southern Beaufort Sea and Mackenzie Delta forms an extensive area of coastal lowlands in northern Canada. This region is underlain by unstable, perennially frozen soils subjected to high rates of marine erosion and deposition.

The nearshore sediments off northern Richards Island are comprised of a wedge of Holocene marine sand, silt and clay underlain by early Wisconsinan sand and clay. The geothermal regime is complex, reflecting deep permafrost conditions established during a period of terrestrial exposure and more recent marine submergence, which results in moderating ground temperatures and creation of a thick thawed layer at the sea bottom. Geotechnical problems encountered in the nearshore area include frost heave of Holocene sediments, thaw settlement related to degradation of ground ice in the early Wisconsinan sediments, and ice push and scour effects.

Onshore sediments consist of early Wisconsinan and older glacial, fluvial and marine sediments overlain by late Wisconsinan and Holocene, eolian and lacustrine sediments. Ground ice, which forms a significant volumetric component of the near-surface soils, occurs as pore ice, wedge ice, pingo ice and as massive bodies of segregated ice of various ages. Geotechnical problems in onshore areas include thaw settlement due to degradation of ground ice, creep of ice-rich soils and frost heave.

Introduction

The Mackenzie Delta and coastal areas of the southern Beaufort Sea form a large region of coastal lowlands in northern Canada. Sedimentary basins underlying this area are rich in hydrocarbon resources and anticipated development has created an unprecedented need for detailed information regarding geological and geotechnical conditions. This need is particularly evident in coastal areas where construction of development-related structures is most likely.

The coastal zone as discussed in this paper includes onshore areas and nearshore areas to approximately one kilometre offshore. Diverse geological materials, complex thermal conditions and the occurrence of various forms of ground ice, present a variety of unique geotechnical problems. As part of ongoing geological and geotechnical studies of the southern Beaufort Sea area, the Geological Survey of Canada (GSC) has carried out a number of regional studies. These have included surficial geology mapping (Rampton, 1982, 1988), geothermal and geophysical studies (Taylor et al., 1982; Hunter et al., 1978), and geotechnical studies (Kurfurst et al., 1984; Kurfurst & Pullan, 1985). Site specific investigations of seabottom and terrestrial sediments have also been carried out along several

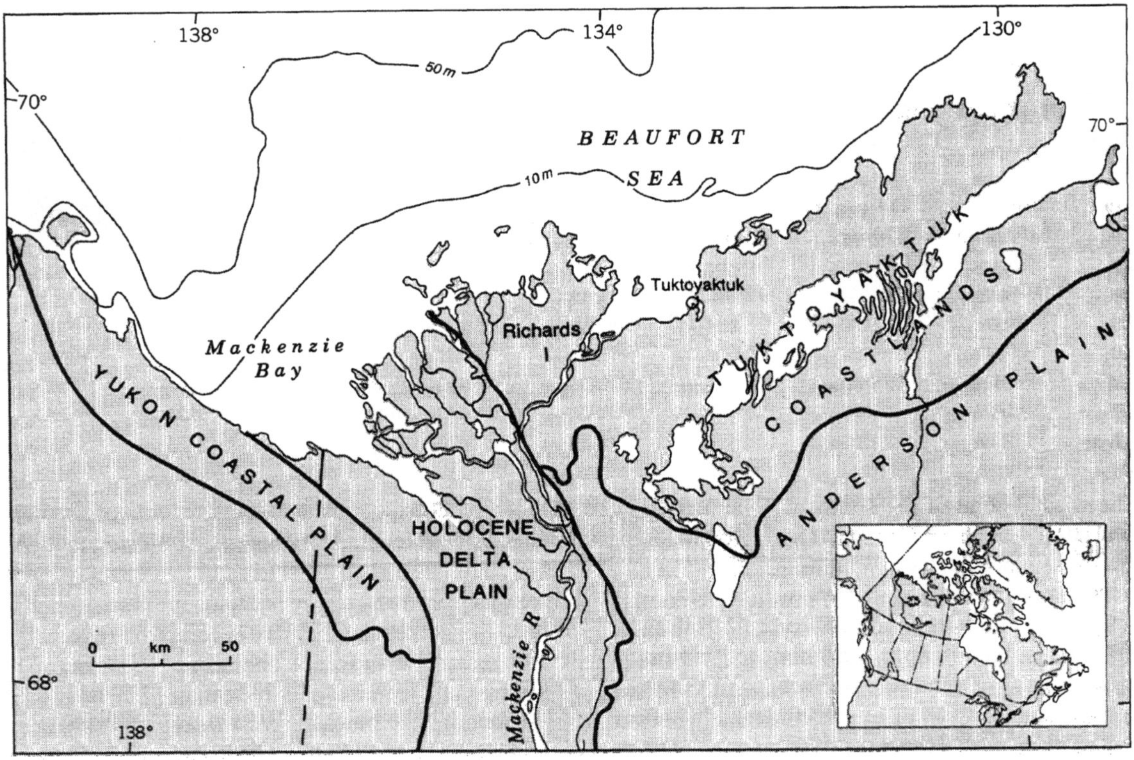

Fig. 1. Physiographic units of the Beaufort Sea coastal zone (after Rampton, 1988).

onshore-offshore transects near areas of particular development interest (Kurfurst, 1984 and 1986; Hill et al., 1986).

This paper discusses the geological conditions of the coastal zone and some of the unique geotechnical problems which are encountered in this arctic environment. Results of recently completed geotechnical investigations along onshore-offshore transects in the vicinity of Richards Island, N.W.T. are discussed in detail.

Geological setting

General

The coastal lowland area of the Canadian Beaufort Sea can be subdivided into three general physiographic units (Rampton, 1988), bounded by upland areas to the south. The Yukon Coastal Plain, the Holocene Mackenzie Delta and the Tuktoyaktuk Coastlands, shown in Fig. 1, form an area over 20 000 km^2 in size. Most of this region is made up of low-lying areas with elevations below 30 m ASL; large areas have elevations below 10 m ASL.

The submarine morphology of offshore areas is complex with a relatively narrow shelf area off the northern Yukon, a deep U-shaped trough in the Mackenzie Bay area and a large shelf area to the east with numerous drowned river channels and other remnant terrestrial features.

Geology of onshore areas

Thick deposits of unconsolidated Quaternary and Holocene sediments with virtually no bedrock exposures characterize nearly all of the onshore coastal zone. The Mackenzie Delta is located in the centre of this zone, forming an extensive low-lying plain with a maze of lakes and channels (Hill, 1987;

Mackay, 1963a). The Tuktoyaktuk Coastlands and the Yukon Coastal Plain have experienced a complicated Quaternay history during which a variety of sediments of various ages have been deposited. Correlation of Quaternary deposits and events is often difficult (Heginbottom & Vincent, 1986). The oldest deposits exposed on the Yukon Coast are thought to be early Wisconsinan or older glacial, marine and alluvial sediments (Rampton, 1982). In the Tuktoyaktuk Coastlands area pre-early Wisconsinan sediments are extensive and occur as a marine clay and as thick marine and alluvial sand units (Rampton, 1988).

Glacial deposits thought to be related to early Wisconsinan ice advances are widespread in the Yukon Coastal Plain and the Tuktoyaktuk Coastlands area. Late Wisconsinan ice is thought to have been confined primarily to the Mackenzie Valley area. Processes affecting the landscape after the retreat of Wisconsinan ice sheets have continued at varying rates to the present. Postglacial and Holocene sediments include widespread alluvial, lacustrine, and colluvial deposits and local eolian and marine deposits.

Geology of nearshore areas

Relatively little is known about the surficial geology of the nearshore zone primarily because of difficulties in collecting geological and geotechnical information in this area of shallow water. A simplified model for the surficial geology of the Beaufort Sea has been presented by O'Connor (1983). Three units are recognized, based on the last marine transgression in the area. The oldest unit consists of sediments which were exposed terrestrially prior to submergence. The second unit was deposited during the marine transgression and the third unit consists of fine-grained marine sediments deposited subsequent to this transgression. In the nearshore area, sediments of the first two units are generally present.

Permafrost and Ground Ice

Permafrost, or perennially frozen ground, is continuous onshore and widespread in offshore areas. The present ground thermal conditions and permafrost thickness reflect a sequence of geologic and climatic conditions which have occurred since early or even pre-Wisconsinan times. These events include fluctuations in sea level, general climatic warming and cooling, and the presence or absence of insulating masses of glacial ice.

At present, permafrost thickness on land varies from less than 100 metres beneath the flood plain of the Mackenzie Delta (Smith 1975) to over 700 metres thickness beneath Richards Island (Taylor et al., 1982). Offshore permafrost is also widespread in certain areas of the Beaufort Sea. In particular, a substantial thickness of permafrost has been detected offshore in the area to the east of the Mackenzie Delta (Hunter et al., 1978).

Nearly all perennially frozen soils in the coastal zone contain some pore water in the form of ice, with the proportions of ice and water being determined by physical and chemical conditions unique to each soil. In addition to pore ice, ground ice also occurs in the form of thin lenses or veins, and as large bodies of nearly pure massive ice. The type and amount of ground ice is highly variable, being determined by the physical characteristics of the enclosing soil material, the hydraulic and thermal conditions experienced during freezing, cover and preservation of glacial ice, and post-ice formation processes such as creep and glacial tectonics.

Coastal stability

Much of the coastline of the Canadian Beaufort Sea is undergoing high rates of coastal retreat with only a few areas of local accretion (Mackay, 1963a, b; McDonald & Lewis, 1973; Lewis & Forbes, 1974; and Harper et al., 1985). Maximum retreat rates of over 20 m/a have been measured in the distal areas of the Mackenzie Delta; however, most coastal sections exhibit retreat rates between 1 m/a and 3 m/a (Harper et al., 1985).

Richards Island

General

Richards Island is situated east of the modern Mackenzie Delta within the Tuktoyaktuk Coastlands physiographic region (Fig. 1). The island is strategically situated for hydrocarbon-related de-

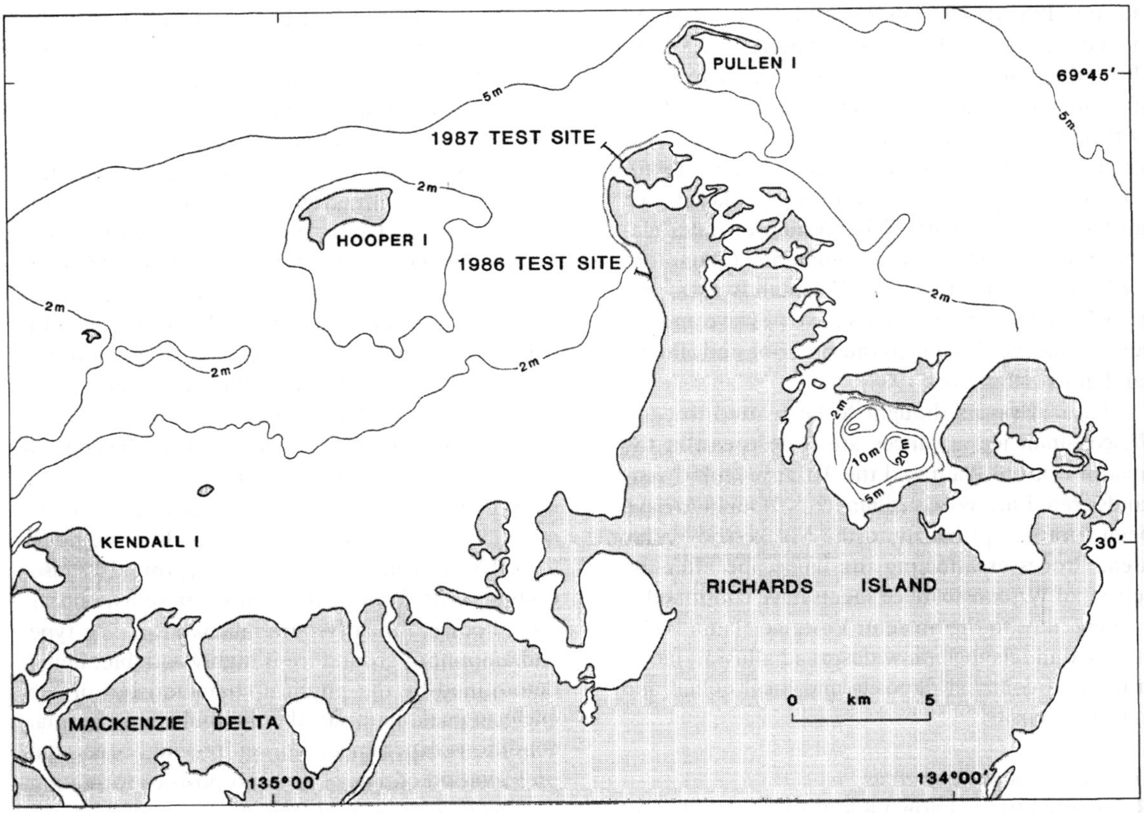

Fig. 2. Richards Island – location of onshore-offshore transects.

velopment with potential production fields locateu at the western end of the island and in areas immediately offshore. At least two proposed pipeline routes cross the island. One of the proposed pipelines would bring offshore oil from the north with an anticipated landfall near the northern tip of the island.

The topography of Richards Island is flat to gently rolling with numerous small lakes, some of which have been inundated by the sea. Land areas in the northern part of the island are covered by a discontinuous veneer of glacial till which overlies older preglacial sands. The coastal areas exhibit a variety of erosional and accretional landforms. As in other areas of the Beaufort Sea coastline, many localities are undergoing rapid coastal retreat while some appear to be relatively stable. Although conditions in onshore and offshore areas of Richards Island cannot be considered as a model for the whole Beaufort Sea region, many of the geotechnical problems encountered in this area are similar to those which might be encountered elsewhere.

For structures such as port facilities or offshore/onshore pipelines, the transition between the offshore and onshore areas is critical. The sites for detailed investigations were therefore chosen to study and document geological and geotechnical conditions in this transitional zone. In order to assess the importance of coastal stability, geotechnical investigations were undertaken in 1986 and 1987 at two sites on northern Richards Island (Fig. 2). The 1986 test site is located in an area of stable coastline, whereas the 1987 test site is characterized by active marine erosion.

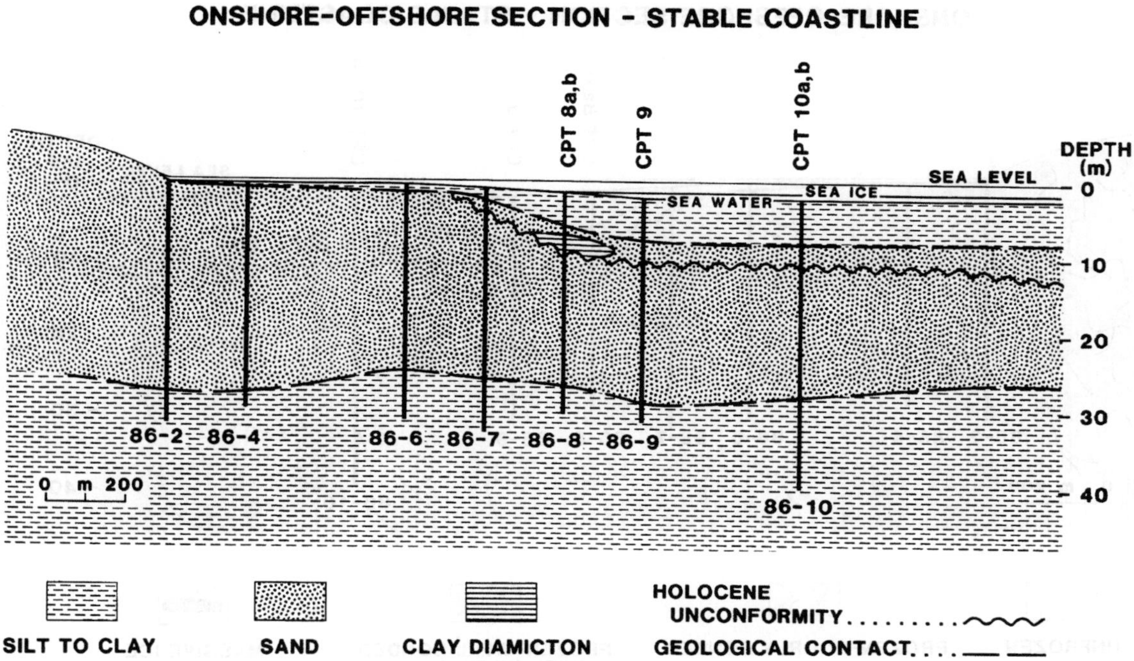

Fig. 3. Stable coastal site – geological conditions.

Stable coastal site

The 1986 geotechnical investigation was undertaken on the west side of Richards Island in an area which has shown no coastal retreat since 1947. This area has a very gentle offshore profile with gradients of approximately 1.0 m/km. The coast has been stabilized by the development of extensive longitudinal sand bars which have been built up by longshore drift from the southwest.

A summary of the geology of the onshore-offshore transect is given in Fig. 3. A seaward thickening wedge of Holocene sediments occurs in the nearshore zone. These sediments represent a transgressive sequence of reworked sands overlain by fine-grained marine silts. The sediments beneath the Holocene unconformity consist of alluvial sands and marine silts of early Wisconsinan or older age. The early Wisconsinan sand is exposed close to the surface near the coast and is present inland where it is overlain by a discontinuous cover of glacial till and colluvial material.

The ground ice and permafrost conditions at the stable coastal site are shown on Fig. 4. The 0° C isotherm, which defines the permafrost table, occurs within 50 cm of the surface at the coast at the end of the thaw season in September. It dips very gently offshore to a depth of less than 2 m, approximately 800 m from shore. Beyond this point between boreholes 86-7 and 87-10, the permafrost table dips down more steeply to approximately 10 m depth beneath the sea floor. The zone of frozen sediments near the shore between boreholes 86-2 and 86-7 is caused primarily by the effect of sea ice freezing to the sea bed during the winter. This exposes the underlying sediments to very cold temperatures (Kurfurst, 1986) throughout much of the year. At 800 m offshore the water column is sufficiently thick that a layer of warm sea water is present year round, causing rapid thaw.

Actively eroding coastal site

The site of the 1987 geotechnical investigation is located at the northern tip of Richards Island. This area is exposed to direct wave action from the

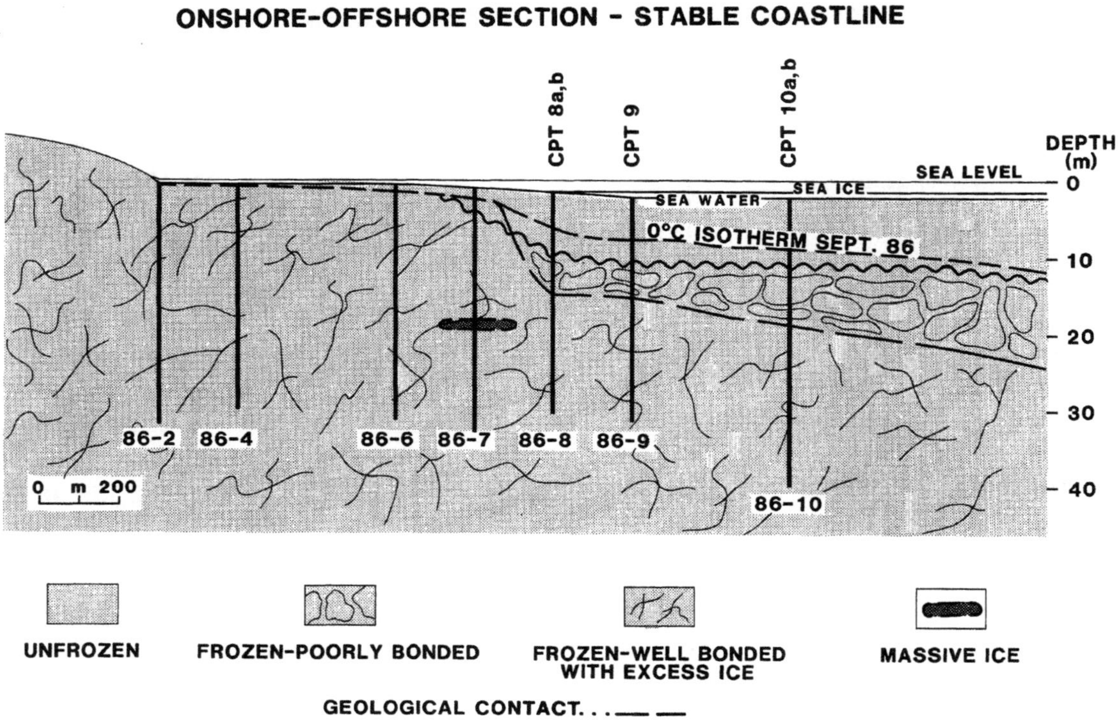

Fig. 4. Stable coastal site – ground ice and permafrost conditions.

Beaufort Sea. Coastal retreat rates measured from sequential air photography since 1947 range from 0.8 m/a to 3.5 m/a. At the site of the onshore-offshore transect, approximately 80 metres of retreat occurred between 1947 and 1985. A nearly vertical coastal bluff approximately 16 m high has developed because of the high rates of erosion.

A summary of the geology at the active coastal retreat site is shown on Fig. 5. Similar to the 1986 site, this site shows a transgressive sequence of re-worked sand overlain by fine-grained marine sediments. The early Wisconsinan sediments, which occur below the Holocene unconformity and are exposed onshore, are more variable than at the stable coastal site. These sediments contain substantial amounts of excess ice in the form of thin lenses and as a pod-shaped body of massive ice exposed just below sea level. It is probable that the higher ice content has contributed to the rapid rate of coastal retreat, accelerating erosion as a result of ablation and reduced sediment deposition offshore.

The thermal and ground ice conditions of the 1987 transect are summarized on Fig. 6. The permafrost table occurs close to the sea bottom only in the nearshore area within 100 m of the cliff. In areas of deeper water further offshore, the permafrost table dips rapidly away from the coast, as it is influenced by relatively warm year-round sea bottom temperatures.

It appears that offshore areas in this region are experiencing thermokarst-like conditions that are common in the adjacent terrestrial environment. During marine transgression, ice-rich early Wisconsinan sediments are eroded to wave base by melting and wave action. Initially, thaw beneath the sea bottom is limited because of the shallow water depths and sea ice freezing to the bottom. As the coast continues to retreat, the water becomes deeper and thaw is accelerated. Where the early

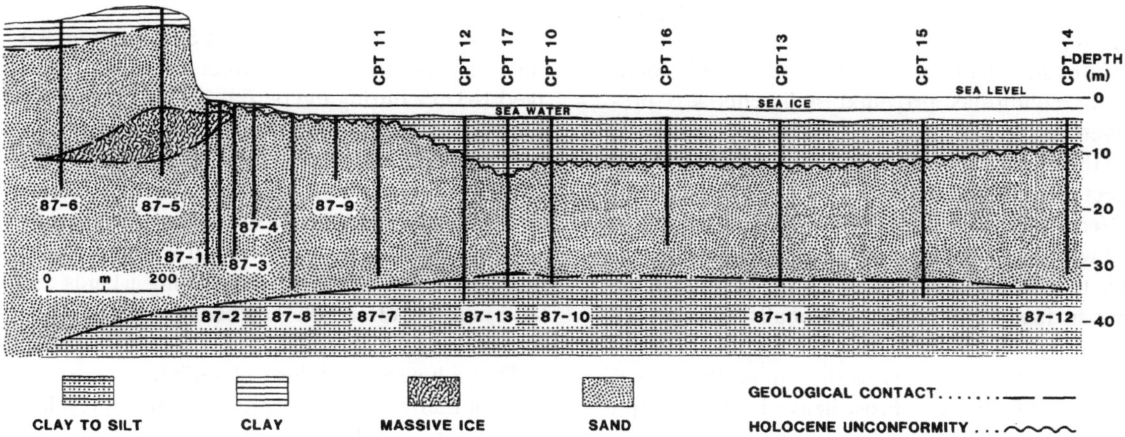

Fig. 5. Actively eroding coastal site – geological conditions.

Wisconsinan sediments are ice-rich, a substantial volume reduction occurs during the thaw process, creating a pitted topography on the sea floor. It appears offshore sedimentation is so rapid that in the vicinity of the 1987 transect the thermokarst depressions are quickly infilled.

Geotechnical considerations

Cone penetrometer tests of Holocene sediments at both sites show that the silts and clays present at the sea bottom are relatively soft materials with low bearing capacities. A further reduction in bearing capacity can be expected in offshore areas where

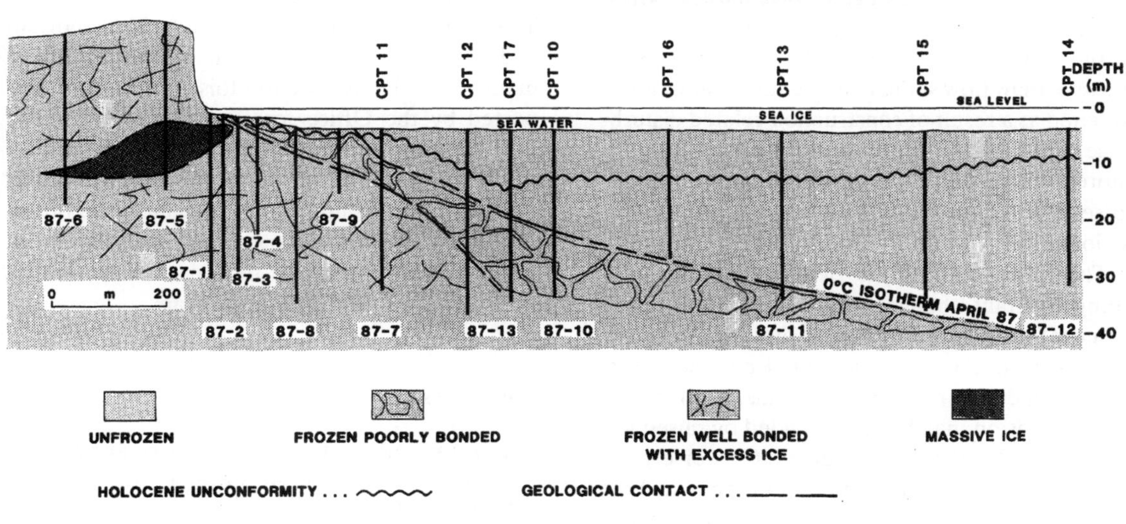

Fig. 6. Actively eroding coastal site – ground ice and permafrost conditions.

ice-rich sediments are thawing at depth. The liberation of water after thawing may also cause high pore water pressures in overlying sediments. Most onshore sediments provide relatively stable foundation conditions provided sediments are preserved in the frozen state. Creep of ice-rich soils and the effect of pore water salinity may negatively effect deep foundations.

Nearshore areas with shallow coastal gradients and water depths less than 1.5 m are likely susceptible to frost heave caused by ice aggradation in the near-surface layer which seasonally freezes and thaws. Further work is being undertaken in the vicinity of the transects to determine the importance of this process. Frost heave is also a major concern in onshore areas, in the active layer, or in areas where drained lakes are experiencing permafrost aggradation.

Thaw settlement caused by melting of excess ice is a major geotechnical concern where engineering structures may alter the existing ground thermal regime. Thaw settlement is also an ongoing natural process in the nearshore area where previously exposed terrestrial sediments are submerged. The amount of settlement expected as a result of man-made or natural thaw is a function of the type and volume of ground ice present, and the physical and hydraulic properties of enclosing sediments.

There are a number of geological hazards which may influence geotechnical design. Although relief in the coastal zone is relatively low, small retrogressive thaw flow slides occur naturally and can be initiated as a result of construction-related disturbance. The weak Holocene sediments in the offshore area may be susceptible to liquefaction due to seismic induced loading or as a result of wave action.

The entire coast of the Beaufort Sea is undergoing constant changes which may affect development facilities well within the lifetime of a particular project. For example, rapid coastal retreat may occur during major storm events. High rates of coastal erosion may be compounded by raised water levels related to storm surges. Conversely, locally high rates of sedimentation in the nearshore zone can affect the construction, operation and maintenance of engineering structures.

Conclusions

Much of the Canadian Beaufort Sea coastline has been undergoing coastal retreat which has averaged 1–3 m/a during the period 1947–87; however, a few areas of local accretion have also been documented. The nearshore areas to approximately 1 km offshore exhibit diverse geotechical conditions which may influence the design of hydrocarbon development-related structures such as pipelines, harbours and artificial drilling islands and have negative effects on their construction and performance.

The submerged terrestrial sediments are in a state of thermal disequilibrium which can result in melting of ice-rich sediments and massive ice, resulting in volume reduction and pitting of the sea floor, reduction of bearing capacity, and thaw settlement. Onshore sediments may be susceptible to frost heave, thaw settlement, and the initiation of small retrogressive thaw flow slides by construction-related disturbance. However, the majority of onshore sediments are considered relatively stable foundation materials provided they are maintained in the frozen state.

Acknowledgment

The authors are grateful to Mr. J.A. Heginbottom and Dr. D.G. Harry for their comments on the manuscript. Fieldwork for this project was supported by the Office for Energy, Research and Development and the Northern Oil and Gas Action Program. The logistical assistance provided by the Polar Continental Shelf Project and the Inuvik Research Center is acknowledged as are the contributions made by all the field participants in the 1986 and 1987 field drilling program.

References

Harper, J.R., Reimer, P.D. & Collins, A.D. 1985 Canadian Beaufort Sea physical shore-zone analysis – G.S.C. Open File Rept 1689: 130 pp.

Heginbottom, J.A. & Vincent, J-S. 1986 (eds): Correlation of Quaternary deposits and events around the margin of Beaufort Sea – Proc., Can. – U.S. Workshop 1984 – G.S.C. Open File Rept 1237: 60 pp.

Hill, P.R. 1987 The Mackenzie Delta and adjacent coastal lowlands – Canadian Beaufort Sea – Symp. on Coastal Lowlands: Geology and Geotechnology, The Hague, May, 1987, Abstracts and Programme: 39.

Hill, P.R., Forbes, D.L., Dallimore, S.R. & Morgan, P. 1986 Shoreface development in the Canadian Beaufort Sea – Proc., Assoc. Comm. Research on Shoreline Erosion and Sedimentation 1986, Burlington, Ontario: 428–448.

Hunter, J.A., Neave, K.G., MacAulay, H.A. & Hobson, G.D. 1978 Interpretation of sub-seabottom permafrost in the Beaufort Sea by seismic methods, Part 1 and 2 – Proc., 3rd Int. Conf. on Permafrost, Edmonton, Alberta: 514–526.

Kurfurst, P.J. 1984 Geotechnical investigation in the southern Beaufort Sea, Spring 1984 – G.S.C. Open File Rept 1078: 142 pp.

Kurfurst, P.J. 1986 Geotechnical investigations of the nearshore zone, North Head, Richards Island, N.W.T. – G.S.C. Open File Rept 1376: 82 pp.

Kurfurst, P.J., Moran, K. & Nixon, F.M. 1984 Drilling and sampling in frozen seabottom sediments, southern Beaufort Sea – G.S.C. Current Research, B, 84–1B: 193–195.

Kurfurst, P.J. & Pullan, S.E. 1985 Field and laboratory measurements of seismic and mechanical properties of frozen ground – Proc. 4th Int. Symposium on Ground Freezing, Sapporo, Japan: 258–262.

Lewis, C.P. & Forbes, D.L. 1974 Sediments and sedimentary processes, Yukon Beaufort Sea coast – Environm. Soc. Comm.-Northern Pipeline, Task Force Rept 74-29: 41 pp.

Mackay, J.R. 1963a Notes on the shoreline recession along the coast of the Yukon Territory – Arctic, 16, (3): 195–197.

Mackay, J.R. 1963b The Mackenzie Delta area – Can. Dept. Mines & Tech. Surv., Mem. 8: 201 pp.

McDonald, B.C. & Lewis, C.P. 1973 Geomorphic and sedimentologic processes of rivers and coast, Yukon coastal plain – Environm.-Soc. Comm.-Northern Pipeline, Task Force Rept 73-39: 52 pp.

O'Conner, M.J. 1983 Distribution of shallow subsea permafrost beneath the Beaufort Sea Continental Shelf – G.S.C. Open File Rept 953: 128 pp.

Rampton, V.N. 1982 Quaternary geology of Yukon coastal plain, Northwest Territories – G.S.C. Bull. 317: 49 pp.

Rampton, V.N. 1988 Quaternary geology of the Tuktoyaktuk Coastlands – G.S.C., Mem. 423: 98 pp.

Smith, M.W. 1975 Permafrost in the Mackenzie Delta, N.W.T. – G.S.C. Pap. 75-28: 34 pp.

Taylor, A.E., Burgess, M., Judge, A.S. & Allen, V.S. 1982 Canadian geothermal data collection, Norman Wells – Earth Physics Branch, Geotherm. Ser. 13: 153 pp.

Geologie en Mijnbouw 68: 131–142 (1989)

Time dependent groundwater flow under river embankments

Christophe M.H. Bauduin & Christinus J.B. Moes
Delft Geotechnics, Earth Structure Department,
P.O. Box 69, 2600 AB Delft, The Netherlands

Received 27 August 1987; accepted in revised form 24 May 1988

Keywords: dike, embankment, groundwater, slope, stability, uplift

Abstract

The Dutch Meuse – Rhine Delta mostly consists of Holocene clay and peat layers, overlying a Pleistocene sand substratum. The lowlands are below MSL, are protected against inundation by riverdikes and are mechanically drained. Time dependent variations of the river levels provoke time and distance dependent piezometric level responses in the Pleistocene sand layers. During storm surges, high piezometric pressures may reduce the bearing capacity of the embankment's foundation.

Predictions of the maximum piezometric pressures are needed for safe and economic design of the dikes, taking into account the unsteadiness of groundwater flow during design storm surge.

An analytical method allows to describe the piezometric response for both a tidal and step (surge) input. This method has been applied for the geo-hydrological conditions of the Meuse-Rhine delta: a pervious aquifer (Pleistocene stratum) overlain by an aquitard (Holocene layers) with time dependent leakage. The model accounts for the possible presence of a silt or mud layer between the river and the aquifer. If the calculated pressure in the sand somewhere exceeds the weight of the Holocene layers, pressure redistribution occurs and an area will be uplifted. The model can be adapted to this non-linearity and then allow to evaluate the time dependent length of the uplifted area and new piezometric levels.

The model parameters can be obtained from measurements made during normal tides.

Introduction

The Lower Meuse-Rhine delta consists mostly of Holocene layers of clay and peat, with a total thickness of approximately 8 to 12 metres, overlying a Pleistocene sand stratum with a thickness of approximately 20 to 40 metres (Fig. 1). The mean soil surface is several metres below mean sea level (MSL). The land is protected against inundation by artificial levees. The groundwater level is maintained below the soil surface by artificial drainage. Time dependent variations of the water level in the sea or river provoke time and distance dependent changes of the pore water pressure in the Pleistocene sand strata.

From time to time, the crests of the levees have to be raised and their slopes have to be fortified. The evaluation of their safety against embankment failure, caused by sliding of the inner slopes as a result of piping in or uplift of the Holocene layers, needs a reliable prediction of the pore pressure in the Pleistocene sand stratum under design storm conditions.

The classical prediction of the pore water pressure in the sand strata during storm surges, based on an assumed stationary ground water flow, yields maximal values. In most of the geo-hydrological situations existing in the Lower Meuse-Rhine delta, however, the period of a storm surge is too short to cause a stationary flow. Taking into account a

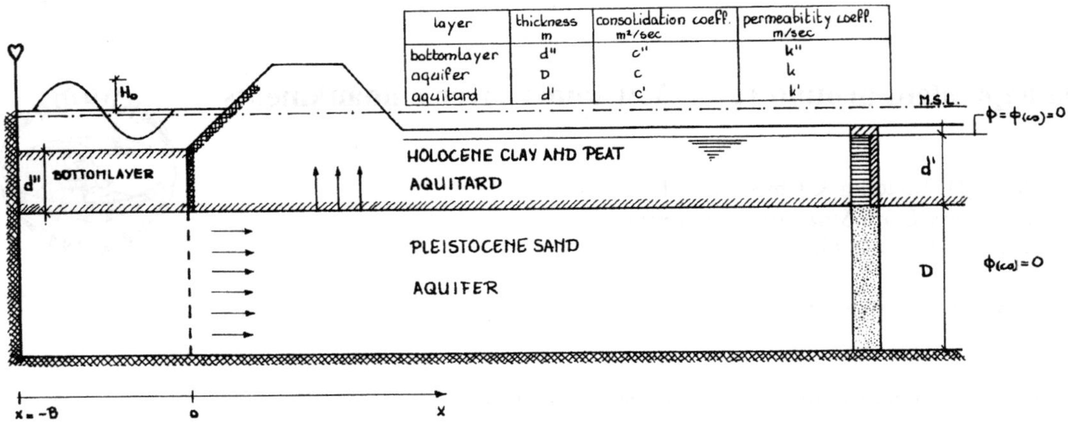

Fig. 1. General geo-hydrological situation in the lower Meuse-Rhine delta.

stationary flow during a storm surge, obviously leads to lower maximum pore water pressure in the Pleistocene aquifer system and consequently to a more economic design.

Therefore an analytical method was developed, capable of describing the piezometric response in the aquifer system, due to both a steady state (tidal) and step (surge) variation of the water level in the river or the sea. It has been applied for the most common geo-hydrological situation of the lower Meuse-Rhine delta, which is shown in Fig. 1: a pervious aquifer (Pleistocene sand) overlain by an aquitard (Holocene layers). The method accounts for the possible presence of a silt or mud layer between the river bottom and the aquifer (bottomlayer).

A geo-hydrological model for the time dependent flow

In general, an aquifer-aquitard model described as above is referred to as a leaky aquifer system. The contribution of the aquitard usually is defined by means of a constant leakage factor (Jacob, 1940). For time dependent flow however, it has been shown that this description is not correct (Barends, 1982). A more accurate description is based on the concept of a time dependent leakage factor. This concept is extended to describe the influence of compressible, low permeable silt or mud layers between the river bottom and the aquifer or the influence of shallow rivers, which do not directly contact the aquifer.

The semi-coupled analytical model presented here is an extension of Barends' model (1982) and links the time dependent storage and flow processes in the compressible bottom layer, the aquifer and the aquitard time dependently. The consolidation processes in the compressible layers and the flow and storage of water in the sand stratum are described by a set of four differential equations for time and place, which are coupled together to form the mathematical model. The coupling between the seepage flow in the aquifer and the consolidation in the bottom layer and in the aquitard is settled by the conditions at their common boundaries (interfaces): continuity of specific discharge and water pressure. The model is completed with the boundary conditions and the demand of compatible initial state conditions. This leads to a rather complicated set of equations. However, they can be solved with the aid of some simplifying assumptions:

- linear flow and linear elastic deformations;
- vertical flow only in the bottom layer and in the aquitard;
- horizontal flow only in the aquifer;
- the bottomlayer, the aquifer and the aquitard are homogenous;

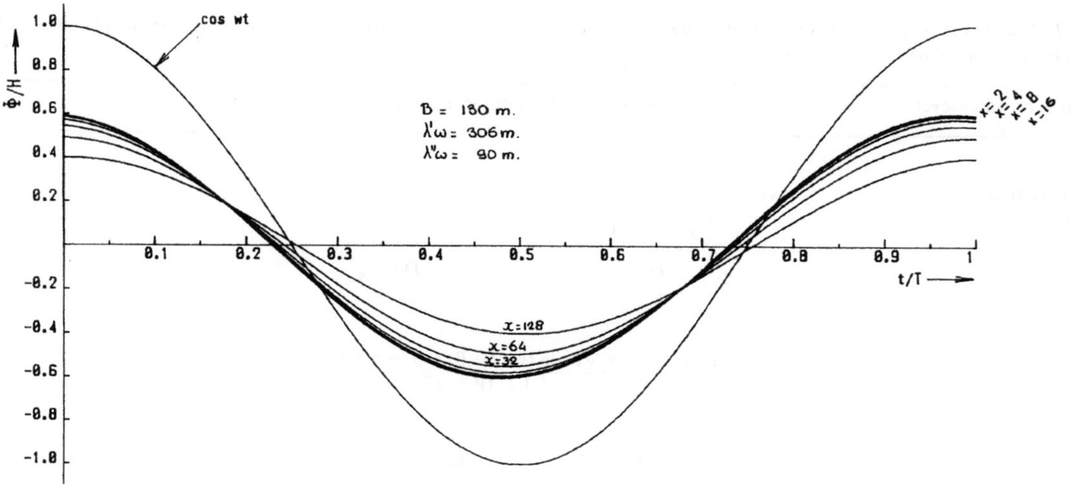

Fig. 2. Water pressure response on tidal variation of the water level at different x-values.

– the level of the phreatic line in the aquitard does not change.

The parameters of the different layers are summarized in Fig. 1. The thickness d' of the aquitard is measured between its boundary with the aquifer and the phreatic level. The riverbottom has a width equal to 2B. In the following, the reference potential head will be taken equal to zero on the phreatic level.

Response of the pore water pressure in the aquifer, due to tidal and surge variations of the water level

The equations describing the model can be solved for a sinusoidal input (tidal wave), using harmonic analysis and for a step input (sudden storm surge), using Laplace transform. The solutions are presented in Table 1, which also gives the solution for the stationary flow for comparison.

For the geo-hydrological conditions prevailing in the Lower Meuse-Rhine delta, the value of ω/c is very low compared to the other ones (Barends, 1982). Assuming ω/c tending to zero (incompressible sand layer), results in the formulas on Table 2. For B reaching an infinite value, the response due to a sinusoidal or step variation of the water level in a sea or a lake can be found.

Analysis of the response on a sinusoidal variation of the water level

As can be seen from equations 2 and 5, the response in the aquifer shows damping and phase shift, compared to the tidal variation of the water level. Both damping and phase shift are due to the consolidation process in the bottom layer, storage in the aquifer and consolidation in the aquitard.

The contribution of the bottom layer to the damping and phase shift, is independent of the distance to the river or the sea, but only depends on consolidation parameters and width and thickness values. The effect of the bottom layer on the phase disappears when the river becomes very wide (sea or lake), but its damping effect remains. It is interesting to observe that for a distance between the river at x = O to x = β (for the value of β see eq. 2 or eq. 5) the pore pressure in the aquifer preceeds the variations of the water level in the case of a river and reacts without time lag in the case of a sea or lake. The response first mentioned has been measured in the Lower Meuse-Rhine delta by De Lange et al. (1986).

The response of an aquifer-aquitard system without bottomlayer can easily be found from the mentioned formulas for f, g and β or λ''_ω equal to 0.

Table 1. Potential head in the aquifer (stationary flow, tide and sudden surge). All potential heads are referred to the phreatic level, which has a potential head equal to zero).

variation of water level	response in aquifer as function of distance and time
stationary flow (eq. 1)	$\varphi(x) = \dfrac{He^{-\frac{x}{\lambda'}}}{1 + \frac{\lambda''}{\lambda'} \coth(\frac{B}{\lambda''})}$ $\quad x>0$ $\lambda' = (\frac{kDd'}{k'})^{0.5} \qquad \lambda'' = (\frac{kDd''}{k''})^{0.5}$
tide (eq. 2)	$\varphi(x,t) = \dfrac{H \cos(\omega t - x \sqrt{r'} \sin(\psi'/2 - \beta)) \exp(- x \sqrt{r'} \cos(\psi'/2))}{[(1+f)^2 + g^2]^{0.5}}$ $\quad x>0$ $r' = [(\frac{k'}{kD} \sqrt{(\frac{\omega}{2c'})})^2 + (\frac{\omega}{c} + \frac{k'}{kD} \sqrt{(\frac{\omega}{2c'})})^2]^{0.5}$ $r'' = [(\frac{k''}{kD} \sqrt{(\frac{\omega}{2c''})})^2 + (\frac{\omega}{c} + \frac{k''}{kD} \sqrt{(\frac{\omega}{2c''})})^2]^{0.5}$ $\psi' = \arctan\left(\dfrac{\frac{w}{c} + \frac{k'}{kD}\sqrt{(\frac{\omega}{2c'})}}{\frac{k'}{kD}\sqrt{(\frac{\omega}{2c'})}}\right)$ $\psi'' = \arctan\left(\dfrac{\frac{w}{c} + \frac{k''}{kD}\sqrt{(\frac{\omega}{2c''})}}{\frac{k''}{kD}\sqrt{(\frac{\omega}{2c''})}}\right)$ $\beta = \arctan(\frac{g}{1+f})$ $f = (\frac{r'}{r''})^{0.5} \dfrac{\cos(\frac{\psi'-\psi''}{2}) \sinh \varrho + \sin(\frac{\psi'-\psi''}{2}) \cos \tau}{\cosh \varrho - \cos \tau}$ $g = (\frac{r'}{r''})^{0.5} \dfrac{\sin(\frac{\psi'-\psi''}{2}) \sinh \varrho - \cos(\frac{\psi'-\psi''}{2}) \sin \tau}{\cosh \varrho - \cos \tau}$ $\varrho = 2B \sqrt{r''} \cos(\psi''/2) \qquad \tau = 2B \sqrt{r''} \sin(\psi''/2)$
sudden surge (eq. 3)	$\varphi(x,t) = \dfrac{H \exp(- x/\lambda_t')}{1 + \frac{\lambda_t''}{\lambda_t'} \coth(\frac{B}{\lambda_t''})}$ $\quad x>0$ $\lambda_t' = [\frac{1}{2ct} + \frac{k'}{kD} \sqrt{(\frac{1}{2c't})} \coth(\frac{d'}{\sqrt{2c't}})]^{-0.5} \qquad \lambda_t'' = [\frac{1}{2ct} + \frac{k''}{kD} \sqrt{(\frac{1}{2c''t})} \coth(\frac{d''}{\sqrt{2c''t}})]^{-0.5}$

<u>Special case:</u> no bottomlayer: $\lambda'' = 0$
$f = g = \beta = 0$
$\lambda_t'' = 0$

Analysis of the response, due to a step variation of the water level

The response due to a step input (see equations 3 and 6), is illustrated in Fig. 3. One can see that the response becomes less with an increasing distance between the considered location and the river (or the sea). The gradient of the time-response curve is also reduced. The response is weaker when a bottomlayer is present because this lowers the discharge to the aquifer. In the case of an incompressible sand skeleton (c tending to an infinite value) the time dependent leakage factors become:

$$\lambda_t' = [\frac{k'}{kD} \sqrt{\frac{1}{2c't}} \cdot \coth(\frac{d'}{2c't})]^{-0,5}$$

$$\lambda_t'' = [\frac{k''}{kD} \sqrt{\frac{1}{2c''t}} \cdot \coth(\frac{d''}{2c''t})]^{-0,5} \qquad \text{(eq. 7)}$$

When moreover the time t is very short compared to the hydrodynamic periods of the aquitard and

Table 2. Potential head in the aquifer (no elastic storage in the Pleistocene sand; all potential heads are referred to the phreatic level, which has $\varphi_{ref} = 0$).

variation of water level	response in aquifer as function of time and distance (no storage in sand)
stationary flow (eq. 4)	$\varphi(x) = \frac{H \exp(-x/\lambda')}{1 + \frac{\lambda''}{\lambda'} \coth(\frac{B}{\lambda''})}$
tide (eq. 5)	$\varphi(x,t) = \frac{H \cos(\omega t - x \tan(\pi/8)/\lambda_\omega - \beta) \exp(-x/\lambda_\omega')}{[(1+m)^2 + n^2]^{0,5}}$
	$m = \frac{\lambda_\omega''}{\lambda_\omega'} \frac{\sinh(2B/\lambda_\omega''}{\cosh(2B/\lambda_\omega'' - \cos(2B \tan(\frac{\pi}{8})/\lambda_\omega'')})$
	$n = \frac{\lambda_\omega''}{\lambda_\omega'} \frac{\sin(2B \tan(\pi/8)/\lambda_\omega''}{\cosh(2B/\lambda_\omega'' - \cos(2B \tan(\frac{\pi}{8})/\lambda_\omega'')})$
	$\beta = \arctan(\frac{n}{1+m})$
	$\lambda_\omega' = \sqrt{(\frac{kD}{k'})} \sqrt[4]{(\frac{c'}{\omega})} \frac{1}{\cos(\pi/8)}$
	$\lambda_\omega'' = \sqrt{(\frac{kD}{k''})} \sqrt[4]{(\frac{c''}{\omega})} \frac{1}{\cos(\pi/8)}$
sudden surge (eq. 6)	$\varphi(x,t) = \frac{H \exp(-x/\lambda_t')}{1 + \frac{\lambda_t''}{\lambda_t'} \coth(\frac{B}{\lambda_t''})}$
	$\lambda_t' = [\frac{k'}{kD} \sqrt{(\frac{1}{2c't})} \coth(\frac{d'}{\sqrt{(2c't)}})]^{-0,5}$
	$\lambda_t'' = [\frac{k''}{kD} \sqrt{(\frac{1}{2c''t})} \coth(\frac{d''}{\sqrt{(2c''t)}})]^{-0,5}$
Special case: no bottomlayer:	$\lambda'' = 0$, $\lambda_\omega'' = 0$, $\lambda_t'' = 0$

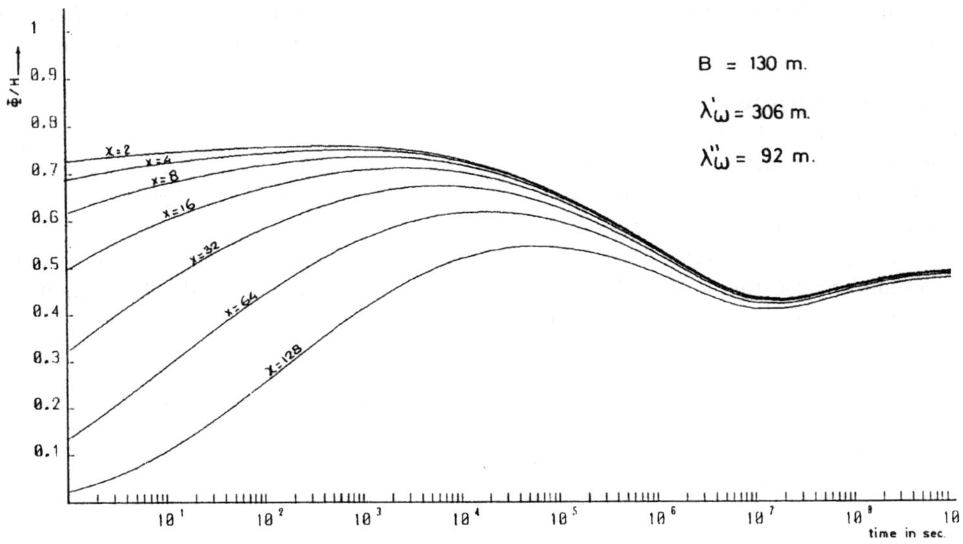

Fig. 3. Water pressure response on a step variation of the water level at different x-values.

the bottomlayer (equal to $\frac{2d'^2}{c'}$ and $\frac{2d''^2}{c''}$), the leakage factors for step variations of the water levels can be well approximated by:

$$\lambda'_t \approx \lambda'_\omega \sqrt[4]{4\omega t} \cdot \frac{1}{1{,}287}$$

$$\lambda''_t \approx \lambda''_\omega \sqrt[4]{4\omega t} \cdot \frac{1}{1{,}287} \qquad \text{(eq. 8)}$$

in which λ'_ω and λ''_ω are the leakage factors for tidal flow.

Response due to an arbitrary variation of the water level

Closed analytical solutions with an acceptable accuracy for the response of the water pressure, due to variations of the water level other than the ones presented above are difficult to derive. However, since the soil parameters (permeability coefficients, consolidation coefficients) are assumed to be constant and the conditions at the boundaries between the layers and the boundary conditions do not vary, the superposition principle is applicable. Using this, one can evaluate the response due to an arbitrary variation of the water level by subdividing it into a series of positive and negative step functions shifted in the time. The superposition of the results of each step function forms the solution. The procedure is overlined in Fig. 4.

Prediction of the pore water pressure in the pleistocene sand substratum during design storm conditions

The objectives of the geo-hydrological model presented, are to provide a reliable way of predicting the water pressure in the soil layers during a design storm and to provide a method to estimate the necessary soil parameters, using measurements of the water pressure response under normal variations of the water level in the river or sea. In the Lower Meuse-Rhine Delta, the maximum water level for design conditions can be subdivided into three components: A stationary one, a tidal one (due to storm tide in the North Sea) and one with a more or less block shape (due to high river discharge). The duration of this combined maximal water level is generally short compared with the

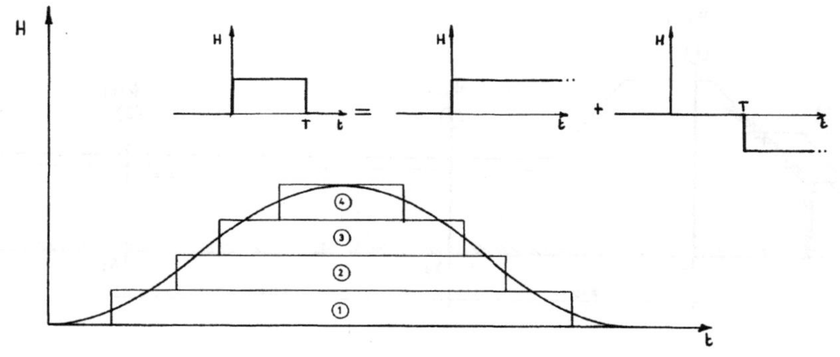

Fig. 4. Superposition principle to evaluate the response, due to an arbritary variation of the water level.

hydrodynamic periods of the bottom layer and the aquitard.

During normal conditions, the water level is composed of a stationary (mean water level) and a tidal component. Based on this, a useful procedure can be outlined (see Fig. 5). Piezometers are installed into the aquifer in at least two, but preferably three locations with sufficient distance from one to the other.

The analysis of the measured response and the prediction of the response under design flood can be performed using the formulas given in Table 2 as the consolidation of the aquitard and the bottom layer are dominating the time-lag observed in the transient response of leaky aquifers (Barends, 1982). Using the damping of the response of the piezometers one to the other (e.g. response of piezometer 2 on piezometer 1 etc.) and equation 5, one can find the leakage factor of the aquitard for tidal variations of the water level λ'_ω:

$$\lambda'_\omega = \frac{x_{ij}}{\ln \frac{\varphi_i}{\varphi_j}} \quad \text{(eq. 9a) or}$$

$$\lambda'_\omega = \frac{\sum x_{ij}^2}{\sum x_{ij} \ln \frac{\varphi_i}{\varphi_j}} \quad \text{(eq. 9b)}$$

in which: x_{ij} is the distance between piezometers i and j and
φ_i, φ_j is the measured piezometric level

Equation 9b follows from the least squares method.

Based on λ'_ω and the response of the piezometers on the tidal variation of the water level, one can find λ''_ω by solving equation 5 (or the equivalent one in the case of a sea). The using λ'_ω and λ''_ω, one can calculate λ'_t and λ''_t at every moment, using equation 8, provided that the duration of the surge is short, compared to the hydrodynamic period. The leakage factors λ' and λ'' for stationary flow, follow in the same way from the measured mean water and piezometric levels. Care must be taken with large scale seasonal effects which can affect the mean piezometric level. Subsequently, with the geo-hydrological parameters found as described above, one can predict the water pressures in the aquifer for the design storm, i.e. for the expected ultimate water level and its duration, using the formulas corresponding to the chosen geo-hydrological model.

Uplift of the aquitard

A basic assumption underlying all above mentioned formulas, is that the water pressure in the aquifer can increase without being limited by the weight of the overlying aquitard. When the water pressure somewhere in the sand layer indeed tends to exceed the vertical pressure exerted on the boundary by the weight of the flood bank and the foundation layers no further development of the

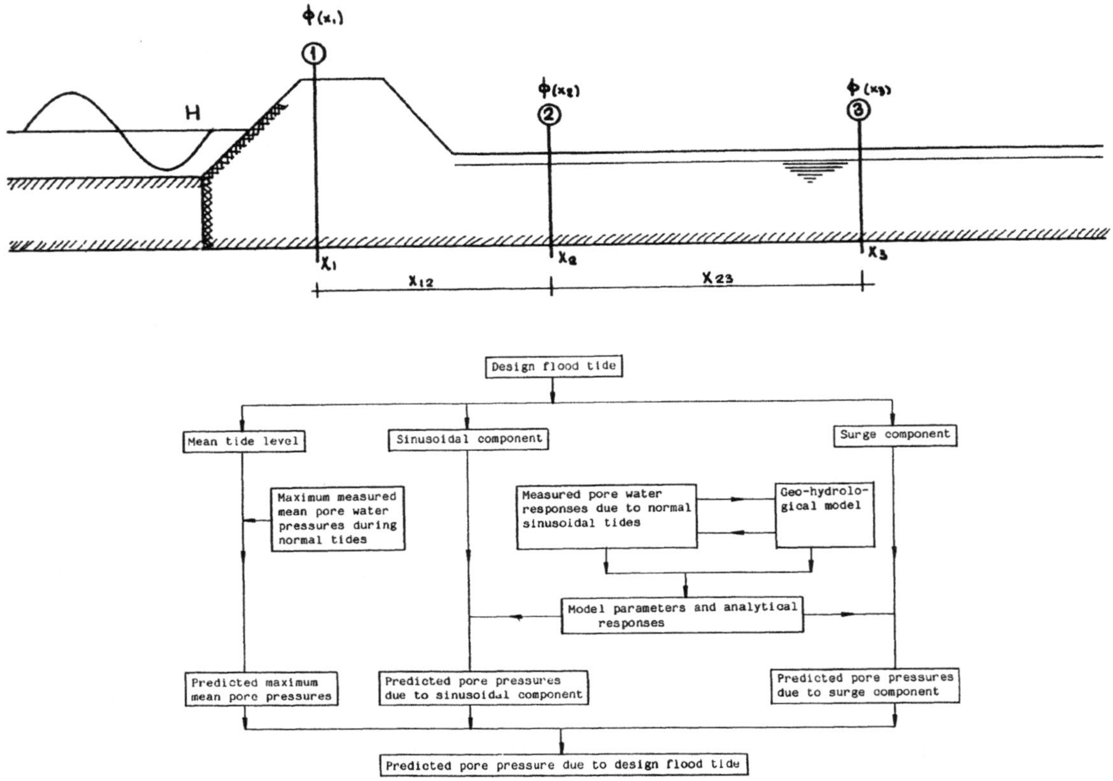

Fig. 5. Outline of the method for prediction of maximum pore pressure, likely to occur during design flood conditions. a. Location of the piezometers. b. Calculation flowchart (after Marsland & Randolph, 1978).

water pressure in this area is possible. In such a case, the formulas are no longer valid because a pressure redistribution will occur. As a result, the water pressure in the aquitard will nowhere exceed the local total vertical stress at the separation between the aquifer and the aquitard.

The area in which the water pressure equals the total vertical stress, the Holocene layers of clay and peat will be uplifted. The significance of uplift on the stability and deformations of levees has been outlined by Marsland (1961), Marsland & Randolph (1978), Padfield & Schofield (1983) and Teunissen et al. (1986): The development of a large uplift aera near the toe of a levee can lead to large deformations or even collapse of the embankment because uplift reduces the shear forces transmitted to the stiff sand stratum (see Fig. 6).

The influence of an uplifted area on the deformations and on the stability of a levee, depends on its location and magnitude, its impact on the existing force equilibrium and the stiffness of the soil near the toe of the levee. The equations mentioned in Tables 1 and 2 give the variations of the piezometric head in the aquitard in response to the variation of the piezometric waterhead of outside the flood embankment. These variations of piezometric heads are independent of the chosen referential piezometric head. As *pressures* have to be compared when considering uplift phenomena, a suitable choice of the referential piezometric head is needed in order to convert the piezometric heads into waterpressures which can directly be compared to the pressure exerted by the Holocene layers on the sand stratum. Taking the reference potential head

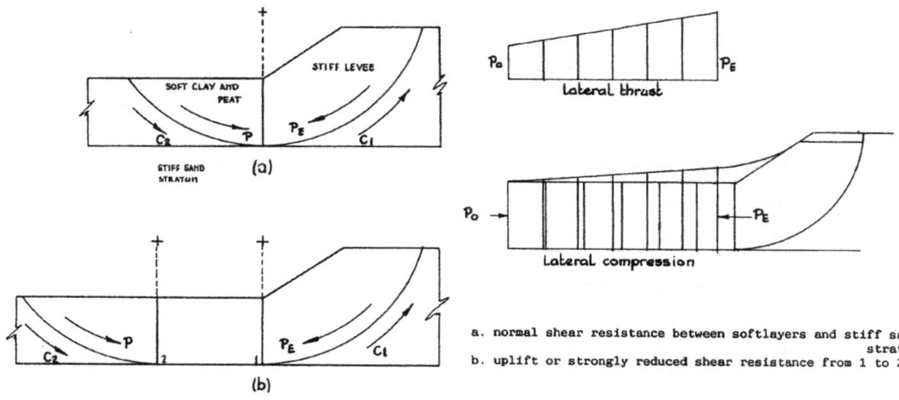

Fig. 6. Dike failure, caused by excessive horizontal deformation of the passive area under uplift conditions (after Padfield & Schofield, 1983).

equal to zero on the phreatic level in the polder, makes direct comparison between the soil pressures and the waterpressures possible: the stationary potential head given in the equation 1 and 4 with $\varphi_{ref} = O$, is also the expression of the stationary waterpressure at the bottom of the Holocene layers if the waterlevel H_1 is expressed in metres above the phreatic level and multiplied by the unit weight of the pore water, γ_w.

As equations 2, 3, 5 and 6 are variations of the potential head, they are also variations of the waterpressure when multiplied by γ_w. Hence, the waterpressure becomes the sum of the stationary pressure referred to above and the time dependent pressure variation. This choice of the reference potential can be made because the phreatic level is considered as being constant during the phenomena which are analyzed. In case of important phreatic storage in the Holocene layers, this would not be true. For typical Dutch dike cross sections in the lower Meuse-Rhine delta, the beginning of the uplifted area (at x = P, see Fig. 7a) can approximately be taken at the x-value where $\gamma_w.\varphi(x, t)$ equals the total vertical stress $\sigma_v(x)$ for the lowest t-value (see the line corresponding with t2 in Fig. 7a).φ (x,t) is calculated with the formulas on Table 1 or 2. The total vertical stress includes the weight of the clay and peat layers and the weight of the flood bank.

The moment the water pressure in the aquifer somewhere equals the vertical stress, the water pressure will no longer increase in that place, but (locally) will remain equal, see Fig. 7b. The uplift area will develop as a function of time, starting from the value of x = P according to the formula given in Table 3. This table also gives the waterpressure in the aquifer from the moment of the beginning of the uplift conditions. The water pressure distribution during the uplift conditions as a function of the distance is given in Table 3 and illustrated by the line corresponding with t3 in Fig. 7a. It is emphasized that the formulas given in Table 3 are approximations. Their accuracy has not been verified yet, because up till now no correct mathematical solution has been found, not analytical, nor numerical. Further research is needed to verify the accuracy of the proposed solutions.

It is interesting to observe that all the geo-hydrological parameters needed to evaluate the magnitude of the uplift area can be obtained from measurements made over normal tides as explained above. The pressure redistribution in uplift conditions however, shows that the determination of λ'_ω and λ''_ω is not possible if uplift conditions occurred during the measurement of the water pressure response due to the (normal) tidal variation of the water level. The superposition principle as explained earlier is no longer more valid when uplift occurs as uplift is a typical non-linearity in the conditions at the separations between the aquifer and the aquitard.

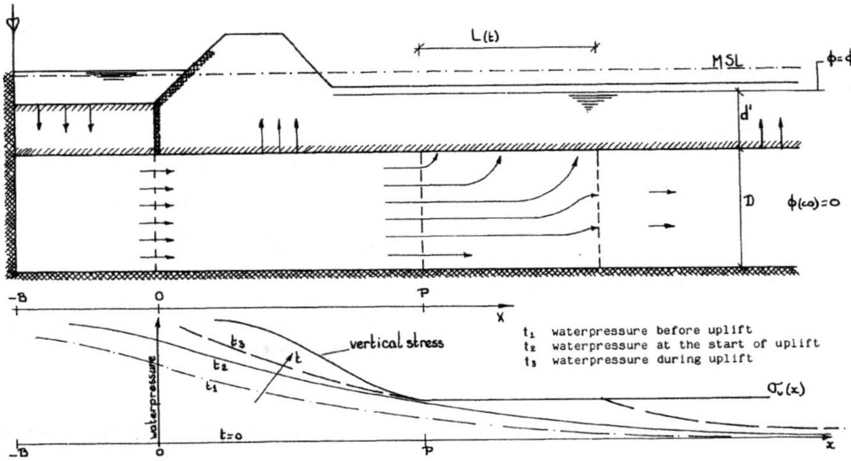

Fig. 7a. Water pressures distribution before and during uplift.

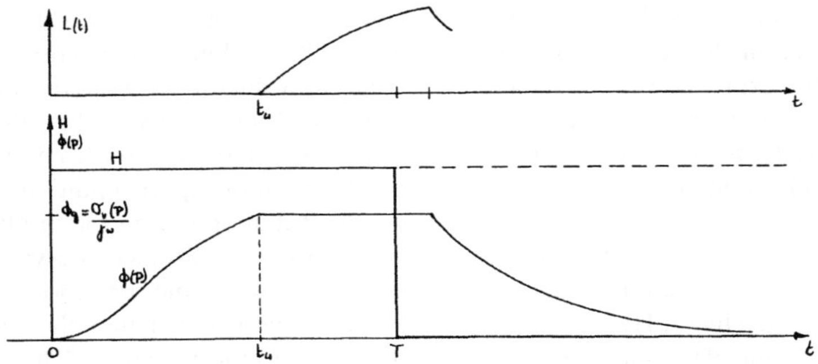

Fig. 7b. Water pressure development and length of the uplift area as a function of time.

Case study

The method outlined above is applied to the measurement of the tidal responses in three piezometers located at 16 m, 60 m and 115 m from the rivershore (Fig. 9).

The actual river width is about 260 m. Using the response of one piezometer to the other (eq. 9), one finds $\lambda'_\omega = 306$ m. Then, by trial and error, one finds the corresponding value of $\lambda''_\omega = 90$ m. The calculated and the measured values of the damping are shown in the Table in Fig. 8. They are fitting very well. These values where used to plot Figs. 2 and 3 (response due to tidal variation and a sudden surge) and Fig. 8 (length of the uplift area). Knowing the total vertical stress acting on the aquifer and the design water level, one can determine the water pressure in the aquitard and determine whether uplift is likely to occur during the design storm. One can calculate the extent of the uplift area using the equations given in Table 3.

piezometer no.	distance from river m.	measured damping -	calculated damping -
1	16	0,727	0,729
2	60	0,636	0,634
3	115	0,500	0,501

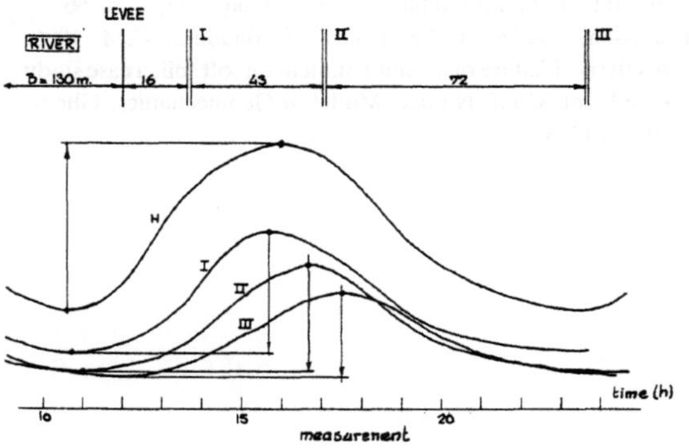

Fig. 8. Case study: Measured and calculated damping ($\lambda_\omega' = 306$ m; $\lambda_\omega'' = 90$ m)

Conclusion

A method is presented here, to assess λ_ω' and λ_ω'', the geohydrological parameters of an aquifer-aquitard system with a bottomlayer between the river and the aquifer. This is done with the aid of the measurement of the response of the water pressure in the aquifer, due to normal tidal variations of the river level. The parameters can be directly used to predict the water pressure during design storm conditions. If the water pressure exceeds the pressure due to the weight of the layers of the aquitard, uplift will occur. The same geohydrological parameters can be used to calculate approximately the length of the uplifted zone and the new distribution of the water pressures.

Table 3. Approximate length of the uplifted aera and potential head in the aquifer during uplift conditions reference potential head is phreatic level, which has $\varphi_{ref} = 0$; incompressible aquifer).

$$L(t) = \frac{2D}{\pi} \text{Arc cosh} \left\{ \frac{1}{\sin \left[\frac{\pi}{2} \frac{\varphi_g \sinh (P/\lambda_t')}{\varphi (x=0) - \varphi_g \cosh (P/\lambda_t')} \right.} \right\} \quad \text{(eq. 10)}$$

$$\varphi(x=o,t) = \frac{H + \frac{\lambda_t'' \coth (B/\lambda_t'')}{\lambda_t' \sinh (P/\lambda_t')} \varphi_g}{1 + \frac{\lambda_t''}{\lambda_t'} \coth (B/\lambda_t'') \coth (P/\lambda_t')}$$

$$\lambda_t' = \left[\frac{k'}{kD} \sqrt{\left(\frac{1}{2c't}\right)} \coth \left(\frac{d'}{\sqrt{2c't}}\right)\right]^{-0,5}$$

$$\lambda_t'' = \left[\frac{k''}{kD} \sqrt{\left(\frac{1}{2c''t}\right)} \coth \left(\frac{d''}{\sqrt{2c''t}}\right)\right]^{-0,5}$$

$0<x<P$

$$\varphi(x,t) = \varphi(x=o,t) = \frac{\sinh\ (P-x)/\lambda_t'}{\sinh\ (P/\lambda_t')} + \varphi_g \frac{\sinh (x/\lambda_t')}{\sinh (P/\lambda_t')}$$

$P<x<P+L$

$$\varphi(x,t) = \varphi_g$$

$P+L<x$

$$\varphi(x,t) = \varphi_g \exp(-(x-(P+L))/\lambda_t')$$

Special case: no bottomlayer: $\lambda_t'' = 0$

References

Barends, F.B.J. 1982. Transient flow in leaky aquifers – Proc. Int. Conf. Modern Approach to Groundwater Resources Management, Capri, 1982: 152–161

De Lange, W.J. & Maas, C. 1986. Over het voorlopen van het grondwatergetij op de getijdebeweging in de Hollandsche IJssel nabij Gouderak - H_20, 2: 24–29

Jacob, C.E. 1940. The flow of water in an elastic artesian aquifer – Trans. Am. Geophys. Union, 21: 574–586

Marsland, A. 1961. A study of a breach in an earthen embankment caused by uplift pressures – Proc 5th Int. Conf. Soil Mech. and Found. Eng.: 663–668

Marsland, A. & Randolph, M.J. 1978. A study of the variation and effects of water pressures in the previous strata underlying Crayford Marshes – Geotechnique 28, 4: 435–464

Padfield, C.J. & Schofield, A.N. 1983. The development of centrifugal models to study the influence of uplift pressures on the stability of a flood bank – Geotechnique 33, 1: 56–66

Teunissen, J.A.M., Calle, E.O.F. & Bauduin, C.M. 1986. Analysis of failure of an embankment on soft soil, a case study – 2nd. Int. Conf. Numer. Model in Geomechanics, Ghent, 1986: 617–628

Proceedings KNGMG Symposium 'Coastal Lowlands, Geology and Geotechnology', 1987: 143–159 (1989)

The development of two major Indonesian river deltas: morphology and sedimentary aspects of the Solo and Porong delta, East Java

P. Hoekstra
Department of Physical Geography, University of Utrecht, P.O. Box 80.115, 3508 TC Utrecht, The Netherlands

Received 25 August 1987; accepted in revised form 15 March 1988

Abstract

In the coastal zone of NE Java, two important rivers debouch into the sea, less than 100 km from each other. The discharge and transport of appreciable amounts of sediment is, due to the monsoonal climate, mainly restricted to the wet season. The input of sediment by both the Solo and Porong rivers has resulted in the rapid outgrowth of two extensive, but morphologically quite different delta systems. The Solo delta is characterized as a highly mud-dominated and rapidly prograding single finger delta, a specimen of a high-constructive, elongate delta type. In spite of several bifurcations in the past, the present Solo delta is made up of one straight, major channel. Lateral migration of the channel and the formation of crevasses is largely counteracted by the huge amount of cohesive and consolidated silts and clays along the channel, partly forming pronounced levees. The subaqueous topography of the delta front displays a number of small-scale bottom features related to post-depositional processes of sediment transport.

The Porong delta is characterized as a lobate multidistributary delta. The input of sand by the Porong river has resulted in the development of a triangular mouth bar complex, with a number of braided, subtidal channels. Both deltas are still subject to further growth and development, though the growth rates for both deltas differ substantially.

Introduction

The Indonesian island of Java (Fig. 1a) is one of the most densely populated regions in the world. The geological backbone of the island is formed by a number of large volcanoes and their fertile, volcanic products are the basis for intensive agriculture. However, under the influence of the humid tropical climate, the relief and the presence of easily erodible soils, soil erosion has become a major hazard (Ongkosongo, 1982). The denudation rates in Java are extremely high, in some watersheds even exceeding those of the Yellow River (Carson & Utomo, 1987). Every wet season, the eroded particles are washed away by local rivers and large volumes of sediment are carried towards the sea. Along the northern coastal plain of the island of Java, a number of important 'monsoonal' rivers

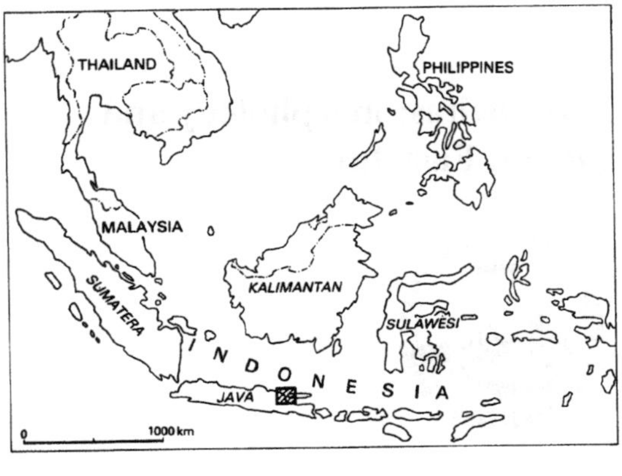

Fig. 1a. The Indonesian Archipelago, part of SE Asia, with 13,000 islands and a total coastline of about 81,000 km. The island of Java is located in Western Indonesia.

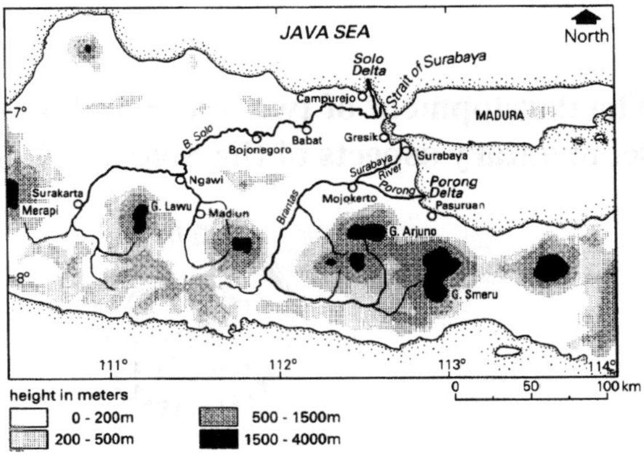

Fig. 1b. Map of Central and East Java and the island of Madura (Western Indonesia), illustrating the topographic setting of the drainage basins of the Solo and Porong rivers. A series of large volcanoes is located along the axis of the island of Java.

debouch into the Java Sea. The input of sediment by these rivers generally exceeds the transport capacity of coastal currents and waves. The surplus of sediment results in a continuously prograding shoreline and the formation and outgrowth of a series of major river deltas (Hollerwöger, 1964; Tjia et al., 1968). New deltaic regions are used for settlements, agriculture and the construction of fish ponds. The deltaic ecosystem is characterized by the presence of delicate natural balances which require a sound coastal management (Bird & Soegiarto, 1979; Coleman & Roberts, 1988, this volume). The deltaic shoreline is a result of a dynamic equilibrium between the input and transfer of sediments. This balance is easily disrupted when the river-network within the delta becomes subject to changes. The natural coastal protection may locally be damaged, thereby causing severe flooding of the low-lying deltaic areas.

Further knowledge with regard to delta-growth and dynamics should contribute to delta-management. Recent studies of delta-development and dynamics have been presented by Coleman & Wright (1975), Allen et al. (1979), Coleman (1982), Coleman & Prior (1982b), Thom & Wright (1983), Wang (1984), Wright (1985), Wright et al. (1986), Terwindt et al. (1987) and Hoekstra et al. (in press).

River outflow, sediment transport, depositional processes and facies and the morphological development of the Solo and Porong deltas (Fig. 1a and b) have been the subject of research during the Indonesian-Dutch Snellius-II oceanographical expedition (LIPI-NRZ, 1984; Hoekstra & Tiktanata, in press). In this paper changes in coastal morphology in the recent past and present are studied, both qualitatively and quantitatively, and explained against the background of the present-day river regime and sediment transport processes.

The island of Java is part of the humid tropics and the climate is characterized by the alternation of two monsoons. The wet, W-NW monsoon prevails from December until March and the dry E-SE monsoon from May to September. Average annual precipitation for East-Java is about 2,100 mm · year^{-1} with minimum and maximum values of 1,500 and 3,000 mm · year^{-1} respectively (data Proyek Bengawan Solo, PBS).

Measurements, observations and sampling were carried out in July-September 1984 of the dry season and in December 1984-February 1985, during the wet season (Hoekstra & Tiktanata, in press).

River basin, river regime and sediment transport

The headwaters of the Solo river drain the flanks of the volcanoes Gunung Lawu and Merapi, in Central Java (Fig. 1b). The sources of the Brantas river are located on East-Java, in the volcanic complexes of the G. Smeru and G. Arjuno. Some general characteristics of the two riversystems and their respective deltas are presented in Table 1. Both the Solo and Porong river are 'monsoonal' rivers and the major part of river discharge takes place during the four or five months of the wet season (Fig. 2, Solo river, data 1984–1985).

Solo

The mean annual flood ($Q_{2.33}$, data PBS) for the river Solo is about 1,350 $m^3 \cdot s^{-1}$ (Location Babat; Fig. 1b). The discharge regime of the Solo is especially characterized by a seasonal flashy discharge (Fig. 2, Table 1 and 2). During periods of high discharge or banjirs, as in February 1984 and 1985, river discharges increase to more than 2,500 $m^3 \cdot s^{-1}$ and may even reach values up to 4,000 $m^3 \cdot s^{-1}$ (Fig. 2). The frequency of banjirs is given in Table 2. The minimum and maximum daily flows during the field campaign in the period of December 1984-January 1985 (wet season) were respectively

Table 1. Characteristic data of river basin, river regime and delta-environment of the Solo and Porong delta.

Data	Solo	Brantas/Porong
Drainage basin		
area (km^2)	16,000	12,000
channel length (km)	550	300
River Slope		
alluvial valley ($m \cdot m^{-1}$)	1.6–4.0 × 10^{-4}	larger 4.1 × 10^{-4}
delta environment ($m \cdot m^{-1}$)	less 0.5 × 10^{-4}	3.0 × 10^{-4}
Discharge[1]		
location (Figure 1)	Babat	Porong
mean annual flood ($m^3 \cdot s^{-1}$)	1,350	± 600
minimum discharge ($m^3 \cdot s^{-1}$)	<10	<1.0
maximum discharge ($m^3 \cdot s^{-1}$)	2,500–4,000	1,250
Sediment transport		
mechanism	suspension load	suspension load and bedload
concentrations dry season (ppm)	25–40	0–50
concentrations wet season (ppm)	1,700–3,200	1,200
total sediment transport ($tonne \cdot y^{-1}$)	19 × 10^6	19.7 × 10^6*
Delta dimensions		
maximum length (km)	12	3.2
area (km^2)	33.6	47.7
volume (km^3)	1.7	0.8
Tidal regime		
character	Diurnal	mixed diurnal-semi diurnal
tidal range (m)	0.9–2.1	0.5–2.7
Waves		
character	seawaves and swell	seawaves
wave approach from:		
dry season	NE	E.
wet season	NW	–
wave height (m)	less 1.5 (90%)	–
	2–2.5 (10%)	–

* based on deltagrowth.

[1] Data Proyek Bengawan Solo, PBS, Surakarta.

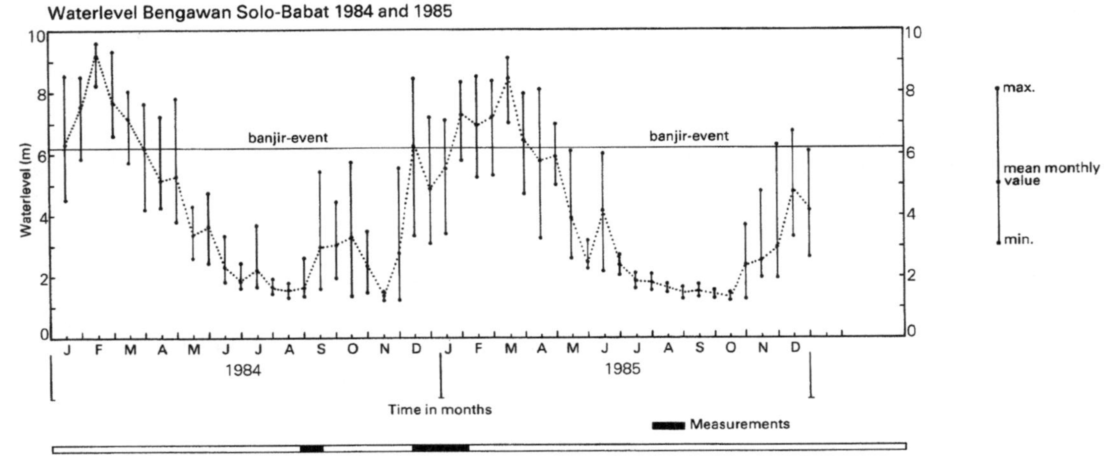

Fig. 2. Discharge distribution of the Solo river as indicated by the mean, half-monthly water levels for the years 1984–1985. Minimum and maximum values are also included. (Data, PBS, Surakarta and Institute for Hydraulic Engineering, Bandung).

$310\,m^3 \cdot s^{-1}$ and $1{,}800\,m^3 \cdot s^{-1}$ with average monthly-values of about 800–$1{,}000\,m^3 \cdot s^{-1}$. In contrast to the wet season, dry season river discharge in the Solo reduces drastically and in August and September 1984 within the river delta, the average daily freshwater discharge was approximately 20–$30\,m^3 \cdot s^{-1}$. Average monthly flow at Babat varied from 80–$250\,m^3 \cdot s^{-1}$.

Porong

Flow in the river Porong is almost entirely controlled by man. In the wet season the Porong has a flood-relief function and up to 80% of the water, supplied by the Brantas river, flows towards the Porong delta. Average wet season flow is about $600\,m^3 \cdot s^{-1}$. In extremely wet years (1984) the discharge increases to about $1200\,m^3 \cdot s^{-1}$. In the dry season river flow is diverted to the town of Surabaya, and flow in the Kali Porong is often reduced to almost zero. In 1972 mean monthly flow in the period June-October was less than $0.6\,m^3 \cdot s^{-1}$ (MacDonald and Partners, 1977). A comparable situation was observed in the dry season of 1984.

Table 2. Frequency of banjirs for a random selection of years. For the Solo river, a banjir is defined as a discharge-event, during which the average daily discharge exceeds values of $1{,}000\,m^3 \cdot s^{-1}$.

Period-year	total of recorded days	banjir days	frequency
1954	306	11	0.04
1955	365	14	0.04
1956	366	15	0.04
1959	365	47	0.13
1960	366	32	0.09
1974	334	57	0.17
1975	332	95 (max)	0.29
1984	366	92	0.25
1985	365	79	0.21

Denudation and Sediment transport

In the upper Solo basin annual denudation rates are high. Estimations and calculations of erosion differ tremendously and there is no consensus about the (average) annual denudation. Carson & Utomo (1987) suggest denudation rates varying from $3{,}000$ tonne $\cdot\, km^{-2} \cdot y^{-1}$ for gently sloping areas (<5%) to $38{,}000$ tonne $\cdot\, km^{-2} \cdot y^{-1}$ for steep mountain slopes (>50%). In the period of November 1983-April 1984 recorded soil losses at Gondek (Solo basin, Central Java) were about 10,000 tonne $\cdot\, km^{-2}$ (Carson & Utomo, 1987).

However, these are extreme values, since the calculations and measurements, made by Proyek Bengawan Solo (PBS) generally result in much smaller estimations with average values of about 2,100 tonne · km^{-2} · y^{-1}. Estimations of denudation rates for the entire drainage basin may give an indication of the total supply of sediment to the river, as well as the potential sediment transport by the river.

Average annual soil losses are estimated to be about 29 – 37 × 10^6 tonne · y^{-1} (Table 3). Accordingly, the amount of sediment transported by the river Solo may vary substantially, e.g. as a result of different degrees of interception of sediment by dams and reservoirs: 6 – 22 × 10^6 tonne · y^{-1} (Table 3).

The lower part of the Solo is essentially a suspended-load river. Suspended sediment concentrations in the dry season are about 25–40 ppm. In the wet season concentrations range between 1,000–6,000 ppm. in the upper-Solo basin while in major mountain tributaries, concentrations of 35,000 ppm. have been measured (Carson & Utomo, 1987). The measured average concentrations in the river delta (December 1984-January 1985) varied from 1,700 to 3,200 ppm. and almost 95% of the suspended matter appeared to have a grain size smaller than 50 micron.

Relations between river discharge and the suspended sediment load are often a more useful tool to calculate average, annual sediment transport rates. The average monthly sediment transport S (in tonne · $month^{-1}$) as a linear function of the mean-monthly total flow Q (in m^3 · $month^{-1}$) at a location close to Bojonegoro (Fig. 1b) is expressed by (Fig. 3):

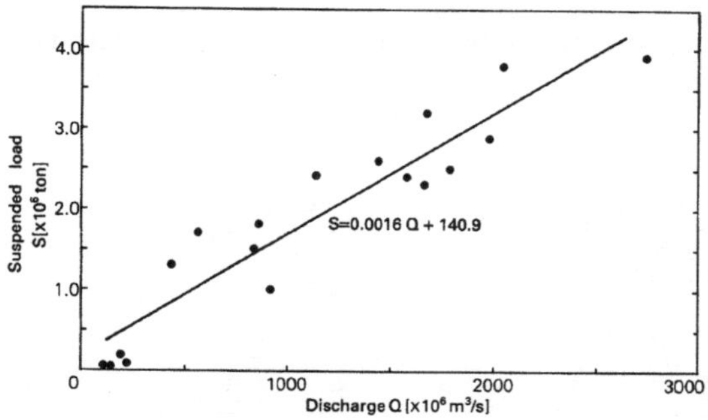

Fig. 3. Relation between total monthly flows and suspended sediment load for the Solo river near Bojonegoro (based on data PBS and MONENCO, Surakarta).

$$S = 0.0016\ Q + 140.961$$

Using the long-term averages of mean monthly flows for Bojonegoro (Hoekstra et al., in press) for an entire year, the total annual, suspended sediment transport by the river Solo is calculated to be approximately 19 × 10^6 tonne · y^{-1}. This figure is comparable to the maximum transport, deduced from denudation rates and interception (Table 3).

Neither denudation rates nor accurate measure-

Table 3. The estimated rates of denudation of the Solo river basin and the potential sediment transport by the Solo river. (Data partly based on OCTA, 1973; MONENCO, 1984 and Yunianto, 1982).

Drainage area	Area (km^2)	Denudation rate (tonne · km^{-2} · y^{-1})	Total soil loss (10^6 tonne · y^{-1})
upper Solo Basin	6,070	2,050–3,000	12.4–18.2
Madiun Basin	3,750	2,000	7.5
Lower Solo Basin	6,200	1,500–1,800	9.3–11.2

Interception of sediment by lakes and reservoirs

Percentage of interception	Potential Sed. transport Solo (10^6 tonne · y^{-1})
80%	5.8– 7.4
70%	8.8–11.1
60%	11.7–14.8
50%	14.6–18.5
40%	17.5–22.1

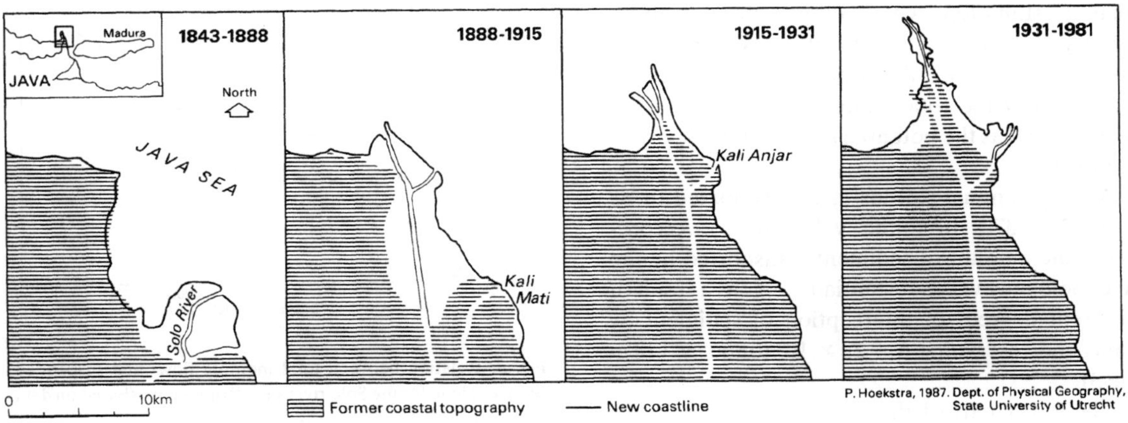

Fig. 4. A reconstruction of the coastal development of NE Java and the Solo Delta since 1843.

ments of suspensian load are available for the river Porong, but suspended sediment concentrations of 50 ppm in the dry season to 1,200–1,300 ppm in the wet season have been measured. The magnitude of bedload transport is almost unknown. In addition, the influence of a series of dams, sluices and reservoirs near Mojokerto (Fig. 1b) has not been measured or estimated recently.

Coastal Developments in the recent past

The most recent topographical maps of the Java coastal regions originated from 1935 and 1942 but Landsat images of 1972 and 1985 indicate that the coastline of NE Java has been subject to drastic changes. Surveying and charting of the present coastline and coastal waters was considered to be a very crucial part of the program to enable navigation and positioning during the field campaign in 1984 and 1985. A study of the present-day coastal morphology and a reconstruction of the former coastal topography was also essential to be able to make predictions about future growth and development. Furthermore, mass-balance or sediment-budget studies for both deltaic-regions required the knowledge of changes in coastal topography and hydrography.

For this study hydrographical (1843, 1846, 1883, 1889, 1908, 1912, 1917, 1980 and 1984) topographical (1915, 1916, 1918, 1931, 1938) and geological (1938) maps have been analysed. Aerial photographs (1943, 1981) and Landsat images have been used for further interpretation. Additional data were obtained from historical publications (Historisch Genootschap, 1925).

Changes in coastal morphology

The Solo Delta. In the mid 19th century the Solo river debouched into the Strait of Surabaya (Fig. 1b, 4). In these days, the original fairway to the port of Surabaya was located close to the former coastline of NE Java and the sediment supply by the Solo river was a direct threat to the navigability of this channel.

At about 1880 the Solo river was diverted by man to flow in a more northern direction, a solution which provided only temporary relief. The remnants of this river branch (Kali Mati or Dead river) still exist as an extremely wide tidal creek, having no connection with the present river system. Finally, in 1890, a channel had been dredged across the tidal flats along the coastline of E Java, to the cape of Pankah. The outflow of the river Solo was thus diverted to the Java Sea (Fig. 4; Historisch Genootschap, 1925).

At the beginning of the 20th century, a new river delta formed at the cape of Pankah. The coastal area on which the new delta developed, consisted of calcareous sandy beaches with beach ridges and

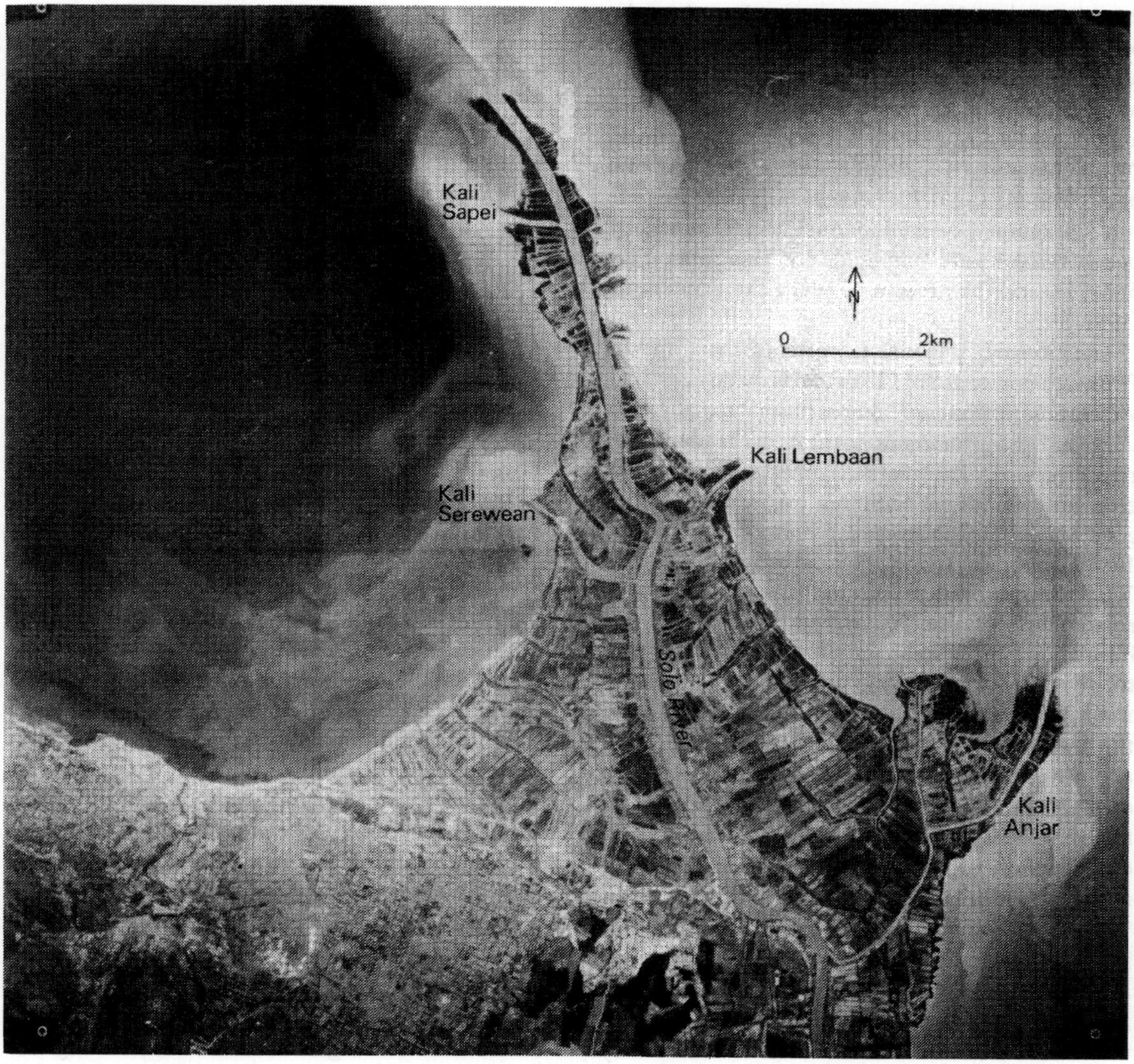

Fig. 5. Aerial photograph of the Solo Delta, showing the main river channel with its prominent natural levees and some smaller outlets on both the east and west side of the delta.

berms. In a seaward direction, the beaches locally passed into calcareous rock, consisting of coral- and shell debris. These beaches are still found along the shoreline west of the delta (Terwindt et al., 1987). However, because of the development of a pronounced mouthbar, the river channel soon bifurcated in two distinct channels, the later Kali Anjar (Fig. 4) and the Solo river. The initial angle between the two branches was about 100–110°. After 1915 the western branch became the main channel. In the period 1915–1931 a similar development was noticed for the Kali Serewean. The growth of the delta continued while the development of a major mouthbar in the mouth of the main channel caused another bifurcation. The angle between the two branches was comparable: 90–100°. Soon after this bifurcation, however, one river branch again prevailed and the influence of the other channel was largely reduced. For that reason, the most western branch, the present Kali Serewean (Figs. 4 and 5) lost its role as a major distributary channel. Along the main river channel delta growth continued, although the delta becomes smaller in a seaward direction. Some minor outlets in the northern part of the delta are man-made e.g. Kali Sapei (Fig. 5; 1974), and the present outlet of the Kali Anjar (1981).

The present Solo delta, north of the Cape of Pankah has a length of 12 km and is made up of one straight major channel, almost normal to the coastline and with prominent natural levees along the river (Fig. 5). The main channel ends in a dip-elongate mouthbar with an elongated middle-groundbar (Terwindt et al., 1987; Wright, 1977, 1985), separating two shallow subchannels (1–2 m). The elongated middle-ground bar at the present river mouth is considered to be a precursor of a new bifurcation in the near future. The distance from the first bifurcation of the Kali Anjar to the one of the Kali Serewean and from the Kali Serewean to the present river mouth is also strikingly the same, about 6 km. Each minor channel has built up its own levees. Channel outlets are characterized by the presence of arcuate mouthbars. The subaqueous morphology in front of the river outlets (Fig. 6) is almost entirely dominated by these mouthbars. The echo sounding profile (Fig. 6, dry season) represents a longitudinal section of the Kali Serewean and shows a rather abrupt decrease in water depth, from 4 m in the channel to less than 1 m above the mouthbar. Part of this mouthbar is covered by unconsolidated muds. The open-bay environment between the river outlets is made up of flat, gradually sloping and extremely muddy tidal flats and mangrove swamps. The transport of sediment from the river outlets to the tidal flats and mangrove swamps has resulted in a widening of the apex of the delta.

The subaqueous delta front has a complex and irregular morphology (Fig. 7). The upper boundary of the delta front is marked by the 2 m depth contour and the front extends to at least a depth of 20–25 m below mean sea level. Although a uniform

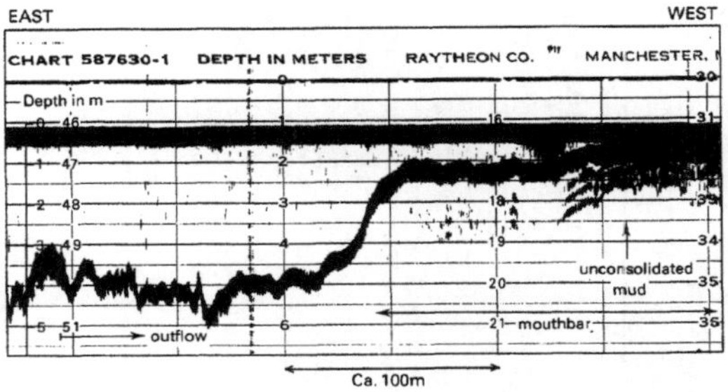

Fig. 6. The mouth bar of the Kali Serewean, partly covered with unconsolidated muds as detected by the echo-sounder (dry season; longitudinal section).

slope is definitely not observed (Fig. 7), average angles of slope are derived from the general bathymetry. Steepest slopes are found along the northwestern and western part of the delta front; angles vary from 0.37° to 0.51°. The northeastern part of the front is characterized by moderate slopes with slope angles of 0.30°–0.47° and lowest gradients (0.20°–0.30°) are observed at the southwest side of the front. Two different morphological types of small scale bottom features are distinguished. The upper part of the subaqueous delta front, to a waterdepth of about 6 m, displays a number of depressions (Fig. 7) of varying width (25–70 m) and depth 0.3–1.0 m), which are characterized by a remarkable flat, and even horizontal bottom.

At greater depths (Fig. 7), the depressions are more irregular and asymmetric. The diameter of these depressions is generally also smaller and the bottom dips in a landward direction. Unfortunately sidescan sonar images of the Solo and Porong deltas are not available and conclusions about the nature and origin of these bottom features are (still) somewhat speculative. Both types of bottom features have a regular and 'fresh' appearance and are almost certainly expressions of post-depositional and gravity-induced sediment movements of various types and origin. Such features and processes are very common in actively prograding deltas with high deposition rates (Prior et al., 1986a, b; Wright, 1985) and are considered to be important mechanisms in the redistribution of sediments.

The rather wide and flat depressions on top of

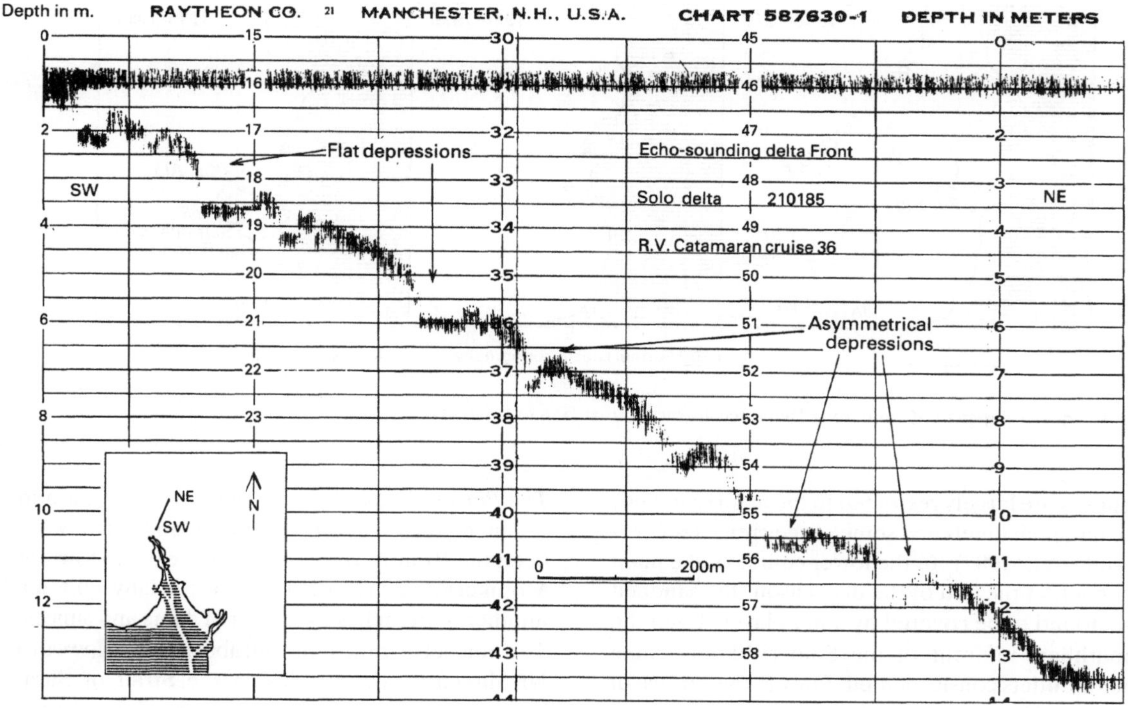

Fig. 7. The irregular morphology of the delta front of the Solo river, indicating sediment transport and mass movements. (echo sounding, section SW-NE).

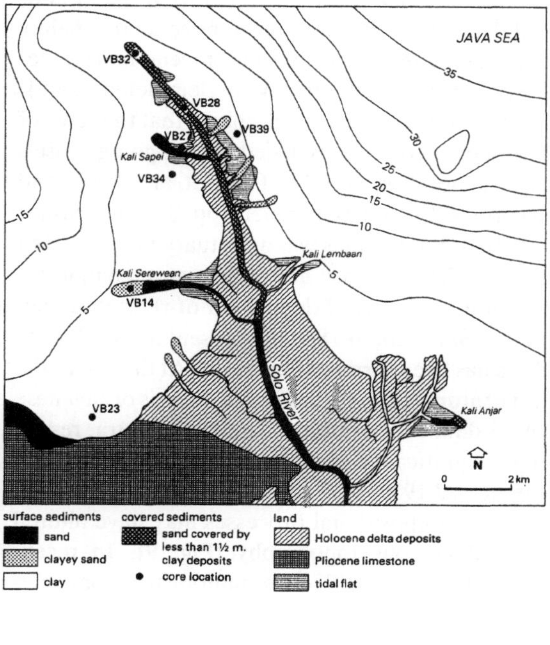

the delta front seem to be related to gravity-induced sediment flows or mud flows, resulting in a pronounced gullying. If these features were the result of a low- or high-density underflow (syndepositional), channel incision is normally observed, and in these conditions a flat and almost horizontal channel bottom is not likely. The small-scale features at the lower part of the delta front definitely indicate the presence of mass-movements, caused by slope instability. The asymmetrical shape of the depressions is almost certainly a result of rotational slumps.

The Solo delta is essentially a mud-dominated system and muddy deltaic sediments make up the seabed in the vicinity of the delta (Fig. 8). Sandy

←

Fig. 8. Lithological map of the distribution of surface deposits of the Solo Delta and adjacent waters.

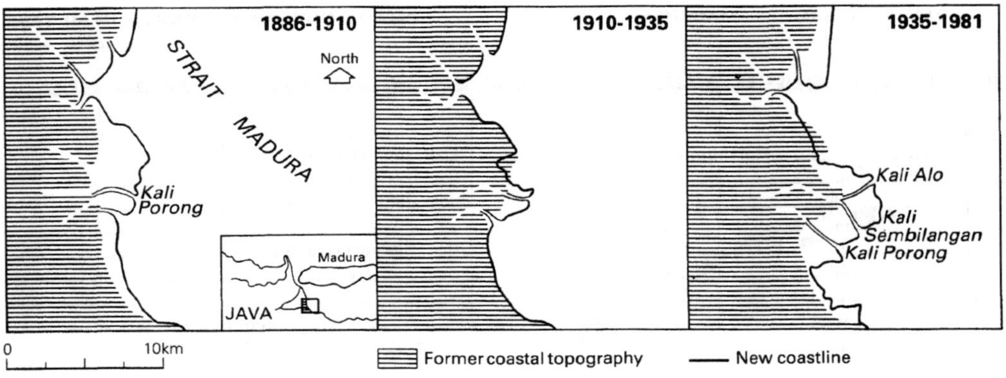

Fig. 9. The development of the Porong Delta, reconstructed from 1886 to 1981/1984.

deposits are largely restricted to the main river bed, the outlet channels and some of the crevasses (Terwindt et al., 1987). In the lower course of the main channel, at the end of the dry season, the sandbed was found to be covered by a mud layer. Even the mouthbars in front of the Solo river and some minor outlets consist of mud. Only at the surface of the mouthbars of the Kali Serewean, Kali Sapei and Kali Anjar some sand and clayey sand is found. These layers only have a limited lateral and vertical extend.

The Solo delta has a morphology which is at variance with the types described in literature so far (Coleman & Prior, 1982a, b; Wright, 1985). The Solo delta definitely has the characteristics of a high-constructive elongate delta, but the present topography makes it less appropriate to classify it as a birdfoot delta. In spite of bifurcations in the past the delta still is an almost single-channel system. This phenomenon may be considered as a logical consequence of the recent age of the system, however, there is serious doubt whether this is the only reason. Especially the lack of natural crevasses in a system which is characterized by frequent floods (Table 2, Fig. 2), suggests that the large amount of cohesive, muddy sediments (Fig. 8) also has a major influence on delta dynamics. It is preferred to characterize the Solo delta as a highly mud-dominated, rapidly-extending, single-finger delta (Terwindt et al., 1987), a rare specimen of an elongate delta.

The Porong Delta. The major source of water and sediment for the Porong delta is and has always been the Brantas river (Fig. 1b). Near the town of Mojokerto, this river has been artificially split in a number of branches by means of dams and sluices. The northern branch, the Surabaya river, flows in a NE-direction, debouching in the Strait of Surabaya. The southern channels of the Porong river and Porong (irrigation) canal flow eastwards but rejoin before entering the Porong delta and the Strait of Madura. The triangular valley between the Surabaya and Porong rivers is covered by alluvial deposits, although a few tens of metres below the present sea level, mid-Pleistocene marine deposits have been found (Van Bemmelen, 1949). From historical sources it is known that the town of Mojokerto could be reached by sea-going vessels until at least A.D. 1396 (Tjia et al., 1968) and according to Van Bemmelen (op.cit.) the former mouth of the Brantas was an estuary until the 10th century. This estuary has disappeared completely because of the rapid deposition of riverborne deposits. Since about 1880 the present active Porong delta has been formed (Fig. 9, 10). At the end of the last century (1881-1910, Fig. 9), a series of Javanese rivers debouched into the Strait of Madura, resulting in significant coastal accretion. Delta growth in the period 1910–1935 seems to have been limited, although depositional processes may have altered the subaqueous topography (Fig. 9). In recent times (1935–1981/1984) a multidistributary net-

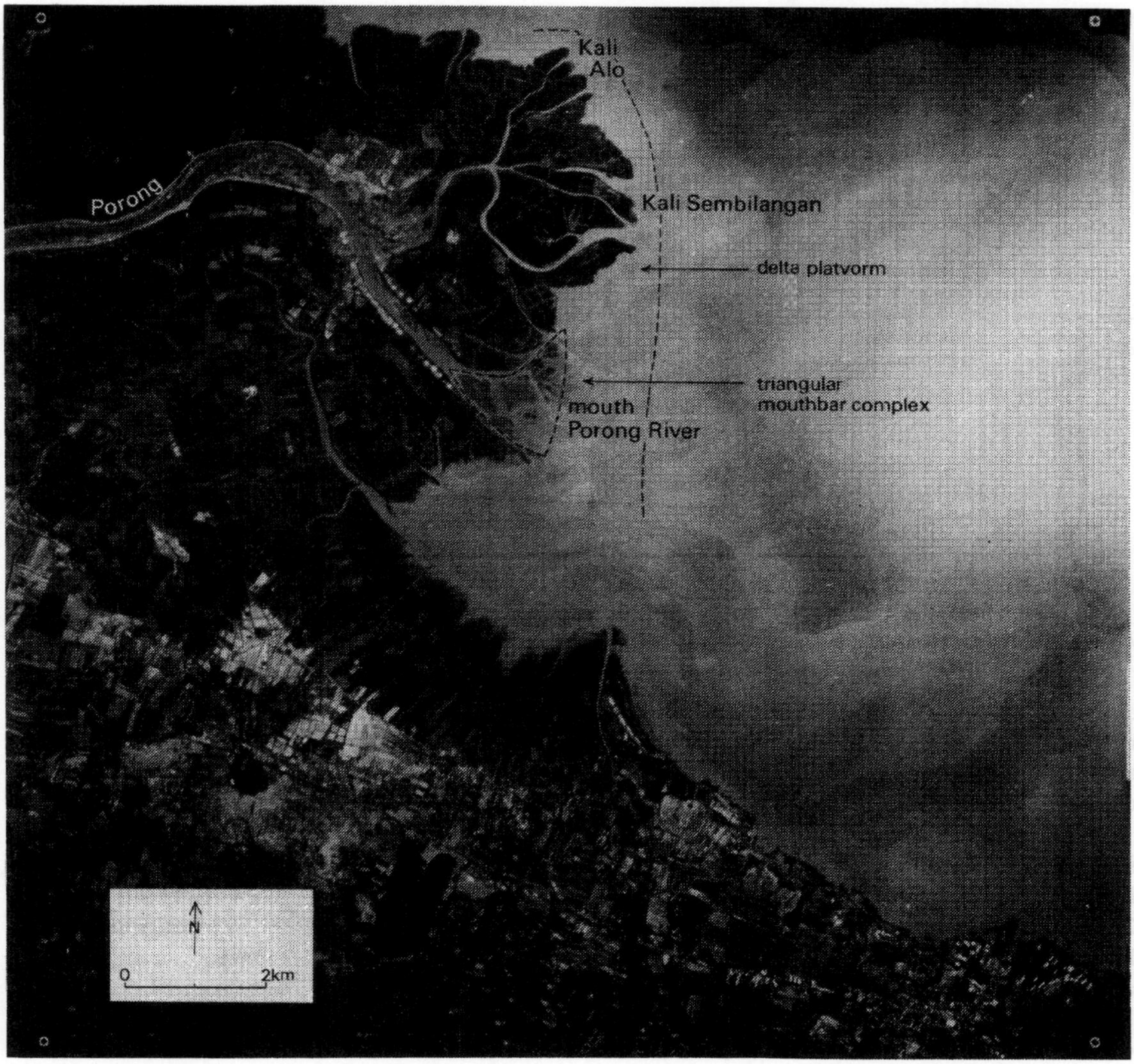

Fig. 10. Aerial photograph of the Porong Delta with the coastline of 1981. Note the triangular mouth bar complex in the main river mouth, consisting of a changing pattern of sandbars, intersected by braided subtidal channels.

work of channels has developed. The river mouth morphology repeatedly resulted in a significant bed friction and a decrease in flow velocities. Consequently, deposition rates at the river mouth increased, creating a mouthbar of substantial dimensions. As soon as the bed friction became too large, because of depositional processes (friction-dominated outflow; Wright 1977, 1985), the channel bifurcated in branches.

The more regular distribution of sediments resulted in a classical shaped delta-lobe. Until the late 1970s the Kali Porong discharged into the Strait of Madura using the channels of the Kali Sembilangan and Kali Alo. In about 1979 an artificial shortcut was made, connecting the Porong river with the coastal waters at the SE side of the delta (Fig. 9). This branch is at present the most active

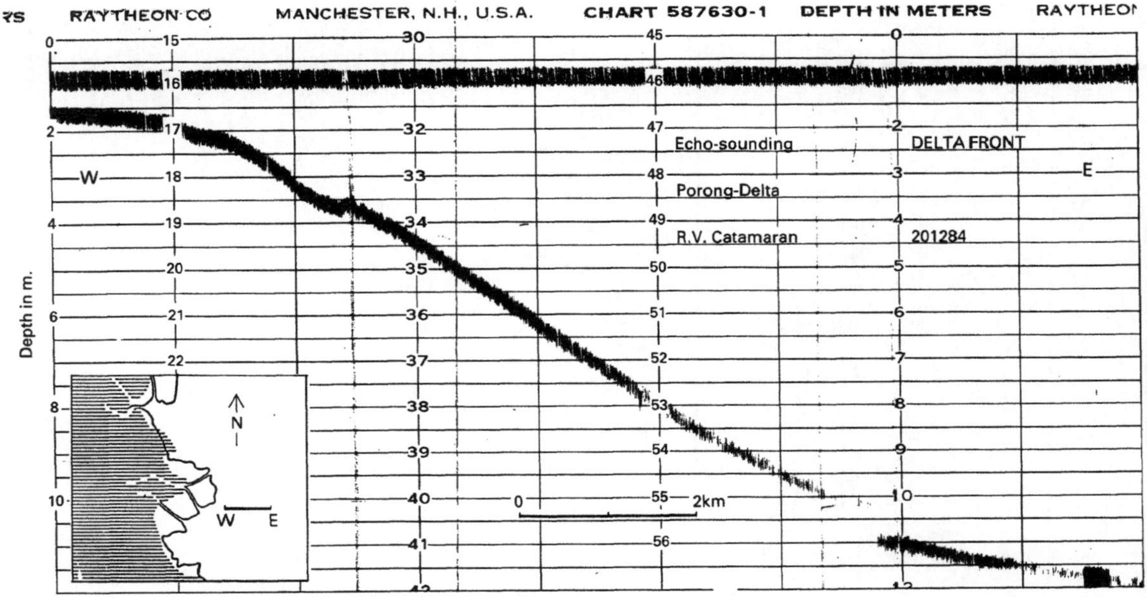

Fig. 11. The smooth delta-front of the Porong Delta; profile measured by echo-sounder (section W-E).

and most important channel within the distributary network. As a consequence, the present coastal accretion is concentrated to the SE side of the delta.

In contrast to the Solo river, the Porong carries a substantial amount of its sediment as sandy bed-load. In the main Porong river mouth (depth 2–3 m), a continuously changing pattern of sand-bars is developed. This pattern is considered to be an extensive triangular mouthbar-complex, cut by braided subtidal channels (Fig. 10). However, river outflow is mainly concentrated in a larger channel, running almost NW-SE/E, being the remnant of a dredging campaign in the recent past.

The delta is surrounded by a submarine deltaic platform at a depth of 1–2 m below mean sea level. In a seaward direction, this delta platform passes into the delta front (Fig. 11). This delta front, extending from a waterdepth of about 2 m to the depth contour of ca. 10 m, has a very regular and smoothly sloping profile. The frontal slope varies from 0.10–0.12° for the northeastern part, to 0.25° along the southeastern section of the delta front. The small discontinuities in the bathymetry (Fig. 11) are artefacts, caused by a temporary halt of the survey vessel. Any evidence for mass-movements or other sediment transport processes along the slope, as observed in the Solo delta, is absent. The interdistributary areas are dominated by extensive tidal flats and mangrove swamps, composed of fine grained deltaic deposits. The Porong delta is morphologically classified as a high-constructive, lobate, multidistributary delta.

Delta Growth

Data on delta growth are generally expressed in terms of linear, areal or cubic growth. Linear growth is an unreliable parameter as coastal accretion may vary substantially within a single delta. Data of areal growth, although better than linear growth (Tjia, 1968; Verstappen, 1968), are still insufficient because they do not take into account changes in the submarine coastal topography. A substantial amount of deltaic sediments is deposited below MSL (mean sea level). Sediment budget studies therefore require the use of cubic growth figures rather than areal growth.

The coastal zone of NE Java, including the Solo

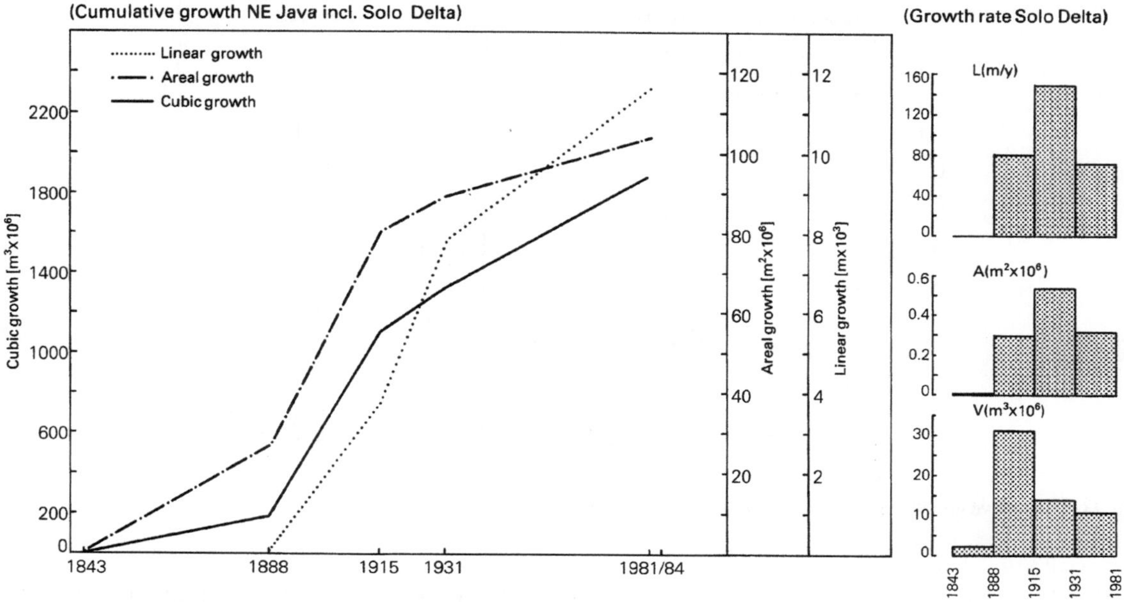

Fig. 12. Cumulative growth and growth rate of the Solo Delta.

and Porong deltas, has been divided in segments, each having a length ($\triangle y$) of 1 to 2 km. Changes in surface area (A) within each segment over a certain time interval have been computed. Only areas which are located above the local highwater-level are taken into account and thus, intertidal areas have been omitted. The average linear growth rate (L) for a segment is:

$$L = \frac{A}{\triangle y}$$

For calculating the cubic growth (V), the changes in cross-sectional profiles (P), perpendicular to the coastline, have been computed. For this purpose, mean sea level is taken as a reference level. The cubic growth is determined by:

Table 4. The longitudinal, areal and cubic growth of the Solo and Porong Delta.

	Length L (m)	L/year ($m \cdot y^{-1}$)	Area A m^2 ($\times 10^6$)	A/year ($m^2 \cdot y^{-1}$) ($\times 10^6$)	Volume V m^3 ($\times 10^6$)	V/year ($m^3 \cdot y^{-1}$) ($\times 10^6$)
Solo Delta						
1843–1888	0	0	0.5	0.01	97.6	2.2
1888–1915	3750	83.8	8.0	0.30	841.0	31.2
1915–1931	4125	152.8	8.7	0.54	223.4	14.0
1931–1981*	3675	73.5	16.4	0.32	545.9	10.9 (17.4 tonne $\cdot y^{-1}$)
Porong Delta						
1886–1910	1690	70.4	25.4	1.1	184.5	7.7
1910–1935	193	7.7	2.9	0.1	59.3	2.4
1935–1981*	1296	28.2	19.4	0.4	570.6	12.4 (19.7 tonne $\cdot y^{-1}$)

* Aerial photographs 1981; hydrographic survey of submarine topography 1984.

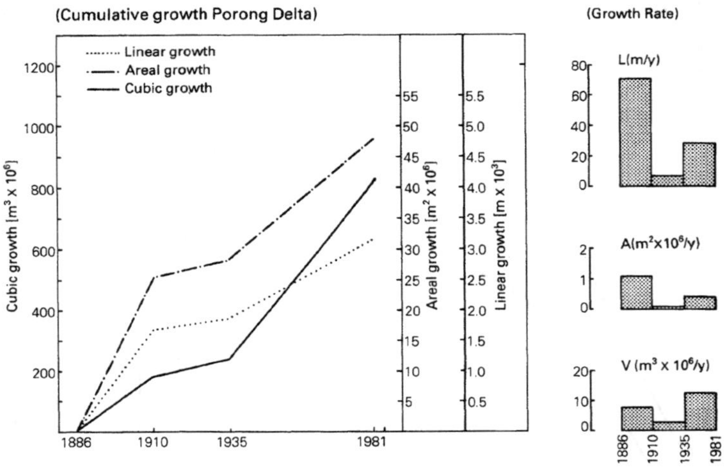

Fig. 13. Cumulative growth and growth rate of the Porong Delta.

$$V = (P) \cdot (\Delta y)$$

The data for linear (L), areal (A), and cubic (V) growth are presented in Figs. 12 and 13 and Table 4.

The Solo delta realized its maximum mean annual cubic growth in the period 1888–1915 (about $30 \times 10^6 m^3 \cdot year^{-1}$). In fact in the period 1843–1915, the foundation was made for the further development of the subaerial part of the delta, which is illustrated by the linear and areal growth figures in the next period ($150 m \cdot year^{-1}$ and $0.54 \times 10^6 m^2 \cdot year^{-1}$). Since 1915 the average volumetric growth has decreased steadily. After 1931 the same is observed with regard to the linear and areal growth. The latter is partly a result of the fact that riverborne sediments have to build up a delta in coastal waters of increasing depth. The delta front has moved to deeper water and river-mouth depositional processes are expected to be more intensively affected by monsoon-induced coastal currents and waves. Data on subsidence rates of the delta cannot be given. However, since the delta covers a stable carbonate platform, subsidence is expected to be less important and mainly a result of consolidation of deltaic sediments.

At present (1931–1981/1984) average rates of deposition are $10.9 \times 10^6 m^3 year^{-1}$ (about 17.4×10^6 tonne $\cdot year^{-1}$). This results in an average longitudinal growth rate of $70 m \cdot year^{-1}$.

In general there is a tendency towards diminishing growth of the Solo delta. The building of numerous dams and sluices as well as the effect of 'check dams' in the upper part of the drainage basin, has reduced the amount of sediment which is transported to the sea. According to estimations based on model calculations by PBS, for certain reservoirs, 80–85% of the sediment transported by the river(s) is retained in these reservoirs (MONENCO, 1984). However, the accuracy of these estimations may be questionable. Some reservoirs which should have been filled up already with sediments, are still in use (pers. comm. PBS, January, 1985).

The maximum average annual linear and areal growth for the Porong delta occurred during the turn of the century (1886–1910) with rates of respectively $70 m \cdot year^{-1}$ and $1.1 \times 10^6 m^2 \cdot year^{-1}$. However, the maximum, mean annual volumetric growth was realized during the last decades (1935–1981/1984). The present amount of deposition is about $12.4 \times 10^6 m^3 \cdot year^{-1}$ (19.7×10^6 tonne $\cdot$ $year^{-1}$) which gives an average longitudinal growth rate of $28.2 m \cdot year^{-1}$ (Table 4, Fig. 11). After the 'slow-down' in growth in the period 1910–1935, the linear and areal growth have also increased again.

The growth of the Porong delta shows a trend which is certainly different from the one observed for the Solo delta, probably because of significant variations in sediment supply and type to both rivers. In the downstream part of the Porong river, even close to the delta, the river is supplied with a considerable amount of sediment from the volcanic Arjuno complex (Fig. 1b).

Volcanic eruptions occasionally result in a tremendous production of loose, highly erodible material. The volcano Kelud in the mid-Brantas Basin has erupted on average once every 15 years over the last 150 years. It is estimated that the eruption of 1951 and 1966 resulted in an annual supply of 2×10^6 tons of sediment to the Brantas river in the period 1950–1970 (Carson & Utomo, 1987).

The headwaters of the Solo river drain a number of volcanoes, but the lower part of the Solo river basin is lacking this high-mountaineous relief. Compared to the lower Brantas and Porong river basin, the total amount of rainfall and rainfall intensities are lower and the floodplain deposits of the river Solo are also less susceptible to erosion (Yunianto, 1982). As a result, the supply of sediment to the Solo is probably smaller. In addition, the influence of agriculture may partly explain the differences in delta growth rates. In the downstream part of the Solo, agriculture seems to be less intensive and soil losses are probably much smaller than in the Porong-alluvial plain or on the mountain slopes.

Finally the gradient of the lower Solo is definitely lower than that of the Porong river. This combined with the effect of man-made constructions further reduces the sediment transport.

Discussion and Conclusions

In the coastal zone of NE Java two important rivers, the Solo and Porong debouch into the sea less than 100 km from each other. During every wet monsoon, the Solo and Porong rivers carry large volumes of sediment towards the sea. The Solo is a dominantly suspension-load river whereas the Porong carries a substantial part of its sediment as bedload. In both cases the input of sediment largely exceeds the transport capacity of coastal currents and waves, resulting in the rapid progradation of two recent but already extensive delta systems. The morphology of both systems is, in spite of the rather small distance between the two deltas, quite different. The Solo delta consists of one straight major channel with prominent natural levees, almost normal to the coastline. The river outlets are dominated by extensive dip-elongate or arcuate mouth bars. Formation of mouthbars has caused several upstream channel bifurcations.

Since the major part of the suspended sediment, transported by the Solo, appears to have a grain size smaller than 50 micron (silt, clay), the Solo delta itself is in fact a huge mud body (Fig. 8). The river is embedded in cohesive, consolidated silts and clays which, in combination with the natural levees, limits the lateral migration of the river channel. For the same reason, natural crevasses are rather scarce.

The high amount of silt- and clay-rich deposits also affects the morphology of the subaqueous delta front of the Solo delta (Fig. 7). Two types of bottom features are distinguished, which are almost certainly reflections of post-depositional sediment transport processes. Their origin is probably related to gravity-induced sediment flows or mud flows (wide and flat depressions on top of front) or rotational slumps (asymmetrical depressions on lower part of front). Several conditions within the Solo delta are in favour for the formation of the various small scale bottom features. First of all, the average, annual transport of suspended matter by the river Solo, as calculated on the basis of discharge and concentration data (PBS) approximates 19×10^6 tonne (Table 1). This figure corresponds with the amount, which may be expected, because of local denudation rates (Table 3). The average, annual sedimentation, based on map analysis of deltagrowth, is about 17×10^6 tonne (Table 4). Although these data represent long-term averages, for the Solo delta it is evident that only a small part of the delivered sediments (about 10%) escapes to the coastal waters. Depositional processes and mechanisms of growth as well as the mutual relations, between delta morphology and hydrodynamics have been discussed by Hoekstra et al. (in press) and Hoekstra (1988, this volume). Since depositional processes in the Solo delta are mainly restricted to the wet season, deposition rates must be high, resulting in an excess pore pressure and a loss of strength. The high amount of silt and clay-rich deposits, which are very sensitive for this effect, enhances the process.

Secondly, the slope of the NE section of the delta front varies from 0.3°–0.5° and these angles are large enough to initiate and stimulate mass-wasting or sediment flows. Finally, incoming waves (from the NE) may supply the essential energy to trigger the processes. The Porong delta, by contrast, is classified as a lobate multidistributary delta. The genesis of this multidistributary network of river channels is a natural response to the fact that the Porong river carries a substantial amount of its sediment as sandy bedload. Rapid deposition of sediments at the river mouth has resulted in the formation of a triangular mouthbar complex, consisting of a changing pattern of sandbars, intersected by braided subtidal channels. Sooner or later, these sandbars are covered by vegetation and emerge as islands, thereby separating the various channels and limiting the braided character. The multidistributary network of channels resulted in a more regular distribution of sediment and a lobate delta. The deltaic platform and the delta front also partly consist of sandy deposits. This is reflected for instance by the regular and smooth topography of the subaqueous delta front (Fig. 11).

Acknowledgements

This research has been carried out as a part of the Snellius-II expedition, organized by the Netherlands Council of Oceanic Research (NRZ) and the Indonesian Institute of Sciences (LIPI). We gratefully acknowledge the assistance of the Indonesian Institute of Hydraulic Engineering and the Indonesian Institute of Geology and Mining.

References

Allen, G.P., D. Laurier & J. Thouvenin. 1979. Etude sédimentologique du Delta de la Mahakam – Notes et mémoires 15, Total (Paris), C.F.P.: 156 pp

Bird, C.F.E. & A. Soegiarto. 1979. Environmental problems related to the coastal dynamics of humid tropical deltas – Proc. Jakarta Workshop Coastal Resources Management. LIPI – The United Nations University: 18–21

Carson, B. & W.H. Utomo. 1987. Erosion and sedimentation processes in Java – Badan Penelitan dan Pengembangan Pertanian dan The Ford Foundation; Jakarta: 34 pp

Coleman, J.M. 1982. Deltas. Processes of deposition and models for exploration – Int. Human Res. Devel. Corp. (Boston): 124 pp

Coleman, J.M. & D.B. Prior. 1982a. Deltaic sand bodies – AAPG short course Education Course Note Ser. 15: 171 pp

Coleman, J.M. & D.B. Prior. 1982b. Deltaic environments of deposition. In: Scholle, P.A. & Spearing D. (eds): Sandstone depositional environments AAPG: 139–178

Coleman, J.M. & H.H. Roberts. 1988. Deltaic coastal wetlands. In: Van der Linden, W.J.M., S.A.P.L. Cloetingh, J.P.K. Kaasschieter, J. Vandenberghe, W.J.E. van de Graaff & J.A.M. van der Gun (eds): Coastal Lowlands: Geology and Geotechnology – Proc. KNGMG Symp. The Hague, 1987 – Kluwer Acad. Publ. (Dordrecht): 1–24 (this issue).

Coleman, J.M. & L.D. Wright. 1975. Modern river deltas: Variability of process and sand bodies. In: Broussard, M.L. (ed.): Deltas: Models for exploration – Houston Geol. Soc.: 99–149

Historisch Genootschap. 1925. Het leven van een vloothouder. Gedenkschriften van M.H. Jansen – Kemink en Zoon (Utrecht): 161–169 and 224–235

Hoekstra, P. 1988. Hydrodynamics and depositional processes of the Solo and Porong deltas, East Java, Indonesia. In: Van der Linden, W.J.M., S.A.P.L. Cloetingh, J.P.K. Kaasschieter, J. Vandenberghe, W.J.E. van de Graaff & J.A.M. van der Gun (eds): Coastal Lowlands: Geology and Geotechnology – Proc. KNGMG Symp. The Hague, 1987 – Kluwer Acad. Publ. (Dordrecht): 161–173

Hoekstra, P., P.G.E.F. Augustinus & J.H.J. Terwindt. (in press). River outflow and mud deposition in a monsoon-dominated coastal environment – Intern. Symp. Physical processes in estuaries, 1986, Noordwijkerhout, The Netherlands – Springer (New York)

Hoekstra, P. & Tiktanata, 1988. Coastal hydrodynamics, geomorphology and sedimentary environments of two major Javanese river deltas. Program and preliminary results of the Snellius-II Expedition (Indonesia) – J. S.E. Asian Earth Sci. (2), 2: 95–106

Hollerwöger, F. 1964. The accelerated growth of river deltas in Java: Mad. Geogr. Indonesia, 4(7): 3–15

Keller, G.H. & D.B. Prior. 1986. Sediment dynamics of the Huanghe (Yellow River) Delta and neighboring Gulf of Bohai, People's Republic of China: Project Overview – Geo-Marine Lett., 6: 63–66

LIPI-NRZ. 1984. The Snellius-II expedition Book 2. Operational Plans – Indonesian Institute of Sciences (LIPI) and Netherlands Council of Oceanic Research (NRZ), Jakarta and Amsterdam: 218 pp. with annex

MacDonald and Partners Cons. Eng. 1977. Lower Brantas pollution study – Government of Indonesia, Ministry of Public Works and Power, Direct. Gen. Water Res. Development, Cambridge: 1–53 with annex

MONENCO. 1984. Jipang project feasibility study. Lower Solo River Development Study, Montreal Eng. Co. Ltd. – Proyek Bengawan Solo, Surakarta, Indonesia: 7–17

O.C.T.A. 1974. Survey and study for the development of Solo River Basin – Supporting Rept. 1: Hydrology, Proyek Bengawan Solo, Surakarta: 1–167

Ongkosongo, O.S.R. 1982. The nature of coastline changes in Indonesia – Ind. J. Geogr. 12, 43: 1–22

Prior, D.B, Z.S. Yang, B.D. Bornhold, G.H. Keller, Z.H. Lin, W.J. Wiseman, L.D. Wright & T.C. Lin. 1986a. The subaqueous delta of the modern Huanghe (Yellow River) – Geo-Marine Lett., 6: 67–75

Prior, D.B, Z.S. Yang, B.D. Bornhold, G.H. Keller, N.Z. Lu, W.J. Wiseman jr., L.D. Wright & J. Zhang. 1986b. Active slope failure, sediment collapse, and silt flows on the modern subaqueous Huanghe (Yellow River) Delta – Geo-Marine Lett., 6: 85–95

Terwindt, J.H.J., P.G.E.F. Augustinus, J.R. Boersma & P. Hoekstra. 1987. Mud discharge, dispersion and deposition in a monsoon-dominated coastal environment. In: Coastal Sediments '87. Proc. Internat. Conf. Coastal Sediments '87, New Orleans, WW Div./ASCE: 1976–1988

Thom, B.G. & L.D. Wright. 1983. The geomorphology of the Purari Delta. In: T. Petr. (ed.): The Purari tropical environment of a high rainfall river basin, Dr. Junk (Amsterdam): 47–65

Tjia, H.D. Sukendar Asikin & R.S. Atmadja. 1968. Coastal accretion in Western Indonesia – Nat. Inst. Geol. Mining, Bandung, Bull., 1: 15–45

Van Bemmelen, R.W. 1949. The Geology of Indonesia – General geology of Indonesia and adjacent archipelagoes, 1A – Martinus Nijhoff (The Hague): 732 pp

Verstappen, H.Th. 1968. Geomorphology in delta studies – ITC, Delft, Ser. B, 24: 3–24

Wang, F.C. 1984. The dynamics of a river-bay-delta system – J. Geophys. Res. 89, C5: 8054–8060

Wright, L.D. 1977. Sediment transport and deposition at river mouths: A synthesis – Geol. Soc. Amer. Bull., 88: 857–868

Wright, L.D. 1985. River Deltas. In: Davis, R.A. (ed.): Coastal sedimentary environments – Springer (New York): 1–76

Wright, L.D., Z.S. Yang, B.D. Bornhold, G.H. Keller, D.B. Prior & W.J. Wiseman. 1986. Hyperpycnal plumes and plume fronts over the Huanghe (Yellow River) Delta Front – Geo-Marine Lett., 6: 97–105

Yunianto, T. 1982. Soil erosion hazard study using aerial photography. Case study of the area east of West Progo Hills – Indon. J. Geogr., 12, 43: 29–50

Proceedings KNGMG Symposium 'Coastal Lowlands, Geology and Geotechnology', 1987: 161–173 (1989)

Hydrodynamics and depositional processes of the Solo and Porong Deltas, East Java, Indonesia

P. Hoekstra
Department of Physical Geography, University of Utrecht, P.O. Box 80.115, 3508 TC Utrecht, The Netherlands

Received 8 September 1987; accepted in revised form 9 March 1988

Key words: river outflow, plumes, depositional processes, deltas

Abstract

The river outflow and depositional processes of the rivers Solo and Porong (East Java, Indonesia) have been studied. The input of sediment into the coastal waters by both the Solo and Porong river has resulted in the rapid development of two-delta-systems. The Solo delta is a mud-dominated, rapidly prograding elongate (single-finger) delta whereas the Porong delta is characterized as a lobate, multidistributary delta. In the wet season, river outflow generally has the character of a buoyant jet (Solo) or a friction-dominated plane jet (Porong). The excessive supply of sediment in the Solo delta in combination with decreasing flow velocities, as a result of entrainment and mixing, cause a rapid settling of suspended matter and deposition. Deposition rates in the Solo delta are also high because of the peculiar fact that maximum river outflow takes place around high tide. As a consequence, riverborne sediment is trapped by tidal and monsoon-induced flow beneath the plume. Bedload transport is of great significance for the Porong delta, and a substantial part of this sandy bedload is deposited in and in front of the present, main river mouth. The river mouth morphology is characterized by a braided complex of sandbars, which in the future may result in new channel bifurcations.

Introduction

The Solo and Porong, two monsoonal rivers on East Java (Western Indonesia, see also map South-east Asia; Hoekstra 1988, this volume, pp. 000), are characterized by a huge sediment load in the wet season. Sediment input to the coastal waters of north-east Java in the last century resulted in the rapid outbuilding of two major river deltas (Hoekstra 1988, this volume, pp. 000). The Solo delta is classified as a monsoonal, highly mud-dominated and rapidly prograding single finger delta (Terwindt et al. 1987), a rare specimen of an elongate delta. Deltaic morphology is characterized by the presence of one straight river channel, almost normal to the coastline, and embedded in cohesive, consolidated silts and clays. The main river mouth and some minor outlets are dominated by the presence of arcuate or dip-elongate mouth bars. The Porong delta, by contrast, has a multidistributary network of channels which resulted in a more regular distribution of sediments. This delta is classified as a lobate, multidistributary delta. However, river outflow in the present delta is mainly concentrated in one river channel and one river mouth. In this river mouth a triangular mouth bar-complex is developed, consisting of a number of sandbars, and intersected by braided, subtidal channels (Hoekstra 1988, this volume, pp. 000). Average, annual deposition in the Solo and Porong delta is about 17×10^6 tonnes and 20×10^6 tonnes, respectively. The two deltas exhibit significant differences in morphology, which are only partly explained by variations in river regime, sediment transport and sediment properties (Hoekstra 1988, this volume, pp. 000).

In this paper river outflow, coastal hydrodynamics and depositional processes are studied in detail, in an attempt to understand the mutual relations between delta morphology and hydrodynamics (Coleman & Wright 1975, Coleman 1982, Wang 1984, Wright 1985, 1987; Keller et al. 1986.

This study was part of the Indonesian-Dutch Snellius II oceanographic expedition. Measurements, observations and sampling were carried out in the months July–September 1984 of the dry season and in the period of December 1984–February 1985, during the wet season (Hoekstra & Tiktanata, 1988). The drainage basin of the Porong river showed a rather quick response to some unusually heavy rain showers in the month of September 1984. As a matter of fact, river outflow and depositional processes did not differ in essence from those observed in the wet season.

River regime and tidal characteristics

The island of Java, located at latitude 7–8° S (Fig. 1), is part of the humid tropics and the climate is characterized by the alternation of wet and dry monsoons. In the months December until March, the wet W-NW monsoon dominates and causes abundant precipitation. The E-SE monsoon between May and September contains less moisture and this period is the dry season. Average annual precipitation for East Java is about $2{,}100\ \mathrm{mm} \cdot \mathrm{a}^{-1}$.

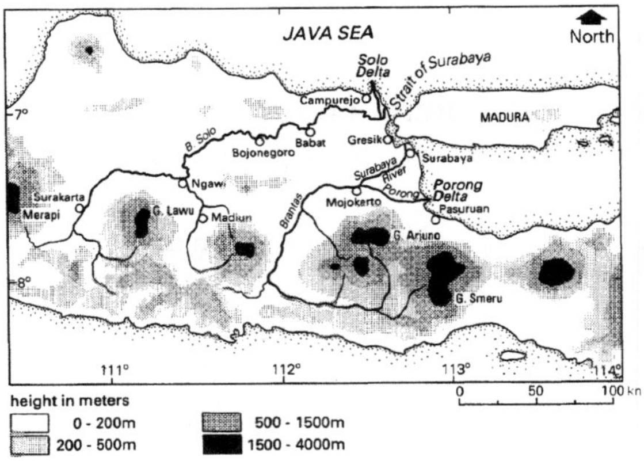

Fig. 1. The drainage basins of the Solo and Brantas river on East Java, Western Indonesia.

Solo river

The sources of the river Solo are on the slopes of the volcanoes G. Lawu and Merapi, near the town of Surakarta in Central Java (Fig. 1). The river has a length of about 550 km and the drainage basin covers an area of ca. $16{,}000\ \mathrm{km}^2$.

Discharge data for Babat (Figs. 1 and 2, data 1984–1985), clearly reflect the succession of wet and dry seasons (Hoekstra 1988, this volume, pp. 000). The minimum and maximum daily flows, recorded (during the field campaign) in the wet season of 1984–1985, were $310\ \mathrm{m}^3 \cdot \mathrm{s}^{-1}$ and $1{,}800\ \mathrm{m}^3 \cdot \mathrm{s}^{-1}$, respectively. Average monthly values varied from $800\text{–}1{,}000\ \mathrm{m}^3 \cdot \mathrm{s}^{-1}$. Daily river flow in the dry season of 1984 (August–September) was often less than $50\ \mathrm{m}^3 \cdot \mathrm{s}^{-1}$. Average monthly flow at Babat ranged from 80 to $250\ \mathrm{m}^3 \cdot \mathrm{s}^{-1}$, however, the average daily fresh-water discharge in the river delta, during measurements, was approximately $20\text{–}30\ \mathrm{m}^3 \cdot \mathrm{s}^{-1}$.

Porong river

The Porong river receives its water from the Brantas river, which drains the volcanic complexes of G.

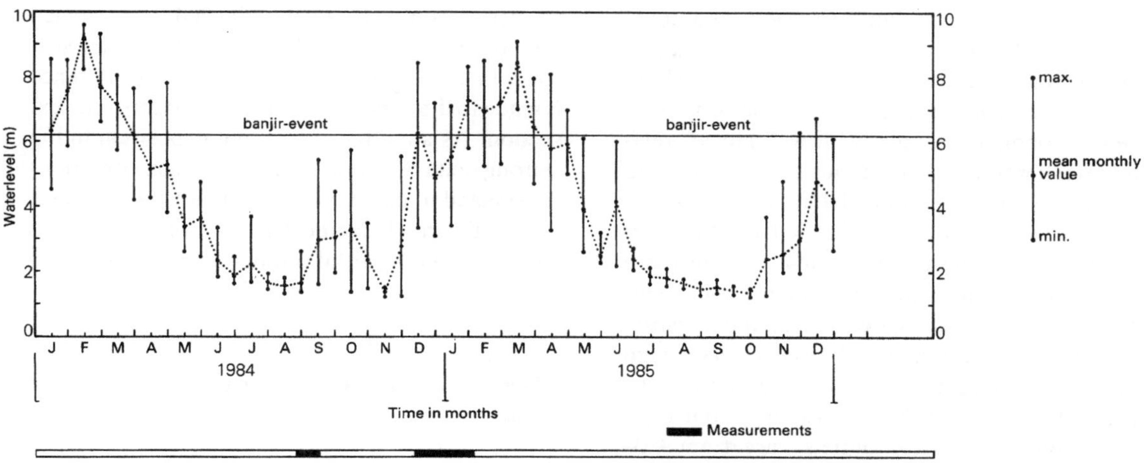

Fig. 2. Discharge distribution of the Solo river, as indicated by the mean, half-monthly water levels at Babat for the years 1984–1985 (data Institute for Hydraulic Engineering, Bandung and Proyek Bengawan Solo, Surakarta).

Smeru and G. Arjuno (Fig. 1). The total length of the river system is about 300 km, covering a drainage basin with an area of ca. 12,000 km^2.

In the wet season, a flood-relief system diverts 80% of the water to the Porong delta. The average wet season flow is about 600 $m^3 \cdot s^{-1}$, but individual floods may result in discharges as high as 1,200 $m^3 \cdot s^{-1}$. In the dry season, river flow is frequently diverted to the town of Surabaya and flow in the Porong river is commonly reduced to almost zero.

Tidal effects

The lower reaches of both the Solo and Porong rivers are subject to tidal influence. The Java Sea near East Java has a diurnal tide with a range of 0.9 m at neap tide and 2.1 m at spring tide. Along the SE part of the Solo delta, the tidal motion is also affected by the diurnal-semidiurnal tide in the Strait of Surabaya, which has a mixed micro-mesotidal range (1.1–2.6 m). Around the Solo delta both tidal regimes interfere and tidal measurements indicate that the horizontal tide generally has a semidiurnal character. A reversal of current direction commonly has no correlation with a high or low water slack period.

In the Strait of Madura and near the Porong delta again a mixed diurnal-semidiurnal tide is observed. Tidal data for Pasuruan (see location Fig. 1) indicate water level fluctuations ranging from 0.5 m at neap tide to 2.7 m at spring tide.

In the dry season, when river discharges are reduced, the alternation of flood and ebb currents generates a bidirectional transport of water and sediment within the deltas.

River outflow dry season

Solo delta

Low river discharges ($Q = 20\text{–}30\ m^3 \cdot s^{-1}$) in the dry season coupled with the small or even negative gradient of the lower part of the Solo river (Hoekstra 1988, this volume, pp. 000), are conditions favourable for the upstream flow of considerable amounts of saline water. In extremely dry seasons, the inflow of saline water is sometimes measurable up to 100 km upstream from the river mouth. Generally, however, seawater and fresh river water are subject to mixing within the delta area. The degree of mixing depends on river discharge, tidal phase, tidal amplitude, density contrasts and the local effects of bottom friction. For dry season conditions, the Solo delta may be considered as a partially mixed estuary (Hoekstra et al., in press). Three different zones are distinguished within this estuarine environment. A zone with well-mixed, relatively fresh water (salinity about 10‰) is observed in the areas upstream of the delta. Downstream of this section, a zone with stratified conditions is developed (bottom salinity 20–30‰). The increase in bottom friction in the river mouth and minor outlets effectively stimulates mixing and a disruption of the stratification, which results in a seaward zone with a well-mixed saline waterbody (30–32‰). Maximum outflow velocities in the main river mouth and minor outlets vary from 0.3–1.0 $m \cdot s^{-1}$. Highest flow velocities were recorded in the smaller outlets because these channels are shorter and have a steeper seaward slope (Hoekstra et al., in press). The river outflow essentially has the character of a turbulent jet (Wright 1977, 1985), although there are local differences. The pronounced mouth bar at the main river mouth may cause the development of a friction-dominated plane (turbulent) jet, instead of a fully turbulent jet.

The Solo river is a suspension-load river and the suspended sediment concentrations in the dry season are approximately 10–40 ppm. In the river outlets and coastal waters, resuspension by currents and waves results in sediment concentrations of 100–300 ppm. The low concentrations in the main river channel, the bidirectional transport and the effects of resuspension delimit the net deposition. According to an estimate by PBS (Proyek Bengawan Solo), the sediment load in the dry season accounts for only 10% of the total average annual input of sediment into the coastal zone. Our own estimates vary from 10–25%, depending on the length of the dry season.

Porong delta

The relatively steep gradient in the lower part of the Porong river ($3.0 \times 10^{-4}\, m \cdot m^{-1}$ or larger) precludes the inflow of large amounts of saline water. Average daily discharges during measurements in the dry season of 1984 were 100–200 $m^3 \cdot s^{-1}$ which are not quite representative for a normal dry season. For a discussion of these flow events reference is made to the conditions in the wet season; see next section. As a matter of fact, net deposition in the Porong delta during average dry season conditions is expected to be nil, since the major part of water and sediment delivered by the river Brantas, is carried to the Strait of Surabaya by the Surabaya river.

River outflow and depositional processes in the wet season

Buoyant plumes of the Solo river

Overall river discharge in the period of December 1984–February 1985 (Fig. 2) was characterized by an alternation of periods of high discharges (1,300–1,800 $m^3 \cdot s^{-1}$) and phases with declining flow (270–300 $m^3 \cdot s^{-1}$). An incursion of salt water was no longer observed and the vertical distribution of salinities or temperatures was homogeneous with fresh river water present in the entire cross section (Hoekstra et al., in press). River outflow processes have been observed in detail during spring tide as well as neap tide conditions.

Simultaneous (tidal) measurements were executed by using two survey vessels, the Indonesian r.v. Hatiga (code HA)and a Dutch catamaran (code CA). Both ships were anchored in front of the main river mouth (Fig. 3) and measurements continued for at least a complete tidal cycle of 25-hours. The results for neap tide conditions (December 17, 1984, Q = 550–650 $m^3 \cdot s^{-1}$ at Babat) are presented in Fig. 4. The maximum river output is at high tide, as indicated by a drop in surface temperature (to 28° C) and salinity (to 15.7‰), whereas increased values are observed with respect to surface turbidity and suspended sediment load (about 300 ppm, Fig. 4). This phenomenon is an unexpected consequence of the interference of the tidal regimes north and south of the delta (Hoekstra et al., in press). Maximum flow velocities in front of the river outlet are not linked to river outflow, but are primarily the result of the mon-

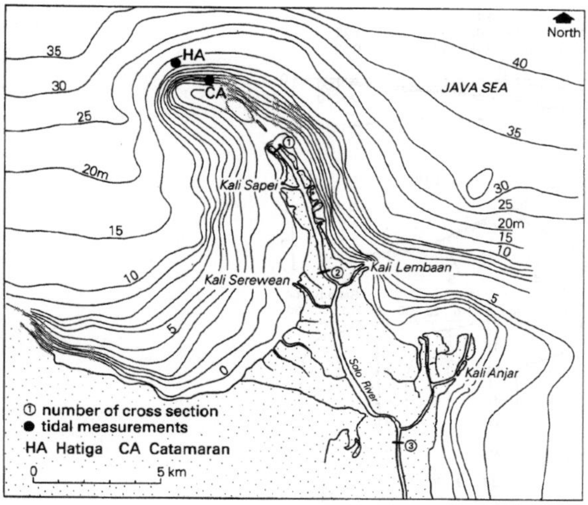

Fig. 3. Bathymetry of the Solo delta with anchor locations of the r.v. Hatiga (HA) and catamaran (CA) during measurements.

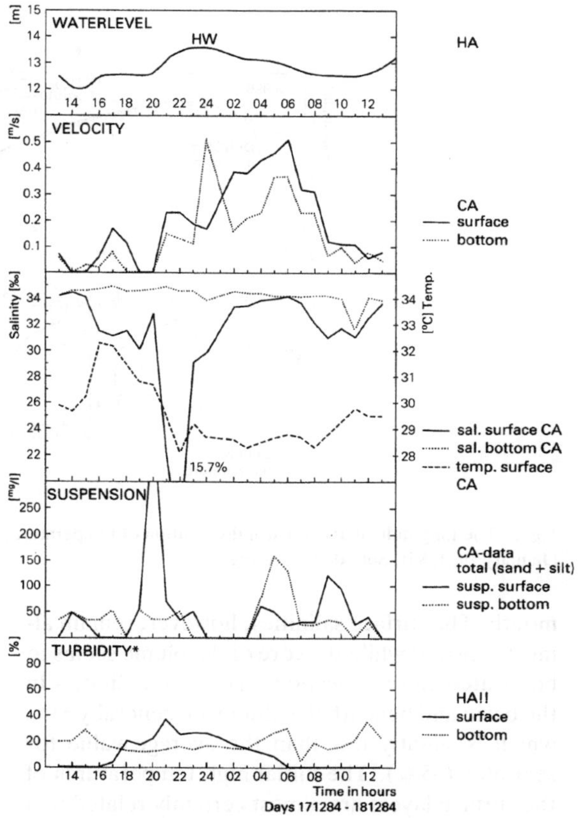

* measurements HA; Loc ± 1200m north of CA!

Fig. 4. The outflow processes in front of the main river mouth of the Solo river, during a tidal-cycle with neap tide (December 17, 1984): water level, flow velocity, salinity, temperature, suspended load and turbidity.

soon-induced eastward drift. Flow patterns around the Solo delta are largely governed by this monsoon-generated coastal drift (Terwindt et al., 1987) and Eulerian residual currents have a strength of 0.14–0.21 $m \cdot s^{-1}$ with a flow direction which varies between NNE and ENE.

The river outflow has the character of a buoyant jet because of the extreme difference in density between the river water and the saline water of the Java Sea (Hoekstra et al. in press, Wright 1977, 1985). The river mouth bar (Fig. 3) affects outflow to a considerable degree. The submarine morphology with a middle ground bar and two bifurcating, divergent channels is typical of outflow conditions, in which frictional forces are predominant (Wright et al. 1980, Wright 1985, 1987). In addition, only thin layers of fresh water are released from the river outlets. In a seaward direction, buoyant forces result in a lateral spreading of the flow. The buoyancy of the effluent fresh water gives rise to the development of buoyant river plumes (e.g. Fig. 5). These plumes are characterized by a high suspended sediment load and the presence of organic, yellow substances, giving the plumes a distinct dark brown colour. Surface fronts are marked by lines of detritus and concentrated foam. The thickness of these river plumes is commonly very limited: average values range from 0.5 to 1.5 m.

The distribution of flow velocities within the plume determines the gradients in sediment transport capacity and thus the suspension outfall. The area and rate of deposition of suspended matter are further governed by the tracks and the shape of the plumes. Figure 5 gives an illustration of such a buoyant river plume, observed at January 19, 1985 in front of Kali Serewean (see Fig. 3; river discharge at the outlet, Q_0 is about 215 $m^3 \cdot s^{-1}$). The thickness of the freshwater layer (density less than 1005 $kg \cdot m^{-3}$) thins rapidly from 2 m in the river channel to 0.3 m at a distance of 4.5 km from the

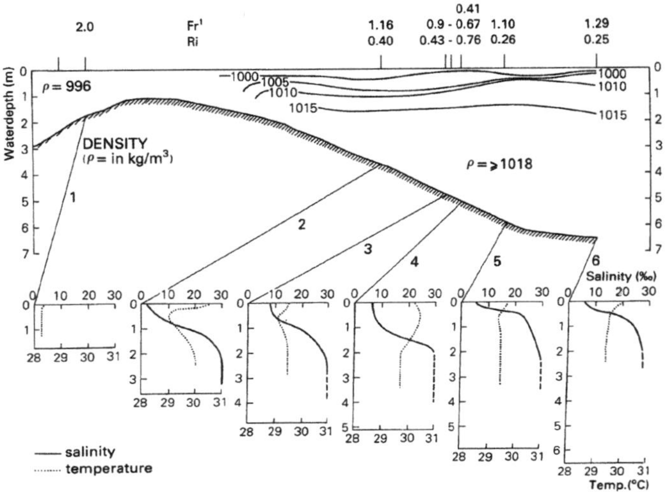

Fig. 5. The longitudinal and vertical distribution of temperature, salinity and density in a buoyant river plume of the Kali Serewean (January 19, 1985), Solo delta (see Fig. 4).

mouth. The surface salinities, however, remain almost constant while the core of the plume seems to be located close to section 4 (Fig. 5). Salinities in the basin and beneath the plume are generally 31‰ which is slightly less than the average value for seawater (35‰). The initial high temperatures of the surface layers are almost certainly related to a secondary circulation in the buoyant plume. This circulation results in an upwelling of relatively warm water along the axis of the plume. Further seaward the vertical temperature profiles become homogeneous (Fig. 5).

The decrease in flow velocities (from $1.1\,\mathrm{m\cdot s^{-1}}$ to $0.3\,\mathrm{m\cdot s^{-1}}$ at 4 km from the mouth), the thinning of the freshwater layer with increasing densities in the lower parts of the plume and the vertical temperature profiles, as illustrated in Fig. 5, stress the role of entrainment and vertical mixing in the transfer of water, momentum and thermal energy.

Plume modelling

Model simulations have been executed to estimate the entire longitudinal flow velocity field within the plume. A relatively simple analytical model, based on research by Bondar (1970) and Wright & Coleman (1971, 1974) has been used. The model assumes that lateral spreading and vertical thinning of the plume are essentially the result of buoyant forces. The lateral flow results in a lateral expansion, which is directly linked with vertical thinning on the basis of continuity. Both lateral expansion and vertical thinning depend on the dynamic buoyant expansion coefficient A, incorporating inertia and the effects of buoyancy:

$$A = \frac{3}{2}\left\{\frac{\frac{4}{3}(2g\gamma)^{1/2}\cdot(1-\gamma/2)\cdot h_0\,(h_0)^{1/2}}{Q_0}\right\} \qquad (1)$$

in which:

g = acceleration of gravity ($\mathrm{m\cdot s^{-2}}$)

γ = $\frac{\varrho_s - \varrho_f}{\varrho_s}$; relative density (–)

ϱ_s = density saltwater ($\mathrm{kg\cdot m^{-3}}$)

ϱ_f = density of freshwater ($\mathrm{kg\cdot m^{-3}}$)

h_0 = thickness of freshwater layer at river outlet (m)

Q_0 = discharge at the outlet ($\mathrm{m^3\cdot s^{-1}}$)

Flow reduction, lateral spreading and vertical thinning, as well as changes in density, e.g. as a result of

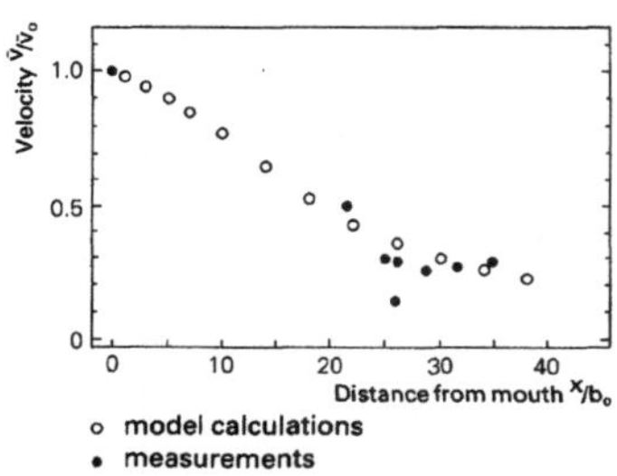

Fig. 6. The distribution of flow velocities within a buoyant river plume: measurements and calculations.

entrainment, affect each other directly and solutions are obtained by iteration. The following boundary conditions were assumed: $Q_0 = 215\,m^3 \cdot s^{-1}$; $b_0 = 125\,m$ and $\gamma = 0.0216$.

The model simulations and measurements (Fig. 6) clearly suggest that in case of a buoyant jet, flow velocities gradually decrease in a seaward direction. Moving away from the river mouth, the behaviour of the buoyant plumes is almost entirely governed by the presence of tidal currents, the monsoon-generated coastal drift, wind and waves. The plumes are carried forward and backward within the coastal environment while the activity of wind, waves and currents tends to destroy the stratified pattern. As a consequence, model simulations become unreliable. However, under certain conditions, plumes which have been discharged during previous outflow phases, are rather stable (see Fig. 4; December 17, 16.00 h.). The river water is still poorly mixed with seawater, which implies little transfer of thermal energy to the basin water. As a result of incident solar radiation, the surface water temperatures rise to extreme values. Meanwhile, during transport in the coastal zone, the suspended sediment gradually settles and surface turbidity is reduced.

Depositional processes in the Solo delta

The suspended sediment concentrations, measured in December 1984–February 1985 reached values varying from 1,700–3,200 ppm. The excessive supply of sediment, in combination with decreasing flow velocities within the buoyant jet, as a result of entrainment and vertical mixing, cause a rapid settling and deposition. The mean vertical suspended sediment concentrations (C_m) decrease exponentially with distance from the mouth:

$$C_m = C_o \cdot e^{\alpha_1 (x/b_0)} \qquad (2)$$

The concentrations, however, are a function of declining flow as given by:

$$\frac{C_m}{C_o} = e^{\alpha_2 (U/U_0) + \beta} \qquad (3)$$

in which:
C_o = concentration at river mouth (ppm);
U = flow velocity at distance x ($m \cdot s^{-1}$);
U_o = flow velocity at river mouth ($m \cdot s^{-1}$);
b_o = width at river mouth (m).
α_1, α_2, and β are empirical values, −0.15, 6.66 and −6.64 respectively, based on the data of January 19, 1985 (Kali Serewean). Since the major part of sediment has a grainsize smaller than 50 μm, the Solo delta is a mud-dominated body (Hoekstra 1988, this volume, pp. 000). The natural levees, which consist of consolidated silts and clays, limit the lateral migration of the river channel and restrict crevasse formation. On several echo soundings, made in the river and in front of the river outlets, the presence of highly turbid bottom waters and layers of fluidised mud with a very low degree of initial consolidation were recorded. These layers indicate the rapid deposition of large amounts of fine-grained sediment. The average annual changes in bed level in the area of the large mouth bar lie in the range of 0.2–0.6 m. These data, based on echo soundings, represent average values for long time intervals. Locally, the annual layers of deposition may therefore exceed a thickness of 1 m.

Rapid deposition, in combination with the high amount of silt and clay-rich deposits, also affects the subaqueous morphology of the delta front, by causing gravity induced sediment flows and mass movements (Hoekstra 1988, this volume).

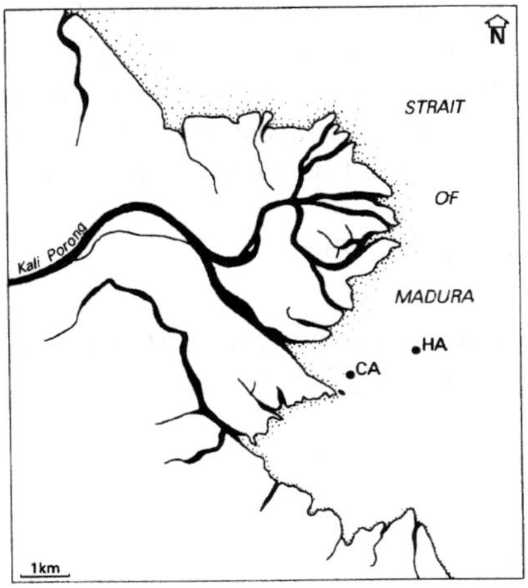

Fig. 7. Porong delta and position of the survey-vessels during measurements.

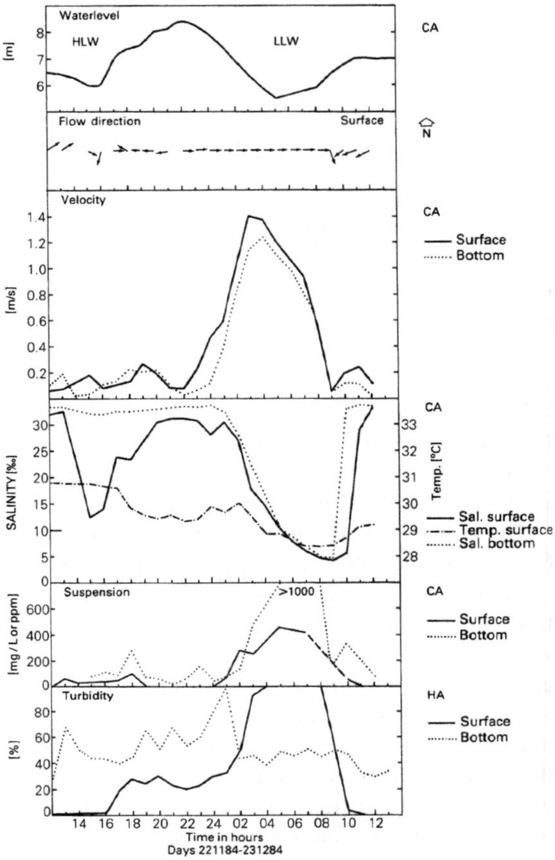

Fig. 8. The outflow processes in front of the Porong river during a tidal cycle, two days after spring tide (December 22, 1984): water level, flow velocity and direction, salinity, temperature, suspended load and turbidity.

Porong delta: friction-dominated outflow

The results of simultaneous tidal measurements in and in front of the Porong delta are presented in Fig. 8 (December 22, 1984; two days after spring tide; Q Porong is about $150\,m^3 \cdot s^{-1}$). The locations of the survey vessels (coded CA and HA) are indicated in Fig. 7. The mixed diurnal-semidiurnal tide is characterized by a high-low water (HLW) and a low-low water (LLW). Two phases of river outflow are observed, both having different lengths and intensities. At HLW river water is still driven up the river by seawater (salinity near the river bed about 34‰), and outflow velocities are less than $0.2\,m \cdot s^{-1}$ (Fig. 8). Directly after HW, outflow recommences and lasts for over eleven hours. The water depth within the outlet channel gradually reduces to about 1 m, and a significant part of the braided complex of sandbars within the river mouth becomes emergent. As a consequence, river outflow is increasingly concentrated in a number of outlet channels. In the meantime, flow velocities increase and have a maximum strength of $1.2\,m \cdot s^{-1}$ at the bed and $1.4\,m \cdot s^{-1}$ at the surface (Fig. 8). At LLW, the minimum salinity is about 5‰ and stratified conditions, as observed at HLW, are not recorded. The maximum suspended sediment concentrations are approximately 400 ppm. Since flow velocities at the bottom are many times higher than the critical flow velocities, required for the transportation of sand grains, sand is easily transported as bedload, but also as suspension load. During measurements of maximum river outflow, about 25% of the suspended sediment load had a grainsize larger than 50 μm.

Since outflow velocities are many times higher than flow velocities at sea (Fig. 8), river outflow has a jet-like character. In contrast to the Solo delta, the submarine topography in and in front of the main river mouth of the Porong delta is expected to result in pronounced bottom friction. The effect of

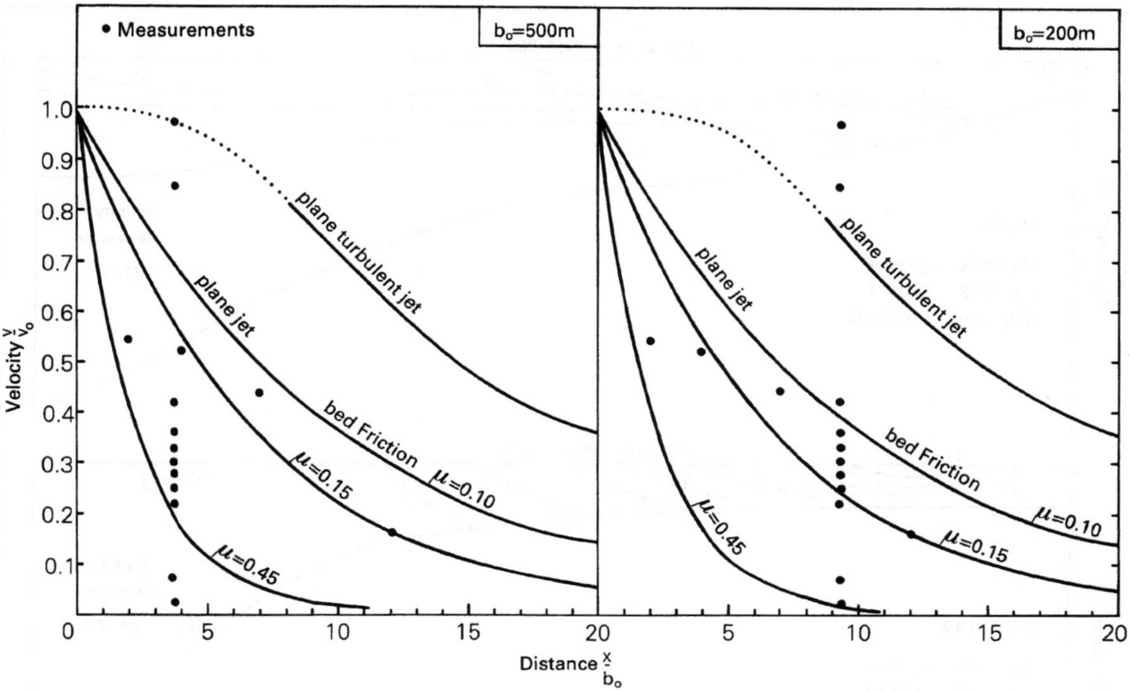

Fig. 9. Measurements and calculations of the velocity distribution in plane jets with varying degrees of bed-friction.

bed friction will increase especially during waning tide, when the water depth steadily decreases. As the solid boundary restricts expansion to the horizontal, outflow ought to have the character of a plane jet. The increased bottom friction, in combination with turbulent jet diffusion normally causes a rapid effluent expansion and deceleration of the flow (Wright 1985, 1987; Wright et al. 1980). Taylor & Dean (1974) and Wright (1985) have proposed formulas for the dimension of, and flow velocities within, a plane jet for different effects of bottom friction:

$$\frac{U}{U_o} = e^{-\mu (x/b_o)} \quad (4)$$

and

$$\frac{b}{b_o} = e^{-\mu (x/b_o)} \quad (5)$$

in which:

μ = friction parameter, related to the Darcy-Weisbach friction factor f_d ($\mu = f_d \cdot h_o/8h$). The estimated values of μ, based on measured changes in plume width, are varying between 0.15 and 0.45.

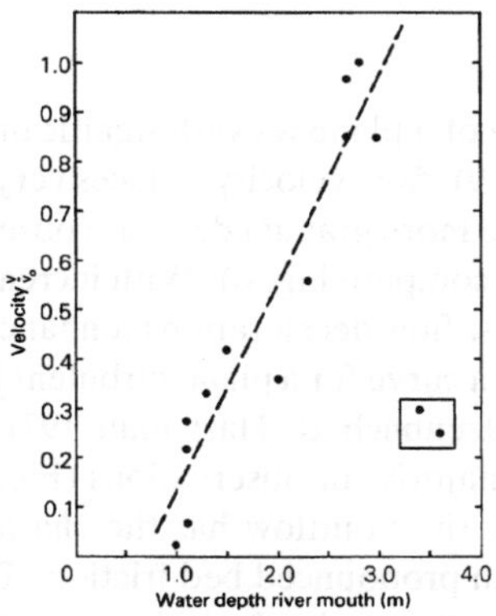

Fig. 10. The effect of a decrease in water depth on the flow deceleration in a plane jet, in front of the Porong delta (measurements December 22, 1984).

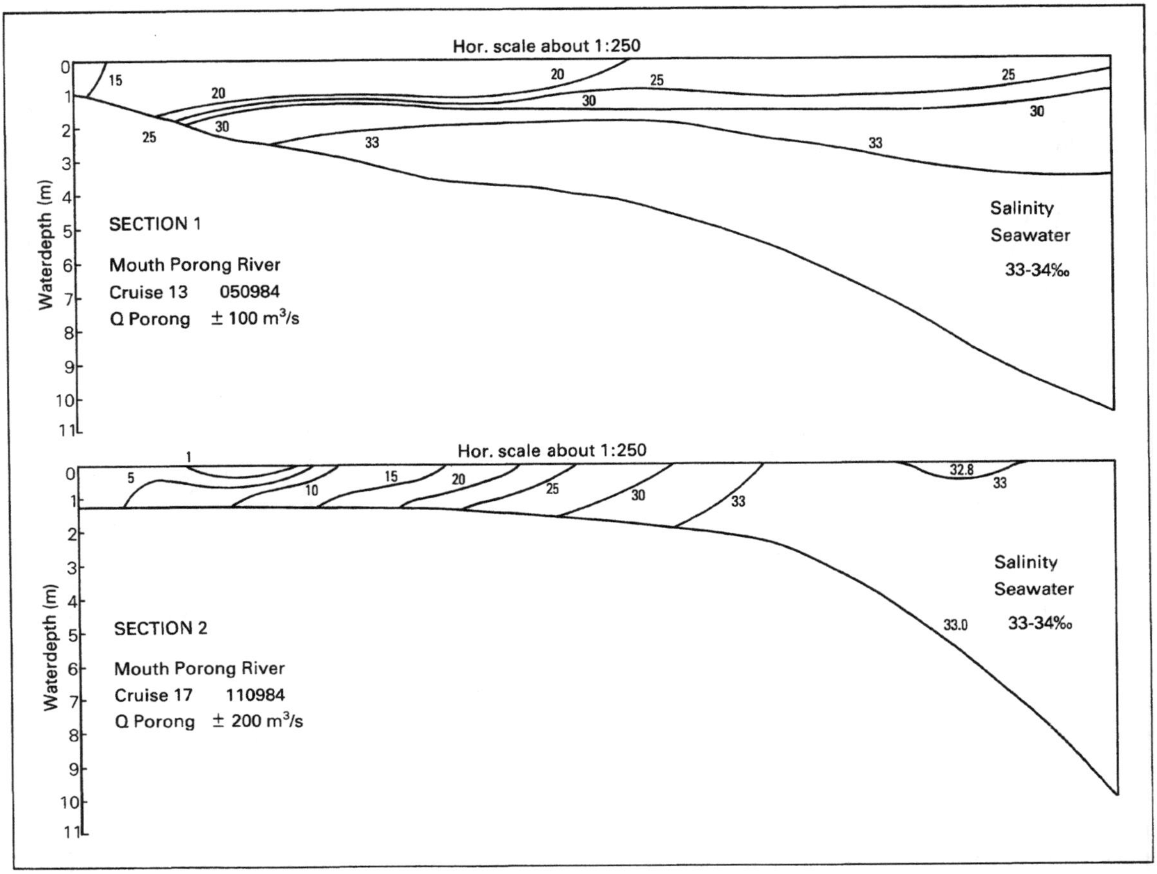

Fig. 11. Mixing and stratification in front of the Porong delta, with variable flow conditions and for a different subaqueous morphology.

In the case of a plane jet with significant bottom friction (Fig. 9), flow velocity reduces very rapidly, instead of the more gradual decrease observed in a buoyant jet (compare Fig. 6). With increasing bottom roughness flow deceleration is enhanced. For a comparison, a curve for a plane turbulent jet is also plotted (Stolzenbach & Harleman 1971, Wright 1985). The majority of observations (Fig. 9) demonstrate that river outflow has the character of a plane jet with pronounced bed friction. The effect of decreasing water depth during waning tide is illustrated in Fig. 10. As soon as the initial water depth in the outlet channel (ca. 3 m) decreases, bottom drag increases and flow reduction is more intense. Hence it appears that the transport capacity of the outflowing effluent is largely reduced.

Mixing and stratification in the Porong delta

Bed friction not only causes a reduction in flow velocities within the outflowing effluent but also promotes mixing, especially in a lateral direction when saline water is encountered. The different degrees and effects of mixing are illustrated by measurements, executed in two radial/longitudinal sections out of the mouth of the Porong river (Fig. 11). In section 1, the water body near the coast and

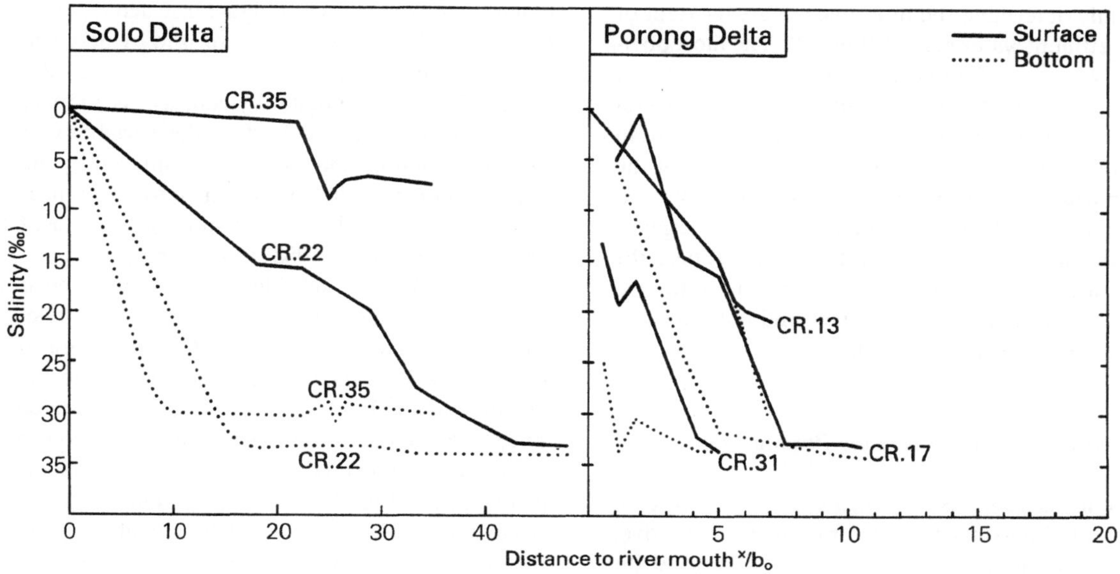

Fig. 12. Longitudinal or radial extend of buoyant river plumes in front of the Solo and Porong delta, as measured during various cruises (Cruise 22, 35: discharge Solo about 1200 $m^3 \cdot s^{-1}$; cr. 13, 17 and 31: discharge Porong 100–200 $m^3 \cdot s^{-1}$).

in shallow water was well mixed (15‰) but still had a lower density in comparison with seawater. Consequently in deeper water, well stratified conditions developed and a buoyant plume was observed. The data of section 2 (Fig. 11) represent measurements recorded a few hours after HLW. The incoming tide drove saline water beneath the plume. The mixing processes, at first during outflow and later on also during inflow, resulted in 'badly-stratified, partially-mixed' conditions. In conclusion, mixing is expected to be less intense in case of (a) low river discharges, (b) low current and outflow velocities and (c) rapidly increasing waterdepth in a seaward direction. However, due to the low current velocities within the plume, the total displacement of the effluent water mass is rather small. If river discharge and outflow velocities increase and the waterdepth is limited over a greater distance (section 2), bed friction and mixing are of major significance. The flow velocities rapidly decrease in a seaward direction and the initial difference in density between the effluent water mass and seawater is reduced by mixing.

For both conditions, the mobility of the partially-mixed or well-mixed water bodies seems to be limited, and the areal extent of the effluent watermasses largely depends on the activity and advective transport caused by wind, waves and currents. Buoyant river plumes, formed in the dry season during periods of high river flow, are generally 'pushed' towards the coast by the Easterly-monsoon. The waves, which strike the delta at almost 90° tend to destroy the density stratification and intensify the mixing processes. Tidal currents in both dry and wet seasons, and in deeper water in front of the delta, are generally weak. The residual currents also have a limited strength with maximum flow velocities of about 0.05 $m \cdot s^{-1}$ and their flow directions vary from N to SE. The buoyant plumes along the Porong delta are often characterized by more than one surface front. A foam front is commonly observed, while in a landward direction a 'colour-front' is recognized, separating water masses of different turbidity and suspended load. These fronts are most probably a result of mixing processes between old and new plumes.

The differences in outflow processes and coastal hydrodynamics between the Solo and Porong delta mainly determine the difference in areal extent of the effluent water masses (Fig. 12). The most obvious reason is, of course the significant difference in average daily discharges between both river systems. The deviating trends in surface and bottom salinities (Fig. 12), however, are indicative of the important role of mixing processes. The buoyant plumes of the Solo delta initially show a divergence of surface and bottom salinities, whereas near the Porong delta, mixing processes operate in the entire vertical. As a result surface and bottom salinities display a similar trend.

Depositional processes in the Porong delta

The measurements of December 22, 1984 (Fig. 8) indicated maximum concentrations near the river mouth of 480 ppm (7% larger than 50 μm) at the surface and 1,100 ppm (10% larger than 50 μm) near the bottom. Over a seaward distance of about 2,000 m these concentrations were reduced to 40–100 ppm and 70–160 ppm respectively. Average concentrations offshore are in general less than 100 ppm. The settling of sediment over the delta front is illustrated by the relatively high values for the bottom turbidity (Fig. 8). However, these turbidity values remain high throughout the tidal cycle which also suggests additional resuspension by currents and waves. The braided pattern of sandbars in the river mouth makes clear that the greater part of the sandy bedload is dropped in the direct vicinity of the outlet. It is concluded that outflow processes in the Porong delta are accompanied by rapid deposition of sediments. The dominant role of bedload transport, the rapid decrease in suspended sediment load due to flow deceleration, and the limited transport by buoyant river plumes, are the main factors, held responsible for this rapid, nearshore deposition.

Discussion and conclusions

The Solo and Porong rivers, two monsoonal rivers in the northern coastal plain of Java, are responsible for the rapid outgrowth of two major river deltas. The Solo delta is a high-constructive, elongate single-finger delta, characterized by only one river channel ('single finger') with some smaller outlets. The Porong delta, by contrast, is a lobate multidistributary delta. Part of the variability in delta morphology and dynamics, displayed by the Solo and Porong deltas, is explained by differences in river regime, sediment transport and sediment properties (Hoekstra 1988, this volume).

Sea level changes, subsidence and compaction as well as tectonic movement commonly are also relevant factors. However, these latter factors are of lesser immediate importance, with respect to the recent Solo and Porong deltas, which were formed since 1880. River outflow processes in relation to wave energy and tidal regime, and the transport processes resulting from coastal currents, definitely have a greater impact on delta morphology. In the dry season, the Solo delta may be considered as a partially-mixed estuary. Low suspended sediment concentrations in the main river channel, the bi-directional transport and the effects of resuspension delimit the net-deposition. In the wet season, the Solo delta is an extremely fluvial dominated system and river outflow has the character of a buoyant jet. Flow velocities in the buoyant plume gradually diminish with distance from the river mouth and, in combination with an excessive supply of suspended sediment, result in an exponential decrease in sediment load and a rapid deposition. Surprisingly, maximum river outflow in the main river channel occurs round high tide. The presence of firmly-built and pronounced natural levees along the main river channel is one of the major consequences of this phenomenon. Moreover, riverborne sediment which settles from the buoyant plume, is retained by the tidal and monsoon-induced drift beneath the plume. All these (hydrodynamic) factors and processes favour limited dispersion of sediments and high deposition rates: almost 90% of the delivered sediments is deposited in the deltaic environment. The morphodynamic system of the Solo delta is partly comparable with e.g. the Mississippi delta (Wright & Coleman 1971, 1974; Wright 1985). However, owing to flow inertia and the fact that the outbuilding river is embedded in cohesive, consolidated silts and clays, natural crevasses are scarce.

For dry season conditions, the Porong delta is classified as an intermediate delta type, between a wave (E-monsoon) and tide-dominated delta system. Net-deposition of sediments in the Porong delta is almost entirely restricted to the wet season. A substantial part of this sediment is supplied as sandy bedload, which is reflected by the braided complex of sandbars in the present river mounth. River mouth morphology causes a significant bottom friction and river outflow has the character of a friction-dominated plane jet. The bed-friction and turbulent diffusion result in a rapid reduction in flow velocities and hence a decrease in transport capacity, but also produce mixing. The transport of the partially-mixed or well-mixed water bodies mainly depends on local coastal currents, which commonly have a limited strength. In the wet season, the Porong delta still is characterized as an intermediate type of delta, but this time between a river and a tide-dominated system.

References

Bondar, C. 1970 Considérations théoriques sur la dispersion d'un courant liquide de densité réduite et à niveau libre, dans un bassin contenant un liquide d'une plus grande densité – Symp. Hydrology of deltas. Publ. 91, AIHS: 246–257

Coleman, J.M. 1982 Deltas. Processes of deposition and models for exploration – Int. Human Res. Devel. Corp. Boston: 124 pp

Coleman, J.M. & Wright, L.D. 1975 Modern river deltas: Variability of process and sand bodies. In: Broussard, M.L. (ed.): Deltas: Models for Exploration – Houston Geol. Soc.: 99–149.

Hoekstra, P. 1988 The development of two major Indonesian river deltas: morphology and sedimentary aspects of the Solo and Porong Deltas, East Java. In: Van der Linden, W.J.M., Cloetingh, S.A.P.L., Kaasschieter, J.P.K., Vandenberghe, J., Van de Graaff, W.J.E. & Van der Gun, J.A.M. (eds): Coastal Lowlands: Geology and Geotechnology – Proc. KNGMG Symp. The Hague, 1987 – Kluwer Acad. Publ. (Dordrecht): 143–159

Hoekstra, P., Augustinus, P.G.E.F. & Terwindt, J.H.J., in press. River outflow and mud deposition in a monsoon-dominated coastal environment – Intern. Symp. Physical Processes in Estuaries, 1986 Noordwijkerhout, The Netherlands – Springer (New York, Berlin)

Hoekstra, P. & Tiktanata, 1988. Coastal hydrodynamics, geomorphology and sedimentary environments of two major Javanese river deltas – Program and Preliminary Results of the Snellius-II Expedition (Indonesia) – J. S.E. Asian Earth Sci. 2 (2): 95–106

Keller, G.H. & Prior, D.B. 1986 Sediment dynamics of the Huanghe (Yellow River) Delta and neighboring Gulf of Bohai, People's Republic of China: Project Overview – Geo-Mar. Lett. 6: 63–66

Stolzenbach, K.D. & Harleman, D.R.F. 1971 An analytical and experimental investigation of surface discharges of heated water – M.I.T., Ralph M. Parson Lab., Water Resour. and Hydrodyn. Dept. Civil Eng. Rept. 135: 212 pp

Taylor, R.B. & Dean, R.G. 1974 Exchange characteristics of tidal inlets – Proc. Coastal Eng. Conf. 14th, Copenhagen: 2268–2289

Terwindt, J.H.J., Augustinus, P.G.E.F., Boersma, J.R. & Hoekstra, P. 1987 Mud discharge, dispersion and deposition in a monsoon-dominated coastal environment. In: Coastal Sediments '87 – Proc. Intern. Conf. Coastal Sediments '87, WW Div./ASCE, New Orleans, 1976–1988

Wang, F.C. 1984 The dynamics of a river-bay-delta system – J. Geophys. Res. 89, C5: 8054–8060

Wright, L.D. 1977 Sediment transport and deposition at river mouths: A synthesis – Geol. Soc. Amer. Bull. 88: 857–868

Wright, L.D. 1985 River deltas. In: Davis, R.A. (ed.): Coastal sedimentary environments – Springer (New York): 1–76

Wright, L.D. 1987 Dispersal and deposition of river sediments in coastal seas: Models from Asia and the tropics – Intern. Symp. Results Indonesian-Dutch Snellius-II expedition, Jakarta: 10 pp

Wright, L.D. & Coleman, J.M. 1971 Effluent expansion and interfacial mixing in the presence of a salt wedge, Mississippi River Delta – J. Geophys. Res. 76: 8649–8861

Wright, L.D. & Coleman, J.M. 1974 Mississippi River mouth processes: effluent dynamics and morphologic development – J. Geol. 82: 751–778

Wright, L.D., Thom, B.G. & Higgins, R. 1980 Wave influences on river-mouth depositional processes: examples from Australia and Papua New Guinea – Estuar. Coast. Mar. Sci. 11: 263–277

Proceedings KNGMG Symposium 'Coastal Lowlands, Geology and Geotechnology', 1987: 175–179 (1989)

Coastal development of Nile Delta

M.A. ElSohby[1], S.O. Mazen[2], M. Abou-Shook[3] & M.A. Bahr[1]
[1] Civil Eng. Dept., Faculty of Engineering, AlAzhar University, Madinet Nasr, Cairo, Egypt; [2] General Organization for Housing, Building and Planning Research, Dokki, Cairo, Egypt; [3] Mining and Geologic Eng. Dept., Faculty of Engineering, AlAzhar University, Madinet Nasr, Cairo, Egypt

Received 9 September 1987; accepted in revised form 29 February 1988

Key words: coast, delta, river, geomorphology

Abstract

This paper reviews the geological and historical development of coastal delta subsoil, littoral processes and geological features of potential engineering significance along a 300 km stretch of coast from Alexandria to Port Said. Particular attention is drawn to the zone comprising Alexandria and Rosetta.

Introduction

The present Nile Delta covers an onshore area of about 25,000 km^2 and an about equal area offshore, down to the 200 m isobath. The coastline of the Nile Delta from Alexandria to Port-Said is about 300 km long. It is roughly straight and has but a few minor offsets. The southern apex of the Delta is at 30° N, some 30 km north of Cairo, where the Nile River splits into the western or Rosetta (Rashid) branch, and the eastern or Damietta (Dymyat) branch (Fig. 1).

The geomorphic features of the Nile Delta coast have undergone considerable changes throughout geologic, historic and recent times. Geologically, the Delta is a lobate structure whose builtup began in Pliocene time. Historically, the several ancient distributaries played a significant role before they disappeared as a result of intensive irrigation constructions in the onshore delta plain to be eventually reduced to the two present branches. Recently, the construction of the Aswan High Dam in the upper reaches of the Nile River produced a significant reduction in outflow while erosion caused a consequent retreat of the delta front by several metres per year. This was studied intensively and is recorded by Orlova & Zenkovich (1974), Sestini (1976), Zaghloul et al. (1982, 1984) and in Reports of the Department for Study of the Side Effects of the High Dam. The present study emphasizes coastal variations in geological and historical times.

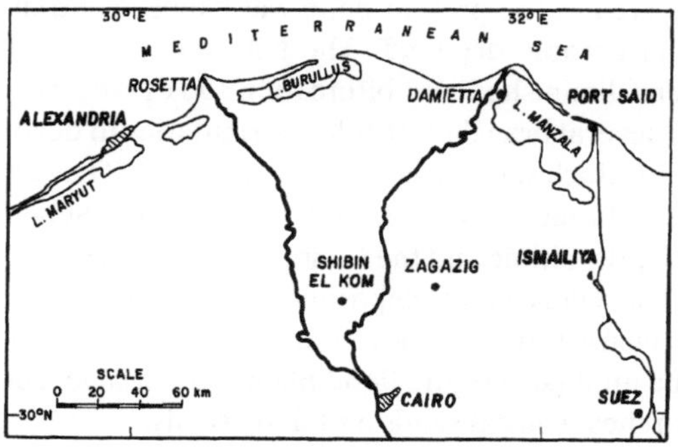

Fig. 1. General plan of the Nile Delta.

Geological evolution of the Nile delta

The Mediterranean Sea flooded the area of the Delta and Nile Valley till Aswan in Middle Pliocene times (Attia, 1954). At that time, the level of the Mediterranean Sea was higher than presently by abbout 180 to 200 metres.

During the Late Pliocene, the sea regressed and the delta area remained occupied by a small embayment whose shoreline was delineated by the line passing through Shibin El-Kom and Abu Hammad, Zagazig (Zaghloul et al., 1977). The sea continued to recede and the embayment began to be successively filled with deltaic sediments carried by the River Nile. According to Ball (1939) near Cairo the river was running in a canyon about 12 km wide and 2 km deep.

During the Pleistocene, the delta continued prograding in stages and at different rates depending upon the sediment load carried by the Nile. At that time the sea level fluctuated; flooding parts of the delta at certain times, retreating from it at others times. As a result, the deltaic sediments, particularly in the north, were subjected to subsidence and intermittent transgression by the sea, causing the deposition of alternating marine sediments and deltaic sands. It was observed (El-Asmar, 1986) that the deltaic sands which form the mouth bar front appear to have developed in the southern area of the delta, whereas clays belonging to the prodelta and the shallow epineritic zone were concentrated in the northern delta area with greater thickness.

During the Late Pleistocence the embayment was completely filled with deltaic sediments. Then the area of the present delta was exposed and the delta became dry land. The Nile water gradient, then, diminished and bifurcation took place causing several branches of Nile distributaries to develop of which seven are historically named. They are the Pelusiac, Tanitic, Mendesian, Bulolic, Sebennitic, Bolbitinic and the Canopic branches (Fig. 2).

The delta cap was deposited in Recent times with its characteristic facies comprising channel fills, Nile mud (silt) of the flood plain, subaerial levees, marches, coastal sands and dune sands.

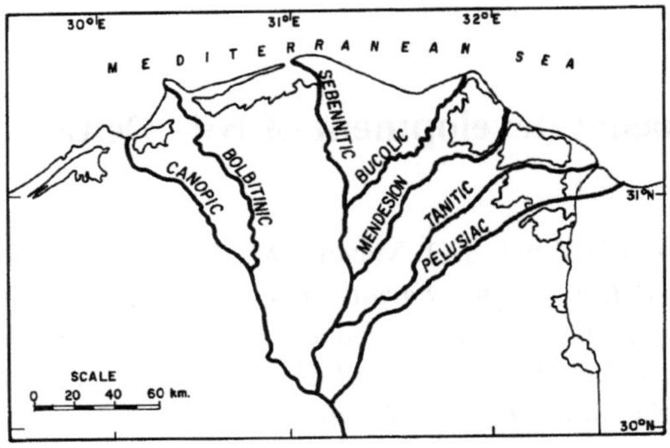

Fig. 2. Old branches of the Nile (after Toussoun; 1922).

Coastal changes in geologic times

In Pleistocene times, sediments were distinctly deposited in cycles. The cyclic sequences are composed of deltaic sands, sometimes pebbly sands, that were formed during regressive phases interbedded with marine clays formed during intermittent periods of marine transgression. The cycles are possibly attributed to the Mediterranean sea level oscillation and climatic changes during the Pleistocene. A genetic relationship between these deltaic cycles and the successive ridges on the Mediterranean coast, west of Alexandria, as well as the River Nile terraces in the valley were studied by Coleman (1975).

The successive shore line progradation and its development have been idealized as shown in Fig. 3.

Coastal changes in historic times

The method adopted in studying the coastal changes in historical times is to compare maps made during the French Expedition of 1798–1801 (Fig. 4) to those made before construction of the High Dam in 1964 (Fig. 5).

By comparison, it was found that major changes occurred in the intervening years. In the present study the changes in the zone commprising Alexandria and Rosetta were studied. The most obvious changes have been: firstly, Maryut lake which be-

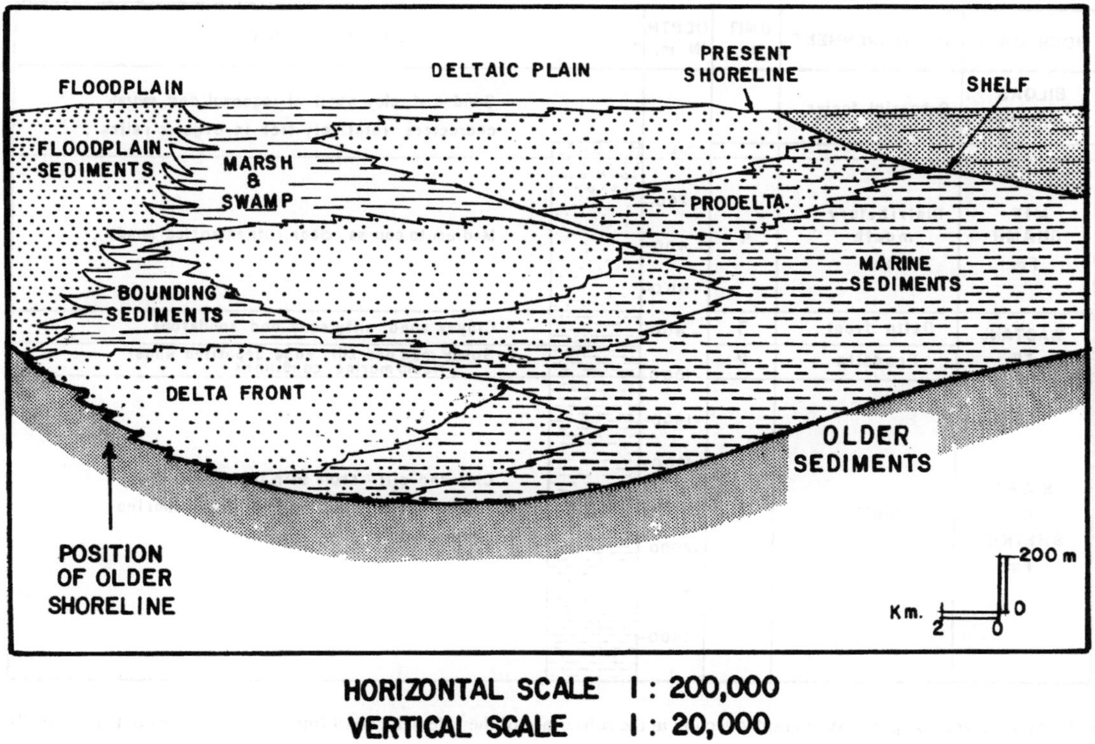

Fig. 3. Coastal changes in geological time (modified after El-Asmar, 1986).

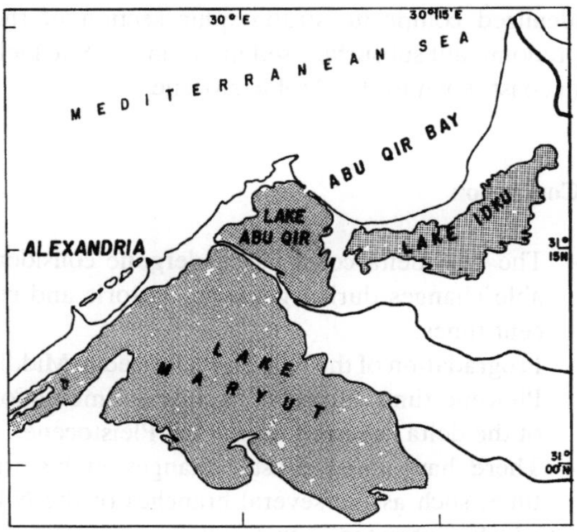

Fig. 4. Map of the Nile Delta during the French Expedition (1798–1801).

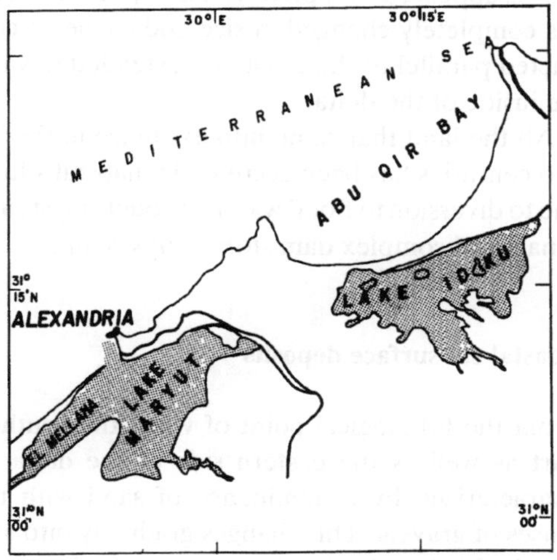

Fig. 5. Map of the Nile Delta before the high dam construction (1964).

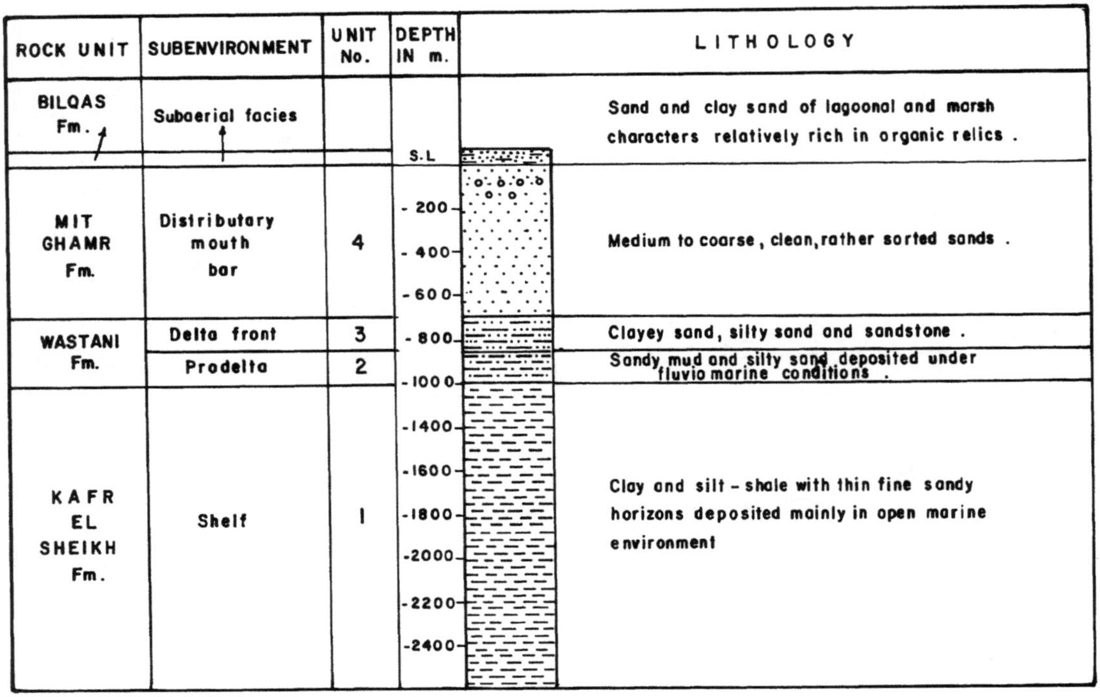

Fig. 6. An idealized composite stratigraphic section in the subsurface of the Nile delta coast (modified after Zahgloul et al., 1977).

came very small when compared with its size during the French expedition; secondly, Abou Qir lake has disappeared completely; thirdly, Idku Lake, has completely changed in size and shape, it contracted parallel to the coast and extended towards the inside of the delta.

All the land that came into existence in the last two centuries has been accreted by natural silting, due to diversion to flood waters through a system of canals and complex dam storage upstream.

Coastal subsurface deposits

From the lithological point of view, the southern part as well as the eastern side of the delta are characterized by a dominance of sand with thin lenses of gravels. This changes gradually into sand interbedded with clay layers towards the north. Occasionally, thin bands of limestone are encountered offshore. The sedimentary pile seems to be thinner and to be wedge-shaped in the south, whereas it is conelike in shape and thicker north of the Nile Delta (El-Asmar, 1986). Figure 6 shows an idealized composite stratigraphic section of the predominant subsurface sediments in the Nile Delta coast down to the PlioPleistocene.

Conclusion

- The Nile delta coast has undergone considerable changes during geologic, historic and recent times.
- Progradation of the Nile Delta started in Middle Pliocene time. However, major sedimentation of the delta occurred during the Pleistocene.
- There have been drastic changes in historic time, such as the several branches of the Nile which contracted to two only.
- A comparison of maps made during the French Expedition and maps made before the High

Dam construction shows that many changes took place and that an area of land was gained.

- After the construction of Aswan High Dam in the upper reaches of the Nile River, it was observed that the outflow was significantly reduced and that the delta front was eroded and retreated.

References

Attia, M.I. 1954 Deposits in the Nile Valley and the Delta – Mines and Quarries Dept., Geol. Surv. Egypt, Cairo, 357 pp.

Ball, J. 1939 Contributions to the geography of Egypt – Geol. Surv. Egypt, Cairo, 308 pp.

Coleman, J.M. 1975 Deltas: processes of deposition and models for exploration – Continuing Education Publication Company, 91 pp.

El-Asmar, H.M. 1986 Lithostratigraphical and sedimentological studies on the subsurface Quaternary succession in the Nile Delta Region, Egypt – M.Sc. Thesis, Fac. of Science, Mansoura Univ., 213 pp.

Orlova, G. & V. Zenkovich 1974 Erosion of the shores of the Nile Delta – Geoforum 18: 68 pp.

Sestini, G. 1976 Geomorphology of the Nile Delta – Proc. seminar on Nile Delta sedimentology, Alexandria, Egypt, 12 pp.

Tousson, Prince Omar, 1922 Mémoire sur les anciennes branches du Nile – Mém. Inst. Egypte, IV, Époque Ancienne, Cairo, 150 pp.

Zaghloul, Z.M., Taha, A.A., Hegab, O. & El Fawal F.M. 1977 The Neogene-Quaternary sedimentary basins of the Nile Delta – Egypt. J. Geol. 21: 1–19.

Zaghloul, Z.M. El-Nasharty F.A. & Isa, I.A. 1982 Post-Aswan High Dam changes of the Nile Delta – Inter. Sympos. Remote Sensing Environment, Cairo, 877 pp.

Zaghloul, Z.M., El-Nasharty, F.A. & Isa, I.A. 1984 Contribution of the coastal changes in the area between Port-Said and Lake Bardawil, Egypt – J. Geol. 28, 1: 25–30.

Proceedings KNGMG Symposium 'Coastal Lowlands, Geology and Geotechnology', 1987: 181–202 (1989)

Development of the Cenozoic Niger Delta in terms of the 'Escalator Regression' model and impact on hydrocarbon distribution

Gordon J. Knox & Ebi M. Omatsola
Shell Internationale Petroleum Maatschappij B.V., Postbus 162, 2501 AN, Den Haag, The Netherlands

Received 10 September 1987; accepted in revised form 13 May 1988

Key words: Niger Delta, Escalator Regression, Cenozoic, stratigraphy, lithofacies, structure, hydrocarbons, shale tectonics, depobelt

Abstract

A conceptual model known as *Escalator Regression* has been developed to explain the particular association of stratigraphy, lithofacies and structure in the central area of the Cenozoic Niger Delta. An important factor is the degree of mobility of the underlying overpressured marine shales which move in response to gravity loading of deltaic sediments. A decrease in mobility caused sand/shale sedimentation to be displaced southwards by continental sand deposits under conditions of lowered submergence rates. A succession of transient arcurate depobelts are defined in a step-wise outbuilding of the delta. Stabilisation of depobelts was followed by slower submergence and deposition of blanket fluvial sands.

The northern delta and delta fringes lie outside the central area where the model was developed. The slope of the Niger Delta is characterised by depocentres separated by shale ridges. The bottom of the slope is marked by imbricated toe thrusts. The slope has continuously built up by sediment addition and shale inversion in response to loading, gradually changing from deep to shallow water environments. Cut and fill channel systems can be related to changes in sea level.

Basement lineaments can be projected under the deltaic prism and do not control arcuate depobelt trends, but some subtle controls are interpreted to have been exerted.

Some consequences for hydrocarbon distribution are discussed. Well temperature profiles show that the alluvial sands carry heat away from the top sand shale section very efficiently. The continental sands stabilise a depobelt temperature profile in existence before these sands advance over the depobelt. Thus a particular stratigraphic level shows only a small increase in temperature when buried to a deeper level, which has consequences for the implications of bacterially degraded oil. Hydrocarbon migration begins at least as early as depobelt stabilisation.

The lithofacies relationship to time shows that several depobelts have been tilted in both dip and strike directions since the passage of the base fluvial sand. Small angles of tilt (1°–2°) could have caused major spillage of hydrocarbons, resulting in important remigration, thus influencing the final hydrocarbon distribution.

Introduction

The Cenozoic Niger Delta is situated at the intersection of the Benue trough and South Atlantic Ocean where a triple junction developed during separation of South America from Africa (Fig 1; Burke et al., 1972; Whiteman, 1982). The delta consists of a sedimentary prism of the order of 12 km overall thickness covering a total area of 140,000 km^2, deposited initially on continental crust and later on oceanic crust. The overall succession represents a gross coarsening upward sequence of a major regressive cycle (Short & Stäuble, 1967; Fränkl & Cordry, 1967).

A large part of the sedimentary package consists of holomarine shales (Akata Formation) which are sandwiched between the underlying basement and overlying paralic and continental sediments. The paralic sequence (Agbada Formation) of alternating sands and shales plays an important role in the synsedimentary structural development. The continental sequence (Benin Formation), a largely continuous sand sheet does not play any significant role in synsedimentary tectonics and merely blankets the structures.

A relationship between synsedimentary growth faulting, structural development and sedimentary facies is recognised. We have developed a conceptual model called 'Escalator Regression' to describe the mechanism of Niger Delta regression. The model is largely based on type sections through the delta (Fig. 2). It combines stratigraphical, structural and lithofacies data to define transient depobelts.

The possible passive effects of basement on the model are discussed as is the response to lowstands of sea level. The model also has important consequences for hydrocarbon habitat which are described later in the paper.

Escalator Regression

Definition
Escalator Regression describes the one-way stepwise outbuilding of the Niger Delta through geological time. The units of these steps are transient depobelts. The shape of the regression line is particularly demonstrated by the change from depobelt paralic sediments (Agbada Formation) to continental sands (Benin Formation). The line broadly represents the coastline position which has exhibited a pulsatory 'slow-fast' advance with time.

Depobelts, as defined here, represent successive phases of delta growth. They are composed of bands of sediments about 40 km wide with lengths of up to 300 km. They contain major fault-bounded sequences which contain a paralic alternating sand/shale sequence limited at the proximal end by a major boundary growth fault and at the distal end by a lithofacies change, a counter-regional growth fault, a major boundary fault of a succeeding depobelt, or any combination of these (Fig. 3). Earlier, Evamy et al., 1978 had defined megastructures which can be subdivided into smaller fault bounded macrostructures, within which seismic reflection horizons are geological time lines.

Along strike, a series of megastructures combine to form a depobelt within which base alluvial sands and top holomarine shale are broadly parallel to time stratigraphy along most of the strike length of the depobelt. Seawards successive depobelts contain sedimentary fills markedly younger than the adjacent ones in a landward direction.

On delta dip sections a relationship is apparent between successive depobelts. The base alluvial sand facies of an updip (older) depobelt is approximately time equivalent to the initiation of the base paralic sand/shale sequence in the down-dip depobelt (Fig. 4a, b). The diachronous sand/shale to alluvial sand lithofacies change at the exterior part of a depobelt is time equivalent to the deposition of the paralic sequence of the next downdip depobelt.

The deposition of paralic sequences within any depobelt is terminated by a rapid advance of an alluvial sand facies over the proximal and central areas of the belt. This advance initiates deposition of the paralic sand/shale sequence in the succeeding depobelt. A paralic sequence develops in this new depobelt, and in the exterior part of the older depobelt, while the continental sands/gravels advance diachronously (Fig. 4a, b). This sequence of events repeated itself five to six times over the last 38 million years to define a series of depobelts

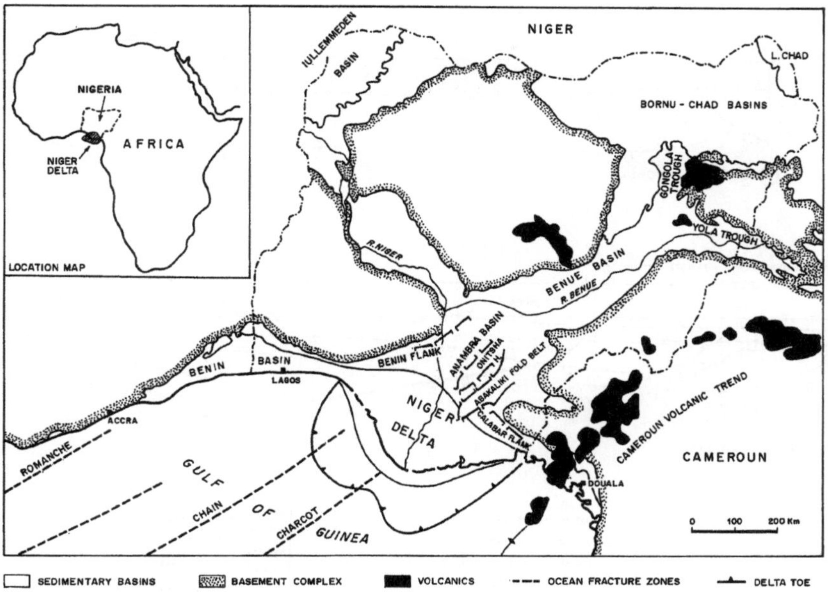

Fig. 1. Regional Setting of the Cenozoic Niger delta. The delta has developed in an embayment defined by the Benin and Calabar flanks at the intersection of the Benue Trough with the Gulf of Guinea part of the South Atlantic Ocean.

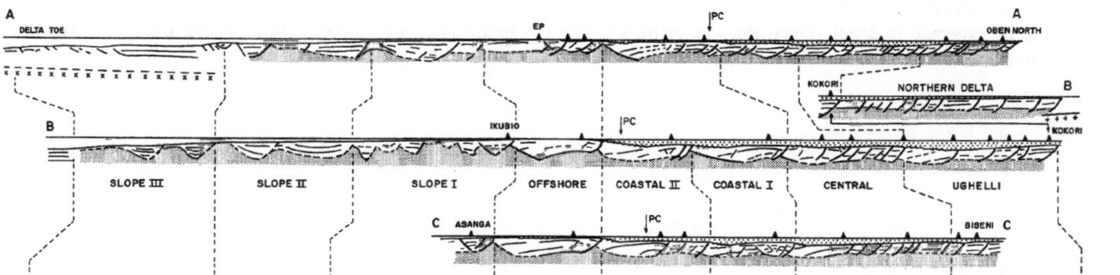

Fig. 2. Schematic dip sections through the Niger Delta. The positions of the depobelts are indicated as are potential future depobelts on the slope. Depobelts on sections A, B, C, and D can be separated but towards the eastern margin (Sections E and F) the belts are no longer individually recognisable. Coarse stipple – continental; fine stipple – marine shales; unstippled depobelts – paralic; unstippled slope – marine to paralic; black – Afam channel.

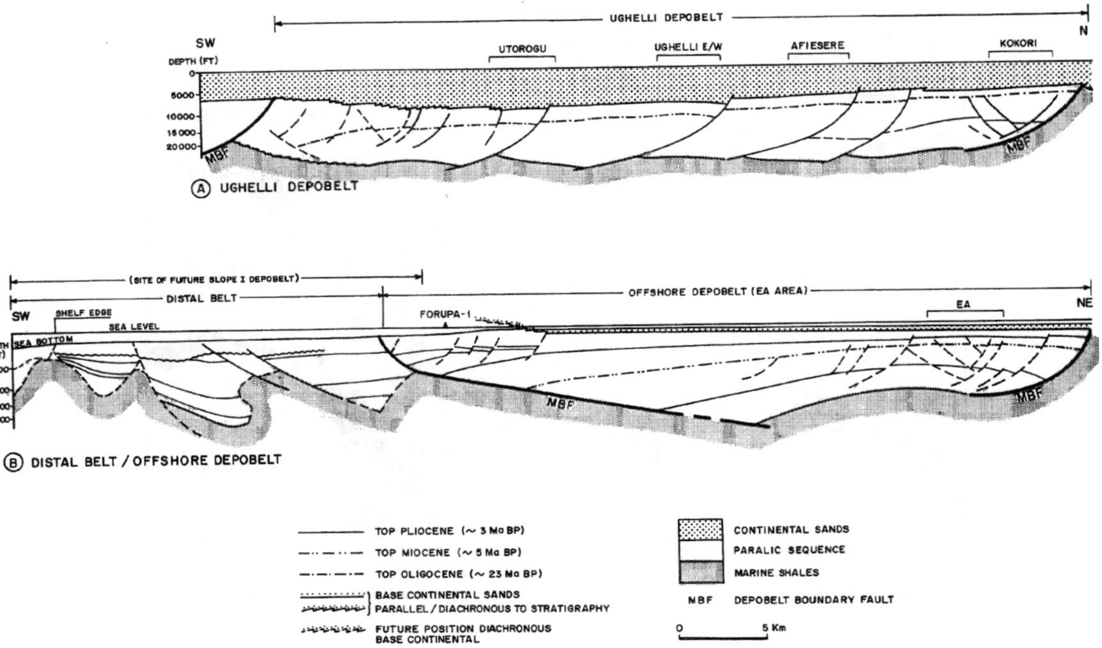

Fig. 3. Relationship of lithostratigraphy and structure in two depobelts. The Ughelli depobelt exhibits a time parallel paralic to continental lithofacies change over its greater part. The EA area is part of the youngest offshore depobelt. In this area the continental sands have only just advanced over the depobelt. Future diachronous advance is indicated. A major boundary fault for a future Slope I depobelt could be expected to develop near the well Forupa-1. The EA field is a collapsed crest structure and has already trapped significant volumes of hydrocarbons. MBF – major boundary fault.

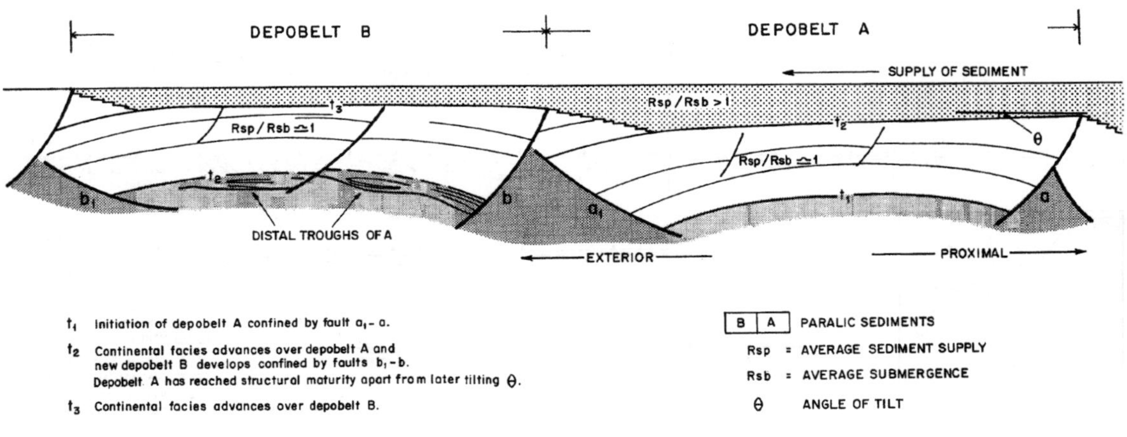

Fig. 4a. Temporal relationship of stratigraphy to time in depobelts. During paralic deposition the rate of average sediment supply was approximately equal to the rate of submergence. At t_2 (depobelt A), t_3 (depobelt B) submergence rate decreased substantially. As rate of sediment supply did not change the continental lithofacies advances rapidly to initiate new depobelt deposition (e.g. depobelt B at t_2).

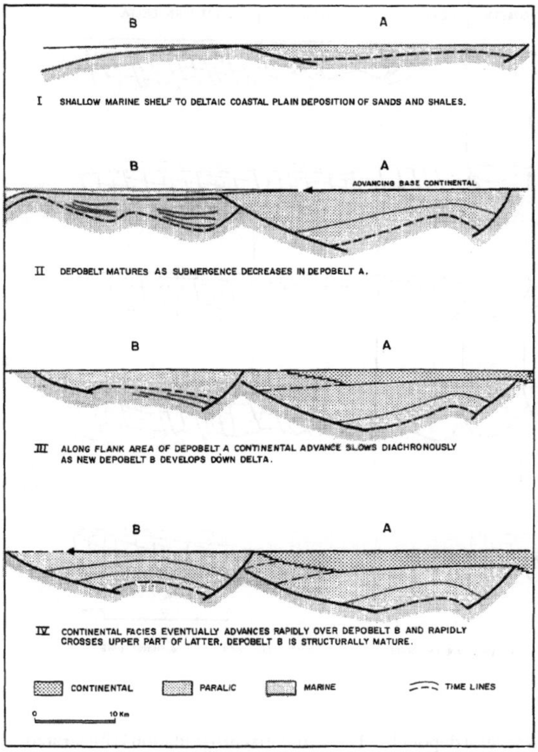

Fig. 4b. Schematic development of a depobelt.

which become increasingly younger seawards forming a more and more convex outer front to the Niger Delta (Figs. 5, 6, 7). The base continental sands which defines the Escalator Regresssion line is recognisable over the last 18 Ma.

Interaction between Sediment Supply and Submergence

Large thicknesses of alternating sand and shale were deposited in the Niger Delta as a series of offlap cythothems (Weber, 1971) 15–100 m thick (Weber & Daukoru, 1975) in a complex of shallow marine, fluviomarine littoral and deltaic plain environments. Average depositional base level was maintained at plus or minus sea level.

The alluvial sand sequence consists predominantly of a series of point bars. The present day topographic gradient of the Niger River from Onitsha (220 km inland) to the coastline is a maximum of 1 in 9.5×10^5 (Nedeco, 1959). This means that, at present, the depositional base level of alluvial sands over an enormous area is slightly above sea level. A similar gradient is likely to have been present over the delta throughout its development.

Using sea level as datum a base level of deposition can be defined and the rate of sediment supply can be related to submergence. Modifying Curtis (1970), the ratio Rsp/Rsb can be defined (Fig. 4a), i.e. average rate of sediment supply over the average rate of submergence relative to sea level. Thus: $Rsp/Rsb \overline{>} 1$; for maintenance of a base-level slightly above sea level. Some sediment (alluvial sand) is deposited to maintain the base level, but the excess is moved forward under conditions of net regression (Fig. 4). $Rsp/Rsb \simeq 1$, the average sediment supplied maintains the base level of a defined area by filling the space created by submergence to create a vertical pile of alternating sand and shale where conditions are alternatively $Rsp/Rsb \overline{>} 1$ (sand) and $Rsp/Rsb \overline{<} 1$ (shale).

During development of a depobelt the average rate of supply did not vary greatly. Climatic history suggests alternating wet/dry periods in the Niger River catchment area (Burka & Durotoye, 1970), resulting in cyclic deposition. These cycles are extremely short compared with the few million years required to develop and fill one depobelt. Thus the major variable controlling depobelt development is more likely to be the rate of submergence, itself ultimately related to the mobility of the underlying overpressured marine shale.

As submergence decreased, towards the end of depobelt development/paralic deposition the supply of sediment exceeded that required to deposit alluvial sands at base level so that a new area of sand/shale deposition develops seaward (e.g. the new depobelt) where the submergence rate is greater and capable of balancing the supply.

It is possible to estimate average rates of submergence. The delta stratigraphy is based on pollen and foraminiferal zones which can be related to the distribution of pelagic Atlantic foraminifera (Berggren, 1972; Stainforth et al., 1975; Saito, 1977) and to an absolute time scale (Fig. 7). The total thickness of paralic sediments can be estimat-

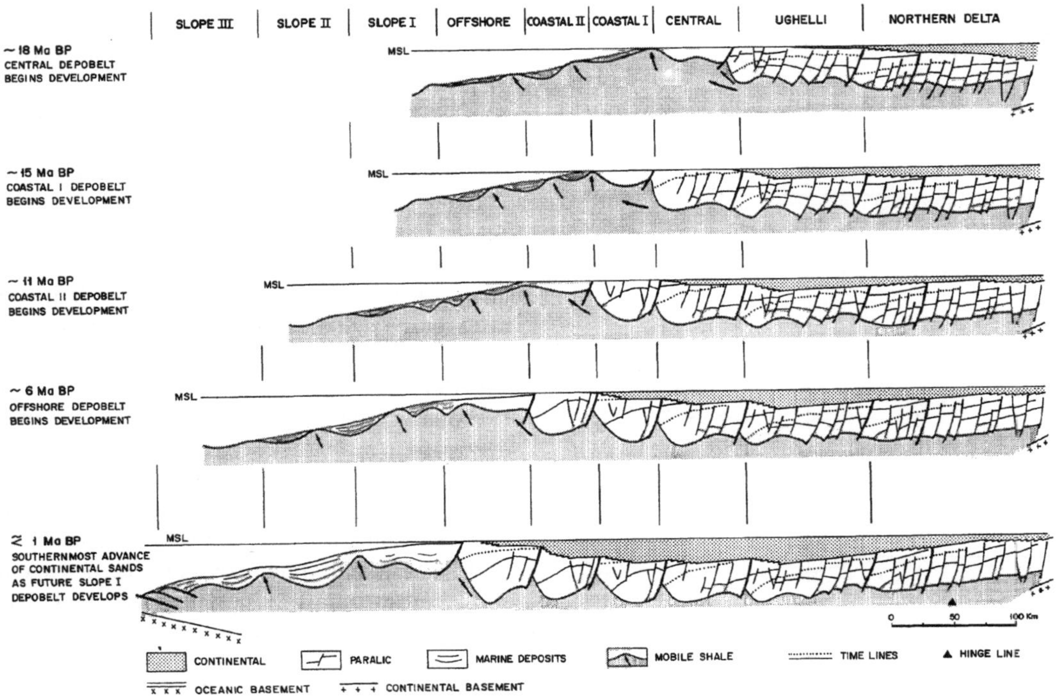

Fig. 5. Conceptual model for forward movement of the Niger Delta across several depobelts. The sequential progradation is illustrated. The shale bulge in front of each depobelt eventually becomes the location of the main boundary and counter-regional fault separating a set of two depobelts. The slope builds up by shale diapirism and deposition. The hinge line represents approximate transition from continental to oceanic crust.

ed within a depobelt. The times of initiation and ending of sand/shale deposition can be approximated by zonal time correlation. Rates of submergence are estimated by dividing thickness of sediment by time required for deposition. The rate of submergence of the base alluvial sands can also be similarly estimated.

The position of base alluvial sand and sand/shale sequence versus time (Fig. 7) can be constructed in a dip direction crossing several depobelts. A picture of stepped progression emerges for about the last 18 Ma of delta history. The approximate submergence rates for various depobelts and subsequent base alluvial sands are plotted on Fig. 7 (inset).

The base level of the alluvial sands has subsided at a rate of around 0.1–0.2 mm/year. The range for the paralic sand/shale of most depobelts is 0.2–1.0 mm/year. Clearly the alluvial sand sedimentation is related to a reduction in submergence rate. The Ughelli depobelt (0.17–0.34 mm/year) and the Northern delta area (0.13–0.16 mm/year) exhibit submergence rates equal to or intermediate between that of the alluvial sand and those of the younger depobelts. These differences suggest that a radical increase in sediment supply took place during the early Neogene, probably related to regional uplift within the African craton, for example, the Cameroon uplift. This is suggested by the total volume of sediment deposited before the Neogene compared with the volume deposited since then. As the time interval of deposition was the same (Fig. 7) the volume of sediment in each period represents the total supply. More than three times the volume of sediment was deposited in the Neogene and younger delta than previously, thus

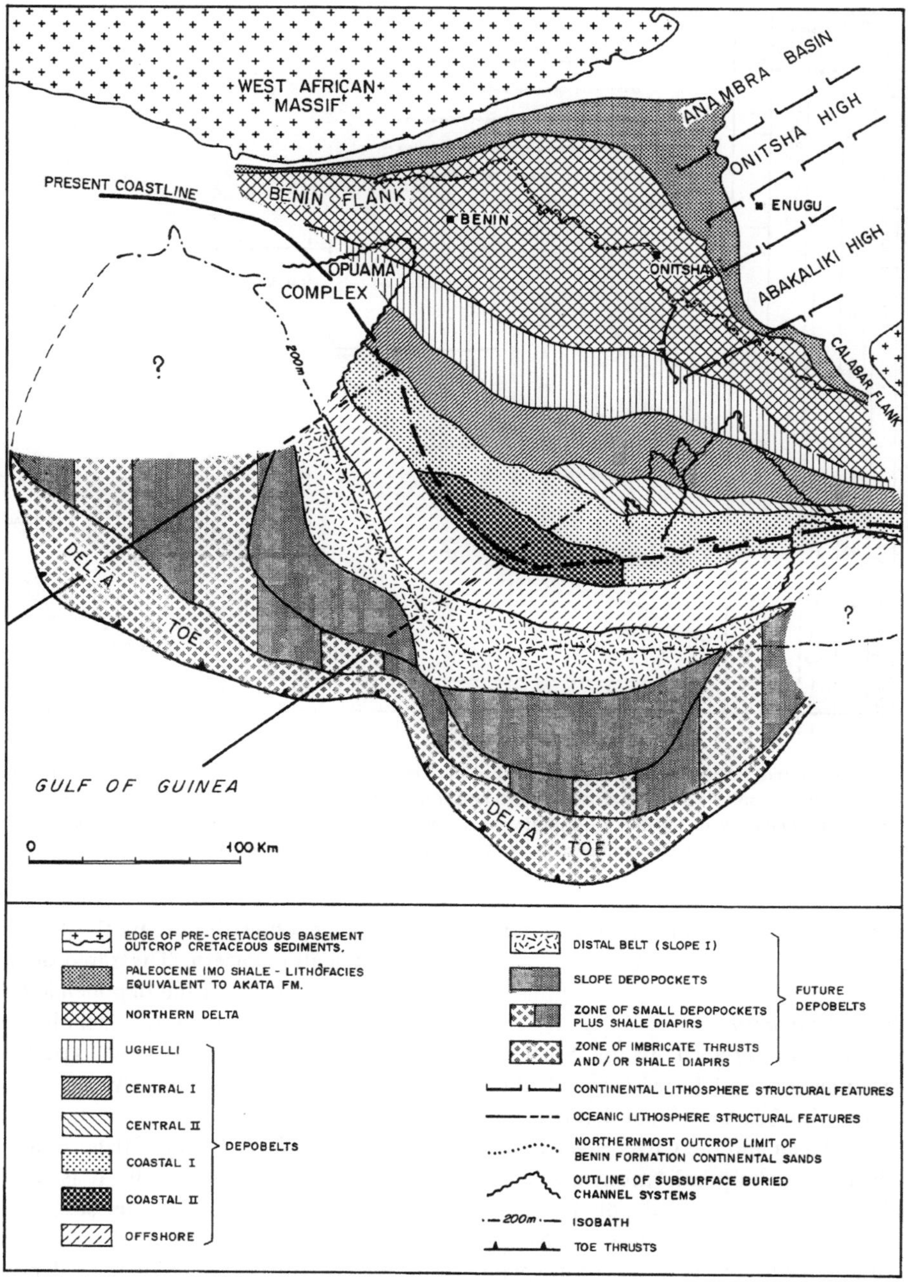

Fig. 6. Regional structural elements and depobelts. The gradual arcuate outbuilding of sequential depobelts is apparent from the Ughelli to offshore belts. The present slope suggests that the delta may be bifurcating into two lobes.

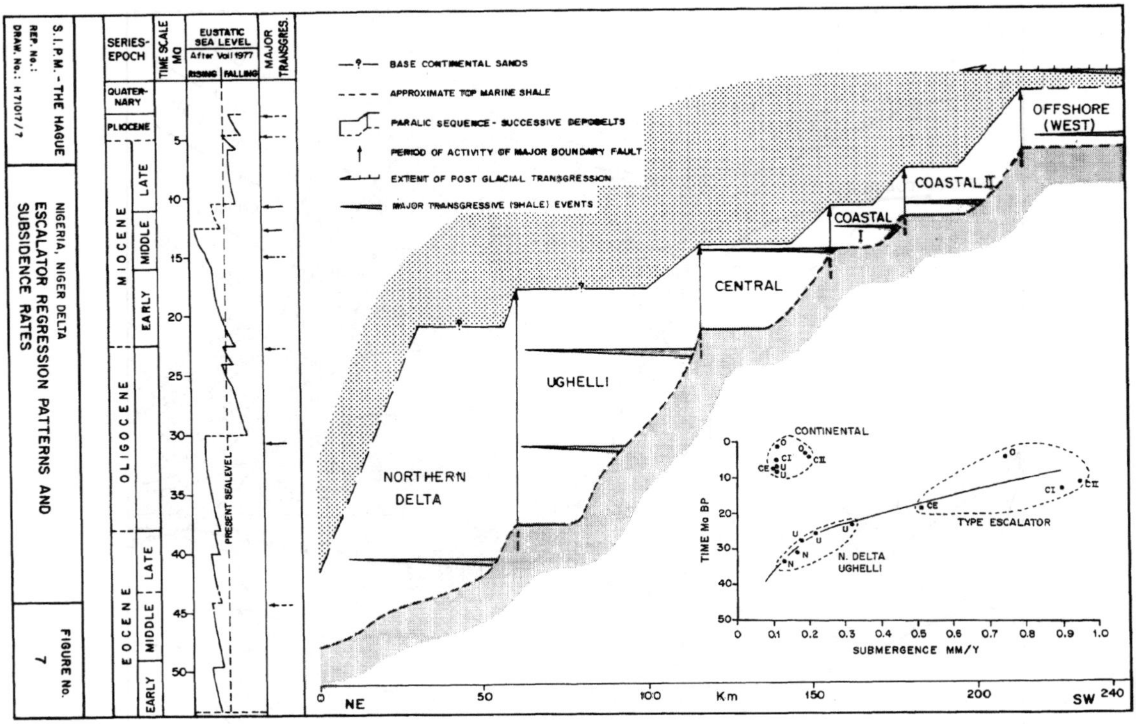

Fig. 7. Escalator regression pattern and subsidence rates. The Escalator regression line is represented by the base continental sands. The Northern delta and Ughelli depobelts developed over a much longer time interval compared to the younger depobelts. Marker shale positions are shown schematically in relation to transgressions. Inset right: Estimated average submergence rate are shown to distinguish the type escalator area from continental sand, Northern delta and Ughelli depobelt sequences.

initiating strong gravity tectonics in the mobile shale substratum.

Variations from the pattern

Northern delta/Ughelli depobelt. In the Northern delta a prograding series of facies occured when Eocene to early Miocene individual deltas in the Lower Anambra basin and the Cross River developed. The Northern delta paralic sequence developed coevally with the Ughelli depobelt. Apart from an initial thin sand/shale sequence, the Northern delta accumulated littoral to lower coastal plain sediments while the Ughelli area developed as a shelf and later as a littoral belt adjacent to the major boundary fault. The base alluvial sand facies eventually advanced rapidly over the Northern delta and a little later over the Ughelli depobelt. The Ughelli depobelt represents the first depobelt of the Escalator pattern. The rapid advance of the base continental over this depobelt led to the initiations of the next (Central) depobelt and subsequently, succeeding depobelts as part of repetitive pattern up to the present day.

Offshore and slope. The Offshore depobelt is the youngest recognisable belt. It is near structural maturity with a diachronous base to the alluvial sands beginning or initiated on the flank of the depobelt. Beyond the Offshore depobelt seismic and well data is limited. Shell/NNPC data and several publications (Beck & Lehner, 1975; Mascle, 1975; Emery et al., 1974; Grunau et al., 1975; Bornhold et al., 1973; Mermey, 1974; Delteil et al., 1974; Houboult, 1969 and Evamy et al., 1978) al-

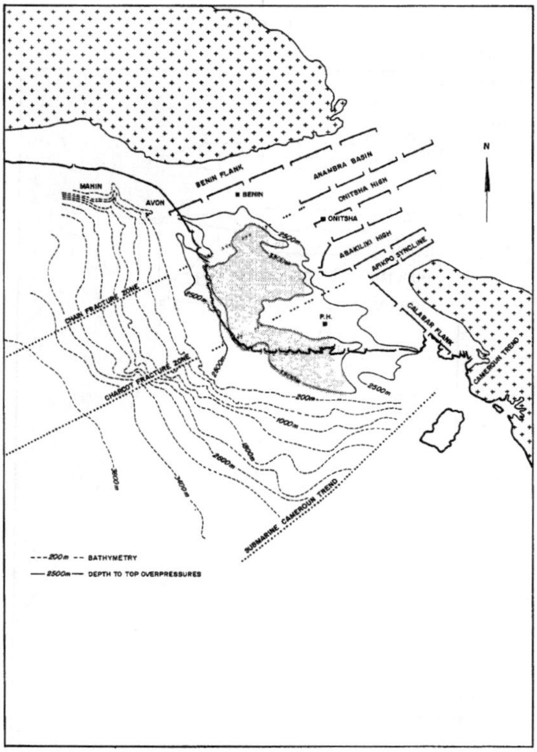

Fig. 8. Offshore bathymetry and subsurface overpressures. The pattern of the latter appears to be influenced by the position of the projection of major oceanic and continental fracture zones. Offshore, the Charcot fracture zone appears to have some effect on the delta slope bathymetry.

low the following composite picture to be built up (see Fig. 2).

The delta slope begins at approximately 200 m and ends at a depth of 3200 m (Fig. 2). An indentation separates two bulges directly southwest of the apex of the present day coastline (Figs. 6, 8). Average gradients on the slope range from 0° 51′ (western bulge) and 1° 16′ (southeastern bulge) to about 3° along the indentation. Locally, sharp changes in topography are present, associated with shale ridges and intervening sediment accumulations, giving gradients of up to 10°, possibly associated with submarine canyons/channels. At both eastern and western limits of the delta, canyons systems are identified, i.e. the Calabar and Mahin Canyon (Burke, 1972) and the Avon Canyon (Fig. 8).

The ocean floor consists of basement reflector overlain by two units:

(a) basal sediments filling basement depressions, followed by

(b) a more widespread layer of pelagic sediments.

The Niger Delta toe marks the base of the delta slope and consists of a zone of imbricate thrusts and or diapirs (Beck & Lehner, 1975). Up the slope, sediments encroach on diapirs and form depositional pockets. These merge into a wider depocentre (Beck & Lehner, 1975; Emery et al., 1974; Mascle, 1975) which is extensive on the two bathymetric delta lobes, but is, however, constricted into small sub-basins between diapirs along the indentation separating the two lobes.

Top of slope diapirs separate pockets of sediments in a Slope I depobelt beyond the offshore depobelt. Well results show that the most proximal pockets contain alternating sand shale sequences. Thus, this depobelt (Slope I) has received sediment and was developing during the Plio-Pleistocene, coincident with the southernmost advance of alluvial sands (Fig. 3). Sedimentation in this depobelt has been interrupted by the post-glacial Flandrian transgression. At present the water depth over the updip area of this depobelt is about 400 m. The maximum extent of this transgression lay inland of the present coastline (Oomkens, 1974). Eventually the Slope I depobelt will receive further sediments as the prograding delta completely overcomes the post-glacial transgression. The positions of potential future Slope II and III depobelts are indicated on Fig. 2.

Delta flanks. The depobelts are not clearly recognisable in the delta fringes (Fig. 2). In the eastern delta, eastern offshore and northwestern edge a thinner paralic section developed, contained in small roll-over structures confined by shale ridges or bulges. These areas submerged less rapidly and were eventually covered by continental sands. These areas merge via structurally intermediate areas into the 'type' escalator depobelts.

Comparison with other deltas. The one-way Escala-

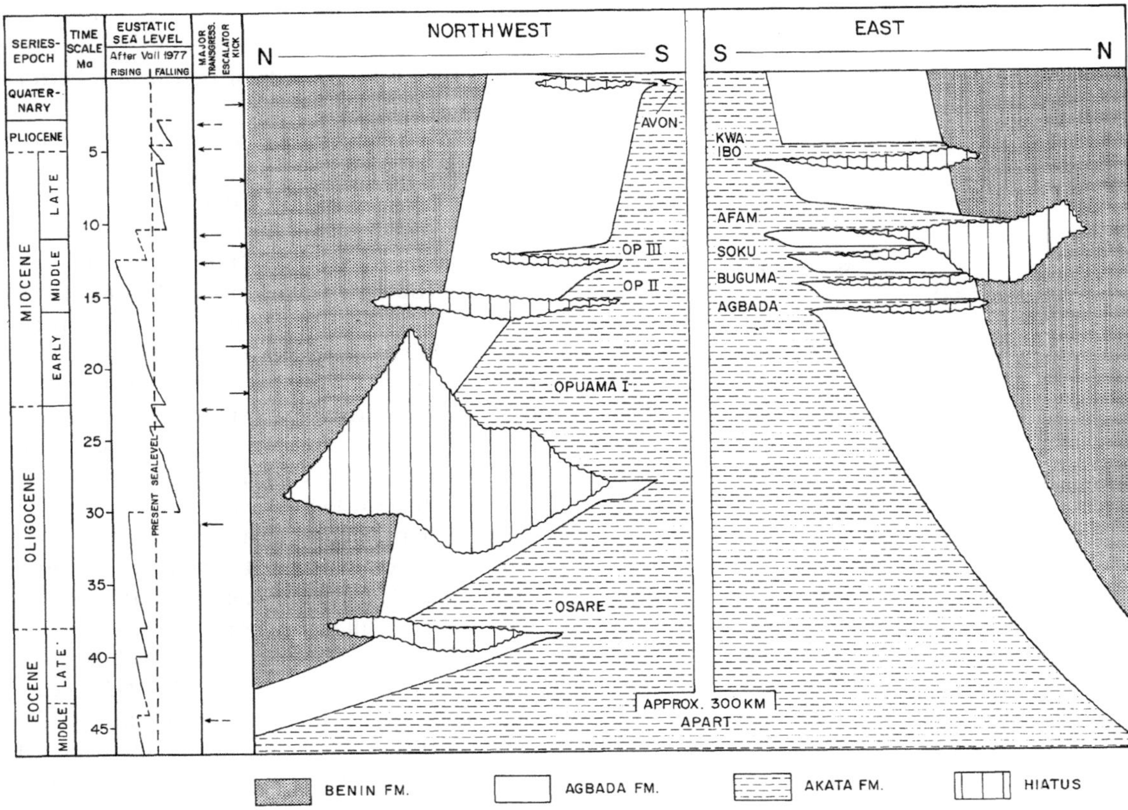

Fig. 9. Stratigraphic position of erosion surfaces and sediment fills in the northwest and east of the Niger Delta. The Opuama complex exhibits repeated channelling at different periods in the same area. In the east the channels occupy different positions broadly within the same area.

tor Regression model is not obviously recognisable in other major deltas. However, Yorston & Miles (1988) has recognised regional depocentres (depobelts), younging seaward in the onshore Gulf of Mexico controlled by salt sectonics including withdrawal and salt wall development. Nevertheless, several major transgressive/regressive cycles climb through the lithostratigraphy (Bruce, 1973). A similar situation applies in the Mackenzie Delta, Canada (Evans et al., 1975; Curtis, 1986). The latter is not a characteristic of the Niger Delta.

The Niger Delta depositional sequence includes a thick marine undercompacted and overpressured shale section which has responded in steps over a series of time transient belts. In an active depobelt the shale has been displaced downward and predominantly seaward. The continuity of these conditions through time has allowed the progressive Escalator Regression pattern to develop in the parts of the basin with a thick shale section (Figs. 2, 3, 6). Towards the edges, where a rising basement and a thinner shale section are present the pattern disappears or becomes highly modified (Fig. 2).

Channels

Several palaeochannel systems are recognised (Figs 9, 10, 11). A major group in the east includes the Agbada, Buguma, Afam (Omatsola & Cordey, 1978), Soku (Murat, 1972), Kwa Ibo and Enwhe channels. On the northwestern edge channelling

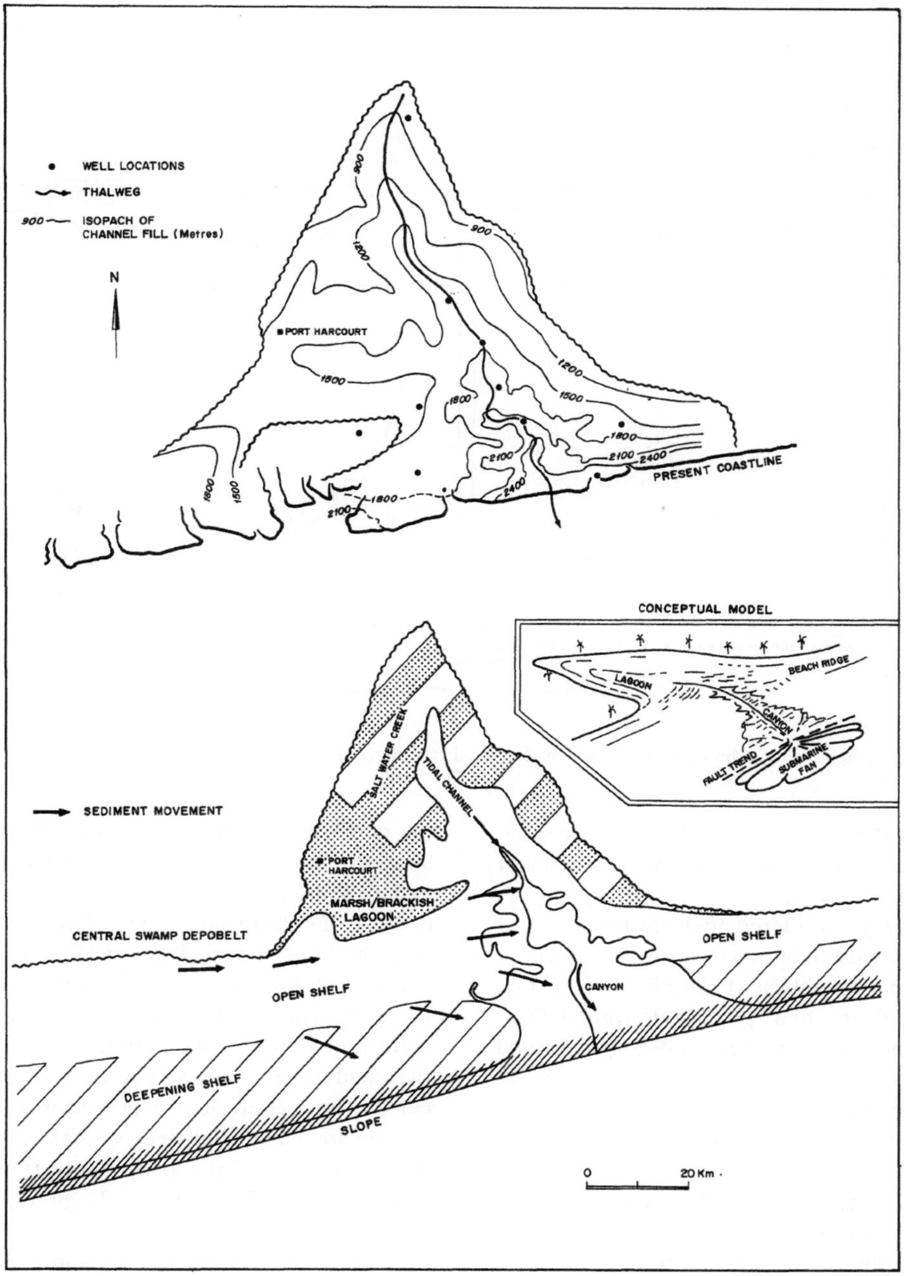

Fig. 10. Afam Channel topography, palaeogeographic setting and conceptual model.

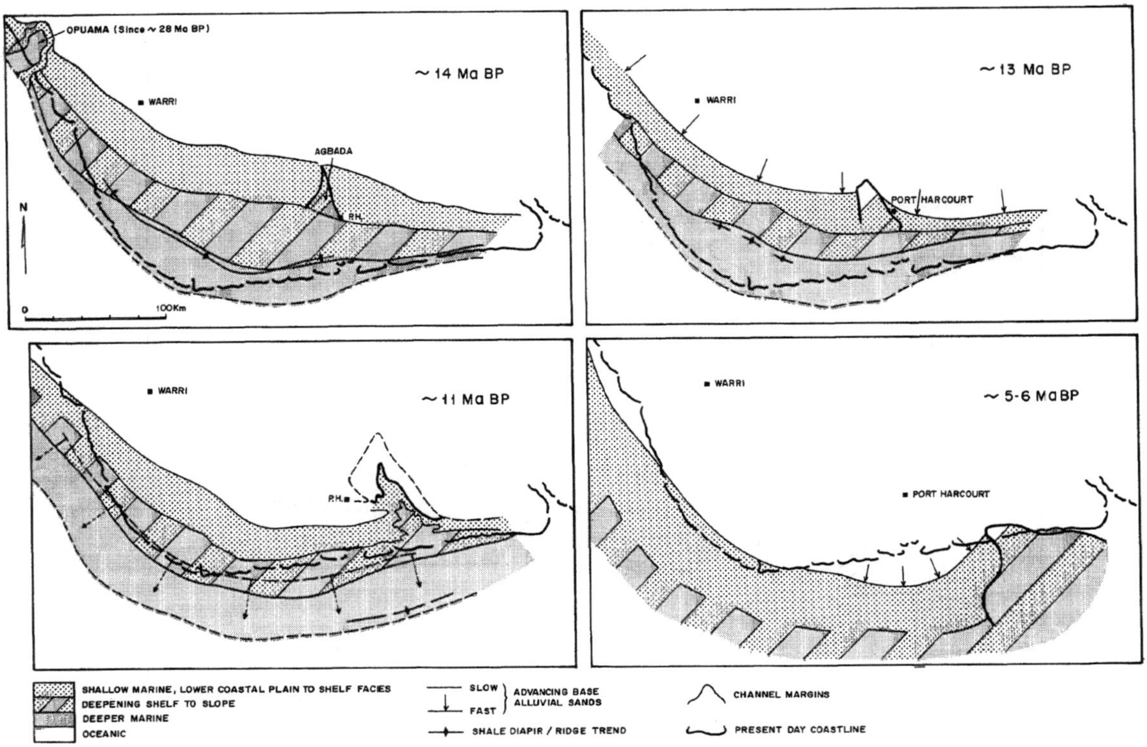

Fig. 11. Palaeogeographic setting through time of various channels in the Niger Delta.

has repeated itself in distinct eposides in the same geographic area resulting in the Opuama channel complex (Peters, 1984).

Murat (1972) considered them to be erosional river gullies filled and buried by transgressive clays. W.A. Knaap (unpublished) described the Afam Channel as a river estuary protected to the south by barrier complexes, while Burke (1972) interpreted it as a submarine canyon. Omatsola & Cordey (1978) interpreted the Afam as an estuary with an offshore submarine canyon. Our interpretation is as follows:

The Afam Channel is the largest channel feature, extending about 100 km northsouth (Fig. 10). The topography of the channel base is generally smooth in the north but becomes more irregular to the south. The original gradient on the base of the channel (using base continental sands on top channel fill as datum, as far as possible) is about 0° 45′ in the north and drops to 0° 16′ towards the centre. Southwards it becomes steeper with a gradient of about 1°. The gradient of this canyon is of the same order as the Zaire canyon (1°; Shepard & Emery, 1973) and the Principe Canyon (1° 27′).

Palaeontological evidence suggests that the northern Afam Channel was a salt water embayment with a central tidal channel contained by barrier sand developments. The incised deeper part cut the time equivalent depobelt as a submarine canyon and channelled deep water sands on to the then palaeo slope at Opobo South (Omatsola & Cordey, 1978). The Agbada, Buguma, Soku and Kwa Ibo Channels (Fig. 11) were also probably transient depobelt-cutting canyons. The larger northwestern Opuama complex consists of a series of cut and fill episodes which include the earlier Osare Channel (Fig. 9).

The channels have developed on the synchronously outbuilding delta front (Figs. 5, 6, 11). They represent clear erosional features which in the case of the larger ones, carried sediments from the depobelt shelves down the thalweg and into a submarine canyon to deposit sediment in deeper water. The erosional origins of the Opuama (Early Miocene, ca. 22 Ma); Afam (Late Mid Miocene, ca. 11 Ma) and Kwa Ibo (Earliest Pliocene, ca. 5 Ma) are clearly related to eustatic low sea level stands (Vail et al., 1977; Haq et al., 1987; Poag & Ward, 1987).

Subsequent transgression led to the deposition of clays which filled these erosional channel systems and then spread over the entire delta. Such clay bodies of individual depobelts are important as regional seals and their base represents stratigraphical events which also form regional seismic marker horizons.

The influence of the basin frame

The Cenozoic Niger Delta fills part of the southern Nigeria sedimentary basin (Murat, 1972) which extends inland to the Benue Trough (Fig. 1). Subsidiary basins relating to different and earlier periods of subsidence are the Benue Trough, Anambra Basin and Abakiliki Trough. The development of the whole basin is structually controlled by lineaments resulting from the opening of the Southern Atlantic Ocean. Important continental structural alignments can be projected into parallel fracture zones on the ocean floor (see Fig. 8).

The continental basement exhibits two main structural elements (Murat, 1972). A NE-SW system defines the northwestern edge of the basin and a NW-SE system defines the northeastern edge (the Calabar hinge line) of the Niger Delta basin (Figs. 6, 8).

On the ocean floor, Emery et al. (1974), Delteil et al. (1974), Mascle (1975), and Beck & Lehner (1975) have recognised extensions of the main ocean fracture zones. West of the delta, Emery et al. (1974) have traced these fracture zones into faults which define the edge of the continental crust. To the east, the Ascension fracture zone runs parallel to the aseismic ridge straddled by islands of the Cameroun volcanic chain.

Beneath the Niger Delta the fracture zones can only be traced for a limited distance under the slope. Between the Chain and Charcot fracture zones the oceanic basement is depressed (Beck & Lehner, 1975). Under the delta, circumstantial evidence suggests that structural lineaments connect oceanic and continental alignments (see also Babalola, 1985).

The Charcot fracture zone appears to coincide with an indentation in the delta toe (Fig. 6) and to affect the shape of the slope above it (Fig. 8). However, no effect is visible above the Chain fracture zone.

Free air (sea) and Bouger anomalies (land) define the sedimentary basin and the presence of basement elements such as the Abakiliki and Onitsha highs on the landward side of the delta (Mascle, 1975; Evamy et al., 1978). Linear magnetic anomalies show that the magnetic basement has elements which parallel the trend of the ocean fracture zones and the conjugate NE-SW and NW-SE continental structural systems. The geometry of the mapped top continuous shales (and overpressures – Fig. 8) and base continental sediments sag between the extrapolation of linear trends bridging the continental and oceanic lineaments. This may be a response to underlying compaction of thickest sedimentary fill developed between the two lineaments (Fig. 8).

Response of the basement to the delta

The basement has subsided to allow a sediment pile of about 12 km maximum thickness to accumulate (Mascle, 1975) in a complex response to depth and temperature dependent mechanical behaviour of the lithosphere. Gravity data (Hospers, 1965, 1971; Ajakaiye & Burke, 1973) has been interpreted to suggest that the basement has exhibited a regional isostatic response by crustal flexure (Walcott, 1972). Under the influence of loading the lithosphere has bent elastically to accumulate the deltaic sedimentary pile in what is only a first order description of the margin's response.

Other contributing processes are no doubt present such as subsidence owing to cooling of oceanic lithosphere (Parsons & Sclater, 1977) and associated continental margin (Sleep, 1971) or down to ocean faulting related to stress differences caused by density-depth changes across the crustal and topographic transition from continental to oceanic lithosphere (Bott & Dean, 1972). However, the relative contribution of each of these processes in the Niger Delta is not known.

Summary of geological history

The Benue Trough developed as a tectonic depression during separation of Africa from South America (Wright, 1968; Burke et al., 1972; Olade, 1975). The fact that the trough margins are parallel to the ocean fracture zones suggests an underlying relationship with the oceanic transforms. Vertical displacements would have been prominent as the developing ocean formed and the early floor subsided.

This is clearly reflected in the sedimentary history of the Benue Trough. From early Albian to possibly Aptian time basic alkaline volcanics were deposited in shallow marine areas (Uzuakpunwa, 1974). They were followed by a series of major siliciclastic transgressive/regressive cycles which were separated by three major structural inversions and one folding episode (Murat, 1972). The last Late Cretaceous/Early Tertiary regressive cycle was the precursor of the Cenozoic delta. It was the areally most extensive cycle and filled the entrance to the Anambra Basin with deltaic sediments. Growth fault trends are recognised, probably related to basement faults, but extensive shale tectonics did not occur, so that equivalent biostratigraphy and lithostratigraphy can be recognised on either side of fault trends. This delta was overwhelmed by the Palaeocene Imo shale transgression which brought deep marine shales to the head of the Anambra Basin.

The Cenozoic Niger Delta advance began in the Middle Eocene. As it developed, growth faulting and roll-over anticlines became more marked. Submergence was slower at first, related to shallower basement, thin mobile shale and less sediment supply than in the later history of the delta. Thus, greater reworking of deposits occurred to form rather sandy sequences which are characteristic of the northern delta.

The Ughelli depobelt expresses the first clear development of the escalator pattern. However, sediment volumes and average submergence rates were less than those of later typical depobelts which formed following a marked increase in supply, related to Neogene uplift of the African hinterland. Submergence rates accelerated on a thicker more mobile shale substratum in response to increased sediment load. The arcuate form of the depobelts suggests that the areas of subsidence were defined by the loading, the severity of which was related to deposition and distribution of the clastic material by the prevailing wave and wind directions, as in the present day (see Pugh, 1954). Shale tectonics developed to form ridges or diapirs along the trend of the prevailing depobelt. As these arcuate trends do not parallel basement fault trends, block faulting of the latter did not have any major influence on deposition.

Deltaic sediment loading caused the underlying overpressured shale mass to squeeze forward and outwards (upwards) on the slope (Merki, 1972). The relationship of the diapirs to the front of the delta, occasional sediment samples, heat flow anomalies (Houbolt, 1969) and low-velocity seismic pull-down effects (Mermey, 1974), where horizontally layered sediments exhibit a synclinal effect on seismic time sections, indicate the presence of overpressured shales.

Rapid depobelt related shale-tectonic-controlled sedimentation overshadowed a continuous slower delta-wide effect which is partly manifested during deposition of the continental sands. Several effects, probably in combination, could have caused this:

- Isostatic adjustment to the sedimentary load would have proceeded continuously by elastic bending of the lithosphere as the delta advanced (Hospers, 1971; Walcott, 1972).
- Cooling of oceanic lithosphere would have caused progressive subsidence away from the mid-ocean ridge (Parson & Sclater, 1977). Following these authors, this effect was probably much more important during the Late Cretaceous/Early Tertiary, than from the Neogene to the present when the heat flow has probably been steady. Up to 5.5 km cumulative subsidence may have occurred based on Parsons & Sclater (1977).

– Hot creep of lower continental lithosphere (Bott, 1977; Bott & Dean, 1972) towards the suboceanic mantle could have caused internal mass changes and thinning of the continental crust. At the same time normal faulting in the brittle upper layers would have caused subsidence and space for sediments. The dominant shale tectonic control on individual depobelts has developed a rhythym (Figs. 6, 7) in the Niger Delta which not only overshadows the complex continental margin lithospheric response but also the effect of eustatic sea level changes. The major channel systems and depobelt-wide shale markers are evidence that the outbuilding depobelts, during fall and rise of sea level, were temporarily interrrupted by incision and subsequent transgression (Fig. 9). However, the effects were largely restricted to the active depobelt and were quickly overcome by renewed regressive advance of the delta. One example of which is the delta's recent recovery and advance following the post-glacial Flandrian transgression (Oomkens, 1974). Despite major sea level rises and falls throughout the Cenozoic there has been little impact on the one-way stepwise advance of successive depobelts.

– Continued overpressure leakage; changes in shale mineralogy and compaction of the sand/shale sequences would have led to further subsidence.

Implications for hydrocarbons

Hydrocarbons are trapped mainly in roll-over anticlines associated with syn-sedimentary growth faults. Evamy et al. (1978) related the physical properties of hydrocarbons particularly to source rock variation and geotemperature. They concluded that late migration occurred near attainment of present day overburden coinciding with source rock maturity.

Variations in oil and gas were ascribed to original source rock differences and later segregation and mixing owing to secondary migration.

Consequences derived from Escalator Regression are as follows:

– Temperature regimes are almost stabilised within the depobelts following passage of continental sands

– Source rock richness and type variations within any depobelt should be broadly similar to that of adjacent updip and downdip depobelts given similar sedimentary environments and rates of deposition.

– Depths to top oil and gas kitchens should have been similar in depobelts developed in an equivalent time period at the time of stabilisation of a depobelt, when continental sand advance occurred. Both Akata and Agbada Formations would contribute as hydrocarbon sources in most of the delta.

– Older depobelts should have incrementally shallower kitchens measured from base continental sands relative to subsequent younger depobelts owing to geological time effect on maturation.

– Significant hydrocarbon migration and accumulation occurs prior to final overburden. In the type escalator area this is as early as deposition of the base continental sands, or even earlier.

– Tilting posterior to and coeval with accumulation in all but the youngest depobelts is likely to have caused remigration of considerable hydrocarbon volumes in updip directions.

Temperature gradients

Temperature gradients average 0.7° C/100 m in the continental sequence, 2.73° C/100 m in the paralic sequence and 5.47° C/100 m in marine shales (Evamy et al., 1978; Weber, in prep.) Very little variation is apparent in the paralic section within a sand percentage range of 35% to 75%. Continental sand gradients are low because of the excellent permeability and convective overturn and mingling of formation waters with near surface waters. Two examples of gradient profiles are illustrated in Fig. 12. Profile A represents a juvenile depobelt where structural maturity has just been attained and profile B represents an older depobelt where 2000 m of continental sequence have been deposited above it. Conversely, profiles A and B can be considered to be from the same depobelt at different stages in its history. Accumulation of continental sands causes the subsurface temperature profile of A to depress to B with only a small rise in temperature in

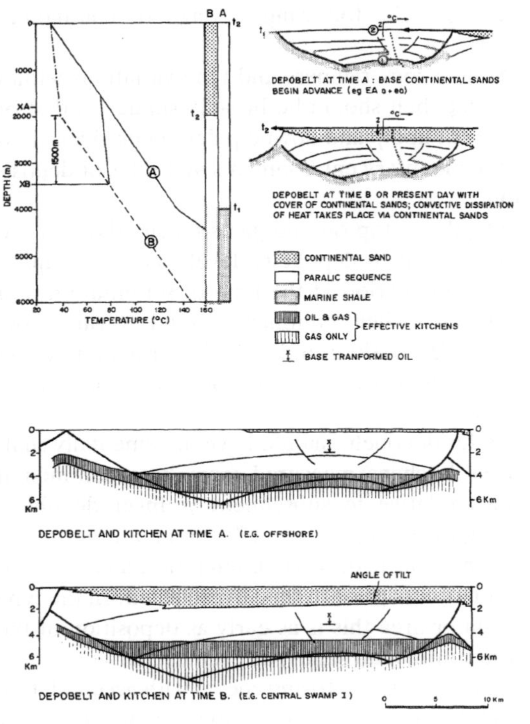

Fig. 12. Schematic relationship of temperature, maturity and degraded oil to depth in depobelts.

the paralic section. The effect of the deposition of a continental sand section is to almost stabilise the temperature regime of the paralic section.

This has implications for conclusions drawn from the presence of bacterially degraded oil which is confined to temperatures less than 65–80° C where the bacterial agents can survive. Evamy et al. (1978) used this as support for late migration when present delta overburdens had been attained. However, the temperature history shown above throws considerable doubt on this conclusion. Transformed oil will occur between approximately 1800 and 1500 m below base continental, or between 3500 and 1800 m from surface in the example shown (Fig. 12). As transformed oils often show a gradation in gravity across the transition depth to normal oil, this may indicate that transformation is curtailed by the small increase in temperature brought about by burial under the continental at a critical temperature between 65°–80° C.

Source rock distribution and maturity

Geochemical studies (Evamy et al., 1978; Weber, 1986) show that the source rocks are dominantly Type II to III, humic to mixed with coaly, lipid and structureless organic components. Lipid components consist of sporomorphs, resins and waxes. Most source rock analyses are derived from the paralic section and little is known of the marine shale source rock potential. The transient repetition of a similar lithofacies geometry in each depobelt means that, whatever the subtleties of source maceral distribution, a broadly similar palynofacies relationship can be expected in each depobelt given a similar rate of sediment supply.

The presence of gas and condensate in distal sediments between shale ridges suggests that hydrocarbons are generated and able to escape from the mobile overpressured section where a hydrocarbon kitchen is already developed. A dilution of the mainly land-derived allochthonous organic matter is likely to take place on the outer shelf and slope where shale deposition was dominant with a consequent reduction in expulsion efficiency. As a result, lighter hydrocarbons may be more prevalent in the distal part of each depobelt. For other parts of the world, Hedberg (1974) has drawn attention to the association of hydrocarbons with mobile shales. In the Niger Delta, Albright (1973) has suggested that offshore seeps are of thermal origin. It is likely that shale diapirs can juxtapose mature section against younger paralic sequences.

Evamy et al. (1978) matched oil and gas generation respectively to temperatures of 116° C and 149° C throughout the delta. However, as the delta is composed of a series of successive time-transient depobelts fossilised beneath continental sands at different periods of time, maturation levels could be expected to show the effect of time as well as temperature (depth). Vitrinite reflectance and other maturity indicators give only a limited picture of maturity as they are derived from wells in structurally crestal positions. As available data show a broad correlation with estimates, a modification of the Lopatin method (Lopatin, 1971) has been used to approximate maturity levels with depobelts.

The offshore depobelt has a considerable volume of sediment within the oil kitchen and part, adjacent to the counter-regional fault, within the gas kitchen. This kitchen is confirmed by several fields including the Meren (Poston et al., 1983) and EA fields. Beyond this depobelt gas condensate has already been referred to above in distal sediment between shale diapirs.

Older depobelts display similar geometry and fill prior to the advance of the base continental sands, therefore given similar time periods of development, sand/shale ratios and therefore temperature gradients will have exhibited comparable maturity profiles at the time of stabilisation. Subsequently the top kitchen will have risen with time with respect to the stratigraphy (base continental) in an individual depobelt (Fig. 12) although it may actually become deeper with respect to the surface.

Calculations indicate that, given similar stratigraphy and temperature gradients, the top oil kitchen should be about 800 m higher with respect to the base continental sands in the Central Swamp I depobelt than in the offshore belt. Thus, both Agbada and Akata Formations contribute as sources for hydrocarbons (see also Ejedawe et al., 1984).

It does not seem necessary to invoke expulsion at low maturity levels from the Agbada Formation (Lambert-Aikhionbare & Ibe, 1984) or particularly from resins in this formation (Ejedawe, 1986). In a study of resin pyrolysis Lewan & Williams (1987) demonstrated that resins transform well within the oil kitchen and produce unusual products which are not normally found in major quantities within crudes. Both the Agbada and Akata Formations can contribute to the Niger Delta crudes at normally accepted maturities. The Akata will naturally provide the earliest hydrocarbons, probably biased towards gas and condensate because of dilution of organic matter and inherent difficulty of expulsion from undercompacted shales (e.g. distal belts and eastern delta). Nevertheless, the mobile character of these shales presents opportunity at an early stage to juxtapose sand-bearing sections in distal pockets against mature shales in diapirs: a process that continues when the diapirs become shale ridges juxtaposed by major boundary and counter-regional faults of the depobelts. Thus, the mechanism envisaged by Weber et al. (1978) whereby seismic pumping of hydrocarbons along fault zones takes place during dilation associated with movement would augment and continue charge from the Akata. Impinging on this and overlapping with it would be hydrocarbon contribution from the Agbada Formation as a depobelt reached its stable form at the time that the continental sequence advanced over it.

Migration and retention

Each depobelt has its own separate integrity. Hydrocarbons rise along flanks from kitchens to structural culminations. The offshore depobelt maturity profile indicates that more gas generation is likely in the exterior part. Augmentation of the kitchen with time and tilting progressively pushes more of the exterior areas of depobelts into the gas kitchens giving a proportionately greater gas charge to the flanks of depobelts.

Since stabilisation of each depobelt, a continental 'piggy-back-basin' has formed above all depobelts with maximum thickness of about 2200 m. This has promoted tilting of the paralic depobelts in both dip and strike directions (Fig. 15). The angle of tilt can be derived in any depobelt by reference to the base continental sequence and present day surface.

Assuming a reservoir geometry of a spherical segment filled to spill-point, say at an intersection with a non-sealing fault (Fig. 13), present day retained hydrocarbons can be estimated as a percentage of original volume, given the present day column height and diameter of the structure at the hydrocarbon-water contact. A nomogram (Fig. 14b) illustrates the relationship between the latter, retained (present-day) hydrocarbons and angle of structural tilt. For example a current closure (especially in areas with the thickest continental sand) with a diameter of 2500 m and column of 25 m (ratio 100) undergoing tilts of 1.0°, 1.5° and 2.0° would now contain 23%, 13.5% and 8% respectively of the originally trapped hydrocarbons. Structures filled to spill point with layered gas and oil would preferentially loose the oil and augment the oil volumes trapped updip of the tilt. Haalebos

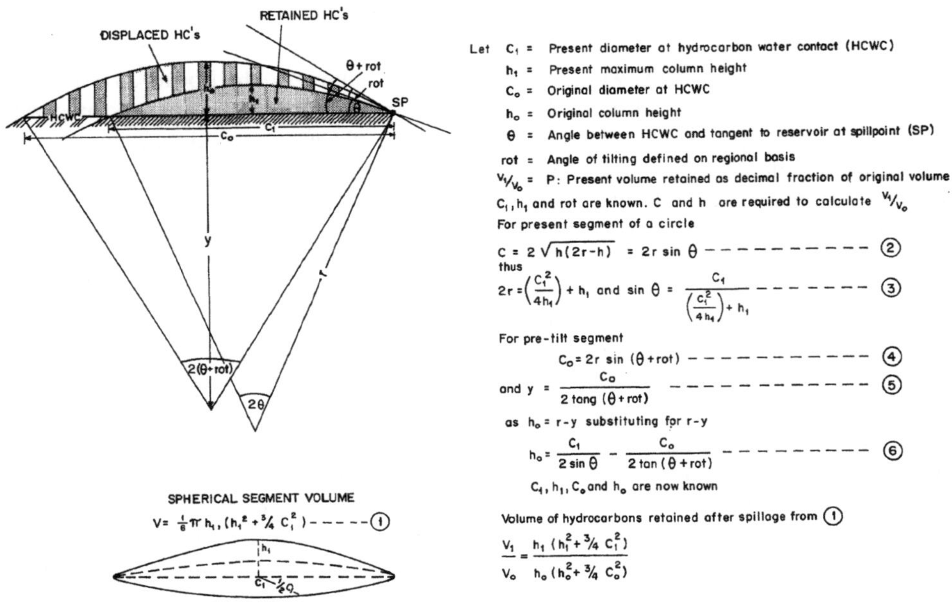

Fig. 13. Model for determination of decimal volume fraction of retained hydrocarbons in a reservoir in the form of a spherical segment which has been structurally tilted about its spill-point.

& Murdock (1981) observed the presence of residual oil in the Oloibiri Field below the oil-water contact which may be suggestive of tilting. This tilt-remigration effect enhances the oil-gas ratio in dip-tilt directions (e.g. Ughelli depobelt, Kokori area), but disturbs it in the strike-tilt directions. The Central and Coastal I depobelts are characterised by the latter and contain gas-prone structures with shallow strike closure in the central, most deeply buried, area. However, updip fields along strike with large volumes of oil and gas are present. In general, hydrocarbon concentration appears to be updip of main tilt direction (Fig. 15, see Evamy et al., 1978 and Ejedawe, 1981 for hydrocarbon distribution).

Conclusions

The Escalator Regression model is a conceptual description of the arrangement of several transient arcuate depobelts defined on the basis of structure, lithofacies and biostratigraphy. One-way advance of the Cenozoic Niger Delta took place principally in a series of steps which can be described by the position of the base continental sands. The model requires the presence of an underlying overpressured mobile shale mass which accommodates depobelts by deformation and displacement. It is apparent that a reduction of the shale thickness restricts the development of the depobelts, particularly towards the edges where the basement frame rises to the surface.

In the type area the dip width of the depobelts shows a surprising uniformity. Neither major changes in sediment supply nor the fact that the delta has spread outwards appear to have had and any great effect on width. The thickness of the prospective sequence is also rather constant. Thinning occurs towards the frame and against basement features such as the Abakiliki and Onitsha Highs. In the type area each depobelt has accommodated a comparable thickness of sand/shale. The importance of the wave destructive delta front

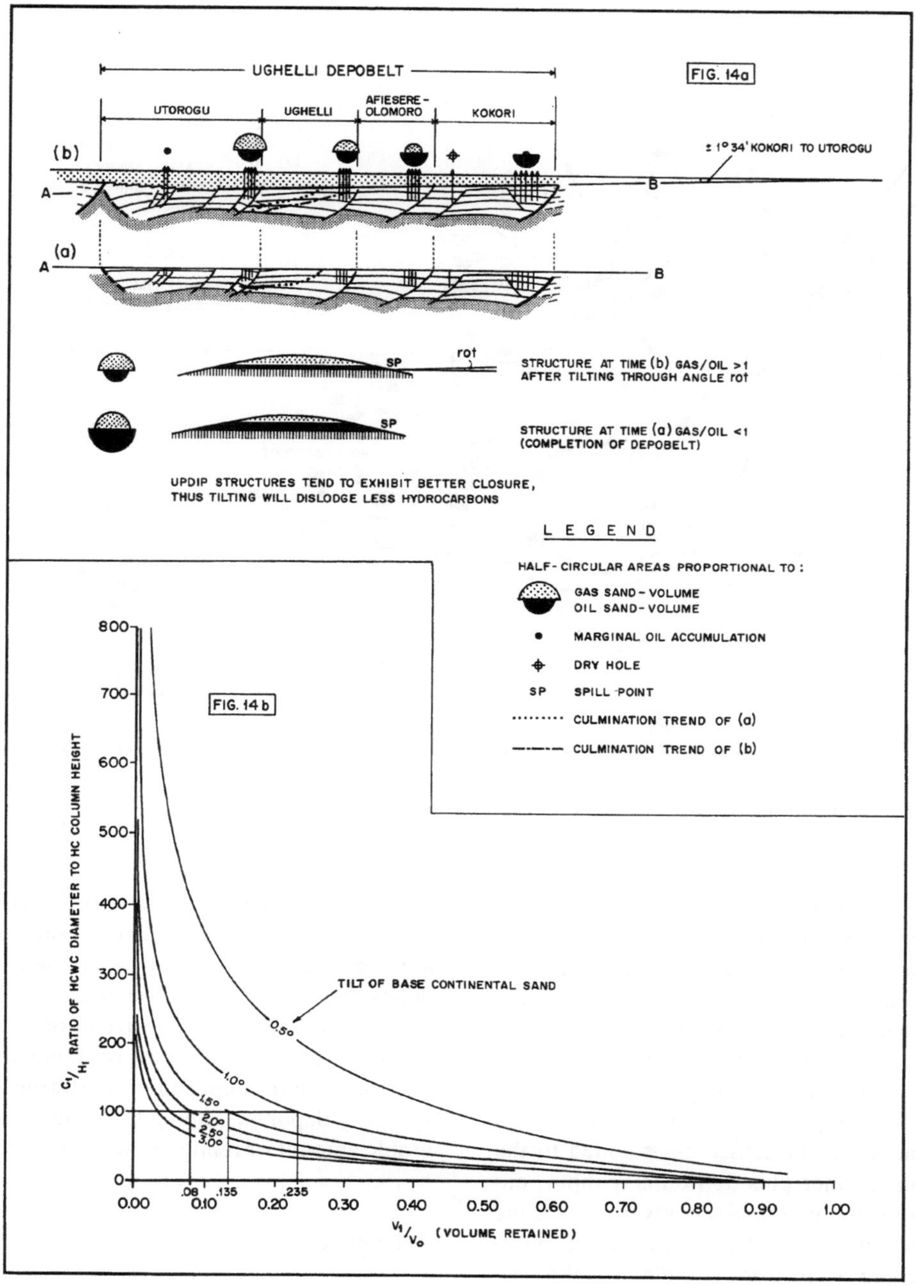

Fig. 14a. Schematic relationship of hydrocarbon accumulation and structural culmination to pre- and post-tilting mode.

Fig. 14b. Nomogram relating present column height, HCWC and tilt of base continental sands for reservoirs originally filled to spill-point.

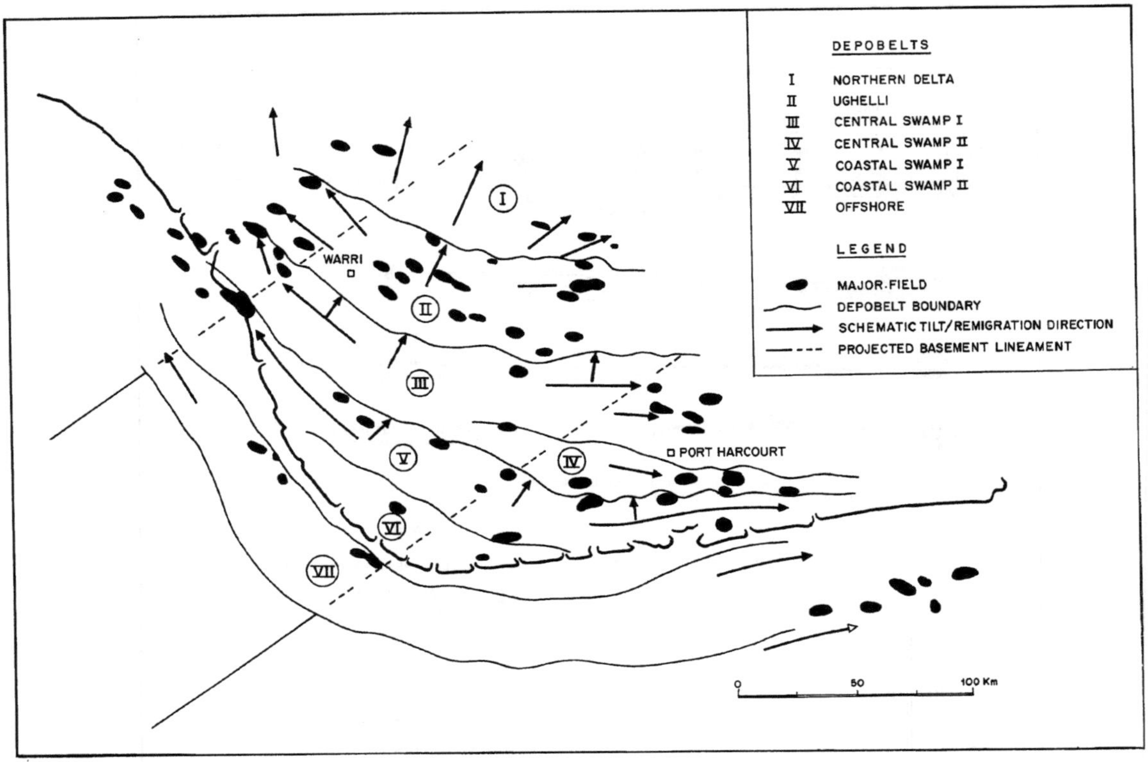

Fig. 15. Major fields, depobelts and conceptual tilt directions.

with lateral sediments dispersal has ensured the continuity of depobelts in a strike direction.

A dynamic geometrical relationship appears to exist between the warped basement, mobile overpressured shale, the developing delta slope and the active depobelt which provide limits to the dip width and thickness of the depobelts. Undoubtedly, depobelt sequences displace shale oceanwards towards the slope. Sediment added to the slope in this manner plus detrital clay input, enables the slope to build up and outwards. At some points an accreting slope balances the depobelt so that submergence slows down and allows the alluvial sand facies to advance.

The consequences of the model would include initial migration and accumulation of hydrocarbons long before final overburden development. The distal and offshore areas are currently in a comparable state to the situation in older depobelts prior to the advance of continental sands. Temperature gradients are broadly stabilised, following deposition of the base continental sands. Maturity models and hydrocarbon distribution indicate that hydrocarbon generation is continuous from an early stage in the depositional and structural history of any depobelt. Tilting has produced a considerable amount of remigration of hydrocarbons.

Acknowledgements

This work was carried out during the authors' employment with Shell in Nigeria. The model was developed in 1976 and has been broadly applied in the intervening years. We would like to thank Mr. Ephraim Umeadi in particular for assistance and colleagues who contributed in discussions to the concepts, particularly W.G. Cordey, P.A. Rowlands and K. Weber. The work also draws on the unpublished efforts and work of numerous Shell Petroleum Development Company geologists and geophysicists. Improvements to the manuscript have benefitted from discussions with H. Doust and R. Eckert.

Permission to publish was generously granted by the Nigerian National Petroleum Corporation and The Shell Petroleum Development Company of Nigeria.

References

Ajakaiye, D.E. & K. Burke, 1973 A Bouger gravity map of Nigeria – Tectonophysics 16: 103–115.

Albright, J.L. 1973 Seismic charts gas seeps – Offshore 33: 123–128.

Babalola, O.O. 1985 Crustal structure and tectonic development of Gulf of Guinea cul-de-sac from integrated interpretation of new aeromagnetic and existing geophysical data – Amer. Assoc. Petr. Geol. Bull. 69: 253.

Beck, R.H. & P. Lehner, 1975 Oceans, new frontier in exploration – Amer. Assoc. Petr. Geol. Bull. 58: 376–395.

Berggren, W.A. 1972 A Cenozoic time-scale, some implications for regional geology and paleobiography – Lethaia 5: 195–212.

Bornhold, B.D., Mascle, J.R. & V. Renard, 1973 Diapiric structures of Niger Delta– Amer. Assoc. Petr. Geol. Bull. 57: 1672–1678.

Bott, M.H.P. 1971 The interior of the Earth – E. Arnold Limited (London): 316 pp.

Bott, M.H.P. & D.S. Dean, 1972 Stress systems at young continental margins – Nature, Phys. Sci. 235: 23–25.

Bruce, C.H. 1973 Pressured shale and related sediment deformation mechanism for development of regional contemporaneous faults – Amer. Assoc. Petr. Geol. Bull. V. 57: 878–886.

Burke, K.C. 1972 Longshore drift, submarine canyons and submarine fans in development of Niger Delta – Amer. Assoc. Petr. Geol. Bull. 56: 1975–1983.

Burke, K.C., Dessauvagie, T.F.J. & A.J. Whiteman, 1972 Geological history of the Benue valley and adjacent areas. In: T.F.J. Dessauvagie, A.J. Whiteman (eds): African Geology – Geol. Dept. Ibadan, Nigeria.

Burke, K.C. & B. Durotoye, 1970 Late Quaternary climatic variation in southwestern Nigeria, evidence from pediments and pediment deposits – Bull. Ass. Sénég. Quatern. Ouest Afr. (Dakar). 25.

Curtis, D.M. 1970 Miocene deltaic sedimentation; Lousiana Gulf Coast – Soc. Econ. Palaeont. Min. Spec. Publ. 15: 293–308.

Curtis, D.M. 1986 Comparative Tertiary petroleum geology of the Gulf Coast, Niger and Beaufort-Mackenzie delta areas – J. Geol. 21: 225–255.

Delteil, J.R., Valery, P., Montodert, L., Fondeur, C., Patriat, P. & J. Mascle, 1974 Continental margin in the northern part of the gulf of Guinea. In: C.A. Burke & C.L. Drake (eds): The geology of continental margins – Springer (Berlin): 297–311.

Ejedawe, J.E. 1981 Patterns of incidence of oil reserve in Niger Delta Basin – Amer. Assoc. Petr. Geol. Bull. 65: 1574–1585.

Ejedawe, J.E. 1986 The expulsion criterion in the evaluation of the petroleum source beds of the Tertiary Niger Delta – J. Petr. Geol. 9, 4: 439–450.

Ejedawe, J.E., Coker, S.J.L., Lambert-Aikhionbare, O.O. Alofe, K.B. & F.O. Adoh, 1984 Evolution of oil-generative window and oil and gas occurrence in Tertiary Niger Delta Basin – Amer. Assoc. Petr. Geol. Bull. 68: 1744–1751.

Emery, K.O., Uchupi, E., Phillips, J., Bowin, C. & J. Mascle, 1974 The continental margin off western Africa: Angola to Sierra Leone (unpublished manuscript) – Contr. Woods Hole Oceanogr. Inst: 1–152.

Evamy, B.D., Haremboure, J., Kamerling, P., Knaap, W.A., Molloy, F.A. & P. Rowlands, 1978 Hydrocarbon habitat of the Tertiary Niger Delta – Amer. Assoc. Petrol. Bull. 62: 1–39.

Evans, C.R., McIvor, D.K. & K. Magara. Organic matter, compaction history and hydrocarbon occurrence – McKenzie Delta, Canada – Proc. 9th World Petr. Congr.: 149–157.

Fränkl, E.J. & E.A. Cordry, 1967 The Niger Delta Oil province: Recent developments onshore and offshore – Proc. World Petr. Congr.: 195–209.

Grunau, H.R., Lehner, P., Cleintuar, M.R., Allenbach, P. & G. Bakker, 1975 New radiometric ages and seismic data from Fluerteventura (Canary Islands), Maio (Cape Verde Islands) and Sao Tome (Gulf of Guinea) In: Borradaile, G.J., Ritsema, A.R., Rondeel, H.E. & Simon, O.J. (eds): Progress in Geodynamics – Royal Netherlands Academy of Arts and Sciences North-Holland Publ. Cy. (Amsterdam): 90–118.

Haalebos, P.E.M. & N.R. Murdock, 1981 New horizons and redevelopment of the Oloibiri field – Soc. Petr. Eng. Conf. Warri, Nigeria: 6 pp.

Haq, B.U., Hardenbol, J. & P.R. Vail, 1987 Chronology of fluctuating sealevels since the Triassic (250 million years ago to present) – Science (in press).

Hedberg, H.D. 1974 Relation of methane generation to undercompacted shales, shale diapirs and mud volcanoes – Amer. Assoc. Petr. Geol. Bull. 58: 661–673.

Hospers, J. 1965 Gravity field and structure of the Niger Delta, Nigeria, West Africa – Geol. Soc. Amer. Bull. 76: 407–422.

Hospers, J. 1971 The geology of the Niger Delta. In: F.M. Delaney (ed.): The Geology of the East Atlantic continental margin: ICSU/SCOR Working Party (31) Symp. – Cambridge Inst. Geol. Sci. Rept: 125–142.

Houbolt. J.J.H.C. 1973 The deep sea canyons in the Gulf of Guinea near Fernando Póo – Verh. Kon. Ned. Geol. Mijnbk. Gen. 30: 7–18.

Lambert-Aikhionbare, D.O. 1984 Petroleum source-bed evaluation of Tertiary Niger Delta: discussion – Amer. Assoc. Petr. Geol. Bull. 68: 387–394.

Lewan, M.D. & J.A. Williams, 1987 Evaluation of Petroleum generation from resinites by hydrous pyrolysis – Amer. Assoc. Petr. Geol. Bull. 71: 201–214.

Lopatin, N.V. 1971 Temperature and geologic time as factors in coalification – Akad. Nauk. SSR Ser. Geol. Izvestiya, 3: 95–106.

Mascle, J. 1975 Submarine Niger Delta: structural framework – J. Min. Geol. 13: 1–2.

Merki, P. 1972 Structural geology of the Cenozoic Niger Delta – In: T.F.J. Dessauvagie & A.J. Whiteman (eds): African Geology – Geol. Dept. Univ. Ibadan, Nigeria: 635–646.

Mermey, P. 1974 Aspect des argiles sous-compactees ou fluantes sur quelques sections sismiques provenant du Delta du Niger – Rev. L'assoc. Franc. Tech. Petròle 228: 21–26.

Murat, R.C. 1972 Stratigraphy and palaeogeography of the Cretaceous and Lower Tertiary in southern Nigeria – In: T.F.J. Dessauvagie & A.J. Whiteman (eds): Geology – Geol. Dept. Univ. Ibadan, Nigeria: 251–266.

Nedeco, 1959 River studies and recommandations on improvement of Niger and Benue – North Holland Publ. Cy: 1000 pp.

Olade, M.A. 1975 Evolution of Nigerian Benue Trough (Aulocogen): a tectonic model – Geol. Mag. 112: 6.575–583.

Omatsola, M.E. & W.G. Cordey, 1976 The Afam Formation, its stratigraphy, microfauna and environment of deposition – Proc. Afric. Micropal. Colloq. 7.

Oomkens, E. 1974 Lithofacies relations in the Late Quaternary Niger Delta complex – Sedimentology 2: 195–222.

Parsons, B. & J.G. Sclater, 1977 An analysis of the variation of ocean-floor bathymetry and heat flow with age – J. Geophys. Res. 82: 803–827.

Peters, S.W. 1984 An ancient submarine canyon in the Oligocene-Miocene of the western Niger Delta – Sedimentology 31: 805–810.

Poag, C.W. & L.W. Ward, 1987 Cenozoic unconformities and depositional supersequences of North Atlantic continental margins: testing the Vail model – Geology 15: 159–162.

Poston, S.W. 1980 G2 and G3 reservoirs, Delta South field, Nigeria: an engineering study pt. 1 – Soc. Petr. Eng. 55th Ann. Tech. Conf./Exhib., Dallas: 1–4.

Poston, S.W., Berry, P.Q. & F.W. Molokwu, 1983 Meren field – the geology and reservoir characteristics of a Nigerian offshore field – J. Petr. Techn. 35: 2095–2104.

Pugh, J.C. 1954 Sand movement in relation to wind direction as exemplified by the Nigerian coastline – Research Notes, Univ. Ibadan, Nigeria 5: 1–14.

Saito, T. 1977 Late Cenozoic planktonic foraminiferal datum levels: the present state of knowledge towards accomplishing Pan Pacific correlation – In: Proc. First Inter. Congr. Pacific. Neogene strat. – Tokyo: 61–80.

Shepard, F.P. & K.O. Emery, 1973 Congo submarine canyon and Fan Valley – Amer. Assoc. Petr. Geol. Bull. 51: 1679–1691.

Short, K.C. & A.J. Stauble, 1967 Outline of geology of Niger Delta – Amer. Assoc. Petr. Geol. Bull. 51: 761–779.

Sleep, N.H. 1971 Thermal effects of the formation of Atlantic continental margins by continental break-up – Geophys. J.R. Astr. Soc. 24: 325–350.

Stainforth, R.M. 1975 Cenozoic planktonic foraminiferal zonation and characteristic index forms – Kansas Univ. Pal. Contr. Art. 62: I, II: 425 pp.

Uzuakpunwa, A.G. 1974 The Abakaliki pyroclastics, eastern Nigeria: New age and tectonic implications – Geol. Mag. 111: 65–70.

Vail, P.R., Mitchum, R.M. jr., Todd, R.G., Widmier, J.M. Thompson, S.III, Sangree, J.B. Bubb, J.N. & W.G. Hatlelid, 1977 Seismic stratigraphy and global changes of sea level. In: Payton, G.E. (ed.): Seismic stratigraphy: application to hydrocarbon exploration – Amer. Assoc. Petr. Geol. Mem. 36: 129–144.

Walcott, R.I. 1972 Gravity, flexure and the growth of sedimentary basins at a continental edge – Geol. Soc. Am. Bull. 83: 1845–1848.

Weber, K.J. 1971 Sedimentological aspects of oil fields in the Niger Delta – Geol. Mijnbouw 50: 559–576.

Weber, K.J. 1986 Hydrocarbon distribution patterns in Nigerian growth fault structures controlled by structural style and stratigraphy – Amer. Assoc. Petr. Geol. Bull. 70: 661–662.

Weber, K.J. & E. Daukoru, 1975 Petroleum geology of the Niger Delta – Proc. World Petr. Cong. 9: 209–221.

Weber, K.J., Mandl, G., Pilaar, W.F., Lehner, F. & R.G. Precious, 1978 The role of faults in hydrocarbon migration and trapping in Nigerian growth fault structures – 10th Ann. Offshore Tech. Conf., Houston, Texas. Proc. 10: 4, 2643–2653.

Whiteman, A.J. 1982 Nigeria: its petroleum geology, resources and potential; I, II – Graham and Trotman, (Edinburgh): 394 pp.

Wright, J.B. 1968 South Atlantic continental drift and the Benin Trough – Tectonophysics 6: 301–310.

Yorston, H.J. & A.E. Miles, 1988 Salt tectonics, patterns of basin fill, and reservoir distribution. Abs. – Amer. Assoc. Petr. Geol. Bull. 72: 261.

Proceedings KNGMG Symposium 'Coastal Lowlands, Geology and Geotechnology', 1987: 203–212 (1989)

The Niger Delta: hydrocarbon potential of a major Tertiary delta province

Harry Doust[1]
[1] Shell Internationale Petroleum Maatschappij B.V., Carel van Bylandtlaan 30, 2506 HR Den Haag, The Netherlands

Received 15 September 1987; accepted in revised form 12 December 1987

Key words: delta, delta stratigraphy, depobelt, gas, hydrocarbon, hydrocarbon source rocks, lithofacies, lithologic variation, macrostructures, Niger, Nigeria, offlap, oil, sand: shale ratio

Abstract

The Niger Delta is one of the largest delta systems in the world, and forms one of its important hydrocarbon provinces. The subaerial portion covers at least 75 000 km^2, and stretches nearly 300 km from apex to coast. An area of even greater size is present below the Gulf of Guinea, where two great lobes protrude a further 250 km from the coast into deep waters. The sedimentary fill of the delta encompasses the Tertiary, during which it has been fed by the drainage systems of the Niger, Benue and Cross rivers. The present-day delta is a complex of individual fluviomarine systems that have succeeded one another in a stepwise fashion as the delta prograded towards the southwest.

Hydrocarbons are trapped in growth fault-related structures throughout the delta both onshore and offshore. The ultimate recovery in existing fields probably totals of the order of 4 billion (10^9) m^3 of oil and there is an underevaluated, but probably vast gas resource base.

In this contribution the impact of oil exploration on the geological knowledge of this delta terrain is discussed. Emphasis is placed on the sedimentary and structural variability inherent in the delta environment and the influence this has on exploration strategies.

Introduction

The Tertiary Niger Delta, which is situated in the Gulf of Guinea in West Africa, has built out into the central Atlantic at the mouth of the Niger, Benue and Cross River drainage systems. The catchment area covers more than a million square kilometres of savannah-covered lowlands, and the delta is one of the world's largest, stretching about 300 km from apex to mouth and covering an area of at least 75 000 km^2 (Fig. 1).

The delta has been in existence throughout the Tertiary, and comprises a regressive sequence of deltaic and marine clastics, thought to be at least 12 km thick in the centre (Fig. 2). It forms one of the world's major hydrocarbon provinces, with a proven ultimate recovery of about 4×10^9 m^3 of oil, as well as an undefined, but probably huge gas reserve.

Exploration for hydrocarbons in this province has been carried out aggressively during the last 30 years and has been based almost wholly on seismic information. Today, it is regarded by many as a 'mature' exploration area, and there has been a drop in recent years both in the rate of new discoveries, and in the volumes of new oil located.

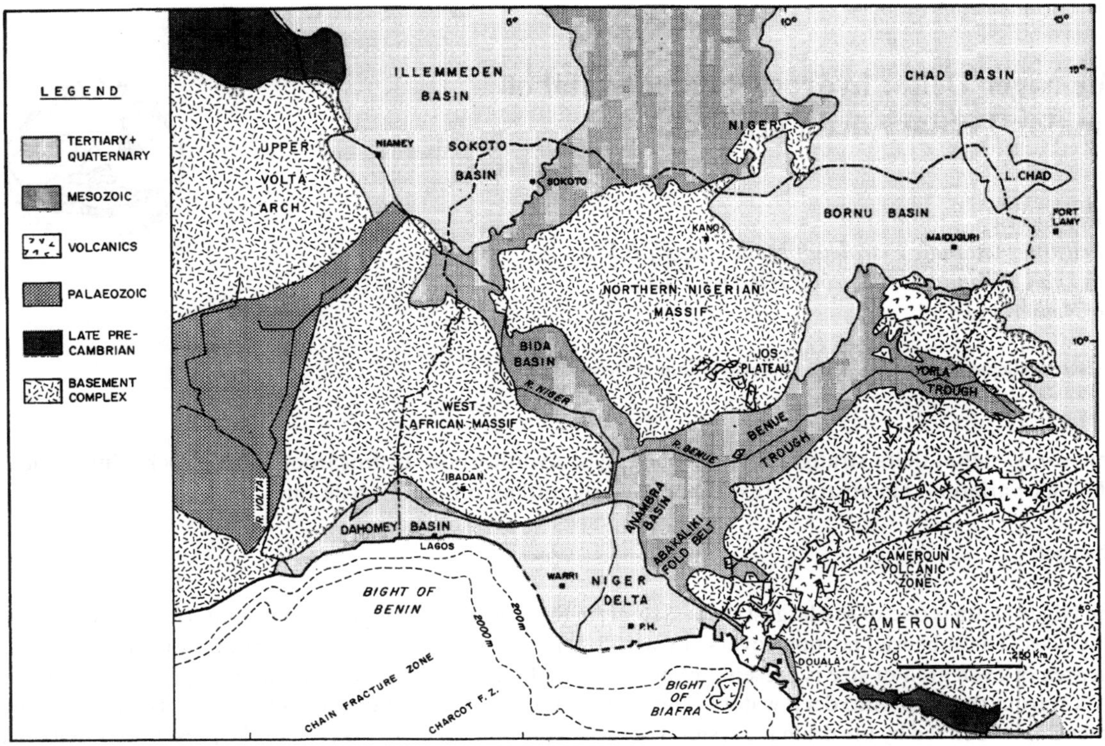

Fig. 1. Simplified geological map of Nigeria, showing the location of the Niger Delta – The catchment area, drained by the Niger and Benue rivers, covers more than one million km² of predominantly lowland savannah. The delta is one of the world's largest, the subaerial portion encompassing about 75 000 km².

In spite of the large number of wells that have been drilled, and their wide geographic distribution – nearly all of the structural complexes in the delta have been penetrated by at least one well – understanding of the habitat of hydrocarbons is rather primitive. Our ability to explain the pattern of oil and gas distribution in fields and discoveries, or to predict what will be found in undrilled traps, is

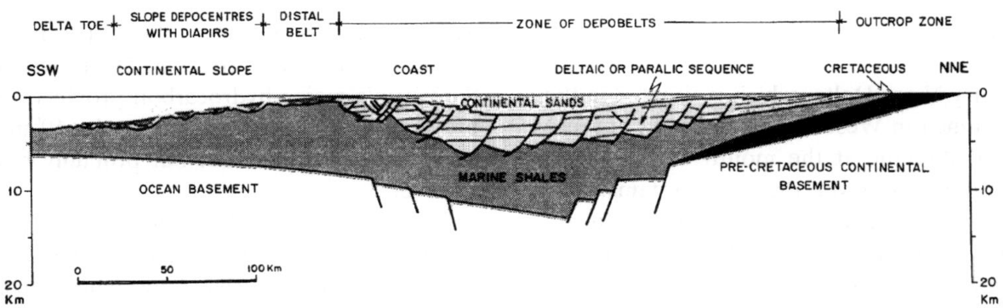

Fig. 2. Generalised sketch section across the Niger Delta – The delta sequence consists of Tertiary sediments with a maximum thickness of about 12 km. It forms a regressive cycle in which 3 basic lithofacies are developed: marine shales, deltaic or paralic sands and shales and alluvial or continental sands.

very circumscribed, and represents a major constraint on exploration strategy.

The wealth of subsurface data available from hydrocarbon exploration should be expected to enable us to appreciate the structural and stratigraphic architecture of the delta – and hence the hydrocarbon habitat; yet explorers are continually confronted with baffling well results, poor forecasting ability and a seemingly bewildering distribution of oil and gas.

In this paper, a number of possible reasons for these problems are discussed, and some consequences for exploration are highlighted.

Geological variability

In deltas like the Niger, the geology is self-contained, and there is relatively little influence from older structural trends, or from subsequent tectonic events. In addition, there is an intimate relationship between structure and stratigraphy – both being dependent on the interplay between rates of sediment supply and subsidence. Structures are synsedimentary and are developed mainly as faulted rollovers on the downthrown sides of growth faults, commonly at the loci of deltaic sand deposition. Between structures lithologic changes take place, mainly affecting the ratio of sand to shale, which reflect the local variations in depositional environment inherent in the deltaic situation.

These changes mean that facies variations in the Niger Delta are much more extensive than in non-deltaic areas, occurring on a much smaller scale. To ignore these variations leads to misinterpretation of the local geological conditions and a consequent obscuring of the underlying patterns. Recognition of local variations, however, is far from simple even with the most extensive data base at one's disposal, and requires a considerable insight into delta structure and stratigraphy.

Stratigraphic variability

The Niger Delta sequence consists of rapidly varying alternations of clastic lithologies, occurring in stacked sections of regressive offlap cycles. The actual lithologies involved are, on the other hand, few – they comprise sandstones, siltstones and shales of great similarity, whatever their age or situation in the sequence. For this reason, it is possible to subdivide the sequence only in terms of sand-shale ratio, or lithofacies.

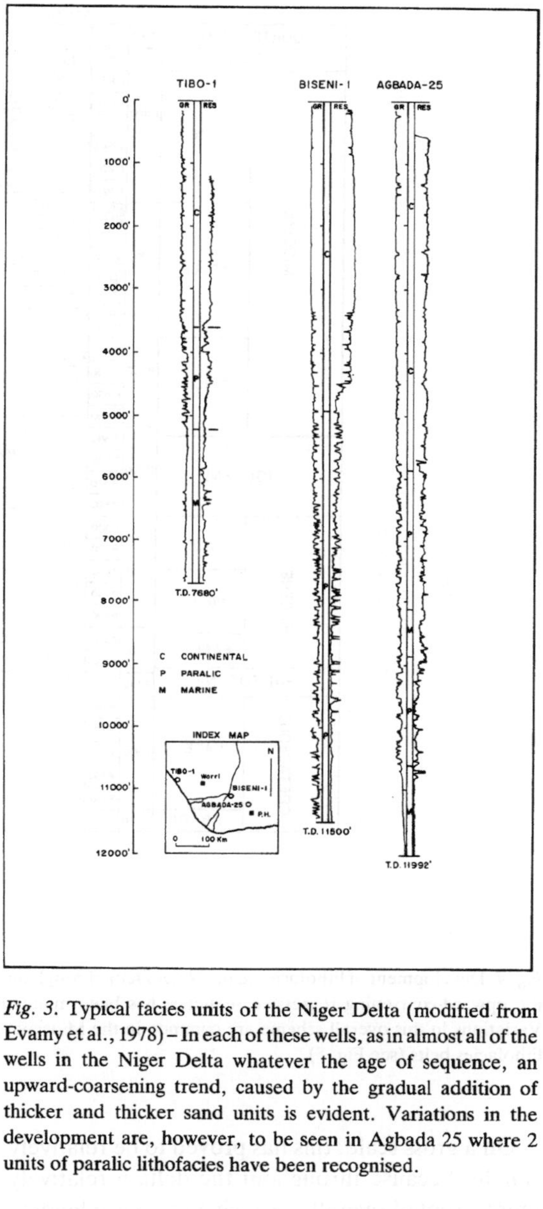

Fig. 3. Typical facies units of the Niger Delta (modified from Evamy et al., 1978) – In each of these wells, as in almost all of the wells in the Niger Delta whatever the age of sequence, an upward-coarsening trend, caused by the gradual addition of thicker and thicker sand units is evident. Variations in the development are, however, to be seen in Agbada 25 where 2 units of paralic lithofacies have been recognised.

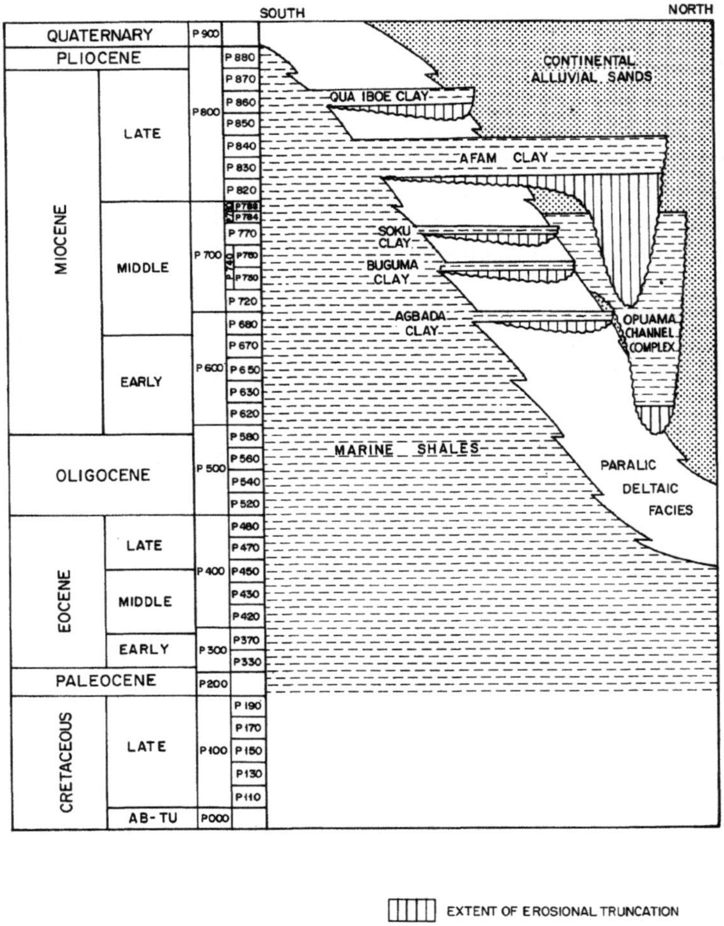

Fig. 4. Development of lithofacies in the Niger Delta through time – Throughout the Tertiary, here subdivided into palynological zones, the upward-coarsening regressive sequence has been maintained, resulting in a southward shift of the lithofacies through time. Variations in this overall scheme are common in the Miocene, particularly on the delta flanks, where channel systems cross-cut the lithofacies belts (see Fig. 5).

On a gross scale, this has proved to be relatively simple, because throughout the delta a relatively uninterrupted overall regressive sequence lending itself to simple subdivision is observed in the subsurface (Fig. 3). Commonly it is defined by three lithofacies: at the base marine shales predominate, and they pass up through the main deltaic section of alternating sands and shales (the 'paralic' sequence of common usage) into a sequence of non-marine alluvial sands (the so-called 'continental'). This pattern has been persistent throughout the Tertiary in the Niger Delta, and is strongly time-transgressive (Fig. 4). It is clearly a pattern of fundamental significance.

A second trend or series of trends of fundamental significance consists of a tendency for the sediments to become finer in a more marine direction. This may be expressed qualitatively as a tendency to 'shale-out' (Fig. 4).

These two trends are inherent in the Niger Delta

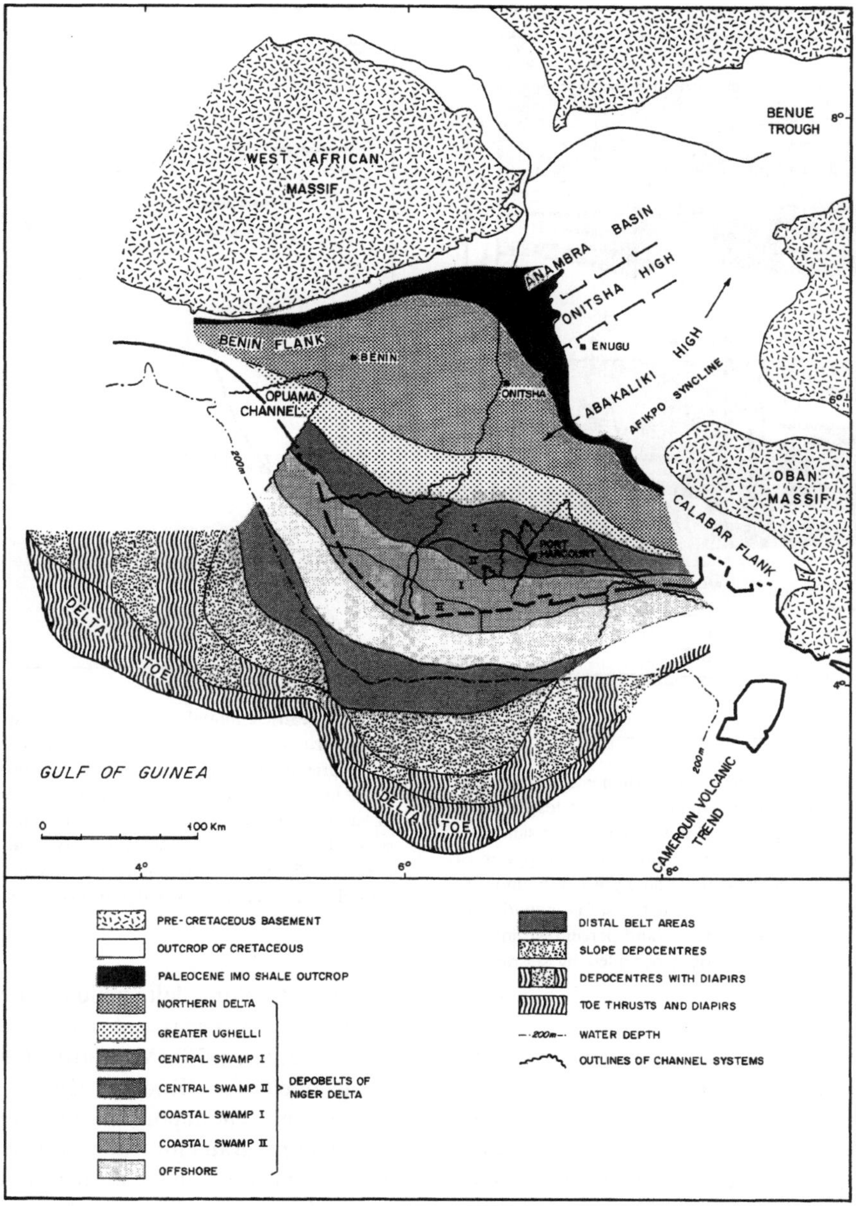

Fig. 5. Niger Delta: regional structural elements and depobelts – Successive phases of delta development referred to as depobelts or megastructures may be distinguished. In each, the deltaic or paralic sequence is distinct in age: thus, for instance, in the Greater Ughelli Depobelt it belongs to the Early Miocene, while in the more seaward Central Swamp Depobelt it belongs to the Middle Miocene. Offshore, the delta forms two large lobes, in which depocentres are located between diapiric shale zones.

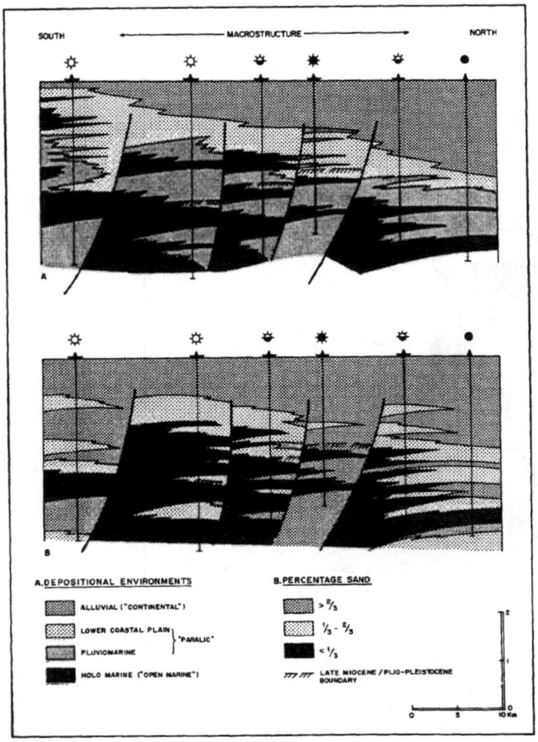

Fig. 6. Distribution of depositional environments and sands in macrostructures offshore – This figure demonstrates the close relationship between depositional environment and sand percentage in the delta sequence. In the faulted zone an uppermost Miocene to Pliocene macrostructural unit is present, bounded to the south and north by other macrostructures. In it a southward shale-out and an upward increase in sand percentage may be seen. Between the wells, control is provided by seismostratigraphic interpretation.

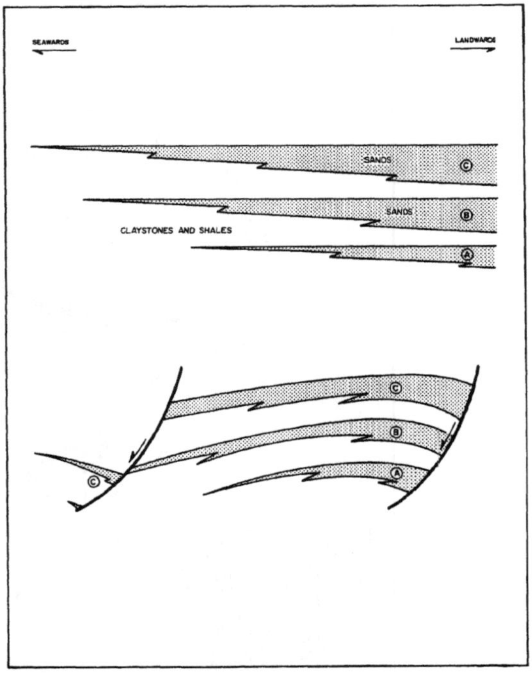

Fig. 7. Idealised model for sand distribution in a regressive offlap cycle within a delta unit. Top: stacked offlap sandy cycles in a regressive sequence. Bottom: distribution of sandy offlap cycles in a macrostructure – The upward and landward increases in sand percentage represent fundamental delta trends, expressed at macrostructural scales (as here) and at megastructural scales (Fig. 8). The distal portion of offlap cycle (c) is represented in the deeper, proximal portion of the succeeding macrostructure.

environment, and are well known in a general sense. They also are manifested on less well-studied scales, however, depending on the local structural and palaeogeographic situation and the ratio of supply to subsidence. To understand this we need to appreciate that the delta is composed of series of relatively independent building blocks, each characterised by its own lithological variations. The important thing to note here is that there are several hierarchies of stratigraphic variability, each of which has to be recognised if the geology and hydrocarbon habitat are to be understood fully.

In the literature, delta building blocks have been referred to as mega- and macrostructural trends (Evamy et al., 1978). Megastructures, also known as depobelts (see Knox & Omatsola, 1988, this issue), represent important stages in the history of the delta progradation – they are bounded by major regional faults and really form independent and integral phases of delta development. Characteristically, the deltaic or paralic sequence in each differs in age from that of its neighbour, and becomes younger towards the sea (Fig. 5).

Macrostructures on the other hand are growth fault-bounded trends that make up magastructures – they form the basic units of delta construction, usually consisting of a rollover feature and a distal

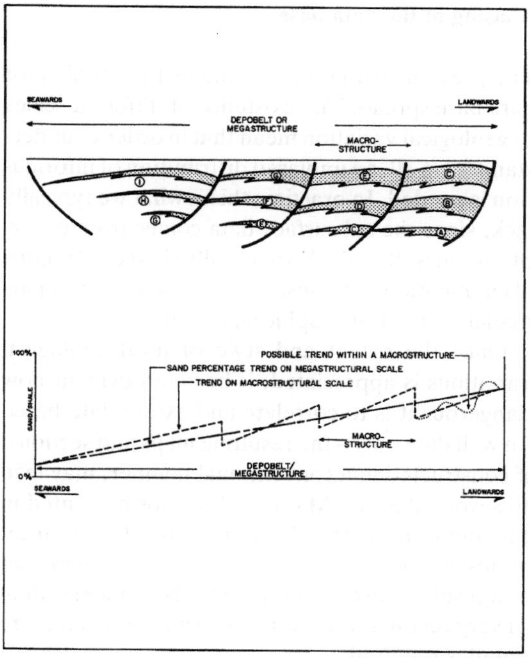

Fig. 8. Idealised models for sand: shale distribution within a megastructure or depobelt. Top: distribution of sandy offlap cycles in a depobelt, showing gradual seaward decrease in sand percentage. Bottom: frequency scales of sand percentage variation in a depobelt – Macrostructural variations (Fig. 7) are here superimposed a lower frequency, megastructural trend and, in turn, may themselves be subject to even higher frequency variations (expressed, for instance, on fieldsized scales; See examples in Weber, 1971). The trends shown here will take place within a framework of delta-wide variations. All geological parameters related to lithologic changes may be expected to vary on different scales in accordance with the sand: shale trend.

flank. There may be several macrostructures in a megastructure, but in some cases (particularly in the younger part of the delta), the subdivision may be obscure. Fig. 6, which comprises a section through a macrostructural trend in the eastern offshore area, shows the environmental and lithologic changes that may occur in such units.

The extent of even smaller-scale variability, found within macrostructures, can be seen in several of the fields described in the classic studies of Weber (e.g. Weber, 1971), and in the Recent subsurface (Oomkens, 1974). In these the complex patterns of sand-shale distribution caused by the juxtaposition of barrier and channel sands are very obvious. At the other end of the scale, delta-wide variations or regional patterns have been described by Evamy et al., 1978 and others.

Recognising the potential for different scales of stratigraphic variation, and their hierarchic nature – the regional, megastructural, macrostructural and local, one can appreciate how important it is to take full account of them. If we do not do so, we cannot recognise any logic in the sedimentary patterns that we see.

Ideally, one might expect the variations in sand percentage, and in other related geological parameters, to lead to the sorts of patterns illustrated in Figs 7 and 8. Here we see macrostructural variations superimposed on megastructural variations of lower frequency. In turn, macrostructural trends themselves are shown to be subject to even higher frequency variations, such as those that occur within a field, due perhaps to the distribution of channel systems (e.g. see Weber (1971)) or sand sources.

In practice, a number of factors tend to obscure these patterns so that they may be locally difficult to recognise; the sites of fluvial delta systems influence sand percentages along strike significantly and, as these shift rapidly, complex patterns may emerge. Other complicating features include the important channel systems that developed on the delta flanks in the Miocene (Figs 4, 5).

Hydrocarbon source rocks and oil and gas mix

An illustration of the problems that arise as a result of stratigraphic variations is provided by the open question as to the source of the oil and gas in the Niger Delta.

Geochemical analyses indicate that the hydrocarbons all belong to one family, and were derived from land plants and amorphous organic matter. There has, however, long been a controversy about the development and distribution of source rocks. Several authors have claimed that the main source lies in shales in the paralic sequence (or Agbada Formation) – for example Lambert-Aikhionbare & Ibe (1984) and Nwachukwu & Chuckwura (1986), while others suggest that it lies below the former in the continuous marine shales (or Akata Formation) – for example Weber & Daukoru (1975) and Ekweozor & Daukoru (1984).

This lack of precision in definition of the most probable source sequence might seem puzzling, but in fact it is understandable since:

1) Hydrocarbons occur throughout the area of the delta, as do both the paralic and marine shale sequences.
2) Source rocks are difficult to locate. They apparently occur in thin and impersistent beds, and can really only satisfactorily be sampled in cores. In ditch cuttings they tend to be so diluted by cavings that they cannot be recognised, while sidewall samples usually fail to locate them.
3) Geothermal gradients measured in wells suggest that considerable parts of the most promising looking source bed sections are immature or barely mature.

At present, we believe that sufficient evidence exists to state that source rocks in the Niger Delta are:

a) indigenous, that is to say within the delta sequence.
b) widespread but locally developed in thin but locally rich beds.
c) strongly environmentally controlled in distribution.

Prediction of either volumes of charge available for trapping, or of ratio of oil to gas charge are at this stage impossible since neither the distribution nor the type or richness of source rocks can be defined. Environmentally-controlled variations in source potential, however, closely linked to changes in the sand-shale ratio, almost certainly do occur on various scales, and in principle could be studied.

It is difficult to be optimistic, however, that source rock studies will ever be able to satisfactorily address the questions of volume generated and proportions of oil and gas charge. Partly this is because the sources are multiple and thin bedded, varying in their development in complex ways; but partly also because there will always be a lack of data from those portions of the delta that lie away from structural culminations. These include distal portions of mega- and macrostructures and other downflank areas.

Aliasing of the data base

The previous paragraph highlights the problem of data base spread. The existence of different scales of geological variation mean that in order to understand them all, an unbiased distribution of information is needed. In practice, this is what we typically lack, since the subsurface data comes from oil exploration wells, which are usually sited on structural culminations – themselves preferred sites of particular tecto-stratigraphic significance.

Once the extent and style of local geological variations is appreciated, it becomes evident how dangerous it is to correlate and extrapolate based on well data only – the resulting maps and sections, if constructed in a conventional manner, may lead to severe aliasing. Many of the maps presented in the literature on the Niger Delta are, for instance, almost certainly aliased to some extent. Their use, as a result, is largely confined to the demonstration of very regional variations – on smaller scales more detail is required.

Comparisons with other deltas

Hierarchical geological variations are characteristic of deltaic systems, and the problems discussed above are common to many areas. The extent and character of the variations, however, differ from delta to delta.

The Niger Delta, in common with the Gulf of Mexico deltaic provinces, is an example of a complex, mature system. It has been in existence throughout the Tertiary and has undergone considerable internal evolution. The main building blocks of the Niger Delta, the megastructures or depobelts, appear to be comparable to the 'progradational wedges' of the Texas Gulf Coast (see Galloway et al., 1982) and the 'blocks' of the Rio Grande Delta described by Busch (1978). Each forms an integral geological province with its own proximal and distal zones, its own characteristic geological variations, and its own individual hydrocarbon habitat.

Some other Tertiary deltas appear to be more simple. In the Neogene Baram Delta of northwest

Borneo (James, 1984) it is not possible to recognise more than one depobelt, and in fact, the entire province seems to constitute one such feature. Both in terms of size and period of development (approximately 15 million years) the Baram is comparable to the Northern Delta depobelt of the Niger.

The hinterland exerts an important influence on the amount of sand that accumulates in deltaic environments. The Niger is a very sandy delta in comparison with the Baram, for instance, thanks to its situation adjacent to ancient shield areas rather than to a young orogenic belt, and it has thicker and more abundant sands, and less abrupt shale-outs. In the Baram, the thick alluvial sand sheet that covers the depobelts of the Niger Delta is absent, probably for the same reason.

Consequences for hydrocarbon exploration

The geological variations discussed above make it extremely difficult to predict hydrocarbon prospectivity at any stage of exploration. It is therefore difficult to define at what stage hydrocarbon exploration in a delta province reaches true maturity.

Studies during the early stages of exploration in Nigeria were directed towards defining regional trends in structural style, source rock maturity, oil properties, and so on (Evamy et al., 1978). At the present stage of exploration, however, these studies are of relatively restricted value because the crucial questions lie on a scale finer than regional. Almost all structural features in the delta have at least one well on them, and exploration has to concentrate on evaluating the potential rewards in increasingly small, complex or deep-lying features. An understanding of the relationship between local prospectivity and local geological environment is critical to such evaluations, and without it, neither expectations from future discoveries nor satisfactory ranking of exploration prospects will be achieved.

Paradoxically, the presence of geological variations of a local nature do mean that even in a thoroughly explored province like the Niger Delta, the chances of locating substantial new reserves are good. They will remain good until every existing potential trap has been tested by the drill, or until sufficient understanding of the local geological environment is achieved to render this unnecessary.

The most suitable way to approach studies leading to an appreciation of the different levels of geological variation is to relate changes in sand-shale ratio to changes in depositional environment in each of the building blocks of the delta. This must be supported by a sound appreciation of delta structure and stratigraphy and must take note of the evolutionary development of the delta. The utmost use of seismic data should be made to infill the geological control provided by wells and thus minimise aliasing of conclusions (Fig. 6). Once these studies have been carried out, the distribution of hydrocarbons may become less bewildering and forecasting ability should improve.

Acknowledgements

I am grateful to many colleagues in the Nigerian oil industry for contributing data that have led to the ideas put forward in this contribution. I am particularly obliged to Ebi Omatsola and Gordon Knox, who critically reviewed the text. The former was kind enough to present the paper at the Symposium in my absence.

Permission to publish was generously granted by the Nigerian National Petroleum Corporation and the Shell Petroleum Development Company of Nigeria Ltd.

References

Busch, D.A., 1978 Influence of growth faulting on sedimentation and prospect evaluation – Am. Assoc. Pet. Geol., Bull. 59: 217–230.

Ekweozor, C.M. & E.M. Daukoru, 1984 Petroleum source-bed evaluation of Tertiary Niger Delta: reply – Am. Assoc. Pet. Geol., Bull. 68: 390–392.

Evamy, B.D., Haremboure, J., Kamerling, J., Knaap, P., Molloy, F.A. & P.H. Rowlands, 1978 Hydrocarbon habitat of Tertiary Niger Delta – Am. Assoc. Pet. Geol., Bull. 62: 1–39.

Galloway, W.E., Hobday, D.K. & K. Magara, 1982 Frio Formation of Texas Gulf Coastal Plain: Depositional systems, structural framework and hydrocarbon distribution – Am.

Assoc. Pet. Geol., Bull. 66: 649–688.

James D.M.D., (ed.), 1984 The Geology and Hydrocarbon Resources of Negara Brunei Darussalam – Muzium Brunei: 164 pp.

Knox, G.J. & E.M. Omatsola, 1988 Development of the Cenozoic Niger Delta in terms of the 'Escalator Regression' model and impact on hydrocarbon distribution. In: Van der Linden, W.J.M, S.A.P.L. Cloetingh, J.P.K. Kaasschieter, J. Vandenberghe, W.J.E. van de Graaff & J.A.M. van der Gun (eds) 1988: Proc. KNGMG Symp. Coastal Lowlands: Geology and Geotechnology. The Hague, 1987 – Kluwer Acad. Publ. (Dordrecht): 181–202.

Lambert-Aikhionbare, D.O. & A.C. Ibe, 1984 Petroleum source-bed evaluation of Tertiary Niger Delta: discussion – Am. Assoc. Pet. Geol., Bull. 68: 387–389.

Nwachukwu, J.I. & P.I. Chukwura, 1986 Organic matter of Agbada Formation, Niger Delta, Nigeria – Am. Assoc. Pet. Geol., Bull. 70: 48–55.

Oomkens, E., 1974 Lithofacies relations in Late Quarternary Niger Delta Complex – Sedimentology 21: 195–222.

Weber, K.J., 1971 Sedimentological aspects of oil fields in the Niger Delta – Geol. Mijnbouw 50: 559–576.

Weber, K.J. & E. Daukoru, 1975 Petroleum geology of the Niger Delta – Proc. 9th World Petr. Congr. 2: 209–221.

Proceedings KNGMG Symposium 'Coastal Lowlands, Geology and Geotechnology', 1987: 213–223 (1989)

Barrier islands, tidal flats, and coastal marshes resulting from a relative rise of sea level in East Frisia on the German North Sea coast

Hansjörg Streif
Niedersächsisches Landesamt für Bodenforschung, Stilleweg 2, D-3000 Hannover 51, F.R.G.

Received 12 January 1988; accepted in revised form 30 June 1988

Key words: Southern North Sea, Holocene, sea-level rise, coastal development, barrier islands, tidal flats, marshes

Abstract

The level of the North Sea has risen about 110 m during the last 15000 years, i.e. during the Weichselian Late Glacial and the Holocene periods. Only the last 46 m of this rise, covering the time interval since 8900 B.P., can be reconstructed on the basis of reliable radiocarbon dates from the southern North Sea and the German Bight.

Along the East Frisian coast, the formation of the present day barrier islands, tidal flats and coastal marshes started shortly after 8000 B.P. when the sea had risen above the −20 m level. Sequences of marine, tidal-flat, shoal and beach sediments, altogether 35 m thick, were formed in the zone of the barrier islands; these were covered by a thin layer of salt marsh deposits and in turn by dunes up to 20 m high. Thick sequences of intertidal and subtidal deposits occur in the seaward part of the tidal flats. In the landward part of the tidal flats and in the coastal marshes, layers of peat are intercalated in the tidal, brackish, and lagoonal deposits. These peat layers were formed since 6500 B.P. and document phases of retardation in the rise of the sea level, from 4800 to 4200 B.P. and from 3300 to 2300 B.P. Fossil soils on brackish sediments and decomposition horizons in the peat indicate that a temporary lowering of the water level took place at about 2700 B.P. and again at 2000 B.P.

Introduction

The German sector of the North Sea is roughly triangular in shape and stretches from the estuary of the River Ems and from the northern end of the island of Sylt in a northwesterly direction as far as the Dogger Bank. The maximum water depth of 70 m is encountered on the northwestern flank of the Dogger Bank; however, the depth is generally less than 45 m in the inner part of the German Bight.

The following tidal zones, using the classification of Hayes (1979), can be distinguished along the southern North Sea coast. Low mesotides with a tide amplitude between 1 and 2 m occur along the coast between the Rhein delta and the island of Terschelling, Netherlands. High mesotides with a tide amplitude of 2 to 3.5 m occur eastwards from Terschelling as far as Wangerooge. Low macrotides with a maximum amplitude of 4 m occur in the innermost part of the German Bight, between the tidal inlet of the Jade Bay and the mouth of the Eider River. The tide amplitude decreases in a northerly direction to that characteristic of high mesotides between Eiderstedt and Sylt, to that of low mesotides between Sylt and Blavandshuk, and to that of microtides to the north of Nymindegab, Denmark.

The present tidal conditions are partly mirrored by the coastal morphology. Continuous barrier systems exist in the zones of small tide amplitudes along the western coasts of the Netherlands and Denmark. Elongated barrier islands occur in the zone which is affected by low and high mesotides, and rounded or sickle-like sandy shoals exist in the macrotidal zone. However, tidal conditions and geomorphological patterns have been much modified under the influence of the rise of sea level and coastal development during the course of the Weichselian Late Glacial and Holocene periods. Moreover, present-day features are only momentary and are in unstable equilibrium with today's hydrodynamic processes.

Sedimentary features and their interpretation

By late Weichselian times, climatic deterioration to fully glacial conditions has caused the sea level in the North Sea area to fall to at least 110 m below the present level. According to Long et al. (1988: Fig. 3), a blanket deposit of Weichselian till extends northeastwards from the coast of East Anglia into the North Sea; it contains boulders derived mainly from the Upper Paleozoic and Mesozoic rocks of eastern England. Its outermost extension reaches as far as the western end of the Dogger Bank. From there the ice margin turns in a northwesterly and then northerly direction and follows the present shoreline of the United Kingdom at a distance of about 50 to 80 km. On the eastern side of the North Sea, the Weichselian end moraines run in a south-north direction through eastern Schleswig-Holstein, Germany, and southern Jutland, Denmark. At Limfjord, the moraines turn west and strike out into the North Sea. Their offshore continuation is uncertain. However, offshore evidence indicates that there was no connection between the British and the Scandinavian ice sheets across the southern North Sea during the Weichselian glacial period (Long et al. 1988: Fig. 3).

During this period the Weichselian rivers discharged into the central North Sea. The most important river system draining the north German hinterland was the so-called 'Elbe-Urstromtal', which acted as an ice-marginal valley during the Weichselian glacial maximum. The course of this 30 to 40 km wide valley system, between the island of Helgoland and the Weiße Bank area has been mapped by Figge (1980) using boomer profiling. During this period, wide areas of the North Sea bottom were exposed to periglacial climatic conditions. Evidence of permafrost is presented by segregated ground-ice fabrics as described by Derbyshire et al. (1985) from the central North Sea, and by fossil ice-wedge structures described by Streif (1985) from the German Bight.

The offshore zone

The earliest Holocene brackish-marine incursion into the southernmost North Sea may have occurred as early as 10000 B.P. (Eisma et al., 1981). With the continuing rise in sea level, tidal flat sedimentation became more widespread between 9000 and 8000 B.P., and fully marine conditions spread out over most of the Southern Bight after 7000 B.P. (Eisma et al., 1981).

Shoreline development in the southern North Sea during the Late Weichselian and early to middle Holocene periods has been reconstructed by various authors (Jelgersma, 1979: Fig. V-11, Oele et al., 1979: colour chart). However, these paleogeographic maps present only hypothetical shorelines, which are mainly constructed on the basis of seafloor morphology and the depth of the present North Sea. In many cases, the thickness of Holocene deposits has not been taken into consideration and the effects of younger erosion, which in places had had a considerable impact, have been neglected.

Only the uppermost 46 m of the 110-m rise of the North Sea in the Weichselian Late Glacial and Holocene periods can be reconstructed on the basis of reliable radiocarbon dates. Most of these offshore dates stem from grass rootlets, roots of trees and from basal peat, all of which grew on a former land surface. This type of material predates the Holocene transgression and provides evidence for the progress of the inundation of the pre-Holocene landscape. A smaller number of dates stems from

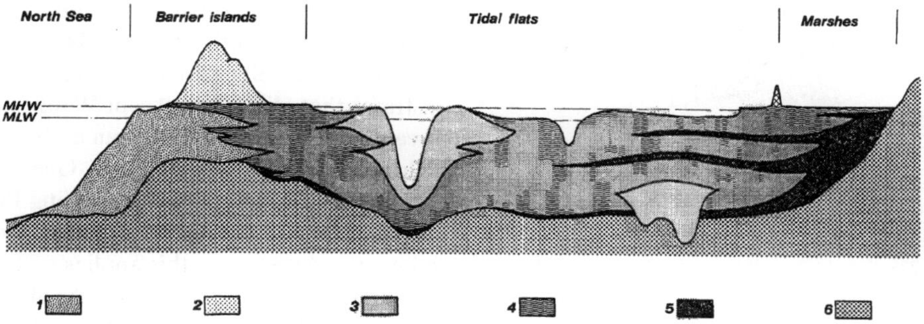

Fig. 1. Schematic cross section through the wedge-like body of coastal deposits from the North Sea to the Pleistocene hinterland. 1 = beach sand, 2 = dunes, 3 = tidal channel infill, 4 = tidal flat deposits, 5 = peat, 6 = Pleistocene deposits.

rootlets and stalks of plants which grew in the basal sequence of brackish-lagoonal deposits, thus indicating the very beginning of the facies changes from terrestrial or semiterrestrial environments to brackish and marine conditions.

Ludwig et al. (1981) compiled sea-level dates and a series of new palynological and radiocarbon dates from the German sector of the North Sea. These were later supplemented by further dates (Behre et al., 1984). These dates point to a rapid and uninterrupted rise of the North Sea from −46 m to −15 m in the time interval between 8600 and 7100 B.P. A great many vibrocores taken from different water depths provide evidence that the ingression of the North Sea occurred in a single transgressive overlap. In many places, Holocene brackish-marine sediments rest with an erosional contact on basal peat or Pleistocene deposits. However, no examples of regressive overlap of semiterrestrial to limnic deposits on top of marine sediments are known from this phase of development of the North Sea.

In contrast to these findings, Kolp (1974, 1976, 1977) published a stepped sea-level curve for the southern North Sea. He distinguished a series of marine terraces at levels of −80, −60, −45 and −30 m and attempted to correlate these with corresponding levels in the Baltic Sea. In this study, no counterpart of the −80 m level exists in the Baltic Sea; however, the −60, −45 and −30 m levels of the North Sea were correlated with the Yoldia, Echeneis and Ancylus terraces, respectively. The younger terrace levels of the Baltic Sea (Mastogloia, Cypleus, Litorina I, Litorina II, and Limnaea terraces) do not seem to be developed or preserved in the North Sea.

Only for the −45 m terrace (Kolp 1976: 7 and plate 1) the genesis and age are known. According to Behre & Menke (1969) and Kolp (1976) this terrace is an accumulation level with a Holocene sequence of freshwater peat of Preboreal age, a hiatus, and an onset of marine sedimentation in Boreal times. These data fit in well with the sea-level curve published by Ludwig et al. (1981: Fig. 2). However, these authors pointed out that all other levels in the southern North Sea are undated planation surfaces which might have formed during the Holocene transgression, but which also might be relict structures belonging to the pretransgression land surface.

The thickness of the brackish-marine Holocene sequences is generally less than 5 m in the German sector of the North Sea. Holocene deposits reach a thickness of more than 15 m in giant-ripple fields and in sediment traps such as the 'Elbe-Urstromtal' and the 'Helgoländer Schlickgebiet'.

The coastal lowland areas

In contrast to the offshore zone, thick sequences of marine-brackish sediments and peat layers occur in the coastal area. The deposits reach a maximum thickness of 35 m in the region of the barrier islands and at the seaward side of the open tidal flat areas. They become thinner towards the landward part of the tidal flats and in the coastal marsh areas, and wedge out against the Pleistocene hinterland, i.e.

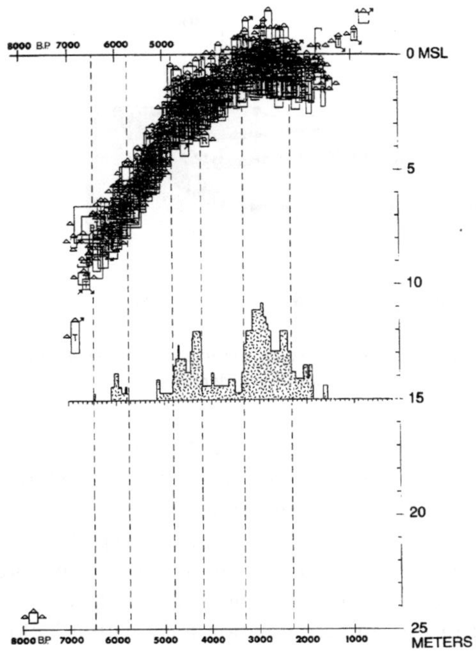

Fig. 2. Time-depth diagram of conventional radiocarbon dates from basal and intercalated peat layers, showing a band-like sea-level curve in the upper and the frequency distribution of samples from intercalated peat layers in its lower part.

the 'Geest'. This wedge-like body of coastal sediments is depicted in Fig. 1, which shows a schematic cross section from the North Sea to the Geest.

The subsurface morphology of the Pleistocene beneath the Holocene deposits is partly a relict of fluvial and terrestrial processes mainly from the Weichselian glacial period. It was partially modified when, under the influence of the rising sea level, tidal currents penetrated into the former valleys and coastal erosion affected progressively higher parts of the submerging landscape. The original morphology was severely modified by marine morphodynamic processes in the seaward part of the present day tidal flat areas. However, it is fully preserved under a cover of marine and brackish sediments and peat in the landward zone of the tidal flats and in the coastal marshes.

The barrier islands. Pleistocene deposits are found beneath the East Frisian barrier islands at −35 m NN (relative to German zero datum) in depressions and at an altitude of −4 m NN in their highest positions. This hilly subsurface morphology demonstrates that all islands except Wangerooge were so-called 'Geestkerninseln' in an early stage of the Holocene coastal development. Geestkerninseln are islands with a core of outcropping Pleistocene deposits mantled by Holocene marine and tidal flat sediments. At present, this applies to Texel in the Netherlands, and to the islands of Sylt, Föhr, and Amrum on the west coast of Schleswig-Holstein. With progressing inundation these cores were fully submerged and the present-day system of barrier islands came into being.

The subsurface morphology of the Pleistocene demonstrates that the islands are not relics of a former eastward-growing spit, and that no general migration of the islands from west to east has taken place. Both views have been stressed by previous investigators. Under these assumptions it should be expected that the elevated parts of the Pleistocene cores were eroded during the horizontal migration of deeply incised tidal inlets. However, this is not the case. In contrast it can be demonstrated that migration of the 12 to 35 m deep tidal channels was restricted to zones 5 to 6 km wide at maximum. In most cases these zones are only 3 km wide. Shallower tidal inlets with a maximum depth of 10 to 12 m may in fact have migrated over greater horizontal distances.

The seaward front of the Pleistocene cores of the islands seems to be steeper than the landward side. This might indicate that marine reworking and cliff formation took place during the course of the Holocene transgression. However, so far there is not enough borehole evidence available for reliable proof of the existence of buried cliffs.

The Holocene sedimentary sequences of the barrier islands begin locally with a thin layer of basal peat from a semiterrestrial or freshwater environment, and/or clayey silty sediments containing roots and stalks of *Phragmites communis* and belonging to a brackish-lagoonal environment. These basal units are overlain by thick sequences of Holocene marine sand, tidal-channel fill, and shoal and beach sediments. The uppermost limit of this marine-littoral sedimentary sequence consists of thin

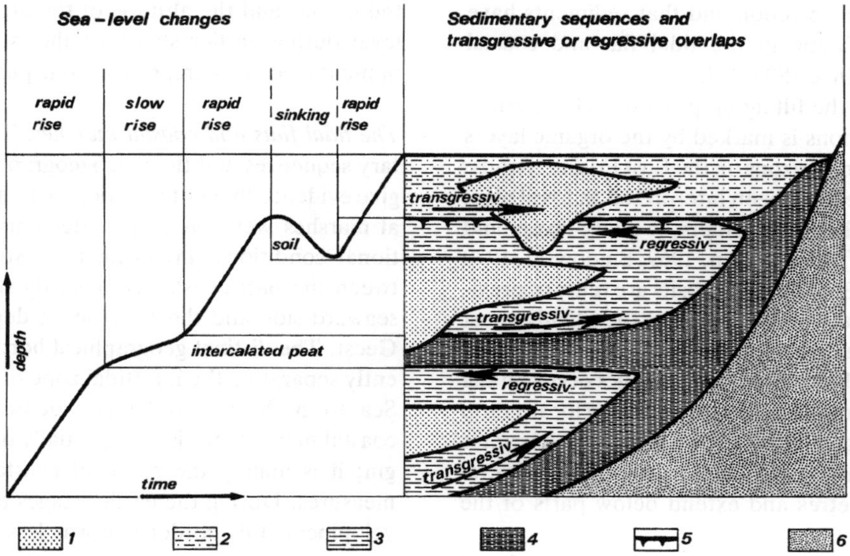

Fig. 3. Processes of interaction between sea-level changes with transgressive and regressive overlaps and the formation of soils on clastic sediments or horizons of decomposition in bog environments. 1 = tidal channel infill, 2 = tidal flat deposits, 3 = brackish and lagoonal deposits, 4 = peat, 5 = soil, 6 = Pleistocene deposits.

layers of salt marsh deposits which occur in the depth interval between the present mean sea level and the mean high tide level. In places these fossil salt marsh deposits are overlain by eolian deposits; these deposits form dunes up to 20 m high which characterize the morphology of the islands.

A series of age determinations is available which has been used to date the formation of the present system of barrier islands. Vibrocore A 10 was collected at a location 6 km offshore from Wangerooge. The upper part consists of a 0.67 m Holocene sequence of marine sand and the lower part of 2.13 m of brackish sediments with rootlets and stalks of *Phragmites communis*. The core did not penetrate the entire Holocene sequence, but it can be assumed that the Pleistocene is very close to the bottom of the borehole.

Radiocarbon dates obtained from the root and stalk material which have been published by Hanisch (1980) and have been used for the construction of the sea-level curve of Ludwig et al. (1981: Fig. 2) are compiled in Table 1. From this information it can be concluded that sea level was close to −24 m at about 7900 B.P. and that the deposition of the brackish sediments took place in a sheltered position, probably behind an older barrier system which was situated further to the north. They also give evidence that the barrier islands were formed

Table 1. Sea-level dates from vibrocore A 10, taken 6 km to the north of the island of Wangerooge at 55° 51,4′ N–07° 50.7′ E. Radiocarbon dates stem from stalks of *Phragmites communis* in brackish sediments.

Depth interval (in m NN)	Laboratory number	Conventional radiocarbon age
−23,08 to −23,24	Hv 8600	7540 ± 80 B.P.
−23,48 to −23,73	Hv 8601	7980 ± 60 B.P.
−24,07 to −24,39	Hv 8602	7960 ± 205 B.P.

in their present position and that sediments have been accumulating in the tidal flat and coastal marsh areas since 7500 B.P.

The end of the filling-up process under marine-littoral conditions is marked by the organic layers and vegetation horizons on the East Frisian islands. Some of these have been interpreted as fossil algal mats (Hanisch, 1980), others as peaty layers and salt marshes (Barckhausen, 1969; Sindowski, 1970 and Streif, 1986). These deposits are relics of previous stages in the development of the islands, most probably formed on the landward side in the shelter of a dune belt. Today, these relics occur on the seaward side of the islands, where they locally crop out on the beach. Borehole evidence demonstrates that they in fact have a wider distribution of several square kilometres and extend below parts of the dune belt.

Radiocarbon dates from these deposits are listed in Table 2 and provide evidence that salt marshes existed within the outlines of the present island of Juist at least since 1965 ± 130 B.P. (Hv 13131). On Wangerooge, clastic deposits with bivalve shells of *Scrobicularia plana* suggest that tidal flat conditions existed in that area from around 1540 ± 75 to around 1450 ± 180 B.P. (Hv 300 and 9257) and that coastal salt marshes developed from about 655 ± 130 B.P. onwards (Hv 8604, 8605, 9169 and 12303). Peaty horizons on Langeoog date from 690 ± 110 B.P. (Hv 1998). These findings can also be used to reconstruct the southward migration of the islands and the altitude of the mean high tide level during earlier stages of the island's development (see last paragraph of this paper).

The tidal flats and coastal marshes. The sedimentary sequences and the paleogeographic evolution give evidence that both the tidal flats and the coastal marshes were built up under uniform depositional conditions, involving the entire zone between the barrier islands or sandy shoals on the seaward side and the Pleistocene deposits of the Geest. The distinct geographical border line presently separating the intertidal zone of the Wadden Sea from the supratidal and densely vegetated coastal marsh lands is only partially of natural origin; it is mainly the result of coastal protection measures. During the earlier stages of coastal development, this border line was less distinct, and from a paleogeographical point of view, it was a very unstable feature which repeatedly migrated landwards or seawards during the Holocene (Fig. 1).

On the basis of the sedimentary sequences, three zones can be distinguished in the tidal flat and coastal marsh areas. On the seaward side thick, continuous sequences of clastic, shallow-marine and intertidal deposits overlie pre-Holocene terrestrial sediments and in some places basal peats. In the zone bordering the Pleistocene hinterland, pure sedentary sequences of peat, often several metres thick, occur and indicate continuous bog

Table 2. Sea-level dates of mollusc shells in tidal-flat sediments, rootlets and peaty layers in salt marsh deposits of the East Frisian barrier islands.

Author		Island	Material	Depth (in m NN)	Laboratory number	Conventional radiocarbon age
Hanisch	1980: 227	Wangerooge	rootlets (salt marsh)	+1.70	Hv 8604	655 ± 130 B.P.
Hanisch	1980: 227	Wangerooge	rootlets (salt marsh)	+1.92	Hv 8605	520 ± 60 B.P.
Hanisch	1980: 227	Wangerooge	rootlets (salt marsh)	+1.85	Hv 9169	580 ± 80 B.P.
Barckhausen	1969: 268	Langeoog	peaty material	+1.53 to +1.58	Hv 1998	690 ± 110 B.P.
Streif	1986: 35	Wangerooge	rootlets (salt marsh)	+1.50 to +1.60	Hv 12303	545 ± 80 B.P.
Streif	1986: 37	Juist	rootlets (salt marsh)	+0.75 to +0.80	Hv 13132	1185 ± 125 B.P.
Streif	1986: 37	Juist	rootlets (salt marsh)	+0.37 to +0.40	Hv 13130	1155 ± 130 B.P.
Streif	1986: 37	Juist	rootlets (salt marsh)	+0.43 to +0.53	Hv 13131	1965 ± 130 B.P.
Sindowski	1969: 21	Wangerooge	*Scrobicularia plana*	at about 0.0	Hv 300	1450 ± 180 B.P.
Hanisch	1980: 224	Wangerooge	*Scrobicularia plana*	−0.10	Hv 9257	1540 ± 75 B.P.

growth. Between these two there is a transitional zone characterized by sequences in which lacustrine deposits and semi-terrestrial peat layers alternate with clastic sediments of marine origin. The cyclic alternation of transgressive overlaps (marine and brackish deposits overlying peat) and regressive overlaps (limnic to semi-terrestrial peat overlying brackish and marine sediments) characterize the zone as being highly sensitive to water-level fluctuations (see last paragraph of this paper).

A great many radiocarbon dates are available from peats from the tidal flat and coastal marsh areas; they provide us with the opportunity of reconstructing the chronology of the coastal development. This is depicted in the band-like sea-level curve in Fig. 2, which is a time/depth graph of conventional radiocarbon sea-level dates. A correction procedure devised by Preuss et al. (1981) was used to minimize the influence of compaction. The vertical width of the band represents the indicative range of the dated material. The true sea-level curve lies within this band, but is masked by the effects of evaluating a multitude of data from a wide coastal lowland area, stemming from different environments, and with various indicative ranges. Minor oscillation of the sea level cannot be deduced from this kind of evaluation, but the general trends become clear.

Samples of basal peats collected from beneath the coastal zone yield results which compare well with, and supplement the results from the offshore area. They also indicate that a unidirectional landward and upward shift of the coastline took place under the influence of a rapidly rising sea level.

As can be seen from the frequency distribution of samples from intercalated peats (Fig. 2), bog growth varied in time. The oldest intercalated peats dating from about 6500 and 6000 B.P. have a very limited distribution. They occur locally in valley-like depressions in the pre-Holocene surface. Ideal conditions for the formation of intercalated peat apparently prevailed between 4800 and 4200 B.P. and between 3300 to 2300 B.P. Intercalated peats from these periods occur throughout the coastal lowland area between the rivers Ems and Elbe. A small number of samples from locally intercalated peat layers yield dates between 2000 to 1600 B.P.

Comparing the frequency distribution of conventional radiocarbon dates given in Fig. 2 with the frequency distribution of radiocarbon ages of the same dendrochronologically corrected set confirms that the peaks and depressions of the histogram are true and do not result from wiggles in the ^{14}C correction curve. Only the small peaks at about 6500 and 6000 B.P. merge into one single peak, but all others remain practically uninfluenced by the dendrochronological correction procedure.

A comparison of the band-like sea-level curve with the frequency distribution of dates from intercalated peat layers shows clearly that frequency peaks of radiocarbon dates coincide with phases in which a gentle rise of sea level took place. In contrast, low frequencies coincide with steep sections of the sea-level curve. Although there is clear evidence of phases with highly favourable conditions for peat formation, a regional evaluation of the dates shows that bog growth was by no means strictly synchronous throughout the coastal lowlands of Lower Saxony. The different coastal areas reacted in their individual ways to the supraregional trends of sea-level changes. Thus, the supraregional trends may be modified by local hydrological and sedimentological conditions in such a manner that in places evidence for them is completely masked. A local sedimentary succession is therefore a record, often an incomplete one, of the interplay between the supraregional changes and the local environment.

Within the general trend of rising sea level, temporary falls can be demonstrated at about 2800 B.P. and 2000 B.P. They are indicated by horizons of decomposed peat, by changes from fen peat to raised-bog peat vegetation, and by soil formation on top of clastic sediments (Preuss, 1979; Behre & Streif, 1980; Behre, 1986). The causes and the processes by which these changes took place are described below.

Course of events and conclusions

On the basis of morphological features it can be assumed that the North Sea was at 110 m below its present level during the coldest period of the Weichselian-Pleniglacial period. No details are known about the processes involved during the rise in sea level from 110 m to 46 m below the present North Sea level. However, taking the end of the Weichselian-Pleniglacial period at about 13000 B.P., the average rate of sea-level rise during Weichselian Late Glacial and early Holocene times amounts to 160 cm per century.

Palynological and radiocarbon age determinations carried out on samples of basal peat from the offshore area indicate that there was a rapid and uninterrupted rise of the North Sea from 46 to 15 m below present sea level in the time interval from 8600 to 7100 B.P. An average rate of sea-level rise of 210 cm per 100 radiocarbon years can be obtained from these figures. The 30 m rise in sea level shifted the coastline landwards over a distance of about 250 km. Borehole evidence from the sedimentary sequences at the bottom of the North Sea indicate that the marine incursion of this period occurred exclusively with transgressive overlaps of tidal-flat and marine deposites over terrestrial sediments, lacustrine muds, and basal peats.

The Holocene sedimentary sequences, which help to build up the system of the East Frisian barrier islands were formed in their present position only after 7500 B.P. Relics of fossil salt marshes within the present outline of the island of Juist indicate that an island existed from 1965 ± 130 B.P onwards (Table 2, Hv 13131). Conclusions can be drawn from the salt-marsh deposits and vegetation horizons as to the distances of southward migration of the islands and the dune belts. Barckhausen (1969, pl. VIII) estimated a southward shift of Langeoog over a distance of at least 2 km in 2000 years. Assuming that dunes also existed in earlier stages of the islands' development, it must be concluded that the dune front on Langeoog prograded at least 500 m southwards over former salt marshes in a 400 year time interval. However, it is misleading to treat this amount as equivalent to the contemporary coastal erosion on the seaward side of the island. This might have been much more, but it could have been a little less than 500 m in 400 years. The island of Wangerooge migrated more than 2 km over its former tidal flat areas in 1500 years, and the southern front of its dune belt prograded at least 600 m over former salt marshes in a 500 year time span. In the western part of Juist, the dune belt prograded at least 1.2 km over former salt marsh areas in 800 years.

Contemporaneously with the deposition of the clastic sedimentary sequences in the region of the barrier islands, cyclic sequences of clastic sediments and intercalated peat layers were formed in the tidal flats and salt marshes. Originally it was assumed that these cyclic facies changes from clastic sediments of marine origin to bog vegetation and vice versa directly mirror vertical changes in sea level. Layers of clastic sediments were looked upon as indicative of a rise in sea level, and peat horizons were taken as indicators of a fall in sea level. However, more detailed investigations demonstrated that this was a rather simplified view of the processes.

The accumulation of thick sequences of shallow-marine and intertidal deposits in the seaward region gives evidence of a general rise in sea level. Hence, comparable conclusions can also be deduced from thick sedentary sequences of fen peat occurring close to the Pleistocene hinterland. Consequently, the cyclic sedimentary sequences in the transitional zone between these two zones can no longer be looked upon as a simple response to variations in sea level. Both clastic sedimentation and bog growth take place in a finely balanced equilibrium which is very sensitive to the direction, amount, and velocity of sea-level changes.

During phases of rapidly rising sea level and dominating marine influence, the zone of clastic sedimentation progrades in a landward direction over former peat areas. In contrast, during phases of slowly rising water level and reduced marine influence, growing bogs expand in a seaward direction over clastic sediments of marine origin. Stable conditions occur when the rate of bog growth can fully compensate for the contemporary accumulation rate of clastic deposits. From this it must be concluded that transgressive overlaps of sediments of marine origin on top of lacustrine deposits or peat occur during phases of rapidly rising sea level. However, regressive overlap of peat on top of brackish or marine deposits does not indicate a temporary fall in sea level but a slowly rising one (Fig. 3).

A temporary lowering of the water level has

other influences on sedimentary sequences. Practically no indications of this can be found in the subaquatic zone. However, in the intertidal zone and especially close to the mean high-tide level, a lowering of the water table initiates soil formation on the surface of clastic sediments. In a bog environment, a lowering of the ground and surface water table often disturbs or interrupts plant growth and leads to oxidation and decomposition at the bog surface (Streif 1982: p 36). These processes of interaction between water-level changes with soil formation on clastic deposits or with decomposition of peat in bog environments is shown schematically in Fig. 3.

In extensive bog areas other effects may also occur as a result of water-level changes. Since in fen-peat areas bog growth is based on the nutrient content of the ground water, a lowering of the ground water level can cut off the nutrient supply for fen peat vegetation and thus initiate a rapid change to raised bog conditions, in which bog growth relies on the nutrient content of the precipitation. Behre & Streif (1980) and Behre (1986: p 47) have dealt with these processes.

The tidal flat and salt-marsh areas offer only limited possibilities of reconstructing the sea-level changes during the course of the last 2000 years, because intercalated peat layers are very rare from this period and because dike construction, the earliest around 1000 A.D., has cut off the protected areas from natural sediment supply. Additionally, drainage measures modified the landscape development and generated intensive compaction processes. All these effects do not occur on the barrier islands or have a much smaller influence there. Therefore, a selected number of the sea-level dates from the islands, given in Table 3, has been dendrochronologically corrected in order to compare geological sea-level dates with historical dates and the results of tide gauge observations.

The *Scrobicularia plana* shells from Wangerooge (Hv 300 and 9257) indicate that the mean high-tide level was at about NN between 410 to 770 AD. On the other hand, the salt-marsh samples from Juist (Hv 13131 and 13130) demonstrate that the mean high-tide level was below +0.4 m in the periods from 105 BC to 210 AD and from 675 to 990 AD. Both groups of dates offer the possibility of establishing the altitude of the local mean high-tide level between zero NN and +0.4 m in the time span between 600 and 700 AD. Comparable conclusions can be drawn from the algal mat and the vegetation horizons on Wangerooge and Langeoog. The algal mat which occurs between +1.22 and +1.35 m on Wangerooge is a very precise indicator of the mean high-tide level. However, palynological investigations have demonstrated a high content of reworked material from older fresh-water and bog environments in this layer. Therefore, no radiocarbon dating has been carried out on the algal mat. The lowermost sample of a vegetation horizon (Hv 12303), collected from only 30 cm above the algal mat indicates that the mean high-tide level was just below +1.5 m in the time interval 1305 to 1435 AD. This result is supported by the sample taken from the 'untere Moorerdebank' on Langeoog (Hv 1998) between +1.53 and +1.58 m, which yielded an age from 1125 to 1395 AD. This second group of

Table 3. Selected sea-level dates with conventional radiocarbon ages and dendrochronologically corrected ages for comparison of geological dates, historical dates and tide gauge observations.

Island	Depth (in m NN)	Laboratory number	Conventional radiocarbon age	Dendrochronologically corrected age
Wangerooge	+1.50 to +1.60	Hv 12303	545 ± 80 B.P.	1305 to 1435 A.D.
Langeoog	+1.53 to +1.58	Hv 1998	690 ± 110 B.P.	1125 to 1395 A.D.
Wangerooge	at about 0.0	Hv 300	1450 ± 180 B.P.	410 to 770 A.D.
Wangerooge	−0.10	Hv 9257	1540 ± 75 B.P.	420 to 600 A.D.
Juist	+0.37 to +0.40	Hv 13130	1185 ± 125 B.P.	675 to 990 A.D.
Juist	+0.43 to +53	Hv 13131	1965 ± 130 B.P.	105 B.C. to 210 A.D.

dates makes it highly probable that the local mean high-tide level was at +1.22 to +1.35 in the time interval 1125 to 1395 AD. This is very close to the present-day mean high-water level which is at +1.29 m on the island of Wangerooge.

Comparison of these geological sea-level dates with the evaluation of historical records of storm-surge levels and tide gauge observations published by Rohde (1975, 1977 and 1985) permits the following conclusions to be drawn. Tide gauge measurements, available from the German North Sea coast for the last 150 years, yielded an average rise of mean high-tide level and storm-surge level of 25 cm per century. Corresponding values can be deduced from historical records of storm-surge marks which are available for the last 300 years. From a combination of both finds, Rohde (1977) concluded that the average rise of the mean high-tide level was 25 cm per century during the last 300 years, which means that the mean high tide in about 1650 AD was 75 cm lower than at present.

On the other hand the geological dates mentioned above indicate that the mean high-tide level on Wangerooge around 1125 to 1395 AD was very close to its present elevation. If both the geological sea-level dates and the tide gauge observations are reliable and can be compared, it must be concluded that a temporary lowering of the mean high-tide level of at least 75 cm occurred between 1125 to 1395 AD and 1650 AD. This period of lowering partly overlaps the period of the little Ice Age (1450 to 1850 AD), during which global temperatures were lower and the circulation pattern was less stable than at present. However, more detailed sea-level studies are necessary to cover the lack of information existing between geological observations and tide gauge evaluations and to shed more light on the relationship between climatic changes and sea-level variations. If these interactions are confirmed, the most probable explanation for such a small scale and ephemeral oscillation of the mean high-tide level is thermal contraction and expansion of the water in the oceans.

References

Barckhausen, J. 1969 Entstehung und Entwicklung der Insel Langeoog – Beiträge zur Quartärgeologie und Paläogeographie eines ostfriesischen Küstenabschnittes – Oldenburger Jahrbuch 68: 239–281.

Behre, K.-E. 1986 Meeresspiegelverhalten und Besiedlung während der Zeit um Christi Geburt in den Nordseemarschen – Offa 43, Festschrift Bantelmann: 45–51.

Behre, K.-E., Dörjes, J. & Irion, G. 1984 Ein datierter Sedimentkern aus dem Holozän der südlichen Nordsee – Probleme der Küstenforschung im südlichen Nordseegebiet 15: 135–148.

Behre, K.-E. & Menke, B. 1969 Pollenanalytische Untersuchungen an einem Bohrkern der südlichen Doggerbank – Beitr. Meereskunde, 24/25: 123–129.

Behre, K.-E. & Streif, H. 1980 Kriterien zu Meeresspiegel- und darauf bezogene Grundwasserabsenkungen – Eiszeitalter Gegenw. 30: 153–160.

Derbyshire, E., Love, M.A. & Edge, M.J. 1985 Fabrics of probable segregated ground-ice origin in some sediment cores from the North Sea basin. In: Broadman, J. (ed.): Soils and Quaternary landscape evolution: 261–280. Chichester, Engl., John Wiley & Sons Ltd.

Eisma, D., Mook, W.G. & Laban, C. 1981 An early Holocene tidal flat in the Southern Bight – Int. Assoc. Sediment. Spec. Publ. 5: 229–237.

Figge, K. 1980 Das Elbe-Urstromtal im Bereich der deutschen Bucht (Nordsee) – Eiszeitalter Gegenw. 30: 203–211.

Hanisch, J. 1980 Neue Meeresspiegeldaten aus dem Raum Wangerooge – Eiszeitalter Gegenw. 30: 221–228.

Hayes, M.O. 1979 Barrier island morphology as a function of wave regime. In: Leatherman, S. (ed.): Barrier Islands: 1–27. New York, Academic Press.

Jelgersma, S. 1979 Sea-level changes in the North Sea basin. In: Oele, E., Schüttenheim, R.T.E. & Wiggers, A.J. (eds.): The Quaternary history of the North Sea – Acta Univ. Ups. Symp. Univ. Ups. Ann. Quing. Cel.: 2: 234–248.

Kolp, O. 1974 Submarine Uferterrassen in der südlichen Ost- und Nordsee als Marken eines stufenweise erfolgten Meeresspiegelanstiegs – Baltica 5: 11–40.

Kolp, O. 1976 Submarine Uferterrassen der südlichen Ost- und Nordsee als Marken des holozänen Meeresanstiegs und der Überflutungsphasen der Ostsee – Petermanns Geogr. Mitt. 120: 1–23.

Kolp, O. 1977 Die Beziehungen zwischen dem eustatischen Meeresanstieg, submarinen Terrassen und den Entwicklungsphasen der Ostsee im Holozän – Z. geol. Wiss. 5 (7): 853–870.

Long, D., Laban, C., Streif, H., Cameron, T.D.J. & Schüttenhelm, R.T.E. 1988 The sedimentary record of climatic variation in the southern North Sea. In: Shackleton, N.J., West, R.G. & Bowen, D.Q. (eds.) The past three million years: Evolution of climatic variability in the North Atlantic region – Phil. Trans. R. Soc. London, B 318: 523–537.

Ludwig, G., Müller, H. & Streif, H. 1981 New dates on Holocene sea-level changes in the German Bight – Int. Assoc. Sediment. Spec. Publ. 5: 211–219.

Oele, E., Schüttenhelm, R.T.E. & Wiggers, A.J. (eds.) 1979 The Quaternary history of the North Sea – Acta Univ. Ups. Symp. Univ. Ups. Annum Quing. Cel. 2: 248 pp.

Preuss, H. 1979 Die holozäne Entwicklung der Nordseeküste im Gebiet der östlichen Wesermarsch – Geol. Jb., A 53: 3–84.

Preuss, H., Streif, H., Tabat, W. & Vinken, R. 1981 Abschlußbericht zum DFG-Forschungsvorhaben 'Meeresspiegelschwankungen der Nordsee im Jungpleistozän und Holozän – Unpubl. Rep. Niedersächsisches Landesamt für Bodenforschung, Hannover: 116 pp.

Rohde, H. 1975 Wasserstandsbeobachtungen im Bereich der deutschen Nordseeküste vor der Mitte des 19. Jahrhunderts – Die Küste, 28: 1-96.

Rohde, H. 1977 Sturmfluthöhen und säkularer Wasserstandsanstieg an der deutschen Nordseeküste – Die Küste, 30: 52–143.

Rohde, H. 1985 New aspects concerning the increase of sea level on the German North Sea coast. In: Edge, B.L. (ed.): Nineteenth Coastal Engineering Conference – Proc. Intern. Conf., Sept. 3–7: 899–907. Amer. Soc. Civil Engineers, New York.

Sindowski, K.-H. 1969 Geologische Karte von Niedersachsen 1 : 25 000 – Erläuterungen Blatt Wangerooge Nr. 2213: 49 pp.

Sindowski, K.-H. 1970 Geologische Karte von Niedersachsen 1 : 25 000 – Erläuterungen Blatt Spiekeroog Nr. 2212: 56 pp.

Streif, H. 1982 The occurrence and significance of peat in the Holocene deposits of the German North Sea coast – In: De Bakker, H. & van den Berg, M.W. (eds): ILRI Publ. 30, Proc. Symp. Peat lands below sea level: 31–41. Wageningen.

Streif, H. 1985 Südliche Nordsee im Eiszeitalter – Überflutungen und Eisvorstöße – Forschung, Mitt. DFG, 1 (1): 9–11.

Streif, H. 1986 Zur Altersstellung und Entwicklung der Ostfriesischen Inseln – Offa, 43, Festschrift Bantelmann: 29–44.

Proceedings KNGMG Symposium 'Coastal Lowlands, Geology and Geotechnology', 1987: 225–235 (1989)

Morphodynamics of the West-Central Florida barrier system: the delicate balance between wave- and tide-domination

Richard A. Davis, Jr
Department of Geology, University of South Florida, Tampa, Florida 33620, U.S.A.

Received 15 September 1987; accepted in revised form 20 January 1988

Key words: barrier island, inlet, tidal delta, tide-domination, wave-domination

Abstract

The barrier island system along the west-central Florida peninsula is probably the most diverse and complicated reach of barrier islands in the world. All types and sizes of barrier islands and inlets persist along a low-energy coast with mean annual wave heights of <30 cm at the shore and tidal ranges <1 m. This mixed-energy barrier coast is bounded at each end by a tide-dominated coast with spring ranges <1.5 m. The general coastal morphology reflects the dominant coastal process in a fashion comparable to the German Bight on the North Sea and the Georgia Bight on the Atlantic coast of the United States.

The morphodynamics of this coast demonstrate well that; 1) the relative effects of coastal processes determine the coastal morphology, and 2) along low-energy coasts there is a very delicate balance between wave- and tide-domination that can be shifted quite easily and thus change morphology.

Introduction

The barrier island system along the west-central Florida peninsula is continuous for nearly 300 km along the eastern Gulf of Mexico. The barrier islands along this low-energy coast extend from Anclote Key in the north to Cape Romano in the south (Fig. 1). This reach of coast includes 29 barrier islands and 30 tidal inlets (Fig. 2). North of Anclote Key the coast becomes quite irregular with the salt marsh environment fronting the open Gulf of Mexico. South of the barrier island system the coast is quite irregular and is dominated by mangrove islands with associated tidal channels (Fig. 1).

This low-energy coastal region represents a very complicated mixing, in both time and space, of tide- and wave-dominated conditions. As such, it affords an opportunity of investigating the cause and effect relationships between sediment supply, tidal energy, wave energy and coastal response. Sea level rise is overprinted on the system as a relatively slow process. In some respects this coastal area is similar to that of the German Bight of the North Sea where high-energy wave and tidal conditions prevail and where coastal morphology responds to the mixed situation. The Georgia Bight along the southeastern coast of the United States shows a similar but less extreme pattern. On the Florida coast the more wave-dominated portion is in the centre of the area in question whereas the opposite circumstances prevail along the German Bight and the Georgia Bight. In each of these areas tide-domination prohibits barriers from being developed but they are present along the mixed energy and wave-dominated reaches.

Because the west-peninsular coast of Florida and its adjacent low gradient shelf represent a general analog to the conditions envisioned for ancient epicontinental and pericontinental seas, there are implications for study of the rock record. A detailed

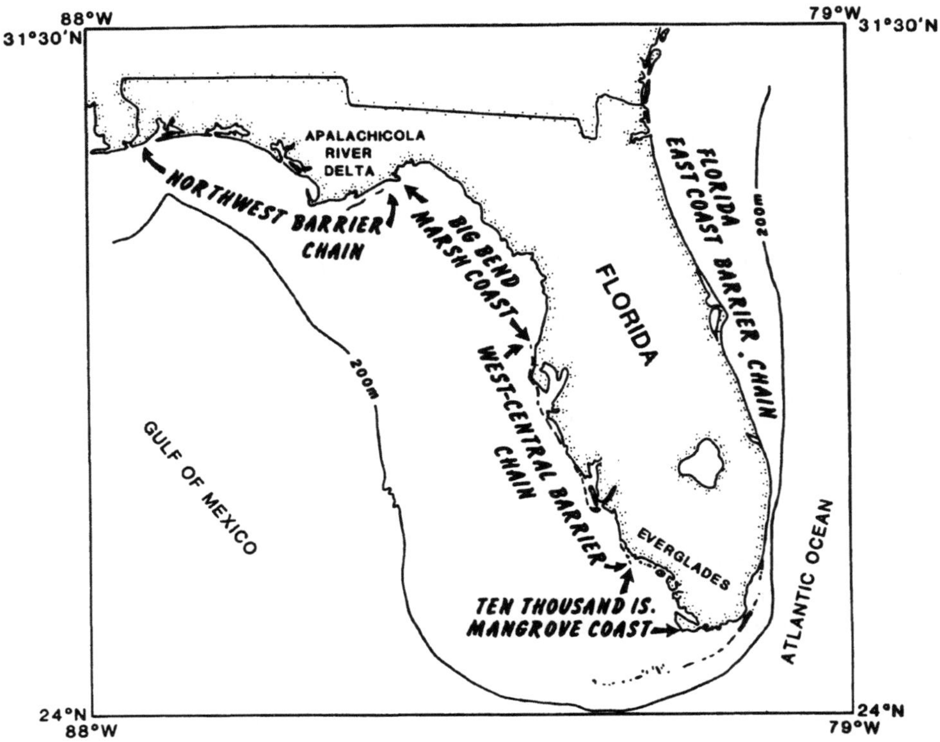

Fig. 1. Location map of the west-central Florida barrier system showing major coastal regions of the State of Florida.

knowledge of the process-response systems operating along a low-energy coast of this type will permit proper assessment of similar coastal sequences preserved in the stratigraphic record as well as provide important data to assist in proper management of coastal environments.

Geologic and geographic setting

The Florida peninsula is a low-relief, generally stable province dominated by Tertiary carbonates and quartz sandstones. The modest drainage systems that emanate from the peninsula carry little sand sized sediment and most of the finer materials are trapped in estuaries that extend throughout the coast. As a result, there is no sand sized sediment that is currently being supplied to the open coast from a landward source.

The Tertiary carbonates are at or near the surface throughout much of the coast in question, especially in the northern and southern portions. These carbonates have had an important effect on coastal morphology by influencing the location of barrier islands. Several islands in the northern part of the barrier system have been shown to be positioned directly above subtle breaks in the carbonate surface only a few metres below sea level (Davis & Kuhn, 1985; Evans et al. 1985). Numerous additional areas have bedrock within less than 3–4 m of present sea level indicating possible influence on coastal morphology.

The shallow bedrock has also had impact on some of tidal inlets that separate the barrier islands. Most of the large inlets are incised into quite resistent strata which cause some stability to the position of the inlet.

The adjacent shoreface in the eastern Gulf of Mexico has a very gentle gradient ranging from about 1 : 1000 off the headland near the north end

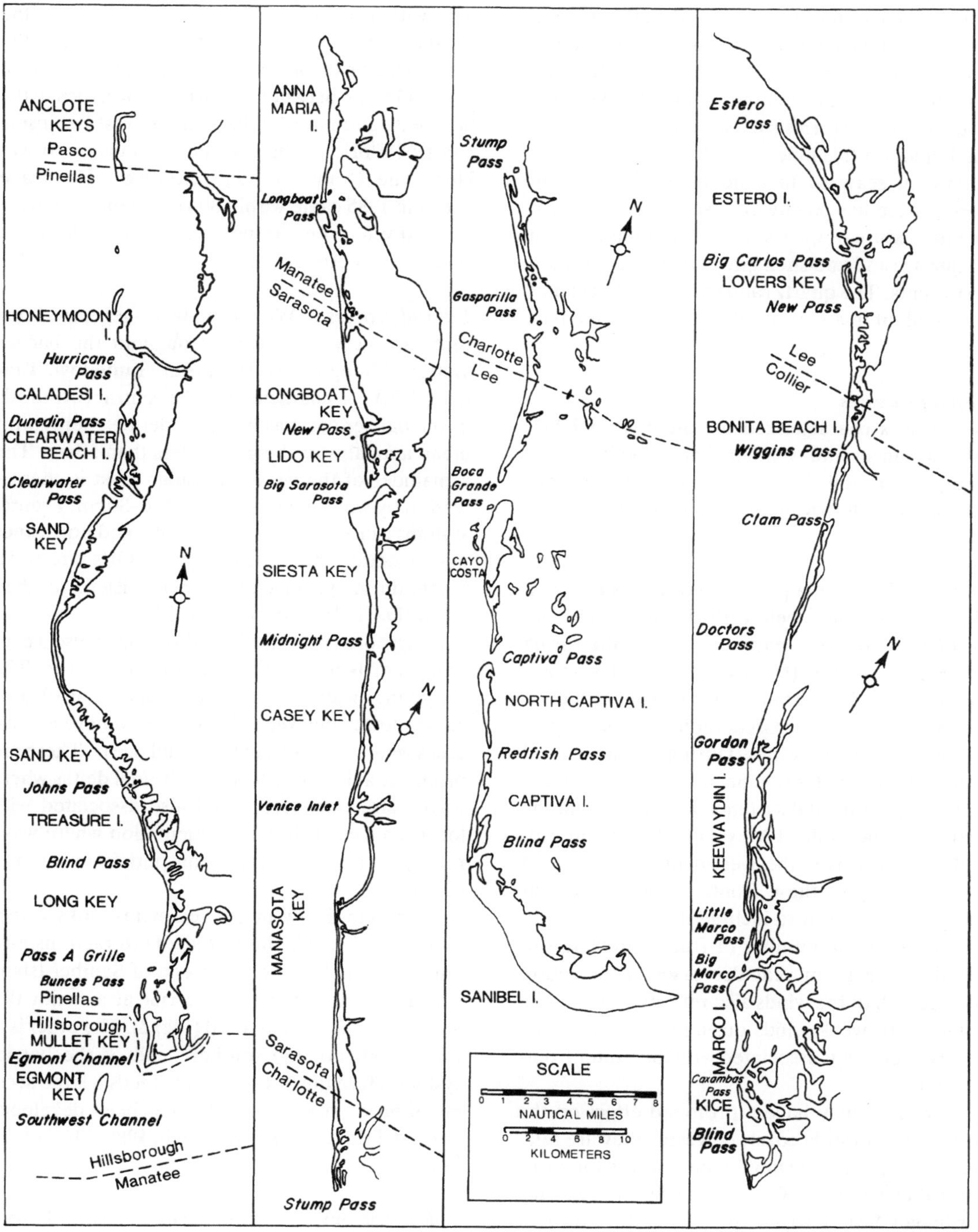

Fig. 2. Strip map of west-central Florida barrier system showing the shape and distribution of each barrier and inlet.

of the barrier system to 1 : 4000 off the mangrove coast south of the barrier system and 1 : 6000 off the salt marsh coast north of the barrier system (Fig. 1). This low gradient causes attenuation of wave energy during major storms such as hurricanes.

Sediment on this inner shelf thins from the barrier system seaward to a depth of about 6–7 m where it becomes patchy or absent. The composition and texture of the sediments are bimodal with fine quartz sand and shell gravel being the primary constituents. The quartz sand component becomes less abundant in a seaward direction from the barriers.

Coastal processes

This coast is characterized by low wave heights, low tidal ranges and weather that is dominated by passage of frontal systems during winter months and occasional hurricanes in late summer and early fall.

Weather. The Florida peninsula occupies part of a subtropical climatic belt with distinctly seasonal and bimodal weather patterns. During the spring and summer months (March–October) this area is dominated by the Bermuda high which produces a clockwise atmospheric circulation resulting in prevailing southeasterly winds. During the late fall and winter months the Florida peninsula is subjected to the passage of frontal systems that move from the north onto the Gulf of Mexico in Texas and proceed eastward over the west-peninsular coast of Florida. Many of these fronts do not extend far enough to the south to include the entire coast in question thus the number of fronts affecting this area during a given year decreases to the south. The strength of the winds also reflects this bimodal pattern with weak winds during the summer and the stronger winds associated with the winter frontal systems.

Hurricanes may have a profound effect on the west-central Florida coast, however, they are relatively infrequent. The most common path for hurricanes that enter the Gulf of Mexico through the Caribbean Sea is to move toward the north or northwest thus bypassing the Florida Gulf coast. These storms that do move in a northerly direction pass within as little as 100 km to the west of the coast and thereby exert significant influence in the form of high wind and wave energy as well as storm surge. On rare occasions a hurricane will turn to the northeast and have landfall on the west-peninsula coast of Florida. The last time this occurred was Hurricane Donna in 1960, however, Hurricane Elena in 1985 came within 80 km of land and had a marked effect on the northern part of the coast (Davis & Andronaco, 1987).

Littoral processes. Waves and wave-generated currents are very important throughout the barrier system although wave heights are quite low. Tanner (1960) has characterized the central two-thirds of the barrier coast as being moderate energy with mean annual wave heights of less than 50 cm. The remainder of the west-peninsular coast is characterized by wave heights of about 10–30 cm. Frontal systems cause breakers of 60–90 cm and hurricanes may generate much larger waves. Offshore wave height during passage of Hurricane Elena reached 2.5 m (Davis & Andronaco, 1987).

Longshore currents along this coast cause a generally north to south net littoral transport. The rates range widely with highest values near 100,000 cubic metres per year. There are numerous local areas where reversals in this trend occur. These are primarily at inlets with large ebb-tidal deltas where wave refraction occurs and also associated with local changes in shoreline orientation where wave refraction results in a net transport from south to north.

There are a few major sediments sinks (other than tidal deltas) in this system of littoral transport. One is located at the south end of Sanibel Island which is a large recurved barrier at a major dislocation in the coastal trend (Fig. 2). The other is at Cape Romano, the southern end of the barrier system. Here the littoral system feeds a large accumulation of quartz sand that has been developed into tidal ridges (Davis & Jewell, 1986; Davis et al. in press).

Tides. The entire west-peninsular coast of Florida is well within the microtidal range. Highest spring ranges are 1.4 m in the Ten Thousand Islands area

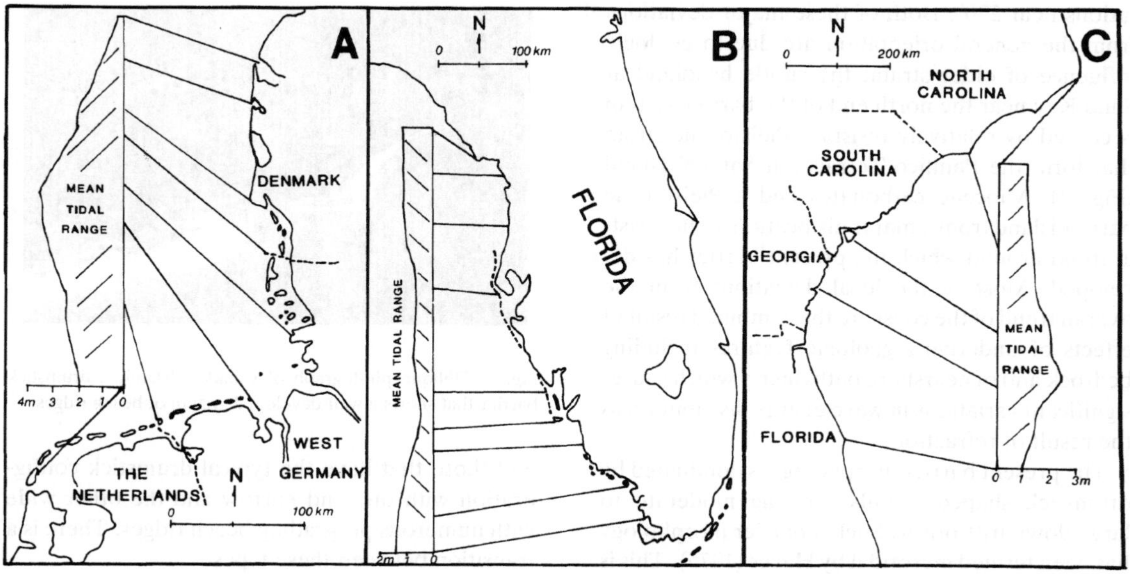

Fig. 3. Generalized outline maps of a) German Bight, b) west-central Florida barrier system, and c) Georgia Bight showing the geographic distribution of mean tidal range for each reach of coast.

south of Cape Romano (Fig. 3b). There is a general decrease toward the north with the central part of the barrier system having spring ranges of only 0.65 m. There is an increase to the north of the barrier system on the salt marsh coast with a maximum spring range of 1.1 m.

This pattern of tidal range is in keeping with the general coastal morphology and with the analogy with the German and Georgia bights (Fig. 3). The Florida situation is one where the lowest tidal ranges are associated with the best developed and most continuous barriers; the same is true for the other two coastal systems. The major difference is in the overall configuration of the coast and position of the barrier islands. On the Florida coast the barriers are in the central portion of the coast whereas the German and Georgia bights have long barriers at either end with a decrease in length toward the centre culminating with an absence of barrier islands where tide range is highest (Fig. 3).

Although tidal ranges on the Florida Gulf coast are low overall, the tidal currents in the inlets may be quite rapid. The combination of large coastal bays and rather widely spaced inlets produces large tidal prisms and consequently also causes high peak velocities of tidal currents. Maximum velocities of over 2 m/sec occur in some inlets.

Coastal morphology

The combination of tidal conditions, wave climate variation and coastal setting produces a varied morphology to the barrier system. There are long, narrow barriers, short, drumstick barriers, some are a few thousand years old, and others have existed for only decades. Tidal inlets include wave-dominated, tide-dominated and mixed morphologies.

General

The reach of coast considered here has a general orientation of northwest-southeast but numerous local changes produce a total range of open coast orientation from about N 20° E on the northern tip of Sand Key to N 45° W on the southern portion of the same barrier (Fig. 2). Sanibel Island curves completely around and has a range of coastal orientations near 270°. Both of these major deviations from the general orientation are due to geologic influence of older strata: the subtle headland at Sand Key near the north end of the barrier system is caused by relatively resistant Pleistocene strata that form the Pamlico Terrace; at Sanibel Island (Fig. 2), Miocene carbonates and a Pleistocene barrier island from a major dislocation in the coastal trend around which the present barrier has developed. Most of the local deviations from the overall trend of the coast are the combined result of effects of underlying geologic features including bedrock and of nearshore bathymetry which causes significant variations in wave energy distribution as the result of refraction.

The present barrier morphology is dominated by drumstick shaped islands showing moderate to large downdrift-offset. Such a barrier morphology has been termed mesotidal by Hayes (1975). This is well shown by Caladesi Island, Siesta Key and Cayo Costa as well as by many other islands (Fig. 2). Several of the drumstick barriers have been subjected to severe building and construction activity which masks their original morphology; examples include Clearwater Beach Island and Treasure Island (Fig. 2).

Barriers with a distinct spit origin are also present. The largest and most obvious is Sand Key in the northern part of the barrier system. This barrier extends to the north and south from the Pleistocene headland at Indian Rocks (Fig. 2). Smaller barrier spit development is evident on the north part of Manasota Key (Fig. 2).

The third prominent barrier morphology is that of long, rather straight and narrow barriers. This is analogous to the microtidal barrier of Hayes (1975). These barriers tend to have a low foredune ridge with a washover apron on the landward side. They typically terminate in a small recurve that may be affected by tidal currents. Examples include Anclote Key, Casey Key and Captiva Island (Fig. 2).

Barrier island morphology

The morphology of the barriers along this coast can conveniently be separated into two broad categories; those that are relatively long and narrow and those that have the typical drumstick configuration with one end narrow and the other wide with numerous prograding beach ridges. There is a transition between these types.

Fig. 4. Oblique photograph of Caladesi Island, a drumstick barrier that shows a well-developed system of beach ridges.

Both upward-shoaling and barrier-spit origins can produce a long and rather narrow barrier island. Three sets of conditions produce this barrier island morphology: 1) longshore transport associated with a headline source, the traditional spit setting, 2) a young barrier formed through upward shoaling, 3) a barrier formed through upward shoaling but with a limited sediment supply. All of these situations also require that wave-dominated conditions prevail.

Drumstick barriers have been described and discussed at great length in the literature since their designation by Hayes, et al. (1974). These barriers tend to be relatively short with a wide updrift end comprised of numerous beach ridges and a narrow downdrift end. The nature and width of the beach ridge complex ranges greatly among barriers. The west-central Florida coast is no exception. Caladesi Island, Anna Maria Island and Siesta Key have a rather simple morphology with numerous beach ridges in a similar orientation (Fig. 4). By contrast, Cayo Costa contains several beach ridge bundles, each truncated by a more recent set (Fig. 5).

Sanibel Island represents a significant deviation from this generalization in that it is a large, relatively old (4000 yrs., Missimer, 1973) example of a spit-type barrier that also has extensive beach ridge accretion sets. This unique morphology to this bar-

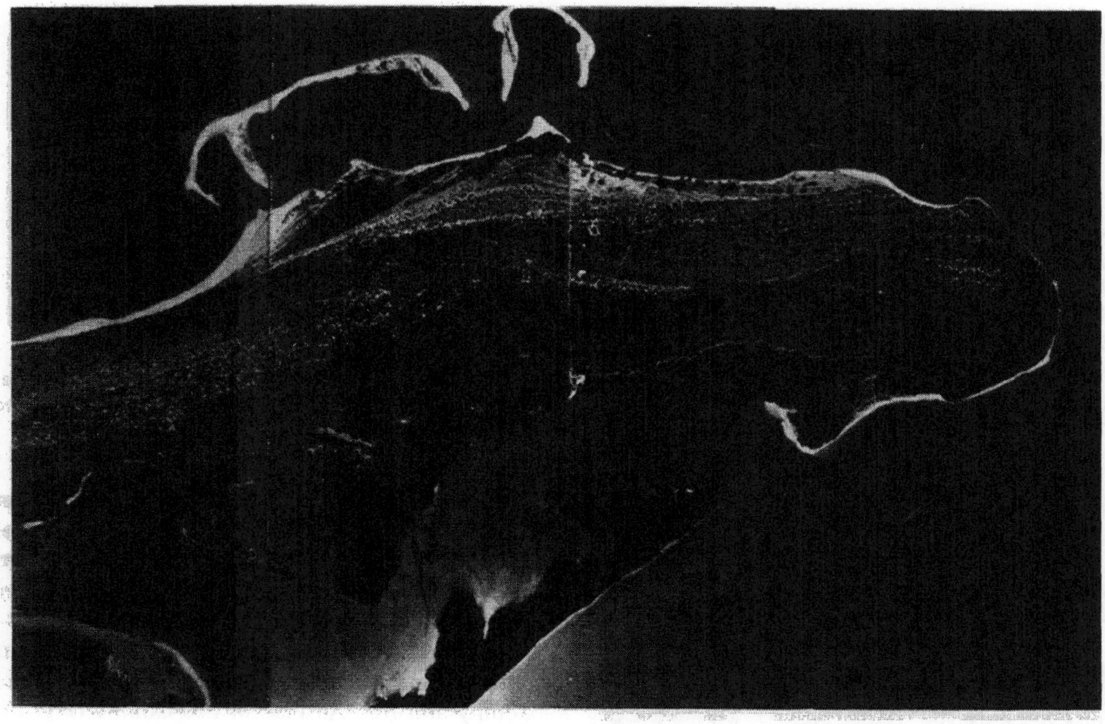

Fig. 5. Vertical aerial photograph of Cayo Costa Island, showing multiple sets of intersecting beach ridges.

rier system is the result of the aforementioned major dislocation in the trend of the barriers and the large recurve associated with it (Fig. 2).

Inlet morphology

The inlets along the west-central Florida coast exhibit great variety in their size and shape. They range from a few tens of metres wide and a metre or so in depth such as Clam Pass in Collier County, up to widths of over a kilometre and depths in excess of 10 m such as Egmont Channel and Boca Grande Pass (Fig. 2). The shapes range widely also with some showing considerable downdrift offset such as Big Sarasota Pass and Gasparilla Pass whereas others have no offset at all, for example Stump Pass and Clam Pass. Generally the larger inlets have the most pronounced offset.

Tidal deltas

The tidal deltas along this coast also exhibit considerable range in size and shape. Their morphology, along with that of the inlet throat can provide a great deal of information about the recent geologic history and dynamics of the inlet system and also about the adjacent barrier island complex.

Flood tidal deltas. Flood deltas are present on the vast majority of the inlets along this barrier system but there are a few that have no apparent landward sediment body associated with the inlet. Those without a flood delta are either the very large inlets such as Egmont Channel and Boca Grande Pass (Fig. 2) or they are ones where development activities have obliterated any discernable flood delta. An example of the latter is Blind Pass in Pinellas County (Fig. 2). The large inlets that lack flood deltas are tide-dominated systems. This morphology is similar to that described in other tide-dominated systems of the Georgia and South Carolina coast (Nummedal et al. 1977; Hayes, 1979). The

Fig. 6. Photographs of a) Redfish Pass and b) Hurricane Pass showing flood tidal deltas. Both of these tidal deltas formed during the 1921 hurricane and have experienced little modification since that time.

large size of the inlet does not present enough of a constriction in the throat and landward deceleration of flooding currents, both conditions that are necessary to produce flood tidal deltas.

The flood deltas that are developed have essentially two different origins although the general morphology is similar. Some originate over a prolonged period of time through tidal processes in the traditional manner of inlet sedimentation. Others

Fig. 7. Flood tidal delta at Midnight Pass which has been vegetated by mangroves. The circular areas are spoil piles which are above high tide and have also been vegetated.

are generated as the result of storms. In the storm situation the barrier is breached and a large fan-shaped sediment body is deposited landward of the breach. These storm-generated flood deltas may subsequently be modified by tidal processes but more commonly they experience little modification after their formation. Hurricane Pass in Pinellas County and Redfish Pass in Lee County (Fig. 2) were both formed by the hurricane on October, 1921. Aerial photographs from shortly thereafter up to the present (Fig. 6) indicate very little change has taken place since that time.

Flood tidal deltas along this coast are typically rather inactive due to the combination of low tidal range and the broad open coastal bays in which they are located. This combination does not produce well-developed channels and overall relief that would result in rapid tidal currents over or around tidal deltas. Consequently, there is little movement of sediment and the surface of the tidal deltas is typically colonized with vegetation. Subtidal surfaces have sea grasses and intertidal surfaces are generally covered with mangroves (Fig. 7) after at least a few years of stability.

Ebb tidal deltas. There is great variation in the size and shape of ebb deltas along this reach of coast. Additionally, the ebb deltas associated with some inlets have undergone drastic changes due to changes in the hydraulic characteristics of the associated inlet. There are a few small inlets that have

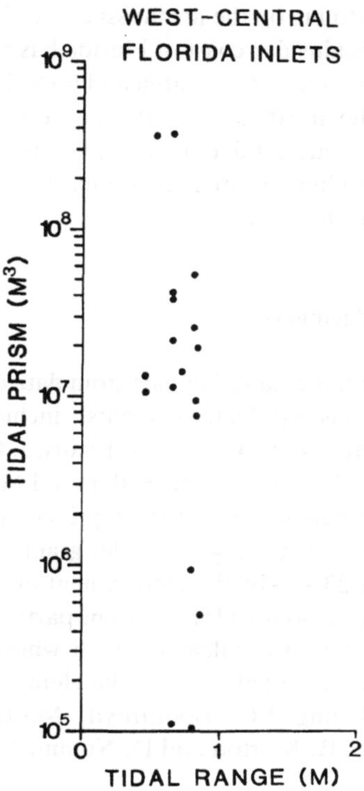

Fig. 8. Plot of tidal range and tidal prism for several inlets in the west-central Florida barrier system. Note that although there is little variation in tidal range, the tidal prism ranges over nearly four orders of magnitude.

no significant ebb delta due to the combination of a very small tidal prism with the wave climate. The result is that waves dominate the coastal area at the inlet and no sediment bodies are permitted to accumulate. Examples include Midnight Pass in Sarasota County and Clam Pass in Collier County (Fig. 2).

The variety of shapes and sizes of ebb tidal deltas reflects the great range in tidal prism over the west-central Florida barrier system. The tidal range shows little difference geographically within the barrier chain with a minimum of near 65 cm and a maximum of near 1 m. Wave energy also varies little with mean annual wave height at the coast being less than 30 cm throughout the area. The fundamental variable in this system is the range in areas of coastal bays that are served by these inlets.

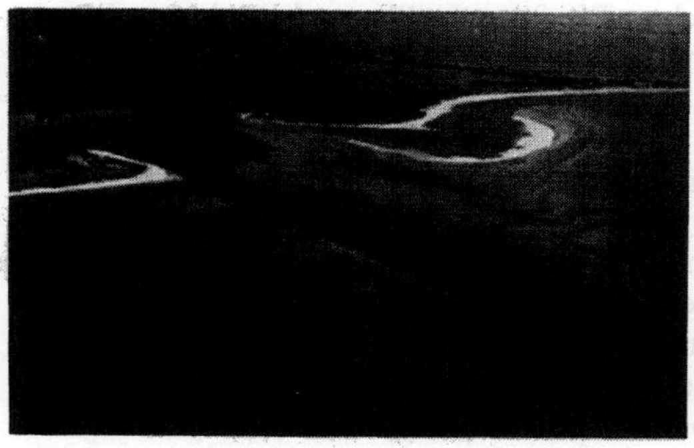

Fig. 9. Oblique photograph of tide-dominated ebb delta at Bunces Pass.

The result is a wide range in tidal prism (Fig. 8) and this tends to control the nature of the ebb tidal delta.

Bunces Pass and Johns Pass in Pinellas County (Fig. 2) have the most tide-dominated morphology with Longboat Pass in Sarasota County and Redfish Pass in Lee County nearly the same. These ebb deltas are characterized by large and elongate sand bodies that are essentially perpendicular to the coastal trend and are adjacent to the main ebb channel (Fig. 9). Although none of these inlets is large, these channel margin linear bars extend 1–2 km beyond the related barriers. These tide-dominated ebb deltas do not have a well-developed terminal lobe because wave energy is ineffective in shaping the tidal delta.

Ebb tidal deltas associated with Egmont Channel at the mouth of Tampa Bay and with Boca Grande Pass at the mouth of Charlotte Harbor (Fig. 2) would also be in the distinctly tide-dominated category because of the large and seaward extending nature of the tidal deltas.

Intermediate or mixed-energy ebb tidal deltas have a morphology that is influenced by both tidal currents and open-water wave energy. These ebb deltas tend to be large and distinct but with a smoothed outer margin due to the well-formed terminal lobe. Examples of this morphology are presently associated with Dunedin Pass and Pass-a-Grille in Pinellas County and Captiva Pass in Lee County (Fig. 2). Ebb deltas associated with major downdrift offsets may also fall into this category such as Big Sarasota Pass and Big Marco Pass in Collier County (Fig. 2).

Summary

The varied nature of the coastal morphology along the west peninsula of Florida represents an excellent example of the relative role between tidal processes and wave-generated processes in shaping the coast. This setting represents perhaps the lowest end of the energy spectrum where this interaction of different coastal processes are manifest. Other coastal settings where these interactive processes have been investigated and discussed are higher energy coasts, especially from the standpoint of tides. Both the German Bight and the Georgia Bight are subjected to tides well into the mesotidal range. By contrast, the west-central coast of Florida experiences tides of just over a metre and less.

Although the regional coastal morphology represents somewhat of a mirror image of the higher energy examples, it shows the same broad coastal types that result from dominance of one primary coastal process over the other regardless of the absolute values of the respective processes. Wave energies in all of the three coastal examples are limited by one or more factors. The west coast of peninsular Florida has a fetch limited by the Gulf of Mexico and, more importantly, has an extremely low-gradient shelf which inhibits wave energy from reaching the coast. The Georgia Bight is also fronted by a broad, rather shallow shelf which limits wave energy. The German Bight has relatively high wave energy during the winter storm season but it is limited by the short fetch of the North Sea.

Tidal energy can be considered as the key to the overall coastal morphology scheme in these situations, both in terms of its absolute magnitude and in its relation to wave energy. On the North Sea coast of Europe the range is from less than 1 m in Denmark to near 4 m in Jade Bay, West Germany and down to about 1 m on the central coast of The Netherlands (Fig. 3). A similar pattern can be seen along the south-eastern coast of the United States where tides range from 1 m in North Carolina up to 3 m in southern South Carolina and back down to 1 m on the northeast coast of Florida. The west peninsular coast of Florida has ranges of about 0.7 m in the central area and up to 1.1 m and 1.4 m at the north and south ends respectively. These quite varied tidal ranges and their accompanying wave climates produce a remarkably similar coastal morphology.

Acknowledgements

This paper has resulted in part from data collected by numerous graduate students, including M. Bland, J. Brame, D. Crowe, M. Evans, J. Gibeaut, J. Gregory, S. Knowles, B. Kuhn, and M. Lynch-Blosse. Funding was supplied primarily by the Florida Sea Grant Program under grants R/OE-17 and R/C-S-23 and by the Department of Geology, University of South Florida. I am particularly indebted to various colleagues with whom I have visited and discussed many of the elements of this coast including J.C. Boothroyd, M.O. Hayes, A.C. Hine, R. Morton and D. Nummedal.

References

Davis, R.A. & Kuhn, B.J. 1985 Origin and development of Anclote Key, west-peninsular Florida – Mar. Geol. 63: 153–171.

Davis, R.A. & Jewell, P. 1986 Low-energy, tide-dominated shelf off southwest Florida – Geol. Soc. Amer. Abstracts with Program 18: 581.

Davis, R.A. & Andronaco, M. 1987 Hurricane effects and post-storm recovery, Pinellas County, Florida (1985–1986). In: Coastal Sediments '87 – Amer. Soc. Civil. Engr.: 1023–1036.

Davis, R.A., Jewell, P. & Sussko, R.J. (in press) Inner continental shelf off southwest Florida. In: Nummedal, D. & Morton, R.H. (eds): Shelf sedimentation, shelf sequences and related hydrocarbon accumulation – Proc. Gulf Coast Section SEPM, 7th Ann. Res. Conf.

Evans, M.W., Hine, A.C., Belknap, D.F. & Davis, R.A. 1985 Bedrock control on barrier island development, Pinellas County, Florida – Mar. Geol. 63: 263–283.

Hayes, M.O. 1975 Morphology and sand accumulations in estuaries. In: Cronin, L.E. (ed.): Estuarine Research – Academic Press (New York) 2: 3–22.

Hayes, M.O. 1979 Barrier island morphology as a function of tidal and wave regime. In: Leatherman, S.P. (ed.): Barrier islands, from the Gulf of St. Lawrence to the Gulf of Mexico – Academic Press (New York): 1–27.

Hayes, M.O., Hulmes, L.J. & Wilson, S.J. 1974 The importance of tidal deltas in erosional and depositional history of barrier islands – Geol. Soc. Amer., Abstracts with Program, 6: 785.

Missimer, T.M. 1973 Growth rates of beach ridges on Sanibel Island, Florida – Gulf Coast Assoc. Geol. Soc. Trans. 23: 383–388.

Nummedal, D., Oertel, G.F., Hubbard, D.K. & Hine, A.C. 1977 Tidal inlet variability – Cape Hatteras to Cape Canaveral. In: Coastal Sediments '77 – Amer. Assoc. Civil Engr.: 543–562.

Tanner, W.F., 1960 Florida coastal classification – Gulf Coast Assoc. Geol. Trans. 10: 259–261.

Proceedings KNGMG Symposium 'Coastal Lowlands, Geology and Geotechnology', 1987: 237–253 (1989)

Pb-210 as a tracer for sediment transport and deposition in the Dutch-German Waddensea

D. Eisma[1], G.W. Berger[1], Chen Wei-Yue[2] & Shen Jian[2]
[1] *Netherlands Institute for Sea Research, Texel, The Netherlands;*
[2] *Institute of Estuarine and Coastal Research, East China Normal University, Shanghai, P.R. China*

Received 29 September 1987; accepted in revised form 9 February 1988

Key words: Pb-210, Waddensea, rate of deposition, resuspension

Abstract

Pb-210 activity has been measured in tidal flat sediments from the Dutch-German Waddensea. It is shown that it can be successfully used to estimate the rate of deposition in areas of intertidal mud deposition, which could be ascertained by absolute dates obtained from soundings and pollen. There is a marked grain size efect and there are indications that Pb-210 and/or Po-210 are enriched in the organic fraction. The degree of reworking was estimated by comparing the Pb-210 age of undisturbed mud layers with absolute ages estimated from repeated soundings. At least 40–90% of the deposited mud was reworked from older sediments. Resuspended old mud acquires within 6.5–9.5 days a Pb-210 activity as is usually found in fresh recent muds. The residence time of Pb-210 in the Waddensea is 43 days.

Introduction

Pb-210, a natural radio-isotope of the U-238 series with a half-life of 22.3 years, has been used extensively for the determination of sediment accumulation rates in relatively quiet environments where fine-grained material is accumulated (Goldberg, 1978; Cochran, 1984; Carpenter et al., 1985). The distribution of Pb-210 in the sediment reflects the depositional history: regular deposition, periods of nondeposition or erosion, mixing by bioturbation (or, in less quiet areas, by stormwaves or bottom trawls), changes in the deposition rate and changes in the sediment source.

Pb-210 measurements can date sediments up to a ca 100 years old and are therefore eminently suitable for the determination of recent depositional rates. A complication is that the mixing of old and young sediment can disturb the chronology. The degree of mixing can be estimated from the sediment structure: a much-used method is to X-ray the sediment so that the structure becomes better visible. When there is little of the original layering left, the mixing rate has to be estimated from the vertical distribution in the sediment of shortlived radio-isotopes such as Th-234 (half-life 24 days) and Th-238 (half-life 1.9 years; Cochran, 1984) or by assuming that the bioturbation process is analogous to a diffusion process (e.g. Officer, 1982). If the sediment mixing rate can be regarded as constant in time, the mixed surface layer will move upwards at the same rate as the deposition rate, which then can be estimated from the Pb-210 activity profile below the surface mixed layer. The relation between Pb-210 activity in the sediment, mixing and sedimentation rate can be described by:

$$\frac{dA}{dt} = D\,\frac{d^2A}{dz^2} - \omega\frac{dA}{dz} - \lambda\, A \qquad \ldots\ (1)$$

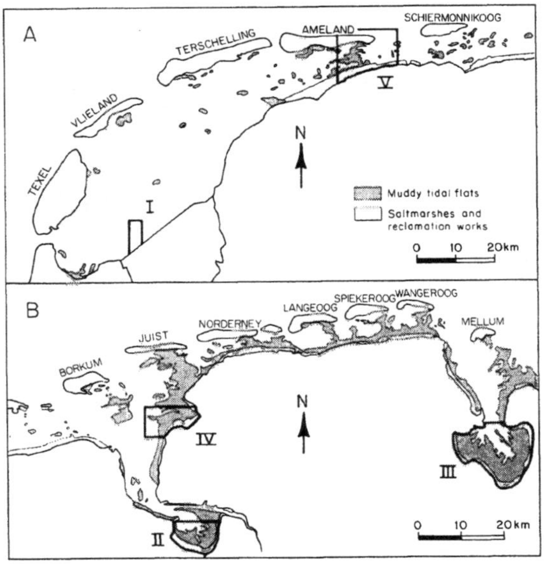

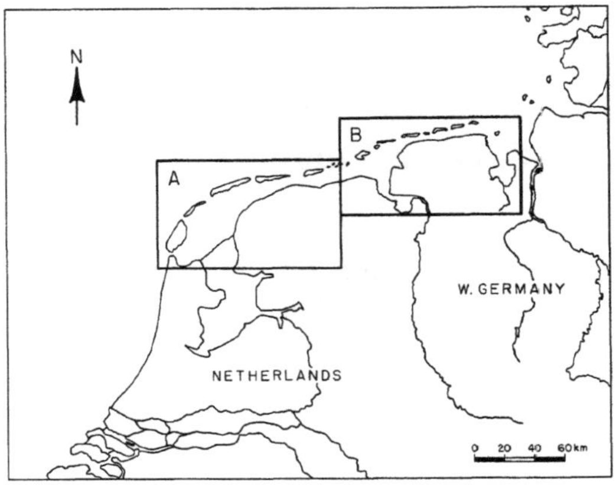

Fig. 1. Principal areas in the Dutch-German Waddensea where cores have been collected for Pb-210 analysis.

where A = Pb-210 activity (dpm/g), t = time (yr), D = mixing coefficient (cm/yr), z = depth in sediment (cm), ω = deposition rate (cm/yr), λ = half-life of Pb-210 (22.3 yr).

The Waddensea is not a quiet environment. Tidal current velocities may reach 1 m/s and large waves during storms (up to several metres high) can have a strong effect on the bottom because of the shallow waterdepth: only in the larger channels is the maximum waterdepth during high tide more than a few metres. Therefore, on the tidal flats erosion dominates during the winter and accumulation during the summer. This is true even for sheltered areas like the Dollard, an embayment at the Dutch-German border, where mud, deposited during the summer, is largely removed during the winter (Bos, 1985). Only salt marshes remain relatively undisturbed but they only cover small areas along the margins of the Waddensea, in total ca 3–4% of the total area (Dijkema, 1987); almost everywhere the Waddensea is bordered by dikes.

To try out the possibilities of using Pb-210 as a tracer for sediment transport and deposition in the Waddensea, short sediment cores were collected in different environments – embayments, filled-in channels, tidal watersheds. A total of 40 vibrocores and boxcores was taken in five separate areas indicated in Fig. 1: the Vlieter (a filled-in channel; area I), the Dollard (area II), the Jade (area III), the Leybucht (area IV) and the tidal flats south of Ameland (area V). The samples were predominantly collected in areas where fine-grained sediment could be expected. The cores were split, described, X-rayed and then sub-sampled for Pb-210 analysis. Additional data on the age of the sediments were obtained from repeated soundings carried out since 1921 in the Dutch Waddensea by the Rijkswaterstaat, and from pollen analysis (in the Dollard area). In addition some laboratory experiments were carried out to estimate the effects of particle size on the Pb-210 activity of a sediment.

Methods

The cores were collected with a vibrocorer of 5 m length containing a PVC liner of 6 cm diameter, and with a boxcorer with boxes of 25 × 40 cm. Coring was done from the R.V. NAVICULA (Netherlands Institute for Sea Research), which was especially designed for work in the Waddensea: it has a draught of 85–100 cm and a flat bottom so that it can stay on the flats during low tide. Positioning was done with Decca Navigation and with radar. On deck the cores from the vibrocore were cut into sections of 120 cm and stored upright. Both ends of each section were closed with a PVC cap. The boxcores were subsampled with PVC pipe of 6 cm diameter. The subsamples were also closed with PVC caps and stored upright. The sediment thickness in the subsamples as well in the original boxsample was measured so that the degree of compaction during subsampling was known.

In the laboratory the vibrocore pipes and the boxcore subsamples were cut into halves and one half was stored. The other half was described (colour, macroscopic features, texture, structure) and X-rayed in a Faxitron X-ray apparatus. Subsamples were taken from undisturbed layers in the cores for Pb-210 analyses and some particle size determinations. Core no. 14 from the Dollard was also subsampled for pollen analysis. The pollen analysis was carried out by H. Heynis (Heynis et al., 1987). Particle size was determined by combined sieve and pipet analysis: samples were split by wet-sieving through a nylon sieve of 63 μm. The smaller fraction was used for pipet analysis, the coarser fraction was wet-sieved. Pb-210 activity was determined by measuring its granddaughter Po-210, which is assumed to be in secular equilibrium with Pb-210. The samples taken in the cores were selected in such a way that they were similar in grain size.

For Pb-210 analysis samples of ca. 2 g (dry weight) were spiked with Po-208 as a yield tracer and digested on an electrical hotplate with a 3 : 2 mixture of concentrated HNO_3 : HC1 and 70% $HC1O_4$. Siliceous residues were removed by addition of a few drops of HF. After evaporation, the Po-210 isotopes were plated on silver discs at 90° C after reduction of Fe^{3+} with ascorbid acid, and analysed using alfa spectroscopy. Calculations were made according to the c.i.c. (constant initial concentration) method (Robbins & Edgington, 1975; Goldberg et al., 1977) with the formula Ti = 1/y*Ln A0/A1 where

A0 = unsupported Pb-210 activity at the sediment surface in dpm/g dry sediment;

A1 = unsupported Pb-210 activity at depth in dpm/g dry sediment;

y = decay constant of Pb-210 (0.03114 year);

Ti = difference in age of surface sediment and sediment at depth in years.

The sedimentation rate was determined from the slope of the least-squares fit for Pb-210 excess values, plotted versus depth.

Sedimentation in an old channel: the Vlieter (area I; Fig. 2)

Before closure of the Afsluitdijk in 1932, the Vlieter was a tidal channel with a maximum depth of ca 12 m, connecting the Texel inlet with the former Zuiderzee (now IJsselmeer). After the closure the tidal currents were reduced and the channel was partly filled in with sand from the surrounding tidal flats and with mud deposited from suspension. The channel was gradually filled until ca 1970 when deposition ended. At present the waterdepth in the former channel is 2–3 m (below spring low tide). Cores were collected in 1982 at locations where the corer was expected to reach into the former channel bottom so that the entire infill since 1932 would be sampled. The selection of the locations was based on the repeated soundings carried out by the Rijkswaterstaat. In a number of cores the former channel bottom, consisting of coarse sand with mollusk shells, was reached. The infill was found to consist of a sequence of virtually undisturbed sandy and muddy layers (Berger et al., 1987), the individual layers having a thickness of up to 3 cm. The freshly sampled sediment showed colour bands of light grey, dark bluish grey and intermediate grey, but these had no relation to the texture of the sediment. The horizontal extent of the layers, as well as of the colour bands, was found to be small: cores collected within 15 m of each other could not be correlated. Core 4, collected where the infill was ca 270 cm thick, was selected for Pb-210 analysis: the Pb-210 activity in the muddy layers decreased regularly with depth in the sediment (Fig. 3), indicating a rate of deposition of 6.5 ± 0.15 cm/yr. Mud sediment thickness in the core was 272 cm; the lowest 8 cm of the core consisted of sand with mollusk shells, indicating the corer had penetrated the infill. Compression of the sediment during coring had been negligible. This gives a duration of mud

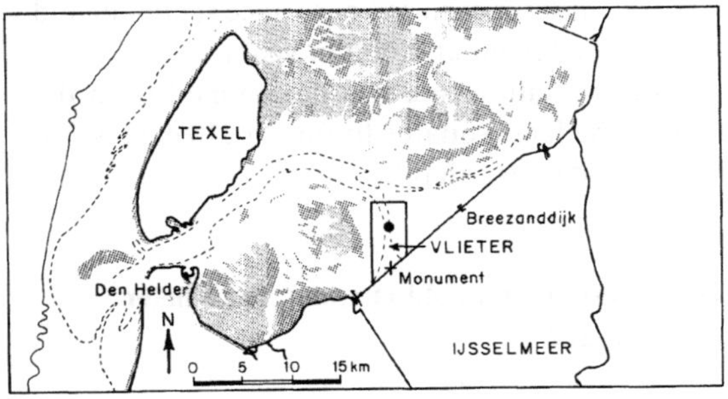

Fig. 2. Location of Vlieter area (area I) in the western Dutch Waddensea and location of vibrocore nr. 4 (black dot). (After Berger et al., 1987).

deposition of 41.8 years. The repeated soundings indicate a depositional rate of 7.1 cm/yr, which implies deposition for a period of 38.3 years. The error in the soundings is of the order of $\pm$ 10 cm (equivalent to a range of ca 3 years of deposition) and in the Pb-210 determinations the overall error is $\pm$ 6.3 cm/yr (equivalent to a range of ca 2 years of deposition). The actual thickness during and shortly after deposition was larger since compaction will have reduced the thickness particularly of the mud layers: continuing compaction after the soundings were made may have contributed to the (small) difference in rate of deposition determined with Pb-210 and estimated from the soundings.

It is important to note that the determination of the rate of deposition with Pb-210 appears to be reasonably accurate but that the ages of the sediment are only relative ages, indicating that the infill was formed in ca 42 years. Owing to the sandy nature of the top sediment and some loss from the top during coring, the initial concentration is not well known. Also the soundings are not explicit since there are indications that the infill started with sand washed in from the sandy flats. Mud deposition may therefore have started later than 1932. The time-span of 41.8 years of deposition, indicating that sedimentation ended around 1974, however, agrees well with the observation from the soundings that deposition stopped around 1970 (allowing for the error of 2–3 yrs in the deposition time estimated from the Pb-210 activity and from the soundings).

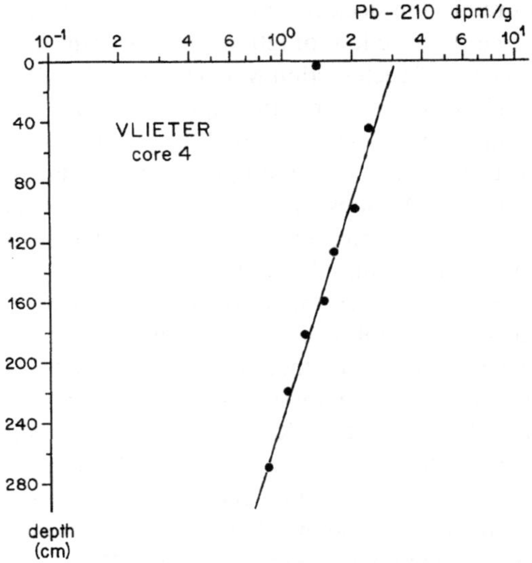

Fig. 3. Pb-210 profile of Vlieter core nr. 4 (from Berger et al., 1987).

Sedimentation on inner tidal flats: the Dollard (area II; Fig. 4)

The Dollard is a large embayment connected with the North Sea by the Ems tidal channel. The sediment on the tidal flats becomes progressively finer in a landward direction. Although the area is sheltered against large waves from the North Sea and the main depositional area is on the inner border along the southern side of the bay, where winds from the west (the dominant wind direction) have little effect, most of the mud deposited during the summer is eroded during the winter. The erosion results at the end of the winter in a surface of consolidated mud, locally covered with some sand and/or plant remains. On this surface, mud is being deposited during spring and summer, resulting in a layer of soft mud on top of the older mud.

Seven cores collected in 1986 were selected for Pb-210 analysis (Fig. 4). Cores 8 and 11 consisted of fine sandy, strongly bioturbated sediment (no. 11 with layers of mollusk shells and shell fragments at 14 cm and 20 cm below the top, no. 8 with muddy layers). Cores 6, 7, 14 and 15 were generally muddy, 7 with some fine sand and traces of burrows, 6

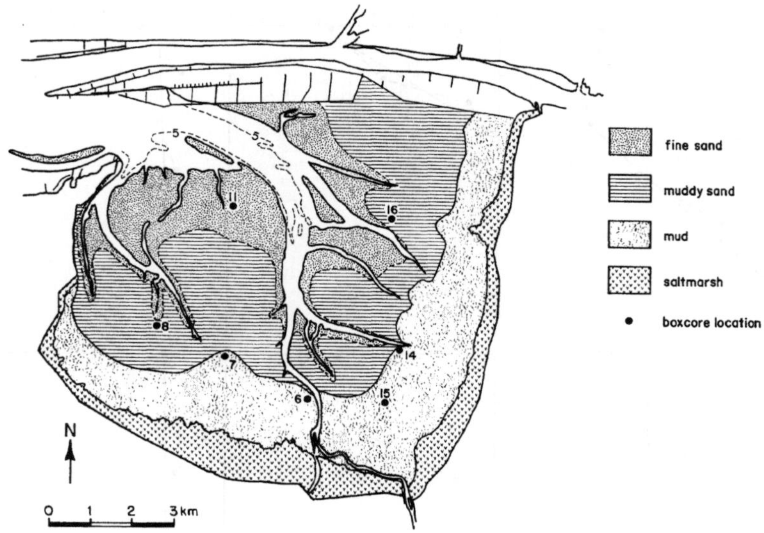

Fig. 4. Map of the Dollard (area II) with locations of boxcores collected for Pb-210 analysis. (from Heynis et al., 1987).

finely laminated with also some burrows, 14 and 15 finely laminated with virtually no bioturbation. No. 16 consisted of fine structureless sandy mud with fine shell fragments and with some shells and large shell fragments at the top.

From the Pb-210 activity distribution in cores 8 and 16 (Fig. 5) it was not possible to estimate a rate of deposition because of sediment mixing and, probably, irregular alternation of erosion and deposition. This may be related to the proximity of small meandering gullies. Core 11 has a surface mixed layer of at least 9 cm; below ca 12 cm the Pb-210 activity rapidly drops to background values. Core 6 has a surface mixed layer of at least 10 cm and at deeper levels a deposition rate of 0.25 cm/yr. Core 15, at 5–15 cm depth in the core, has a deposition rate of 0.14 cm/yr. The surface mixed layer is ca 5 cm thick, which is approximately the layer of freshly deposited, unconsolidated mud. Core 7 also has a deposition rate of 0.14 cm/yr and core 14 a rate of 0.27 cm/yr. In core 14 there were intervals with a somewhat lower (near to the top) and a somewhat higher rate (at 16–19.5 cm; Fig. 6A) of deposition.

Core no. 14, with virtually no bioturbation and with a regular sedimentation rate, was selected for pollen analysis (Heynis et al., 1987). The top 4 cm consisted of soft unconsolidated mud, overlying more consolidated mud with some plant remains at the contact between the two. This sequence reflects the winter erosion followed by deposition during the less stormy spring and summer (the cores were collected in April/May; Bos, 1985). Pollen analysis showed a marked peak of Aster type pollen at ca 33 cm below the top and a significant number of Zea mais pollen in the top 7 cm (Fig. 7). The sudden abundance of the Aster type pollen is the result of the reclamation in 1862 of part of the salt marshes. A new saltmarsh was quickly formed in front of the dike and overgrown with Aster tripolium. The Zea pollen reflect the strong increase of Zea cultivation after 1972. The pollen data provided two fixed dates indicating a deposition rate of 0.28 cm/yr, which compares well with the rate of 0.27 cm/yr based on the Pb-210 activity. The cumulative total pollen curve (Fig. 6B) shows similar fluctuations as the Pb-210 curve with an interval at 16–19.5 m depth in the core when deposition was more rapid, and an interval at 4–6 cm when sedimentation rates were lower. The rate of subsidence in the Dollard is

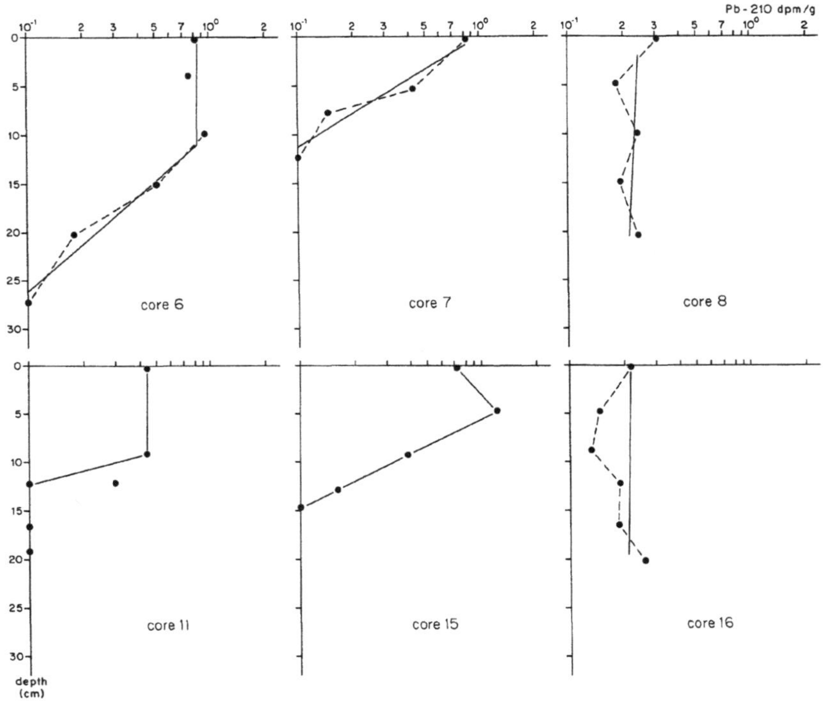

Fig. 5. Pb-210 profiles in six boxcores collected in the Dollard. Numbers indicate core numbers (after Heynis et al., 1987).

ca 0.20 cm/yr (Bouwsema et al., 1986). The deposition rates are generally higher, which is in agreement with the Dollard being an area of mud accumulation. The period of more rapid deposition was ca 1929–1942 on the basis of the cumulative pollen curve, and ca 1925 based on the Pb-210 activity curve which is less precise. It is probable that this period of higher deposition rates was the consequence of the last large reclamation along the margins of the Dollard resulting in the Carel Coenraadpolder, which was completed in 1929. Previous experience has shown that the enclosure of a saltmarsh in this area results in a period of rapid sedimentation on the tidal flats in front of the new dike (K.S. Dijkema, 1987, pers. comm.).

Sedimentation on inner tidal flats: the Jade (area III; Fig. 8)

The Jade is a large embayment connected with the North Sea by one single large tidal channel. Fine-grained sediment is deposited on the inner margins (Fig. 8). Muddy sediments mainly occur on the western and southern side; on the eastern side, which is more exposed to westerly winds and therefore experiences stronger wave action, the sediments are more sandy. The muddy deposits show a layering of sandy and muddy sediments, probably reflecting an alternation of more and less stormy periods. Samples for Pb-210 were taken from eight cores (Fig. 9). All cores show some degree of bioturbation. Three cores on the western side (24b, 25, 27) have a surface mixed layer of 4–6 cm, the others show a more or less regular decrease in Pb-210 activity from the top downwards. One core (26) was very sandy and mixed with shells and shell

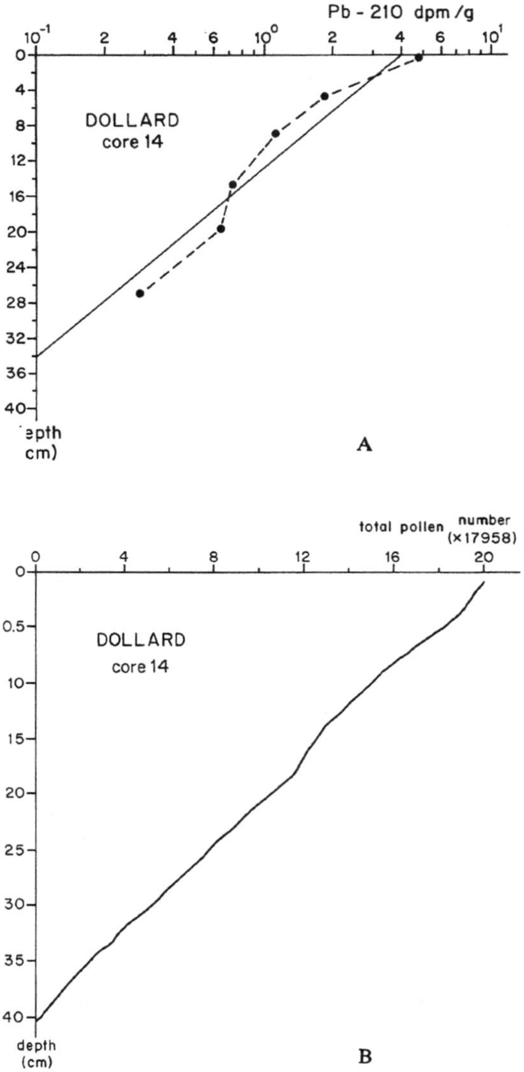

Fig. 6. A. Pb-210 profile in core 14 in the Dollard; B. total pollen profile (after Heynis et al., 1987).

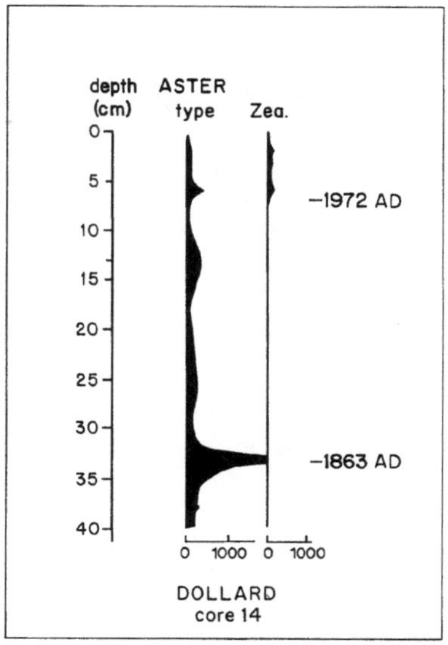

Fig. 7. Profiles of Aster type pollen and Zea mais pollen in core nr. 14 in the Dollard (after Heynis et al., 1987). Horizontal scale indicates number of pollen per cm³.

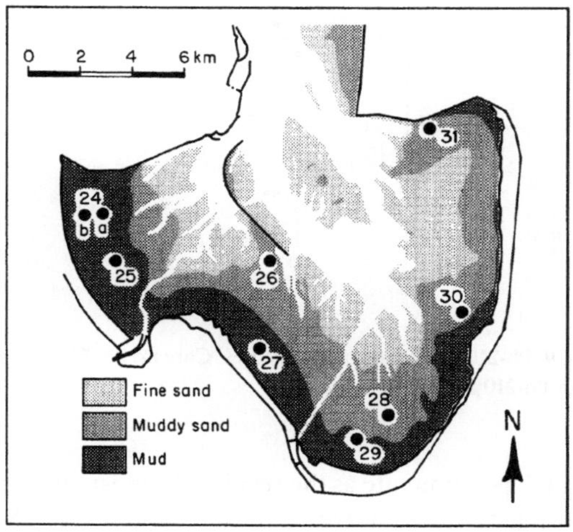

Fig. 8. Map of the Jade (area III) with location of boxcores collected for Pb-210 analysis.

fragments over the entire depth so that no Pb-210 analysis was attempted. Depositional rates were found to vary from 0.25 cm/yr to 1.12 cm/yr. Low depositional rates (0.25–0.27 cm/yr) were indicated where bioturbation was small (cores 24a, 27, 28), higher rates in cores with more bioturbation (0.38–0.49 cm/yr; cores 24b, 25) and highest rates in strongly bioturbated cores (1.03–1.12 cm/yr; cores 29, 30, 31). For cores 29 and 31 different interpretations are possible as indicated in Fig. 9. The rates in the bioturbated cores are considered unreliable since the effect of bioturbation is to bring younger material down into the sediment with a seemingly

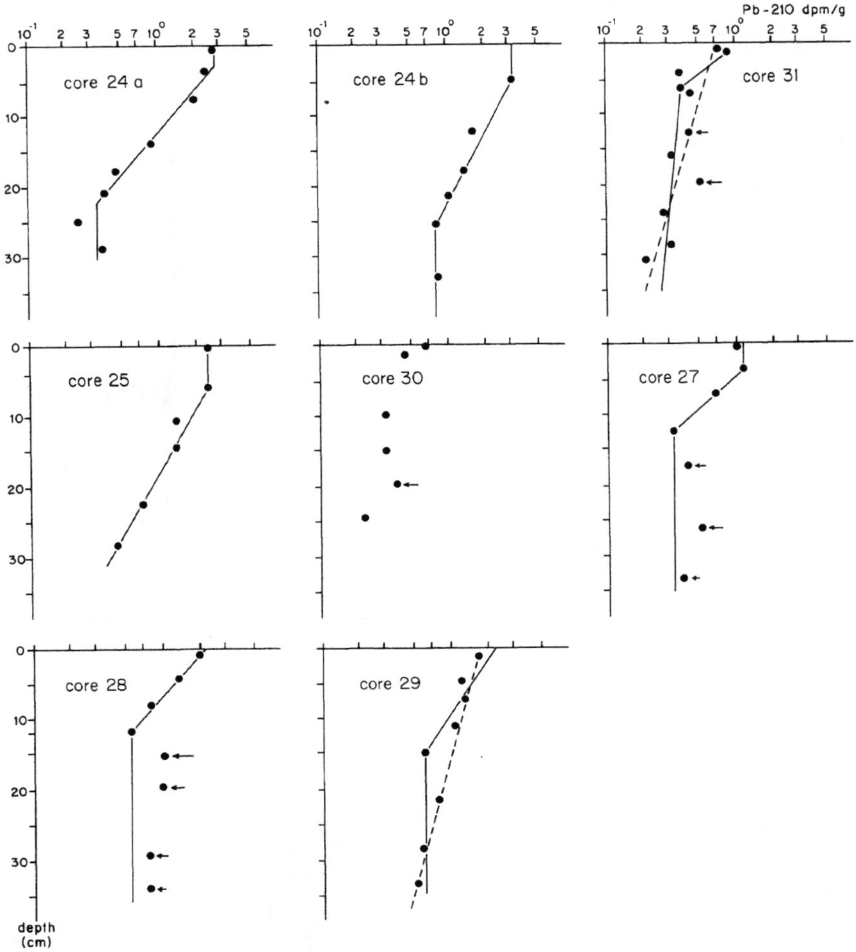

Fig. 9. Pb-210 profiles of Jade boxcores. Core nr. 26 (Fig. 8) was too sandy for Pb-210 analysis. Arrows indicate samples with relatively high Pb-210 activity.

high deposition rate as the result. The most reliable rates are those of cores 24a, 27 and 28 (0.25–0.27 cm/yr). They are only slightly higher than the rate of sea level rise, which is 0.25 cm/yr (measured at Wilhelmshaven in the period 1860–1950; Rohde, 1977).

Another feature is the occurrence of relatively high Pb-210 activities in very fine-grained layers. This is particularly clear in cores 27, 28, 30 and 31 (high activity samples indicated with an arrow), Nittrouer et al. (1979) found a fair correlation between Pb-210 activity and grainsize for sediments on the Washington shelf. For samples from the Waddensea a relation was found as indicated in Fig. 10A. As this was based on a comparison of bottom sediment samples, some experiments were carried out with selected samples (median diameter varying from 30 to 360 micron), that did not contain unsupported Pb-210 (i.e. no Pb-210 from the atmosphere, but only some Pb-210 – 'supported Pb-210' – from the uranium in the sediment particles). Amounts varying from 0.3–2.5 grams were

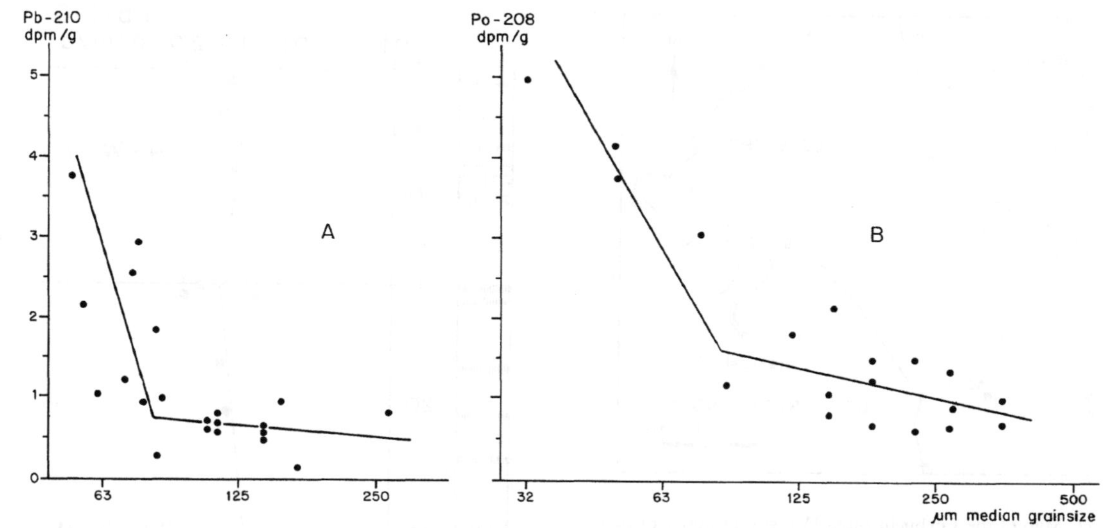

Fig. 10A. Pb-210 activity in the top of bottom sediment samples from the Dutch-German Waddensea in relation to the median grain size. *Fig. 10B.* Relation between Pb-208 activity and particle size as found in adsorption experiments. Differences in suspended matter concentration and time of contact result in variable Pb-208 activities but all results show a sharp increase below ca 90 μm.

shaken with sea water containing 25 mg/l dissolved Po-208 during 1.3, 2.6 and 21.5 hours respectively. Relating the amount of adsorbed Po-208 (in dpm/g) to the particle size shows a similar sharp increase in activity at particle sizes smaller than ca 90 micron as found for the bottom samples. At large sizes the adsorbed amounts are relatively small (less than 2 dpm/g in the experiments, less than 1 dpm/g in the bottom samples), showing a tendency to decrease towards larger sizes. The shift towards higher adsorbed quantities in the experiment as compared with the bottom samples is related to the high dissolved Po-208 concentrations in the experiments. Fig. 10 indicates that in muddy sediments considerable variations in Pb-210 activity are possible, depending on grain size, so that the results have to be normalized with respect to grain size. Applying such a normalization to the relatively high points in cores 27, 28, 30 and 31 brings them on or near to the line that indicates the average decrease in Pb-210 activity with depth in the sediment.

Cores from the Leybucht-Ems (area IV; Fig. 11)

Two cores, collected in this area (Fig. 11), showed a regular layering and very little bioturbation. The deposition rate was found to be 0.12 and 0.14 cm/yr respectively which compares well with the average subsidence rate of ca 0.15 cm/yr. From the water content of the sediment, depositional rates of 0.13 and 0.17 g/cm^2/yr could be estimated. This confirmed that particularly in the mud accummulation areas in the Waddensea, which are not much disturbed and comprise ca 12.4% at most of the total Waddensea area, Pb-210 can be used to determine depositional rates in g/cm^2/yr.

Cores from tidal watersheds and adjacent flats: the Dutch-German Waddensea (area V and the German Waddensea between Ems and Jade)

17 cores collected in the Dutch-German Waddensea on tidal watersheds and adjacent tidal flats were analysed for Pb-210 (Rebers, 1985; Chen & Shen, 1986). Two cores show a regular decrease of Pb-210 activity with depth in the sediment. The

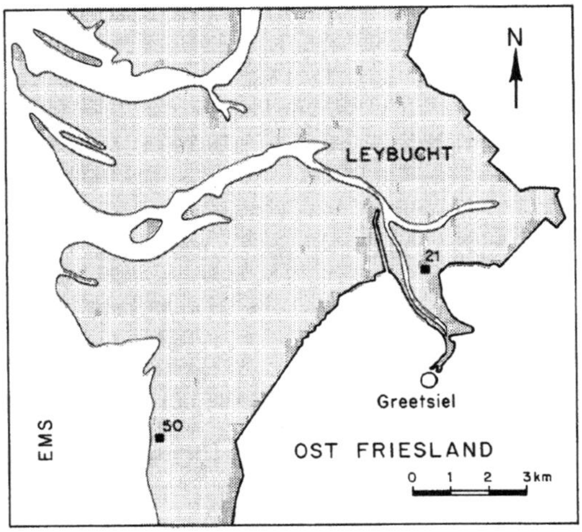

Fig. 11. Map of the Leybucht (area IV) with locations of two boxcores collected for Pb-210 analysis. Shaded area: tidal flats.

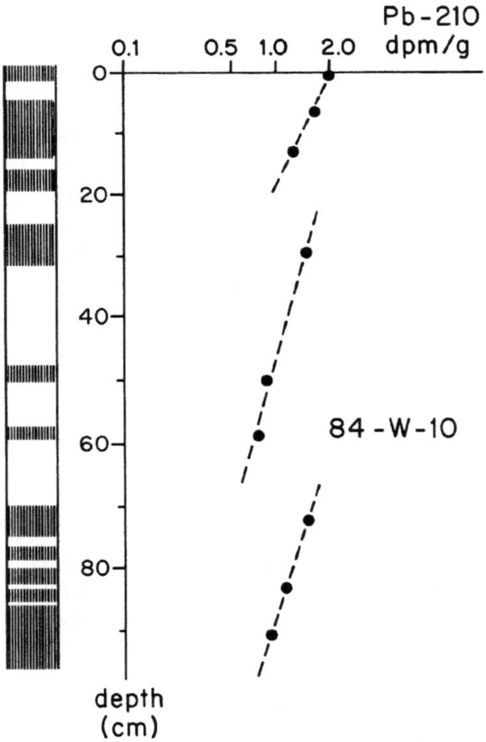

Fig. 12. Pb-210 profile of vibrocore 84-W-10, collected south of Ameland island (area V; Rebers, 1985).

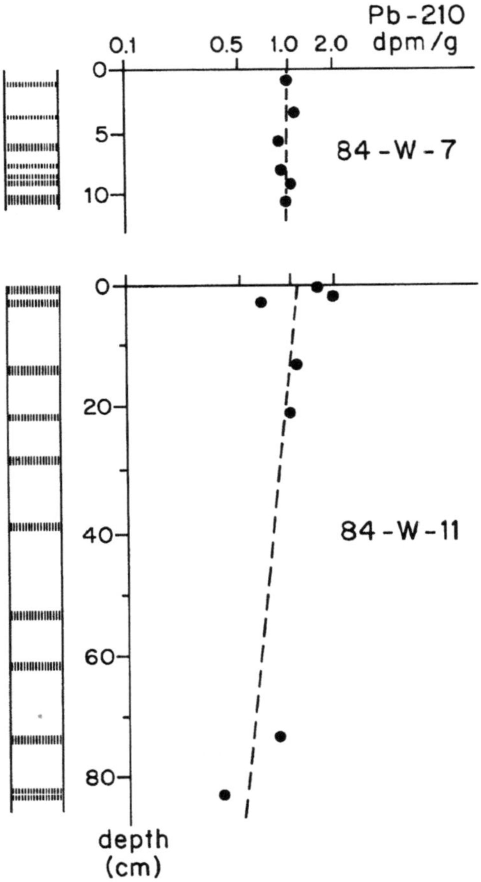

Fig. 13. Pb-210 profiles of vibrocores 84-W-7 and 84-W-11, collected south of Ameland island (area V; Rebers, 1985).

highest depositional rate (0.44 cm/yr) was again found for the most bioturbated core, the lowest rate (0.22 cm/yr) for the least bioturbated. In one core the Pb-210 decrease was more irregular and allowed different interpretations. One core (84-W-10) showed three separate sediment layers with a regular Pb-210 activity decrease with depth in each layer, but not in the core as a whole (Fig. 12). Two cores were completely or almost completely mixed (84-W-11 and 84-W-7; Fig. 13) while three other cores showed a surface mixed layer. All other cores showed an irregular distribution of Pb-210 activity with depth in the sediment. These irregularities cannot be explained by grain size effects only. To understand them, it is important to have a better insight into the Pb-210 system in the Waddensea, including the time scales of supply from the atmosphere, adsorption onto suspended particles, removal from the water column and deposition. This is attempted in the next section.

The 'transfer time', 'adsorption time', 'removal time', and the 'residence time' of Pb-210 in the Waddensea

For a better interpretation of the Pb-210 behaviour in the Waddensea the following concepts were introduced:

- the *'transfer time'*, the average time needed for the transfer of the Pb-210 in solution in the water onto suspended particles;
- the *'adsorption time'* (or *scavenging time*), the time needed for suspended matter without adsorbed unsupported Pb-210 to acquire the Pb-210 activity normally found in bottom sediments;
- the *'removal time'*, the average time needed to remove the Pb-210 in suspension from the water column (either by exchange with another watermass, or by deposition on the bottom);
- the *'residence time'*, the average time during which Pb-210 remains in the Waddensea.

Particularly the capacity of the particles to adsorb Pb-210 from the water as well as the possibility to do so are of importance in large parts of the Waddensea where bottom sediment is frequently resuspended and redeposited.

The Pb-210 technique for determining sedimentation rates is based on the assumption that the flux of Pb-210 to the sediment-water interface remains constant. For samples collected in the Waddensea it was found that 95.5% of the total amount of dissolved and particulate Pb-210 in the water is present in the particulate fraction (Berger & Eisma, 1986). The influence of grain size on the distribution of Pb-210 over the particulate size fractions has been discussed above. To calculate the residence time of the total amount of Pb-210 in the water, a balance equation can be used (Shannon et al., 1970; Nosaki & Tsunogai, 1973; Schell, 1977):

$$dN / dt = I + P - (k + u)\, M \qquad \ldots . (2)$$

where N (dpm/cm^2) is the amount of Pb-210 in the surface water, I ($dpm/cm^2/yr$) is the input rate of Pb-210 from the atmosphere, P ($dpm/cm^2/yr$) is the production rate of Pb-210 from its progenitor Ra-226, k (0.0311/yr) is the radioactive decay constant of Pb-210, u (1/yr) is the removal rate constant of Pb-210 and its reverse (1/u) is defined as the residence time of Pb-210 in the water.

The following equations can be derived, on the assumptions that the exchange of Pb-210 happens only between water and the particles in suspension, and that the Pb-210 from Ra-226 is entirely carried by the particles:

$$d\, Nw / dt = I - k.Nw - Uw.Nw \qquad \ldots . (3)$$
$$d\, Ns / dt = Uw.Nw - Us.Ns - k.Ns + P \,..\, (4)$$

where Ns, Nw (dpm/cm^2) are the amounts of Pb-210 in the particulate fraction (s) and in the soluble fraction (w; Ns + Nw = N), Us and Uw (1/yr) are the removal rate constant of Pb-210 in the water and the transfer constant of Pb-210 from the water onto the particles, their reverse values (1/Us, 1/Uw) being defined as the 'removal time' (Ts) and the 'transfer time' (Tw) of Pb-210.

At steady state conditions the following relations hold:

$$d\, Nw/dt = 0,\ d\, Ns/dt = 0,$$
$$dN/dt = d\, (Nw + Ns)dt = 0$$

and

$$TW = 1/Uw = Nw / (I - k.Nw) \qquad \ldots . (5)$$
$$Tw = 1/Us = Ns / (Uw.Nw - k.Ns + P) \,..\, (6)$$

Adding eq. (3) and eq. (4) and equaling to eq. (2) gives:

$$T = 1/U = (N / Ns)\, Ts \qquad \ldots . (7)$$

where T/Ts increases as N/Ns increases.

In the Waddensea the Po-210 activity in the particulate matter (<45 micron) is ca 0.32 dpm/1 and of the dissolved Po-210 ca 0.015 dpm/1. The amounts of Pb-210 are almost the same since the Pb-210 is in equilibrium with the Po-210 (the ratio Po-210/Pb-210 is around 1). N can be calculated to be 0.067 dpm/cm^2, Nw = 0.003 dpm/cm^2 and Ns = 0.064 dpm/cm^2, assuming that the average depth equals 2 m on the tidal flats, 0.51 $dpm/cm^2/yr$ is used for I (the input of Pb-210 from the atmosphere) according to the observations at Milford Haven, U.K. in 1962–1964 (Pierson et al., 1966). P is estimated to be 0.06 $dpm/cm^2/yr$, being an order of magnitude lower than the atmospheric input. From eq. (5), (6) and (7) the 'transfer time', the 'removal time' and the 'residence time' can be estimated: the 'transfer time' (Tw) is 2.15 days, the 'removal time' (TS) is 41.1 days, and the 'residence time' is 43.0 days. The 'adsorption time' can be estimated from Uw. The average amount of Pb-210 transferred from the water onto the suspended sediment in the Waddensea was found to be 0.001397 $dpm/cm^2/day$ (mean input from the atmosphere being 0.0014 $dpm/cm^2/day$). If the average depth on the flats is 2 m and the average concentration of suspended matter is 15 mg/l, it will take 6.5 – 9.5 days for one gram of suspended matter to adsorb 2–3 dpm of Pb-210 (this amount being normally found in freshly deposition fine grained bottom sediment).

The mean residence time of Pb-210 in the Waddensea (43 days) is very short compared with 58 days in Juan de Fuca Strait, 128–163 days off Cape Flattery, 2.6 years near Bikini (Schell, 1977), and 5 years near Cape Good Hope (Shannon et al., 1970), which means that the removal of Pb-210 from the water column is very fast. This is in agreement with the rather rapid exchange of water and suspended matter with the North Sea and the frequent settling and resuspension of suspended matter.

Us (1/Ts) is the estimate of the rate at wich Pb-210 is being removed from the water in particulate form. The average Pb-210 activity yearly removed from the water through exchange or through deposition can be calculated to be 365/43 × 0.064 $dpm/cm^2/yr$ = 0.576 $dpm/cm^2/yr$. Depositional rates in the muddy sediments are in the order of 0.15 cm/yr, compensating the sea level rise (except in the Dollard where more is deposited). In freshly deposited mud a Pb-210 activity of 2–3 dpm/g is measured. At a water content of 50% and a density of 2.65 of the dry sediment, 0.4–0.6 $dpm/cm^2/yr$ is deposited in the areas of mud deposition.

These areas, mainly along the inner margins of the Waddensea and on the tidal watersheds, cover in the Dutch Waddensea ca 12.4% of the total tidal flat area (2155 km^2; Rijkswaterstaat, 1981a). It follows from these figures that ca 10–15% of the particulate Pb-210 removed yearly from the water is deposited and that the remainder, or 85–90%, is removed towards the North Sea. Assuming the Pb-210 homogeneously distributed over the suspended particles, this also implies that 10–15% of the suspended matter coming into the Waddensea through the inlets from the North Sea, is retained in the Waddensea. This does not fit, however, with the estimates based on transport and depositional rates. The total amount of suspended matter deposited in the mud-deposition areas in the Dutch Waddensea is ca 0.5×10^6 t/yr (at a deposition rate of 0.15 cm/yr, a sediment water content of 50%, a s.g. of 2.65 – as it is largely mineral matter –, and a total area of 12% of 2155 km). Adding the Dollard (0.6×10^6 t/yr; Eisma, 1981) gives 1.1×10^6 t/yr.

The total amount entering yearly through the tidal inlets (3620×10^6 m^3, Rijkswaterstaat, 1981b; and a concentration 10 mg/l) is 26.5×10^6 t/yr. It follows that 4.2% of the suspended matter that enters is retained in the Waddensea through deposition. This agrees well with the estimate of Postma (1981) for the entire Waddensea (ca 5%): with a large area of only sandy tidal flats in the western Waddensea the percentage for the Dutch Waddensea should be somewhat lower than for the Waddensea as a whole. That on the basis of the Pb-210 activities in freshly deposited sediment (2–3 dpm/g) a retention of 10–15% is obtained, suggests that the assumption that Pb-210 – or Po-210 – is homogeneously distributed in the suspended matter and the freshly deposited mud, is not valid, the mineral particles (with a s.g. of 2.65) having a lower Pb-210, or Po-210, activity (in the order of 1–2 dpm/g). A few samples of suspended matter collected in the Waddensea had a Po-210 activity of ca 8 dpm/g

(Berger & Eisma, 1986), which implies a reduction in Po-210activity from ca 8 to 2–3 dpm/g after deposition. The most likely explanation is that Po-210, and possibly also Pb-210, is preferentially adsorbed onto the organic fraction, as was found by Spencer et al. (1980) for the northern North Sea. Organic matter is lost from the freshly deposited sediment through consumption by organisms whereby probably most of the Po-210 is mobilized and lost from the sediment, resulting in lower Pb-210 activities, but there are not sufficient data at present to support this. It implies, however, an imbalance between Po-210 and Pb-210 in the freshly deposited mud and a deviation of the estimated Pb-210 activity in the fresh mud from the Pb-210 activity profile in the older sediment. This is often found. Since most of the mud deposited during calm weather in the summer is resuspended again during stormy weather in the winter (resulting in the observed low long-term average deposition rate of 0.15 cm/yr), a year is probably long enough to reach an equilibrium.

In the Waddensea, we have, however, not only material in suspension coming from the North Sea, but also resuspended material that has entered and has been deposited in earlier years, and that is being reworked. This resuspended older material will acquire the same Pb-210 activity as the recent material entering through the inlets within a short time (6.5–9.5 days: the adsorption time); but if it is redeposited within that time, it still will have the character of 'older' mud. The contribution of locally reworked mud in the mud deposits could be estimated in some cores collected south of Ameland (region V).

Mixing of old resuspended mud with recent mud

Resuspension of bottom sediment is a frequent process in the Wadden Sea. Reworking, resulting in changing patterns of channels and gullies, is a continuous process (Wiemer & Peersmann, 1986). Suspended matter concentrations are higher during and shortly after storms and the regular distribution pattern, related to the tides, is disturbed (Fig. 14, after Postma, 1980).

To estimate the contribution of reworked mud, Pb-210 ages of mud in cores from the tidal flats south of the island of Ameland, were compared with ages estimated from soundings (Wiemer & Peersmann, 1986). The soundings that were used, were made by Rijkswaterstaat in 1959, 1967, 1971, 1973, 1975, 1978, 1981 and 1986. Older soundings were considered less accurate. The error in the measured depth was ca 10 cm before 1980, afterwards ca 5 cm (Rijkswaterstaat, 1981a). The horizontal error, due to positioning, was a few metres in the older soundings, and in the order of 0.5 m in the later soundings. Projection of the soundings from different years on the same profile made it possible to determine the period in which a sediment layer was formed (Fig. 15). It should be realised that the soundings are a momentary record of a process of cutting and infill that is continuous so that a deposit may have been formed very rapidly or very slowly in the interval between two soundings. From seven vibrocores (86-1, 86-4, 86-5, 86-9, 86-14, 86-15 and 86-17; locations see Fig. 15) the Pb-210 activity of 19 mud layers was compared with the age based on the soundings. All mud samples were much older according to their Pb-210 activity than indicated by the soundings. Allowing for the fact that the time-interval between the sucessive soundings is several years and that the reworked mud may have adsorbed some Pb-210 while in suspension, the minimum amount of reworked mud in the deposits was estimated. In these estimates the reworked mud was assumed to contain only supported Pb-210 and the oldest possible age according to the soundings was taken. The Pb-210 activity during the time of deposition (indicated by the soundings) was calculated on the basis of the measured Pb-210 activity and the ratio between (at that time) 'recent' mud with a Pb-210 activity of 2.5 dpm/g and reworked mud with only supported Pb-210 was estimated. It was found that the minimum percentage of reworked mud varied from ca 40 to 90%. These percentages and the large variation in percentage between the mud layers make it clear why so many cores collected on the tidal flats show very irregular Pb-210 activity profile with depth in the sediment.

It is also implied that resuspension of the re-

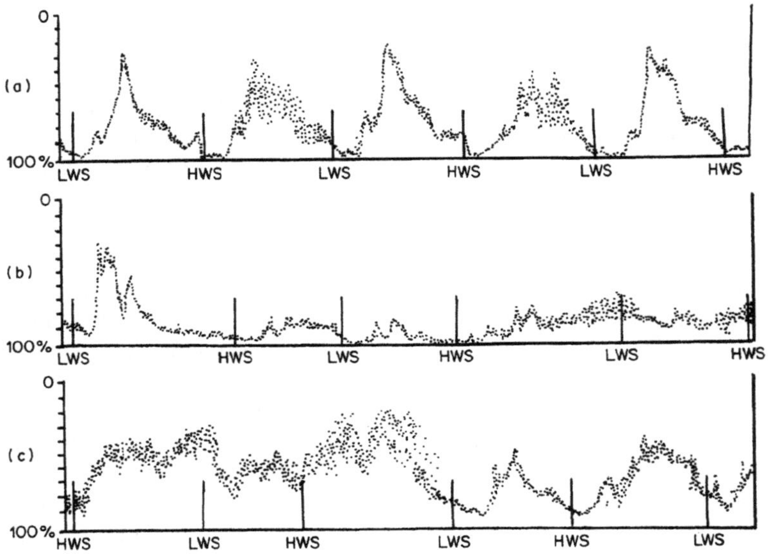

Fig. 14. Transparency (in %) of the water near a tidal inlet in the Waddensea (from Postma, 1980). a) normal tidal variations; b) a storm from W pushes clear North Sea water into the inlet; c) after the storm has subsided turbid water from the inner Waddensea, with mud resuspended during the storm, arrives at the inlet.

worked mud was too short to acquire a 'recent' Pb-210 activity: the period of suspension must have been less than ten days. Considering the thickness of the mud layers (several cm up to more than 10 cm) it is likely that large amounts of bottom mud were resuspended at the same time and quickly redeposited. This happens particularly during storms when large amounts of bottom sediment are

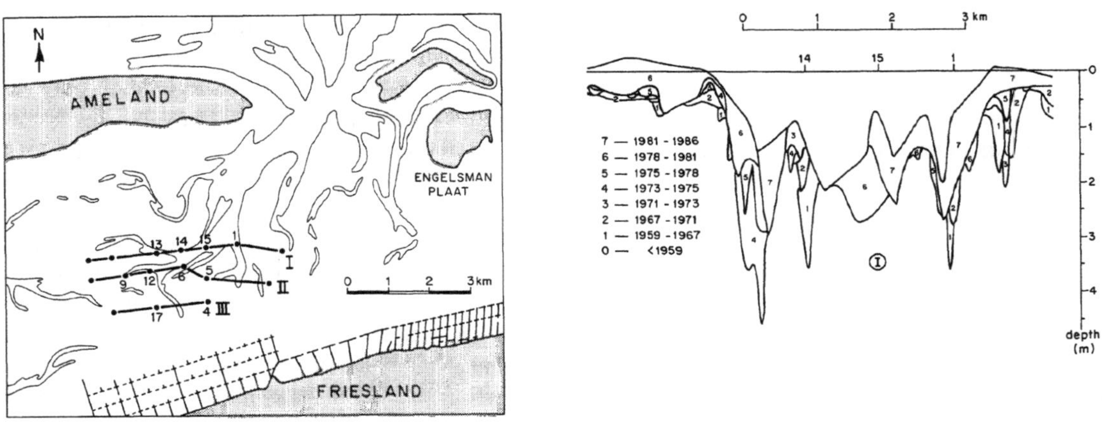

Fig. 15. A. Transects in area V along which sounding data from different years have been compared. Dots indicate stations where vibrocores have been taken. The cores collected at the numbered stations have been used to compare Pb-210 sediment ages with ages estimated from the soundings.

B. Deposition along transect I during seven periods between 1959 and 1986, as estimated from repeated soundings (data Rijkswaterstaat; Wiemer & Peersmann, 1986).

reworked by waves and small channels and gullies are quickly filled in. This leads to the rather paradoxical conclusion that the mud layers are an indication for storm activity. This, however, is not necessarily always true. Low areas on the tidal flats can be gradually filled with muddy sediment, which has even been found in the Texel inlet (Sha 1987, pers. comm.). There has to be, however, a short period between reworking and deposition, which makes storm reworking likely.

The high percentages of reworked material in the muddy layers south of Ameland, made it probable that also the more landward situated muddy deposits contain an admixture of reworked material. They may have been longer in suspension than 10 days, so that the reworked material has acquired the activity of 'recent' mud and therefore is not recognizable any more as reworked. This probable admixture implies a correspondingly lower estimate of the total amount of suspended matter entering the Waddensea through the tidal inlets, as calculated above, so that this estimate should be regarded as a maximum value. Over a longer period, however, and assuming a steady state, the maximum value will apply as the temporary deposition and resuspension can be regarded as an intermediate state between supply through the inlet and final deposition along the inner margins of the Waddensea.

Conclusions

In saltmarshes deposition rates have been successfully estimated from Pb-210 activity profiles with depth in the sediment (Pheiffer Madsen & Sorensen 1979; Gardner et al., 1987). Saltmarshes, however, occur in the Dutch Waddensea only in small areas, together comprising a few percent only of the total area of tidal flats. In this paper it is shown that Pb-210 can be successfully used to estimate depositional rates in areas of intertidal mud deposition in the Dutch-German Waddensea, which in the Dutch part cover ca 12.4% of the total tidal flat area. In a number of cores mixing through bioturbation is small with no effect on the Pb-210 activity profile, which could be ascertained by absolute dates obtained from soundings and pollen. In other cores bioturbation results in seemingly high deposition rates, and a correction has to be applied.

There is a marked grain size effect on the measured Pb-210 activity and Po-210 and/or Pb-210 are not homogeneously distributed in the suspended matter and in freshly deposited sediment; enrichment in the organic fraction is indicated. There is also considerable reworking of older sediment that is mixed with suspended material entering through the tidal inlets. By comparing the Pb-210 age of undisturbed muddy layers in sandy tidal flat sediments south of the island of Ameland, with absolute ages estimated from repeated soundings, it could be estimated that at least 40–90% of the material in these layers had been reworked from older deposits. The 'adsorption time' or 'scavenging time' needed for suspended material without unsupported Pb-210 to acquire the Pb-210 activity normally found in freshly deposited bottom sediments was found to be 6.5–9.5 days, so that the reworked material was quickly deposited again. The rapid redeposition and the quantities involved make it likely that the reworking took place mainly during storms. It is likely that also other mud deposits contain reworked material: this mud may have been in suspension longer than 10 days and therefore acquired the same Pb-210 activity as recent material. By considering the 'transfer time', the 'removal time' and the 'residence time' of Pb-210 in the Waddensea, the time scales of transport and deposition could be estimated. The 'transfer time' (the average time needed for dissolved Pb-210 to become adsorbed onto particles) is 2.15 days, the 'removal time' (the average time needed to remove particulate Pb-210 from the water column) is 41.1 days, the 'residence time' (the average time during which Pb-210 remains in the Waddensea) is 43.0 days. The residence time is very short compared to the residence times calculated for other areas, ranging from 58 days in Juan de Fuca Strait, to 5 years off Cape Good Hope. This reflects the rapid exchange of water and suspended matter between the Waddensea and the coastal North Sea and the frequent settling and resuspension of suspended matter.

Acknowledgements

This study was carried out in cooperation with the following students: M. Rebers (Agricultural University Wageningen), H. Bos (Amsterdam University), H. Heynis (Free University, Amsterdam), M. Peersmann and J. Wiemer (Utrecht State University). Their contributions – reports or papers – are cited in the text. J. Schilling, F. Parlevliet, J. Kalf and J. van Iperen assisted with coring and sampling. To all of them, and to Skipper C.H. Wisse, R.J.R. Anthonijsz and J.C. Tuntelder of the NAVICULA we are indebted for a period of interesting research and pleasant sampling in the Waddensea. To Dr. Georg Irion (Senckenberg Institut, Wilhelmshaven) we are indebted for his help and hospitality when sampling in the German Waddensea, to the State Geological Survey, Haarlem, for the use of vibrocorer, and to Rijkswaterstaat for the use of sounding data. Last but not least, we are much indebted to the China Commission of the Netherlands Academy of Sciences for financing a ten month stay of Chen Wei-Yue and Shen Jian at the Netherlands Institute for Sea Research.

References

Berger, G.W. & D. Eisma 1986 Po-210 metingen in water en gesuspendeerd materiaal in de Nederlandse kustwateren in januari 1986 – Voorlopig Verslag N.I.O.Z.: 23 pp.

Berger, G.W., D. Eisma & A.J. Van Bennekom 1987 210-Pb derived sedimentation rates in the Vlieter, a recently filled-in tidal channel in the Dutch Waddensea – Neth. J. Sea Res. 21 (4): 287–294.

Bos, H.R. 1985 A descriptive study of the sediments in the Dollard estuary with special emphasis on the occurrence of bio-geological structures in the water-sediment transition zone – Rept NIOZ-Mar. Geol. Dept.: 76 pp.

Bouwsema, O., J.H. Bossinade, K.S. Dijkema (ed.), J.W.Th.M. van Meegen, R. Reenders & W. Vrieling 1986 De ontwikkeling van de hoogte en van de omvang van de kwelders in de landaanwinningswerken in Friesland en Groningen – Rapt Rijkswaterstaat Dir. Groningen en RIN, Texel. NOTA ANA-86.05/rapt 86/3: 58 pp.

Carpenter, R., W.L. Peterson & J.T. Bennett 1985 ^{210}Pb derived sediment accumulation and mixing rates for the greater Puget sound region – Mar. Geol. 64: 291-312.

Chen Wei-yue & Shen Jian 1986 Pb-210 dates of sediment cores from the Dutch and German Wadden Sea – Rept Mar. Geol. Dept. NIOZ: 20 pp.

Cochran, J.K. 1984 The fates of uranium and thorium decay series nuclides in the estuarine environment. In: J.F. Kennedy (ed.): The estuary as a filter – Acad. Press.: 179–220.

Dijkema, K.S. 1987 Changes in salt-marsh area in the Netherlands Wadden Sea after 1600. In: A.H.L. Huiskes, C.W.P.M. Blom & J. Rozema (eds): Vegetation between land and sea – Junk, (Dordrecht) Ch. 4: 42–49.

Eisma, D. 1981 Supply and deposition of suspended matter in the North Sea – Int. Ass. Sediment. Spec. Publ. 5: 415–428.

Gardner, L.R., P. Sharma & W.S. Moore 1987 A regeneration model for the effect of bioturbation by fiddler crabs on ^{210}Pb profiles in salt marsh sediments – J. Environm. Radioact. 5: 25–36.

Goldberg, E.D. (ed.): 1978 Biogeochemistry of estuarine sediments – Proc. UNESCO/SCOR Workshop, Melreux, 29 Nov.–3 Dec. 1976: 293 pp.

Goldberg, E.D., E. Gambles, J.J. Griffin & M. Koide 1977 Pollution history in Narragansett Bay as recorded in its sediments – Estuar. Coast. Mar. Sci. 5: 549–561.

Heynis, H., G.W. Berger & D.Eisma 1987 Accumulation rates of estuarine sediment in the Dollard area: comparison of ^{210}Pb and pollen influx methods – Neth. J. Sea Res. 21(4): 295–301.

Nittrouer, C.A., R.W. Sternberg, R. Carpenter & J.T. Bennett 1979 The use of Pb-210 geochronology as a sedimentological tool: applications on the Washington continental shelf – Mar. Geol. 31: 297–316.

Nosaki, Y. & S. Tsunogai 1973 Lead 210 in the North Pacific and transport of terrestrial material through the atmosphere – Earth Planet. Sci. Lett. 20: 88–92.

Officer, Ch.B. 1982 Mixing, sedimentation rates and age dating for sediment cores – Mar. Geol. 46: 261–278.

Pheiffer Madsen, P. & J. Sørensen 1979 Validation of the lead-210 dating method – J. Radioanal. Chem. 54 (1–2): 39–48.

Pierson, D.H., R.S. Cambray & G.S. Spicer 1966 Lead-210 and polonium-210 in the atmosphere – Tellus 18: 427.

Postma, H. 1980 Sediment transport and sedimentation. In: E. Olausson & I. Cato (eds): Chemistry and Biogeochemistry of Estuaries – Wiley, New York: 153–186.

Postma, H. 1981 Exchange of materials between the North Sea and the Wadden Sea – Mar. Geol. 40: 199–213.

Rebers, M. 1985 Sedimentologie en fauna in de Nederlandse Waddenzee – Rapt NIOZ-Mar. Geol. Dept.: 41 pp.

Robbins, J.A. & D.N. Edgington 1975 Determination of recent sedimentation rates in Lake Michigan using 210-Pb and 137-Cs – Geochim. Cosmochim. Acta 39: 285–304.

Rohde, H. 1977 Sturmfluthöhen und säkularer Wasserstandsanstieg an der deutschen Nordzeeküst – Die Küste 30: 52–143.

Rijkswaterstaat 1981a De nauwkeurigheid van lodingen – Nota WWKZ-82.H010. Hoorn: 19 pp.

Rijkswaterstaat 1981b Zandwinning in de Waddenzee. Resultaten hydrografisch-sedimentologisch onderzoek – Directie Friesland, Leeuwarden: 48 pp.

Schell, W.R. 1977 Concentrations, physico-chemical states and mean residence times of ^{210}Pb and ^{210}Po in marine and estua-

rine waters – Geochim. Cosmochim. Acta 41: 1019–1031.
Shannon, L.V., R.D. Cherry & M.J. Orren 1970 Polonium-210 and lead-210 in the marine environment – Geochim. Cosmochim. Acta 34: 701–711.
Spencer, D.W., M.P. Bacon & P.G. Brewer 1980 The distribution of ^{210}Pb and ^{210}Po in the North Sea – Thalassia Jugoslavica 16 (2–4); 125–154.
Wiemer, J.W.J. & M.Rhe Peersmann 1986 Sedimentary processes Waddensea – Rept NIOZ and Dept. Sedimentology, State Univ. Utrecht: 60 pp.

Proceedings KNGMG Symposium 'Coastal Lowlands, Geology and Geotechnology', 1987: 255–260 (1989)

Transfer/noise modelling in groundwater management: an example

Frans C. Van Geer
TNO-DGV Institute of Applied Geoscience,
P.O. Box 285, 2600 AG Delft, The Netherlands.

Received 1 June 1987; accepted in revised form 15 March 1988

Key words: groundwater, trendanalysis, transfer/noise modelling

Abstract

In groundwater management insight is needed into the effects of human activities. These effects can not be measured separately from variations due to other factors.

Therefore a method is required to decompose the measured groundwater level into the various components, corresponding to the causes that influence the groundwater. In this paper transfer/noise modelling is shown to be a useful method to decompose the groundwater level. Transfer/noise modelling is discussed with the help of a case study, where the objective was to detect the occurrence of a trend in a groundwater measurement series.

Introduction

In lowland areas, which are often densely populated and intensively used for agriculture, groundwater is of vital importance. Small changes in the groundwater regime can result in serious crop damage and can have extensive consequences for water supply and stability of constructions. In order to enable a sensible groundwater management, there is a need for information about the behaviour of the groundwater. In particular, the changes in the groundwater regime due to human activities are of utmost importance.

The actual groundwater level is simple to measure. In the Netherlands a Primary Monitoring Network is developed, consisting of many thousands of observation wells. Because the observation wells are measured at regular times, the measurements of each observation well form a time series. In this paper only the measurement frequency of 24 times per year is under consideration, being the most common measurement frequency of the Primary Monitoring Network.

A time series of groundwater measurements often shows variations, caused by many different influences, natural (e.g. precipitation) as well as artificial (e.g. abstraction). Unfortunately, the contribution of each influence to the groundwater variations can not be measured separately. In this paper a method is presented by which the contribution to the groundwater level of a number of known influences can be estimated from measurements. This method is known as transfer/noise modelling (Box & Jenkins, 1976) and is a representative of the class of time series analysis methods. Although transfer/noise models can be applied to a broad class of geohydrological problems (Van Geer, 1987), in this paper they are discussed with the help of a case study. In section 2 the objective of this case study is given. The transfer/noise model is discussed in section 3. In section 4 the calculation results are presented and in section 5 conclusions

Fig. 1. Groundwater is influenced by many causes.

about the applicability of transfer/noise models to groundwater problems are drawn.

Objective

During the last 40 years, groundwater in the Netherlands has been influenced by human activities in many ways. The most important causes that influence groundwater are:

- abstractions of groundwater, for public water supply as well as for industrial purposes,
- drainage, that is achieved by a system of drains, ditches and canals,
- urbanisation,
- large scale regulation of the surface water discharge and recharge,
- land reclamation, in particular in the former Zuiderzee.

Apart from the above mentioned artificial causes, the groundwater level is subject to climatic fluctuations. The groundwater in relation to some natural and artificial causes is sketched in Fig. 1.

Mostly, the effect of an artificial cause on the groundwater level is a decline of that level. In extensive areas the decline of the groundwater level due to each cause is of the order of magnitude of centimetres, while the natural fluctuations are in the order of metres. However, an accumulation of causes can result in a decline of the (average) groundwater level of 10 to 40 centimetres, which may have a considerable effect on vegetation.

The Ministry of Traffic and Public Works (1985) signalized that in large parts of the Netherlands the average groundwater level in the period 1973–1977 was lower than in the period 1965–1960. This decline was assumed to be non-natural. The presumption was made that if the future groundwater management does not change, a further decline of the groundwater level has to be expected.

Before drawing any conclusions, this large scale decline of the groundwater level is analyzed in more detail. The case study presented in this paper is a result of the first phase of that analysis sponsored by the Ministry of Traffic and Public Works.

The objective of the first phase was:

Verification whether the decline of the groundwater level is really caused by non-natural influences.

It was assumed that the natural variations in the groundwater level are caused by variations in precipitation excess. Of course these natural variations are dominated by a periodicity of one year, but also the sequence of wet and dry years causes considerable variations in groundwater level through the years.

With transfer/noise modelling the component of the groundwater level variation due to the precipitation excess was estimated. If there was a decline in the groundwater level, after accounting for the natural variations, this decline was assumed to be non-natural. In a time series of groundwater measurements a decline of the average groundwater level appears as a trend. In the remainder of this paper the term trend is used. An important aspect in this study was to give an indication of the reliability with which the trend can be detected.

The trend analysis, described in this paper has been applied to six groundwater level time series in the Netherlands (Van Geer, 1986). The results of one of these measurement series, the series P-180, is given here.

Methodology

To detect the influence of the most important causes on the groundwater level the relations between these causes and the groundwater level have to be known or estimated. Generally, the relations in

Fig. 2. Black box model.

reality are very complicated and they therefore have to be simplified. The 'classical' deterministic approach in geohydrology is to schematize the soil to a system of aquifers and aquicludes. Then the relations are expressed with the help of Darcy's law and the equation of continuity. The parameters in these relations are calibrated with the available measurements.

In transfer/noise models the soil is treated as a black box (see Fig. 2).

The output of the box is a time series of groundwater measurements. The input series are the causes, natural as well as artificial. The relations between the groundwater measurement series and the input series are estimated from the correlation structure of the input and output series. In transfer/noise modelling the input series are devided into two groups. One group consists of the dominant causes, which are supposed to be measured. The other group consists of all minor (unknown) causes which are combined to one cause: the noise. The basic assumption in transfer/noise modelling is that the groundwater level is the sum of components, each of which corresponds to a dominant cause or the noise. The relation between each dominant cause and the corresponding component of the groundwater level is a linear model, called a transfer model. The relation between the noise and the corresponding component of the groundwater (the noise component) is called the noise model. The total of transfer models and the noise model is called the transfer/noise model.

The coefficients of all models are estimated simultaneously from the measurements and the reliability of all estimated coefficients is calculated.

In the case study presented here the hypothesis is that the groundwater level time series (observation well P-180) is the sum of three components:

- a precipitation excess component,
- a trend component,
- a noise component.

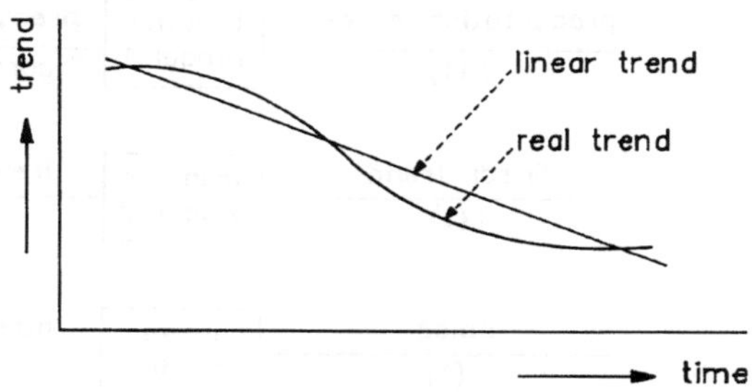

Fig. 3. Linear trend and 'real' trend.

The trend component is assumed to be the result of human activities. In this case study the trend was assumed to be linear. The usefulness of the linearity assumption is motivated as follows:

The objective of this study was to detect whether or not a trend has occurred. This means that there is no need to detect the real form of the trend in the groundwater level. Suppose that there exists a trend in reality; then a significant linear trend can be estimated from the measurements. In the absence of a trend in reality, the estimated linear trend will not differ significantly from zero. Therefore, a hypothetical linear trend can be used to detect the presence of a trend in reality (see Fig. 3).

The transfer/noise model used for modelling the groundwater time series P-180 is schematically shown in Fig. 4.

A general form of the transfer model for the component of the groundwater level due to the precipitation excess is given by:

$$h_1(t) = \delta_{1,1}*h_1(t-1) + \ldots + \delta_{1,k}*h_1(t-k) + \omega_{1,0}*N(t) - \ldots - \omega_{1,m}*N(t-m) \quad (1)$$

where:

$h_1(t)$ is the component of the groundwater level at time t due to the precipitation excess.

$N(t)$ is the precipitation excess during the time interval between t and t-1.

$\delta_{1,1}, \ldots, \delta_{1,k}$ are the autoregressive coefficients.

$\omega_{1,0}, \ldots, \omega_{1,m}$ are the moving average coefficients.

By analogy a general form of the transfer model for

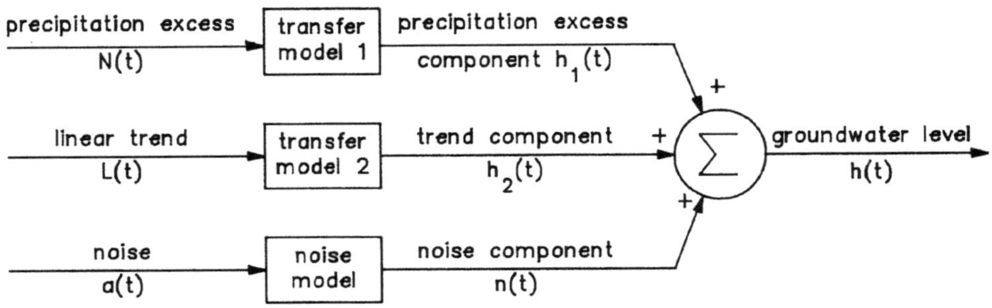

Fig. 4. Transfer/noise model for groundwater series P-180.

the linear trend component is given by:

$$h_2(t) = \delta_{2,1}*h_2(t-1) + \ldots + \delta_{2,r}*h_2(t-r) + \omega_{2,0}*L(t) - \ldots - \omega_{2,s}*L(t-s) \quad (2)$$

where:

$h_2(t)$	is the linear trend component at time t.
$L(t)$	is the linear trend.
$\delta_{2,1}, \ldots, \delta_{2,r}$	are the autoregressive coefficients.
$\omega_{2,0}, \ldots, \omega_{2,s}$	are the moving average coefficients.

In the noise model the input series is assumed to be zero mean white noise, with unknown variance.

The components of the transfer models (1) and (2) are assumed to have a zero mean. Because the groundwater level itself is, generally, non-zero a constant is added to the noise model. The noise model is scaled in such a way that the first moving average coefficient equals 1. The general form of the noise model is given by:

$$\{n(t) - C\} = \varphi_1*\{n(t-1) - C\} + \ldots + \varphi_p*\{n(t-p) - C\} + a(t) - \ldots - \theta_q*a(t-q) \quad (3)$$

where:

$n(t)$	is the noise component.
C	is a constant, to be estimated.
$a(t)$	is the white noise.
$\varphi_1, \ldots, \varphi_p$	are the autoregressive coefficients.
$\theta_1, \ldots, \theta_q$	are the moving average coefficients.

The groundwater level is obtained by adding the three components:

$$h(t) = h_1(t) + h_2(t) + n(t) \quad (4)$$

where:
$h(t)$ is the groundwater level at time t.

The modelling procedure includes the choice of the order of the models (the values of k, m, r, s, p and q) and the estimation of the values of the coefficients. All model coefficients are estimated simultaneously with a standard deviation. The most important quantities in this case study were the coefficients of transfer model 2 and the standard deviation of these coefficients, for they determine the occurrence of a trend and the reliability of the trend detection.

Calculation results

The series P-180 is measured in the province of Noord-Brabant (see Fig. 5). The measurements were taken during the period 1951–1984, with a measurement frequency of 24 times per year. The precipitation and evapotranspiration during this period were determined from measurements at the monitoring station Gemert (also indicated in Fig. 5).

The precipitation excess was calculated by:

$$N(t) = P(t) - 0.8*E_o(t) \quad (5)$$

where:

$P(t)$	is the measured precipitation.
$E_o(t)$	is the potential evapotranspiration, calculated with the formula of Penman.

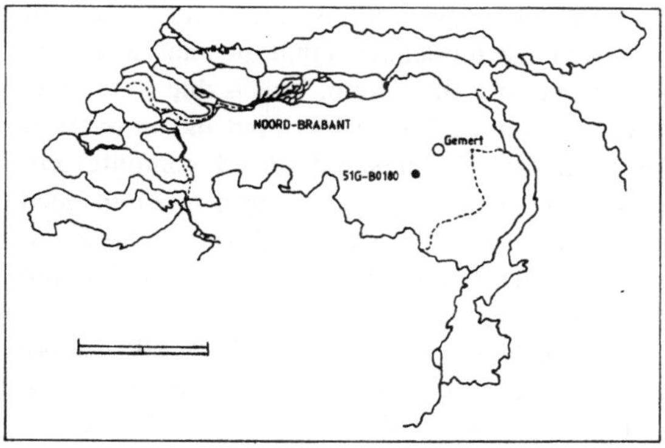

Fig. 5. Location of the groundwater observation well P-180 and the meteorologic monitoring station Gemert. Length of bar is 50 km.

The coefficient 0.8 has proved to give acceptable results in the Netherlands. The measurements used to calculate the precipitation excess were totals over periods of 10 days (decades). To have an input series for the transfer model with the same measurement frequency as the output series (groundwater level) the precipitation excess was converted to a series with measurement frequency of 24 times per year by calculating the intensity of the precipitation excess per decade and taking the mean intensity in the periods corresponding to the groundwater measurement times.

The input series for the trend component is a linear time series. Therefore a series was generated with constant increments. The value of the increment can be chosen arbitrarily, because the coefficients in transfer model 2 are scaled in such a way that the trend component $h_2(t)$ has the appropriate value. To have the coefficients of all model in the same order of magnitude the increment of the linear trend L(t) was chosen to be 0.1 cm per year. The calculations have been performed with the help of Genstat computer programs on a VAX-750 computer. The transfer/noise model for the measurement series P-180 was determined to be:

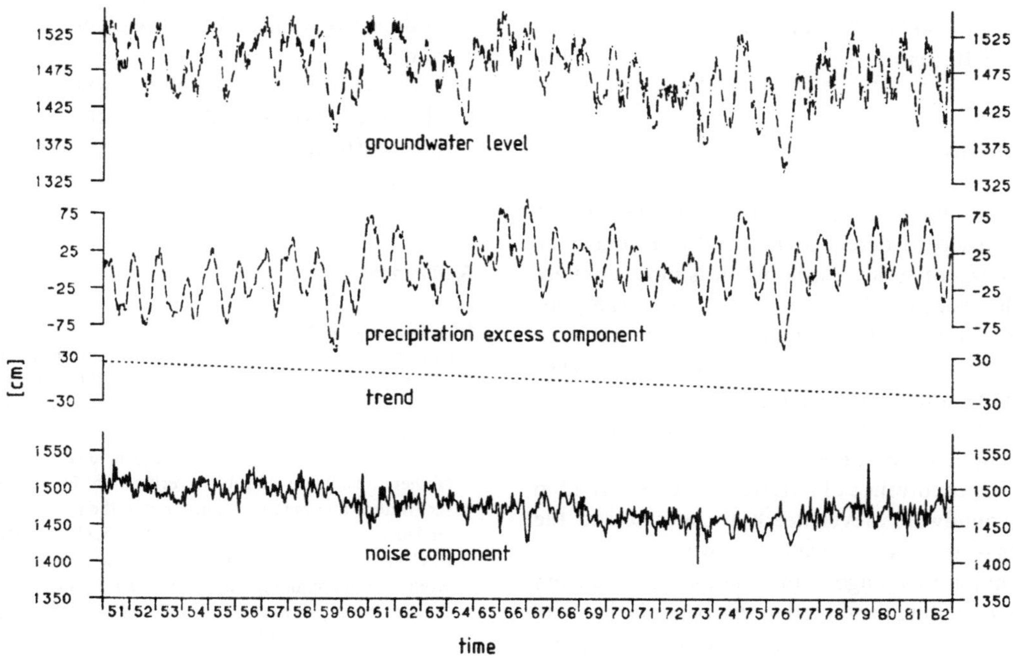

Fig. 6. Result of transfer/noise modelling of P-180.

transfer model 1:

$$h_1(t) = 0.88*h_1(t-1) + 0.42*N(t) \quad (6)$$

transfer model 2:

$$h_2(t) = -0.58*L(t) \quad (7)$$

noise model:

$$\{n(t) + 0.1\} = 0.47*\{n(t-1) + 0.1\} + 0.28*\{n(t-2) + 0.1\} + a(t) \quad (8)$$

The observed groundwater level, the precipitation excess component, the linear trend component and the noise component are shown in Fig. 6.

The standard deviation of the coefficient in transfer model 2 was estimated to be 0.07. The number of time steps over the period 1951–1982 was 24*32 = 768. From (7) it follows that the linear trend over the period 1951–1982 was:

$$-0.58 * 768 * 0.1 \text{ cm} = -45 \text{ cm} \quad (9)$$

Assuming that the probability distribution of the coefficients Gaussian (Box & Jenkins, 1976), the 95% confidence interval of the linear trend over the modelling period is given by:

$$-45 \pm 1.96 * 0.07 * 768 * 0.1 = -45 \pm 10.5 \text{ cm} \quad (10)$$

Conclusions

This case study shows that transfer/noise models can be used to detect a trend in a groundwater measurement series. Although it is most unlikely that the real trend in the groundwater level is linear, the value of the estimated linear trend is an indication of the order of magnitude of the trend. Such indications about changes in the geohydrologic regime are valuable in groundwater management. In this case study the objective was to determine the occurrence of a trend. If the real trend in the groundwater level needs to be assessed the causes of that trend must be known.

A major advantage of the analysis with transfer/noise modelling presented in this paper is that with the decomposition of the groundwater level, also the reliability of this decomposition is indicated by the standard deviation. Another advantage of transfer/noise modelling is that the analysis can be performed in a short time. This is due to the fact that the observed data series (input and output) can be used directly in standardized computerprograms and detailed geohydrologic research need not to be done. Of course overall knowledge about the physical relations between the input and output series is necessary.

Transfer/noise modelling can be used in more situations than the case study presented here. Often in groundwater the causes that influence the groundwater levels are known and measured. There are many situations where transfer/noise models have been applied succesfully, for example:

- to detect variations of the groundwater level due to climatic fluctuations. In this case the groundwater level may be divided into a precipitation excess component and a noise component.
- to detect the influence of changes in surface water level. This may be important close to a river or in areas where the drainage system has been changed. The observed river levels are the input series of a transfer model. A change in the drainage system can be modeled as a step function.
- to detect the drawdown of the groundwater level as a result of groundwater abstraction, where the abstracted volume of groundwater is the input series of a transfer model.

References

Box, G.E.P. & G.M. Jenkins. 1976. Time series analysis; forecasting and control – Holden Day. (San Francisco): 575 pp

Ministry of Traffic and Public Works. 1985. The watermanagement of the Netherlands 1984 (in Dutch) – Staatsuitgeverij, (The Hague, NL): 253 pp

Van Geer, F.C. 1986. Decline of the groundwater level in the Netherlands. Phase 1: Problem verification. Trend analysis of six groundwater measurement series in the period 1951–1984 (in Dutch) – DGV-TNO Rept.: OS 86–25, (Delft, NL): 26 pp

Van Geer, F.C. 1987. Design of groundwater monitoring networks around pumpingstations. Phase 1: Evaluation of existing monitoring networks (in Dutch) – DGV-TNO Rept.: OS 87–39, (Delft, NL): 80 pp

Proceedings KNGMG Symposium 'Coastal Lowlands, Geology and Geotechnology', 1987: 261–266 (1989)

An organisation scheme for the operation and management of the ground water level monitoring network in The Netherlands

M.J. Van Bracht
Dienst Grondwaterverkenning, Institute of Applied Geoscience TNO, P.O. Box 285, 2600 AG Delft, The Netherlands

Received 14 September 1987; accepted in revised form 14 April 1988

Key words: ground water level, monitoring network, organisation scheme, data base

Abstract

As The Netherlands is a low-lying country, piezometric heads of ground water are found at shallow depths. Slight changes in ground water level (a few centimetres) produced by natural or human influence may already have caused considerable damage to agriculture, the natural environment, foundations of buildings, etc.

This vulnerable situation with regard to ground water created the need for monitoring ground water levels. Ground water data are collected by several institutions. The institutions involved decided to co-operate in several ways to create the most effective approach to collect and process the data and to maintain the monitoring network. The TNO-DGV Institute of Applied Geoscience plays an important role in this.

This paper describes an organisation scheme showing the different stages of the data flow from observation wells to the use of data in the central data base of TNO-DGV. The organisation scheme also illustrates among other things, feedback between the different stages in the data flow. In the last chapter the costs of the ground water level network and the data base are described.

Introduction

As the Netherlands is a low-lying country (about one-third of its present area has been reclaimed from the sea), piezometric heads are found at shallow depths. Changes of only a few centimetres in ground water level, induced by natural or human influence, may cause considerable damage to agriculture, the natural environment, the foundations of buildings, etc.

This vulnerability to ground water fluctuations created the need for monitoring ground water level data and gave rise to the founding in 1948 of the Archives for Groundwater Levels TNO. In 1967 these Archives were incorporated into the newly-established Institute of Applied Geoscience TNO.

The main tasks of the Archives for Groundwater Levels are to operate and maintain a nation-wide observation network, to collect, process and evaluate observations, and to make the related data available to interested public and private parties.

Since its foundation the observation network has expanded into a system of over 16 000 observation wells, half of which have two or more observation screens. Fig. 1 shows the number of observation wells of the nation-wide monitoring network of the TNO Archives, from 1948 to 1986.

About half of the observation wells are maintained by TNO. Some 8 000 observation wells are situated in phreatic aquifers, the others in deeper aquifers. There are two kinds of observation wells. The wells of the first category, – primary observa-

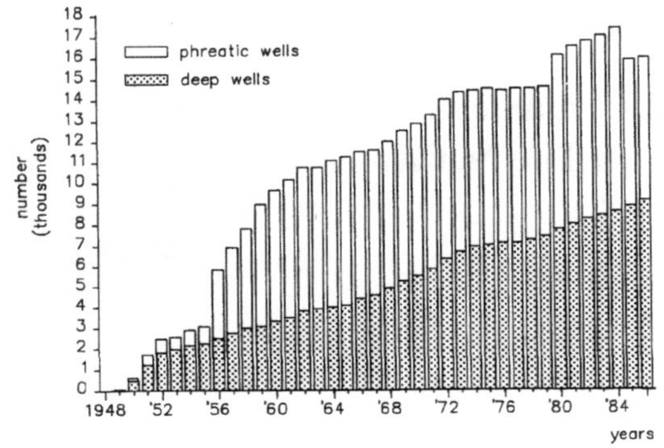

Fig. 1. Number of observation wells of the TNO Archives of Groundwater Levels.

tion wells –, are closely related to the aims of the Dutch provincial ground water management and control. They give a regional picture of the ground water system as a whole, with the emphasis on the natural situation, and also act as a reference for the so-called secondary observation wells. Wells of the latter type are installed for a specific purpose (e.g. to evaluate artificial influences, to monitor local-scale geohydrological systems, etc.).

In 70% of the observation wells, ground water levels are measured fortnightly (on fixed dates). 10% of the recordings are made monthly and 20% quarterly (in April, August, October and December). The observations are carried out voluntarily and free of charge by 4000 observers, including about 2000 private individuals. The others are employees of provincial and municipal institutions, water supply companies, etc. Each year about 500 000 observations are carried out.

Since 1948 more than 8 million water level observations have been collected from all over the country and from various depths and aquifers. All the observations are sorted in the magnetic disc memory of a TNO computer and kept available on-line. Roughly 2 million observational data are supplied to interested parties each year. Most information is used for water management, civil engineering and agricultural purposes. It can be presented on paper (in the form of tables, graphs, contour maps, etc.), on magnetic tapes, or on-line on a computer terminal or micro-computer.

Main activities and organization

The network is managed according to the organization scheme shown in Fig. 2. This scheme comprises:

- three 'main types of activity': network maintenance, data acquisition and processing and data base management;
- five 'organization blocks', containing several related activities;
- twentythree 'elements', each containing one defined activity;
- three 'main elements', elements for co-ordinating other elements in one or two blocks.

The scheme illustrates two kinds of data flow: a regular 'primary data flow', showing the different stages of data from observation well to the final files in the data base, and an irregular data flow which, among other things illustrates feedback between organization blocks.

Network maintenance

The controlling and co-ordinating element for block 1 is the element management of the maintenance.

There are two types of maintenance: 'minor (preventive) maintenance' (a yearly check of the technical state and geohydrological representativeness of the observation well) and major (repair) maintenance (repair of a seriously damaged observation well).

Data acquisition and processing

The first block of this main type of activity contains a number of elements concerning data acquisition (blocks 2 and 3). The controlling element for those two blocks is data acquisition management (creating observation schemes, instructions for observers, etc.). All observations are noted on standard postcards or forms and sent to the Archives of Groundwater Levels. Here all the observations are collected, roughly checked for remarks of the observers and reminders are sent to observers who have not responded.

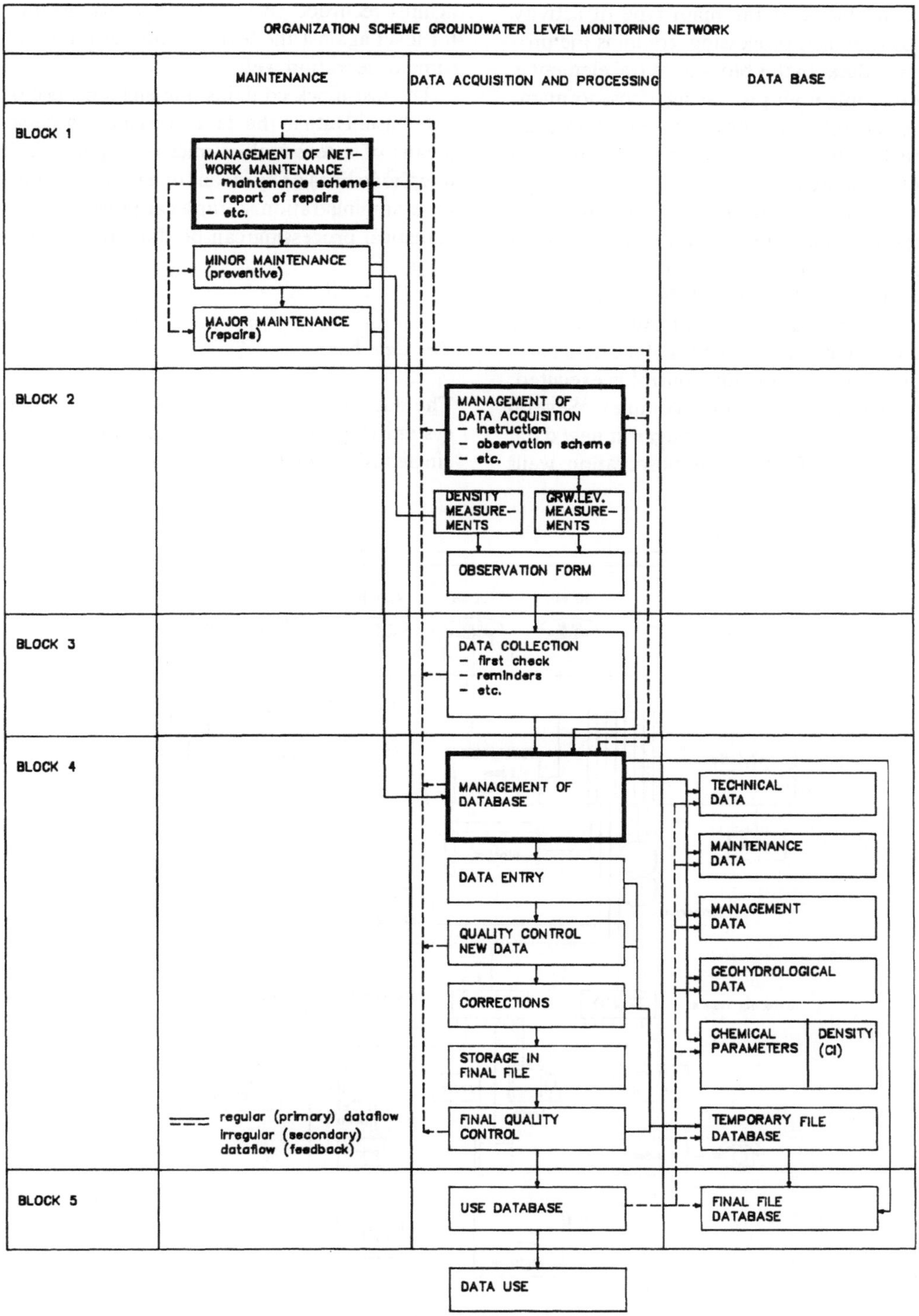

Fig. 2. Organization scheme of Dutch ground water level monitoring network.

The second block of the main type of activity 'data acquisition and processing' contains the processing of the data. In this block a central element is designed to control all processing steps, to maintain all the data in the database, to arrange proprietory protection of the data, to communicate with other central elements, etc.

In the element 'data-entry' all the data are stored in a temporary file, and a quality control, using several automated tests, is carried out.

In the following elements all the recorded data are checked, errors are corrected and the data are finally stored in the definitive files of the data base. Once a year a second quality control is executed. This time the tests are more complex. Several cross-checks are made: old data against new data, the technical specifications of observation wells against recorded ground water levels, ground water level against the level in another screen of the same observation well, etc.

The last block contains one element: use of the data base. Data in the data base can be accessed by means of a terminal or micro-computer (using a telephone line) or by means of a request to the data base administration. Graphical presentations and statistical processing can be done in a number of ways.

The data base

The data base is stored on a VAX 8530 computer system (Fig. 3). This computer is dedicated to a single task, namely management of the ground wa-

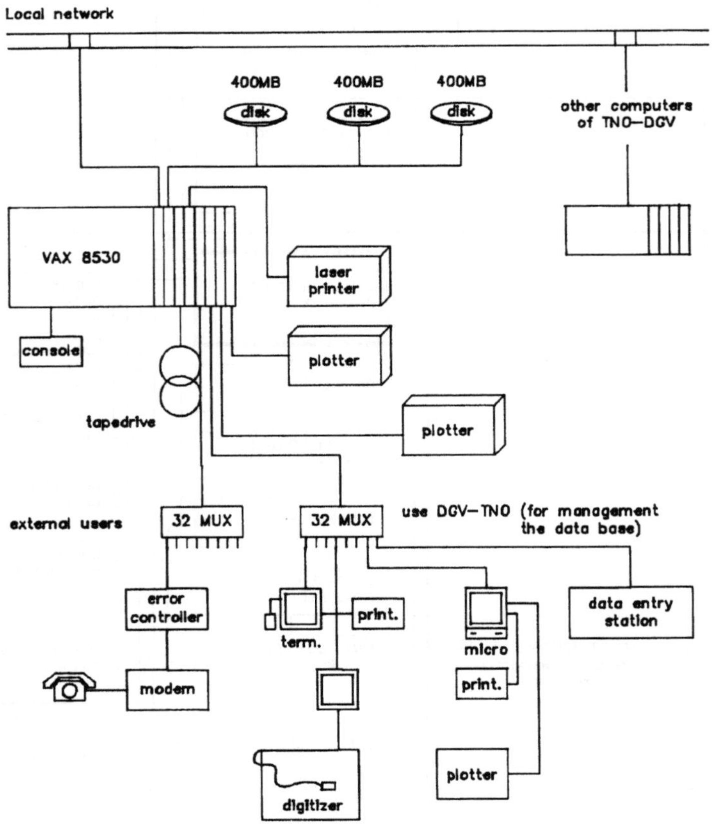

Fig. 3. Hardware configuration data base ground water levels.

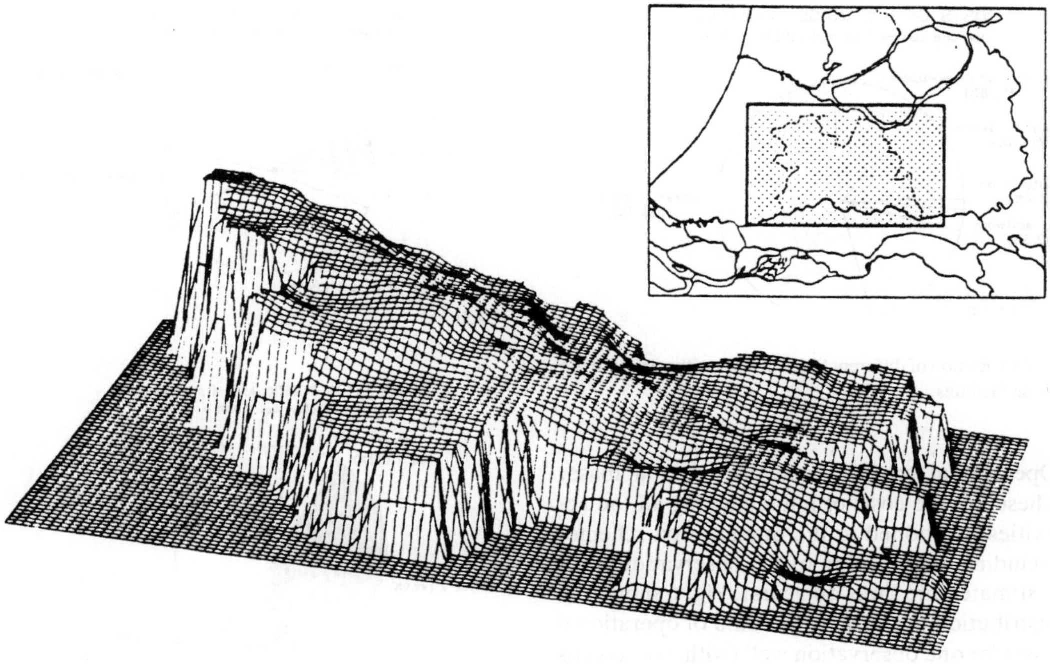

Fig. 4. Three-dimensional diagram of the piezometric level of the province of Utrecht in the Netherlands.

ter level data base. The VAX system is part of a network of other TNO computers and micro-computers. Apart from the ground water levels the data base contains five other types of data:

- technical specifications of observation wells: co-ordinates, elevation of top of tubing or casing with reference to mean sea level, depth of screens, etc.;
- maintenance data: kind of maintenance, kind of repairs, material used, etc.;
- administrative data: province name, district name, address of observers, location of observation well, owners of the observation well, etc.;
- geohydrological data: code of measured aquifer, characteristics of the aquifer, geological description, etc.;
- chemical composition of ground water.

Selections can be made in several ways. The most common are selections made with the help of the identification code of an observation well, selections on a specific aquifer, a period of time (or date), a certain area (name of area or co-ordinates) and selections on combinations of these items.

The data can be displayed as a so-called 'view' on a screen of a terminal, stored on a specific 'work-file' for further processing, or presented in several other ways (time series, contour maps, three-dimensional diagrams etc.) using plotters and printers (for an example, see Fig. 4). Furthermore, some simple derivative values can be directly obtained; for instance, minimum and maximum values, mean values, standard deviations, etc.

Costs of the ground water level monitoring network

Two kinds of costs can be considered:

a) Costs related to the data base facilities (Fig. 5): depreciation and interest on the investment in hardware and software; maintenance of hardware; maintenance of software; costs of the computer room (air-conditioning, etc.); management of the hardware and the support for data base users.

The annual costs of the data base facilities for one observation well amount to some Dfl. 67,- (price level 1987).

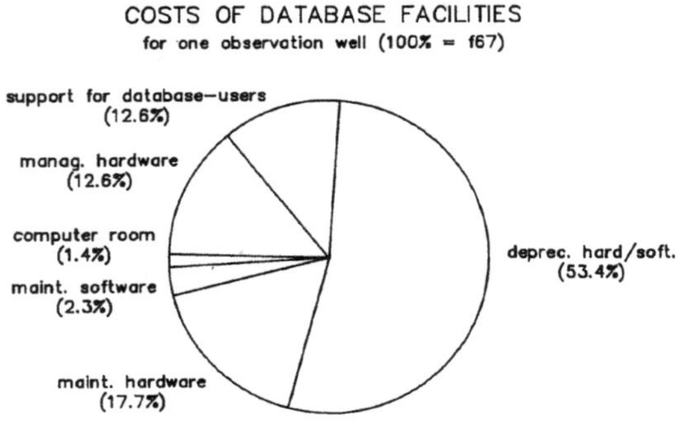

Fig. 5. Distribution of different kinds of costs with respect to the data base facilities.

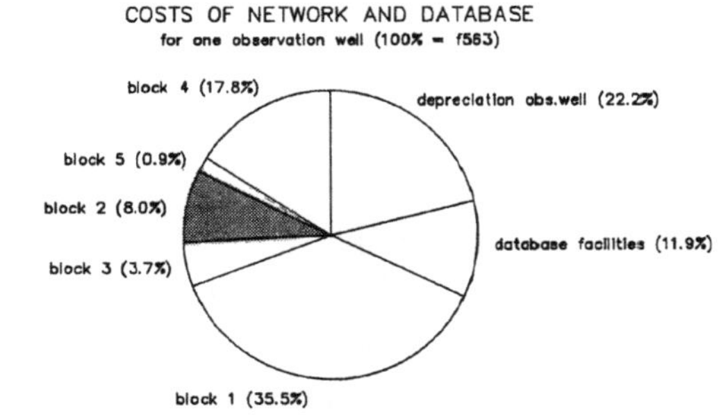

FIG. 6A

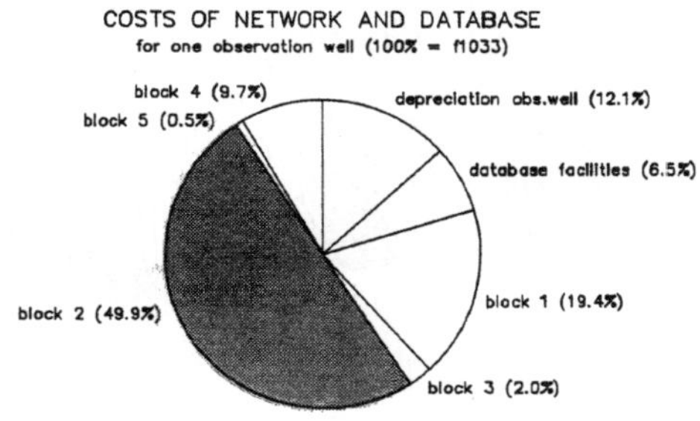

FIG. 6B

Fig. 6. Distribution of costs for one observation well (Fig. 6A using unpaid volunteers as observers, Fig. 6B with paid observers).

b) Operational costs of network and database: These are the costs of implementing all the activities mentioned in the organization scheme, including depreciation of the observation wells (estimated life time: 50 years). Fig. 6 shows the distribution of the different kind of operational costs for one observation well (with two screens and a depth of 25 m) together with the already mentioned costs for data base facilities.

Fig. 6A shows the distribution of costs when the observations are carried out by unpaid volunteers (currently the case in the Dutch network). Fig. 6B shows the distribution of costs when the observation wells are recorded by paid employees. It is obvious that recordings by unpaid volunteers considerably influence the total costs of the ground water level monitoring network.

Proceedings KNGMG Symposium 'Coastal Lowlands, Geology and Geotechnology', 1987: 267–271 (1989)

Study to forecast and to prevent damage resulting from reclamation of the Markerwaard polder

An integrated earth-scientific approach

F.A.M. Claessen
Public Works Department, Rijkswaterstaat, Institute for Inland Water Management and Waste Water Treatment, Lelystad, The Netherlands

Received 1 September 1987; accepted in revised form 26 April 1988

Keywords: geohydrological countermeasures, environmental impact of reclamation, deltaic areas

Abstract

The reclamation of the new Markerwaard polder may have detrimental effects on the eastern part of the province of North Holland due to changes in the groundwater regime. An integrated study was organized to examine possible countermeasures. One of the most important goals of the study was the development of a methodology to forecast the geohydrological changes in the area, their influence on magnitude and rate of settlement of the subsoil and the probable damage caused by these geohydrological and geotechnical changes to buildings, infrastructure, agriculture and ecological aspects of the area.

A method was furthermore developed to compute the lay-out and the scale of magnitude of artificial groundwater recharge works to compensate for land subsidence in those parts of North Holland most susceptible to detrimental effects.

Main results and conclusions of the study and some remarks about recent developments, after this study was ended, will be given.

Introduction

General

The reclamation of the Markerwaard polder will probably be the last phase of the IJsselmeer development project. (Fig. 1). This project started in 1930 with the enclosure of the former Zuiderzee, now Lake IJssel, and in that same year with the reclamation of the Wieringermeer polder (area: 200 km^2). In 1942 the North Eastern polder was reclaimed (480 km^2), in 1957 the Eastern Flevoland polder (540 km^2) and in 1968 the Southern Flevoland polder (440 km^2). The study area lies in the north-western part of the Netherlands (Fig. 1). The Markerwaard polder, covering an area of 410 km^2, is to be reclaimed in the south-western part of the Lake IJssel. The reclamation of this polder induces a drastic drop of the phreatic level in the polder and causes a substantial change in the groundwater regime in the polder and the adjacent areas.

The study was focused on the changes in the groundwater regime of the province of North Holland, since there various types of damage may occur as a result of changes in the piezometric head of the groundwater in the shallow aquifers. Research conducted by the Public Works Department in 1976 showed that the main geohydrological changes would occur in the boxed section of Fig. 1. An area of 500 km^2 in North Holland (the hatched section) was considered most likely to be affected by the predicted changes.

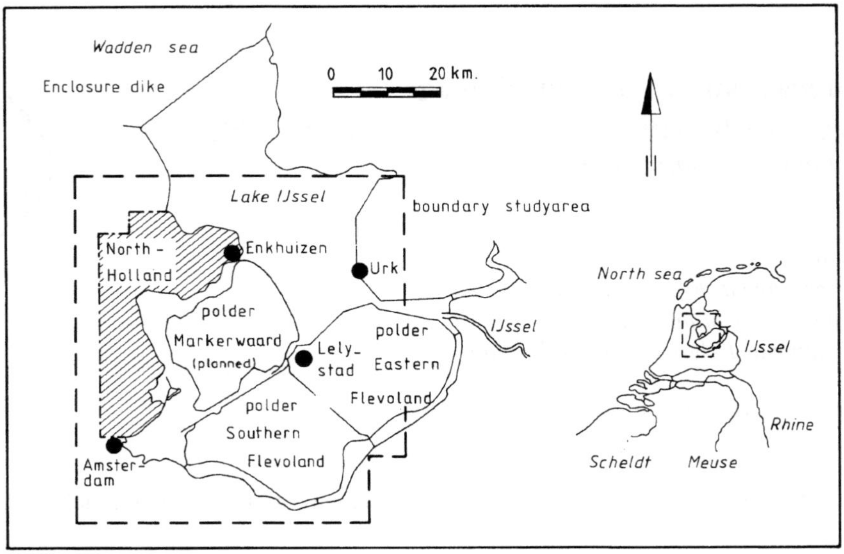

Fig. 1. Area of study.

The problem and the aim of the study

The following harmful effects may occur in North Holland as a result of a change in the groundwater regime in the Markerwaard polder and its surroundings:

- damage to agriculture, through a reduction of crop yields, caused by a drop of the phreatic level and by increasing soil moisture deficiency in the growing season;
- damage to animal and plant life through changes in the abiotic conditions resulting from land subsidence and a drop in the phreatic level in the growing season;
- damage to buildings and physical infrastructure through settlement of the Holocene compressible deposits.

The study was designed to determine the extent of the various negative effects and their spatial pattern, the time span over which they would occur, the countermeasures (e.g. recharge wells and infiltration grooves in the border lakes) required to prevent damage. Potential positive effects as a result of the reclamation were not investigated.

A feasibility study of countermeasures was carried out. It was not restricted to questions of technical feasibility but it also looked at economic viability by comparing the cost of such measures with the cost of damage. Consequently, the hydrogeological and geotechnical aspects were studied together with the agrohydrological, ecological and civil engineering aspects, thus enabling a forecast to be made of the damage to buildings and objects in urban areas, to agriculture and animal and plant life.

Use of the same groundwater models has shown where and how much water has to be injected into the aquifers to prevent or reduce drawdown of the piezometric level in those areas vulnerable to damage. The costs of various countermeasures were also estimated.

Background to and reasons for this study

Hitherto, there have been no systematic inquiries into damages caused by settlement in urban areas, nor any quantitative study to forecast possible damage. Further reasons for this study were:

- at the time this study started the Government had not yet decided wether or not to reclaim the Markerwaard polder. A fresh round of public participation and consultation was launched by the Government in 1981 with the object of reconsidering all relevant aspects so as to arrive at a balanced judgement;

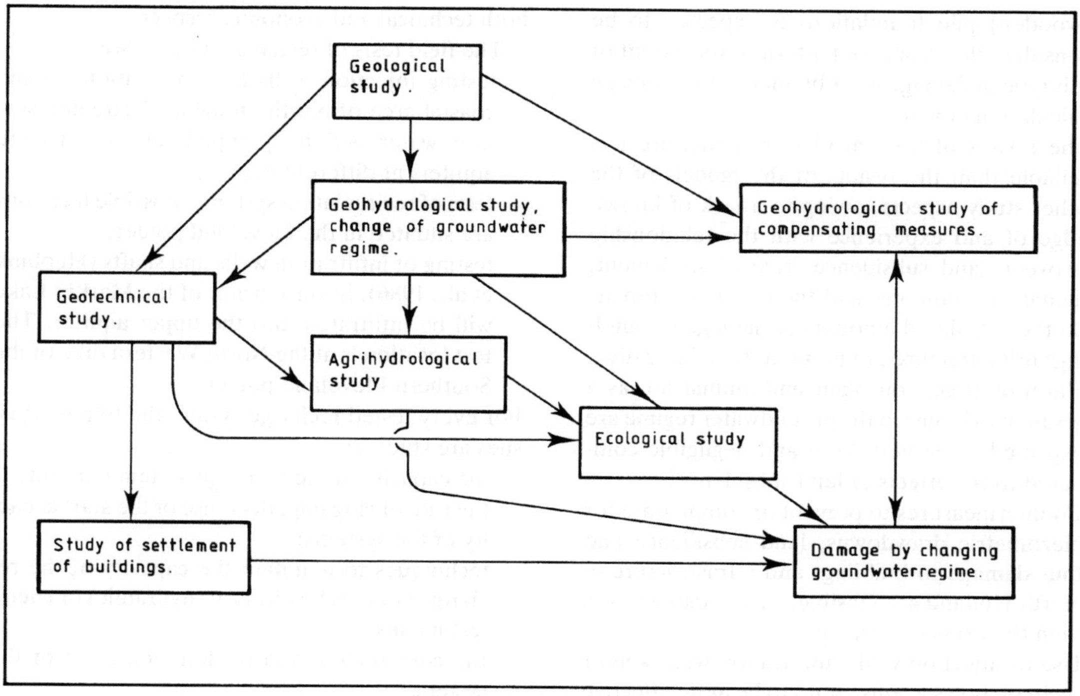

Fig. 2. Relations between different study aspects.

- numerous new data have been obtained on the hydrogeological and geo-technical condition of the area since 1970;
- various numerical calculation models facilitating extensive, detailed and rapid calculations have recently become available. It was therefore possible to approach this problem on a more precise and comprehensive basis.

Structure of the integrated study

The main objective of the study was to weigh the possible damages against the cost and effort of preventing it. This information could then be used when deciding whether or not to reclaim the Markerwaard polder. Since the study embraced various disciplines, it was divided into seven sections. Fig. 2, a diagram of the interrelationship of these various sectoral studies, illustrates their interdependence, viz. the extent to which the results of one may influence those of others.

The various sectoral studies were started concurrently, with the exception of the geological and geohydrological studies, which had been started earlier. They were conducted by six organizations, which worked together on the project under the supervision of the Public Works Department, Rijkswaterstaat.

Main results and conclusions

- The detailed results of the studies, set out in Claessen (1988, this issue), Claessen et al. (1988, this issue) and Satijn (1988, this issue), show that the study achieved its goal of presenting a spatial picture of the various possible types of damage and the specific countermeasures that were considered feasible.
- If no countermeasures are taken, damage to buildings and infrastructure as a result of settlement and additional negative skin friction on

(wooden) pile foundations is expected to be considerable. A spatial pattern of the extent of this type of damage was obtained with a damage calculation model.

- The results of this calculation model are less reliable than the results of the models of the other study aspects, owing to a lack of knowledge of and experience with the relationship between land subsidence, rate of settlement, damage to buildings and their costs for repair. So the calculated amounts of damage to buildings infrastructure are no more than indicative.
- Harmful effects on plant and animal life as a result of a change in the groundwater regime are expected to be very local and negligible compared to the effects of land subsidence.
- Countermeasures to prevent or compensate for piezometric drawdowns, land subsidence and thus damage to buildings and infrastructure in North Holland are feasible. Their costs are less than the expected damage.
- Use of injection wells, infiltration wells and/or recirculation systems seems to be more effective and gives more operational flexibility than use of infiltration grooves, dredged in the peripheral lake. These recharge systems could be gradually phased out in a period of 30 to 50 years.
- The integrated calculation method developed makes it possible to rapidly compare a wide variety of alternatives with diverse basic premises and enables policy makers, among others, to choose optimum measures and their location.

Recent developments

After finishing this integral study in 1983, the National Government decided in 1985 to reclaim the Markerwaard polder. As a result of this decision the Public Works Department started research for setting up an optimal system of geohydrological countermeasures.

The research consists of (i) several pilot projects of recharge systems; (ii) further field investigations for additional geohydrological and geotechnical data; (iii) detailed calculations with mathematical models for optimization of recharge systems in both technical and economic respect.

The field tests of recharge systems are:

- testing injection wells at two locations in the coastal area of North Holland. Pretreated surface water will be pumped into Pleistocene aquifers at different depths;
- tests of recirculation systems. Possible locations are situated in the Flevoland polder;
- testing of infiltration wells and shafts (Hebbink et al., 1986). Surface water of the Marker Lake will be infiltrated into the upper aquifer. The test location is at the North Western dike of the Southern Flevoland polder.

For every tested recharge system the following issues are studied:

- the capacity of the recharge system per unit;
- the rate of clogging, decrease of the start capacity of the systems;
- techniques to maintain the capacity of the recharge systems by physical (hydraulic) or chemical means;
- the construction and exploitation costs of the systems;
- the effectivity of the recharge systems as countermeasures.

Beside the above mentioned research program, at the same time a study by the Public Works Department started for the design of a monitoring system to measure the probable geohydrological and geotechnical alterations in the areas around the Markerwaard polder. One objective of this monitoring system is to distinguish changes in the groundwater regime, and in water pressure and effective stress in the Holocene layers as a result of the construction and reclamation of the new polder, from changes caused by other geohydrological and geotechnical processes.

Data networks will be set up of:

- piezometric levels;
- phreatic levels;
- settlement and subsidence of the Holocene layers;
- subsidence and deformation of buildings and other constructions.

Therefore the number of piezometres, phreatic observation wells, bench marks etc. in the areas surrounding the Markerwaard polder are adjusted for this ends.

In 1986 the National Government postponed the decision about the starting date of the construction of the new polder, for at least several years. So for the time being most above mentioned research activities have been postponed.

As a result of the integral study described above a good insight has been obtained in the feasibility and kind of geohydrological countermeasures which can be applied to prevent detrimental effects for different interests by the construction of the Markerwaard polder. Also a good tool has been established and is operational for this kind of problems in other places and countries.

References

Claessen, F.A.M., van Bruchem, A.J., Hannink, G., Hulsbergen, J.G., de Mulder, E.F.J. 1988. Secondary effects of the reclamation of the Markerwaard polder – In: W.J.M. van der Linden, S.A.P.L. Cloetingh, J.P.K. Kaasschieter, J. Vandenberghe, W.J.E. van de Graaff & J.A.M. van der Gun (eds): Coastal Lowlands: Geology and Geotechnology – Proc. KNGMG Symp. The Hague, 1987 – Kluwer Acad. Publ. (Dordrecht): 283–291

Claessen, F.A.M. 1988. Geohydrological effects of the reclamation of the Markerwaard polder – In: W.J.M. van der Linden, S.A.P.L. Cloetingh, J.P.K. Kaasschieter, J. Vandenberghe, W.J.E. van de Graaff & J.A.M. van der Gun (eds): Coastal Lowlands: Geology and Geotechnology – Proc. KNGMG Symp. The Hague, 1987 – Kluwer Acad. Publ. (Dordrecht): 273–282

Satijn, H.M.G. 1988. Countermeasures to prevent detrimental effects: a feasibility study – In: W.J.M. van der Linden, S.A.P.L. Cloetingh, J.P.K. Kaasschieter, J. Vandenberghe, W.J.E. van de Graaff & J.A.M. van der Gun (eds): Coastal Lowlands: Geology and Geotechnology – Proc. KNGMG Symp. The Hague, 1987 – Kluwer Acad. Publ. (Dordrecht): 301–309

Hebbink A.J., Menting G., Schultz, E. 1986. Infiltratieputten in de randmeren van de Markerwaard – H_2O, 21: 512–518

Proceedings KNGMG Symposium 'Coastal Lowlands, Geology and Geotechnology', 1987: 273–282 (1989)

Geohydrological effects of the reclamation of the Markerwaard polder

F.A.M. Claessen
Public Works Department, Rijkswaterstaat, Institute for Inland Water Management and Waste Water Treatment, Lelystad, The Netherlands

Received 1 September 1987; accepted in revised form 26 April 1988

Key words: ground water modelling, reclamation effects, deltaic plains

Abstract

The reclamation of the Markerwaard polder will cause piezometric and phreatic drawdowns in the province of North Holland. The spatial pattern of the ultimate drawdown and ultimate changes by the reclamation in the ground water balance of the study area was calculated using numerical ground water models.

Introduction

By construction of the Markerwaard polder with an area of 410 km^2 the present water level in the area concerned (on average 0.25 m MSL) will drop 4.5 to 6.5 metre. As a result, ground water flow will be induced towards the new polder. This will mean a considerable change in the ground water regime underneath the new polder and its surrounding area. Preliminary studies showed that the main geohydrological changes would occur within the boxed section of Fig. 1. An area of 500 km^2 in the province of North Holland (the hatched section in Fig. 1) was considered most likely to be affected by the predicted changes. The province of North Holland can be characterized as an old polder area with a surface level varying from 1 to 4 metres below sea level. The geohydrological schematization of the subsurface is derived from the results of the geological investigations (Westerhoff et al., 1986). Figure 2. presents a geohydrological cross section of the Holocene and Pleistocene deposits in the Markerwaard area and North Holland.

Three aquifers can be distinguished which are sometimes completely, sometimes partly separated or covered by poorly permeable layers.

The geohydrological processes that will take place after the construction of the polder are illustrated in two schematic geohydrological cross sections of the coastal area and the Markerwaard area (Fig. 3). Fig. 3a shows the present ground water flow pattern. Figure 3b shows the predicted ultimate ground water flow pattern, after the reclamation. After reclamation ground water will predominantly flow from the borderlake and from the province of North Holland through the aquifers to the new polders. The horizontal component of the ground water flow direction underneath the borderlake will be reversed.

After reclamation, the upward seepage to the deep polders in the province of North Holland will diminish or turn into infiltration. It is very important to assess the expected drawdown of the piezometric head in the aquifer below the Holocene compressible layers, because this drawdown may have several effects, such as:

- a drawdown of the phreatic level.
- an increase of effective stress in the compres-

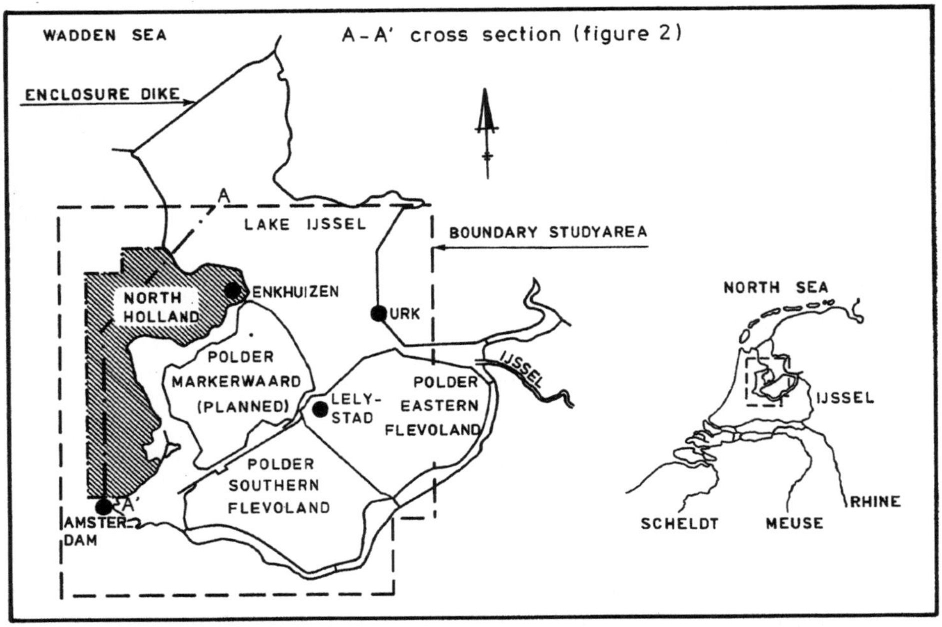

Fig. 1. Area of study.

sible coverlayers, resulting in settlement and land subsidence.

Because of the complexity of the geohydrologycal structure of the subsurface, together with strongly varying boundary conditions, the calculations of the ground water flow had to be carried out by using numerical models. The geohydrological features of the models were derived from the geolog-

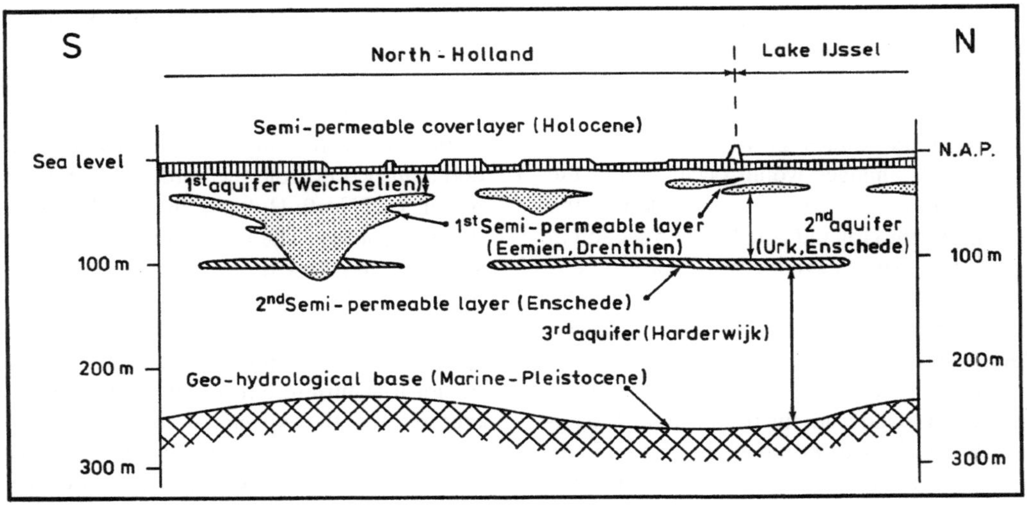

Fig. 2. Geohydrological cross section. Position of section marked on Fig. 1 as A-A'.

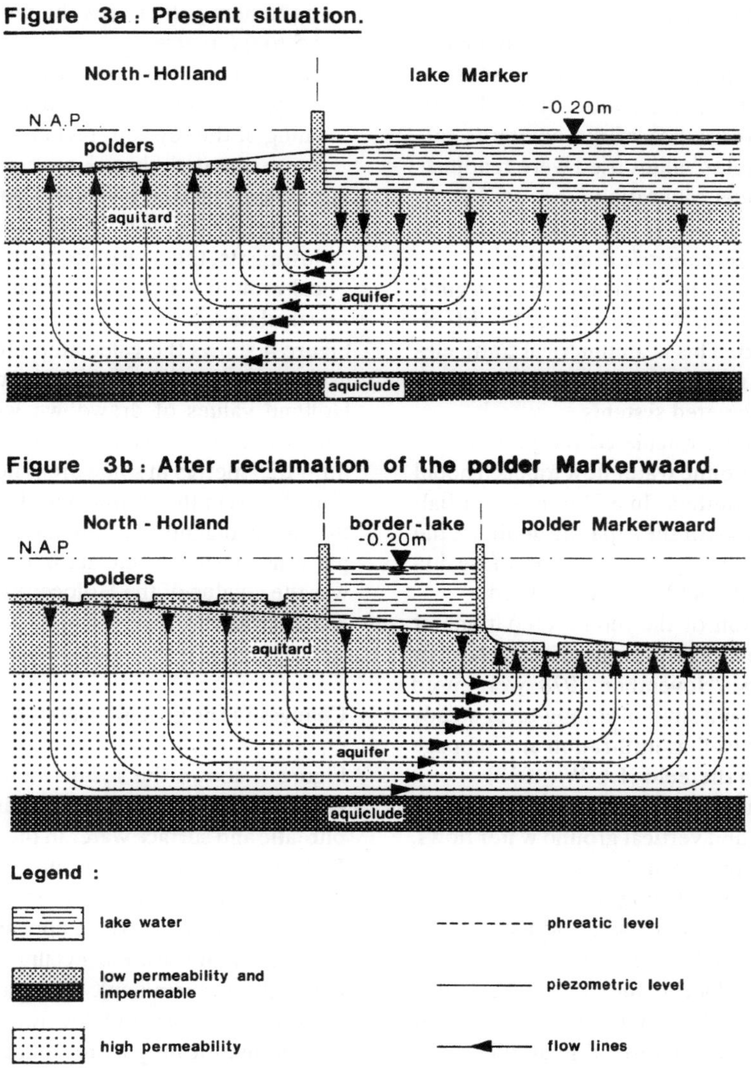

Fig. 3. Schematic geohydrological cross section.

ical and hydrological conditions of the study area.

For the determination of the drawdown of the phreatic level two separate models were used. The calculated piezometric drawdown in the upper aquifer is used as input for the determination of the changes in the phreatic level.

The piezometric drawdown

Geohydrological schematization

For the geohydrological calculations it is necessary to subdivide the surface into aquifers and aquitards, based on the geological conditions. On the basis of a geological framework a number of geohydrological maps were prepared. These maps incorporate the preliminary values of the geohydrological parameters needed for the model computations, such as values of horizontal transmissivity of the aquifers for ground water flow (T) and vertical hydraulic resistance of the aquitards (c). The original parameter values used were readjusted by means of model calibration.

Calculations of the piezometric drawdown
To forecast the changes in the ground water regime after reclamation of the Markerwaard, the FIESTA model was used. This is a numerical finite element model which solves steady ground water flow problems in multi-layered systems.

The FIESTA model calculated the piezometric levels and vertical ground water flow for each nodal point in the three aquifers. In addition, water balances were calculated for the separate aquifers, the complete hydrological system and for those hydrological units that could be distinguished.

For the calibration of the model (= parameter adjustment) the average piezometric levels over the period October/November 1978 were used (these being representative of a long-term steady state situation), as well as infiltration and seepage values measured in the field (I.C.W., 1982).

The output of the model calculations, in terms of piezometric levels and vertical ground water flows, should agree, within certain limits, with the measured data. The geohydrological parameters (T- and c-values) in the model were adjusted to an optimum value by a selective trial and error method (IWACO, 1981). During the calibration process it became clear that the ground water flow in the area is most sensitive for vertical hydraulic resistance of the Holocene aquitard and the surface water levels. After calibration, the forecasting ability of the model was verified with data, different from those used for the calibration. In 1967 the Southern Flevoland polder (Fig. 1) was reclaimed. This resulted in a piezometric drawdown in and around that polder. The actual extent of the drawdown was compared with the calculated drawdown, with fairly similar results.

To indicate the ultimate situation (steady state) after the reclamation of the Markerwaard polder, the input data of the model were adjusted as follows:

- in the new polder area the phreatic level drops 4.5 to 6.5 metres;
- in the same area the vertical hydraulic resistance of the Holocene aquitard decreases through ripening of the top soil, reclamation works, etc. It was estimated, based on experience in other polders, that the c-values would decrease from about 27000 days to 4000 days. After these adjustments, the new piezometric levels and vertical ground water fluxes in the study area were calculated.

Figure 4 shows the calculated ultimate drawdown in the upper aquifer. In the coastal areas of North Holland values of drawdown were calculated of −0.25 m at 5 till 10 km from the coast up to −1.0 till −1.25 metres in some areas near the coast. The drawdowns in the second and third aquifer were of the same magnitude and have the same spatial patterns. An uncertain factor in the FIESTA model was the c-value of the Holocene layer in the polder Markerwaard.

Changes in the ground water balance

Figure 5 shows the results of the change in the vertical flow between the upper aquifer and the phreatic and surface water in the province of North Holland. Negative change means less upward seepage (in deep polders) or more downward seepage (in less deep polders). It appears that the largest increase in infiltration occurs in areas near the coast (0.2–0.4 mm/day). The spatial distribution of areas with upward and downward seepage will not change significantly by reclamation.

Figure 6 shows the net change in the ground water balance between the present situation and the situation after reclamation. The left border of the balance area in the figures, represents the western and northern border of the hatched area in Fig. 1, the right border represents the southern and eastern borders.

The net change in the vertical ground water flux is 14 million m^3 of water per year. The largest changes occur in the third aquifer, i.e. the thickest and most permeable aquifer.

The net change in flux in the hatched area of the

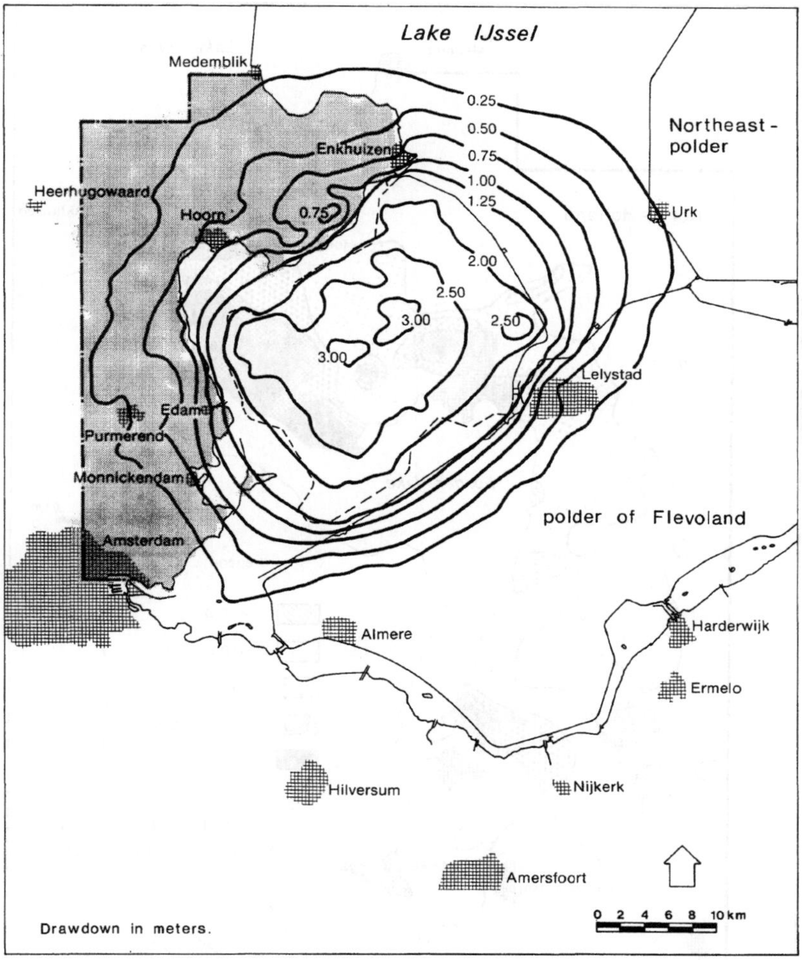

Fig. 4. Ultimate drawdown in upper aquifer.

province of North Holland is about 10% of the annual net precipitation (annual precipitation minus annual actual evapotranspiration). After reclamation, in an average year, the quantity of additional surface water needed for water level control in the province of North Holland will be no more than 3.5% in respect to the present situation.

The rate of piezometric drawdown

A first impression of the rate of the piezometric drawdown by construction of the new polder can be derived from an analysis of well hydrographs of piezometers in and around the Southern Flevoland polder. This polder was reclaimed in 1967 and is situated close to the Markerwaard area. The hydrology and geology of the Flevoland area and the Markerwaard area have much resemblance. Figure 7 shows the hydrographs of some representative wells and the change in the surface water/phreatic level in the Southern Flevoland polder, during and after the reclamation.

As the drop in the surface water level in Fig. 7

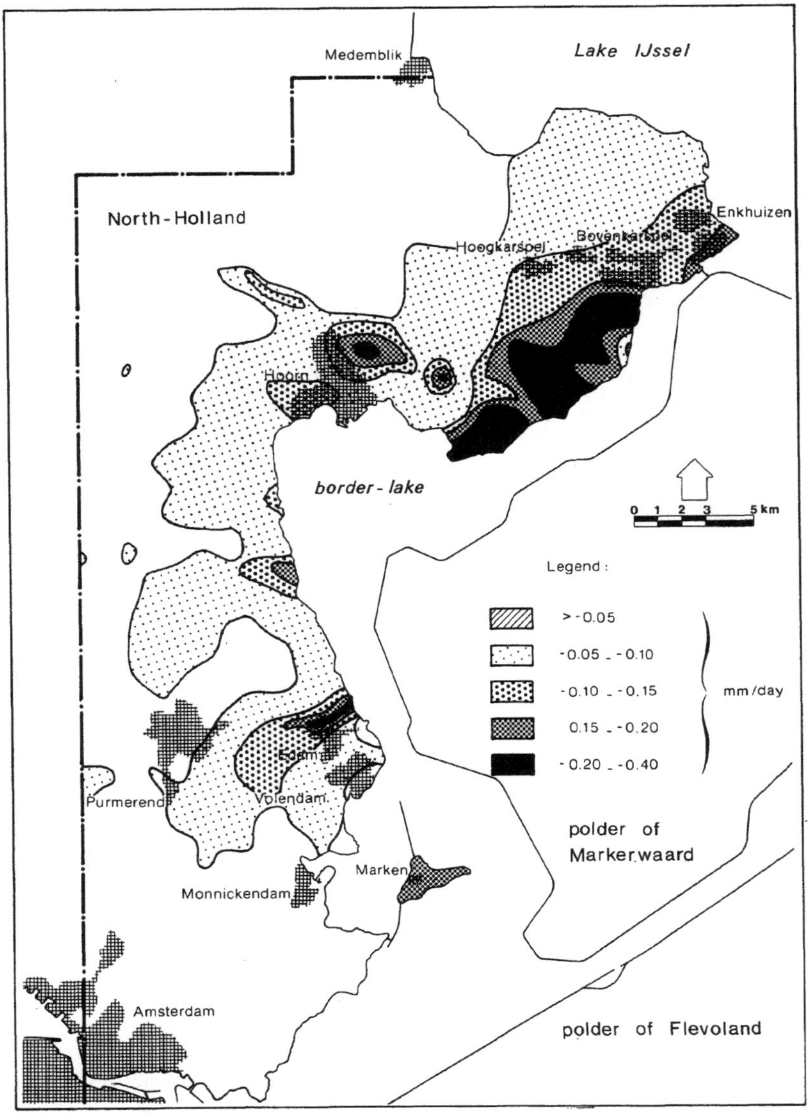

Fig. 5. Change of upward or downward seepage by reclamation of the Markerwaard polder.

shows, this polder was pumped dry in seven months; further reclamation took place during about five years. The diagram shows that within a period of 5 to 7 years after the start of pumping, the major part of the ultimate drawdown occurred.

Calculations were made to estimate the rate of the piezometric drawdown in relation to the rate of pumping dry in the new polder. These calculations were made by using an analytical non-steady ground water flow model (Barends et al., 1984).

It appears from these calculations that under the Markerwaard and North-Holland the piezometric level will decline at the same rate as ascertained under the north-western part of Southern Flevo-

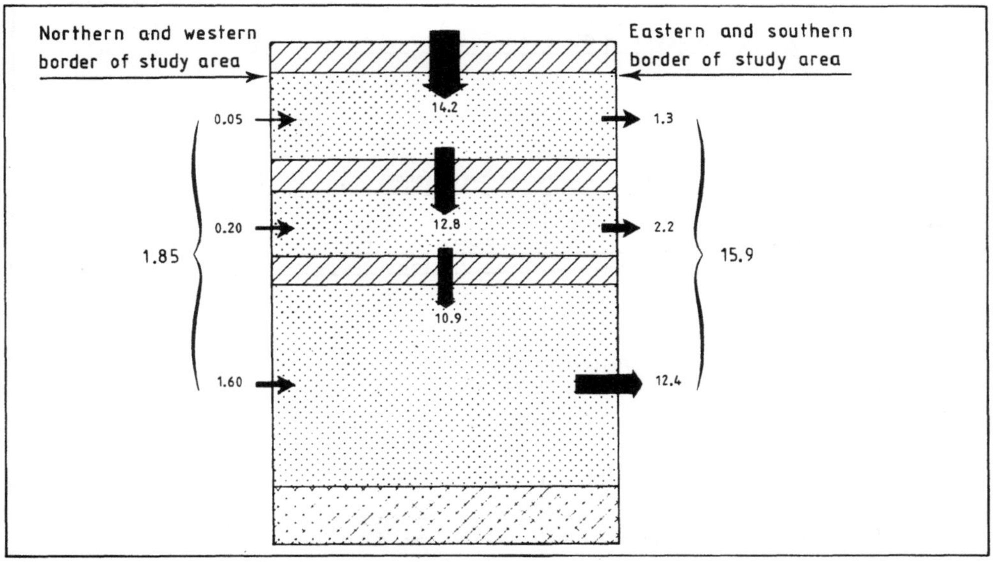

Fig. 6. Net change in ground water flow in the study area as a result of the reclamation (in millions m³ per year).

land (Fig. 7) during and after the reclamation of that polder. If the Markerwaard is pumped dry at the same rate as Southern Flevoland, it is to be expected that after 5 to 7 years about 90% of the ultimate drawdown under and around the Markerwaard will be reached.

The phreatic drawdown

Introduction

As a result of the piezometric drawdown also the phreatic level will decline in the coastal areas of North Holland. In general the phreatic level, –

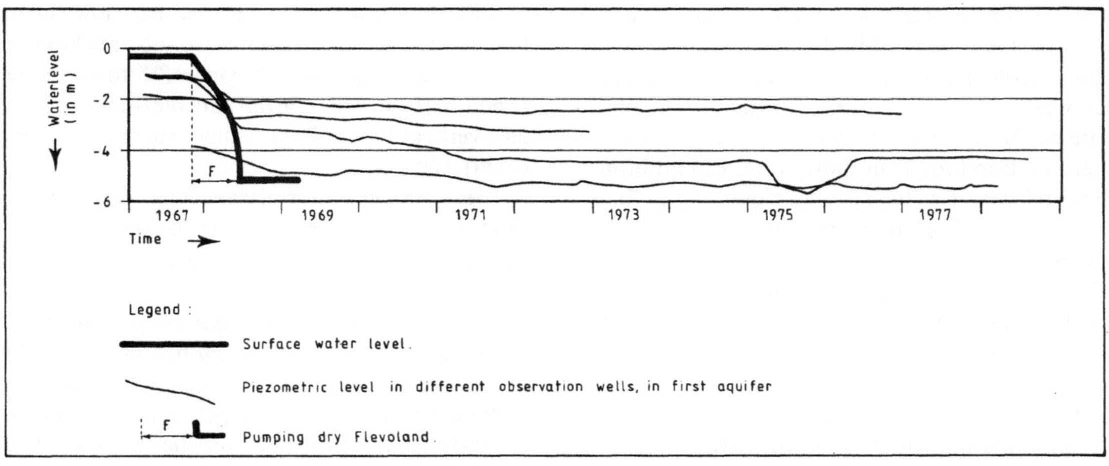

Fig. 7. Observed piezometric drawdown after reclamation in and around the Southern Flevoland polder.

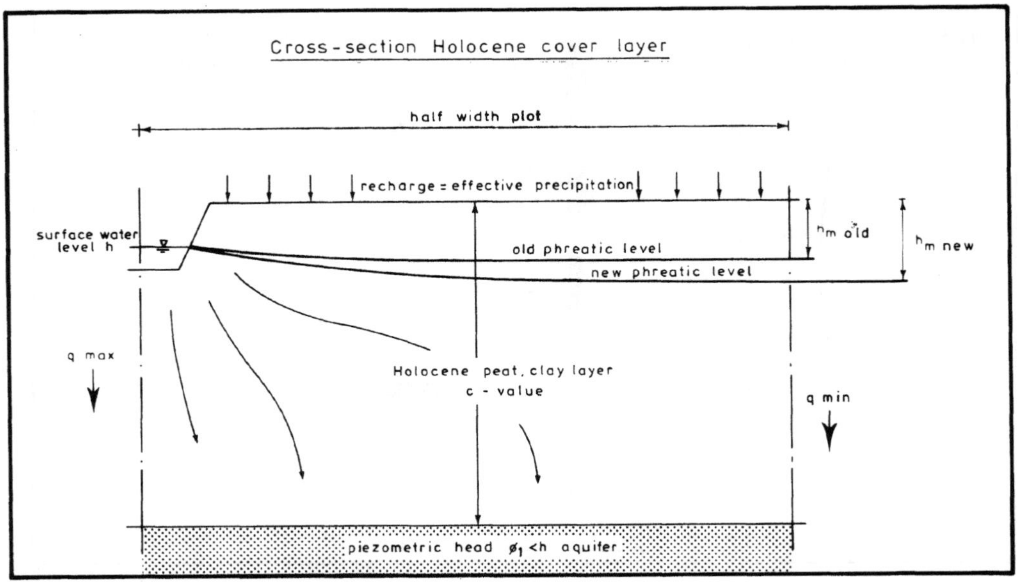

Fig. 8. Hydrological components determining the phreatic level.

averaged over one standard plot of land –, is determined in a polder by (Fig. 8):

- the surface water level and dimensions of and distances between the surface water conduits in the polders;
- the vertical and horizontal permeability of the Holocene coverlayers;
- the net precipitation and evapotranspiration;
- the piezometric level in the upper aquifer.

If the piezometric level drops, at most during summer when net precipitation is negative, at some distance from the water conduits phreatic water level can drop too. Thus more water can infiltrate from surface waters towards the plot or less ground water can discharge to the surface water and atmosphere.

The phreatic drawdown can result in a decrease in the amount of moisture available for plants during the growing season.

Calculation of the phreatic drawdown

The phreatic drawdown was calculated with two ground water models (Mooy & Vinkers, 1982). These models calculate the average drawdown during the summer half year for a standard plot of land, for 9 different representative subareas.

The so called MOTGRO model for vertical-2D saturated ground water flow, (Van der Veer, 1978) determined the interrelationship between the ground water discharge from the plot to the ditch, and the phreatic level in the central part of the plot, for different values of piezometric drawdown in the first aquifer. Each representative sub area has specific values of drainage distances, differences between surface water level and piezometric head and vertical hydraulic resistance of the Holocene coverlayer.

These specific interrelationships were used as input for a second ground water model: the model of Rijtema and De Laat (De Laat, 1980).

This model calculated the vertical ground water fluxes through the root zone and the phreatic level, for time steps of 10 days. Futher input for this model were: 10 days-values of: precipitation and evapotranspiration, crop types, moisture content characteristics and characteristic ground water depth values; all over a period of one year.

The fact that the phreatic drawdown partly is

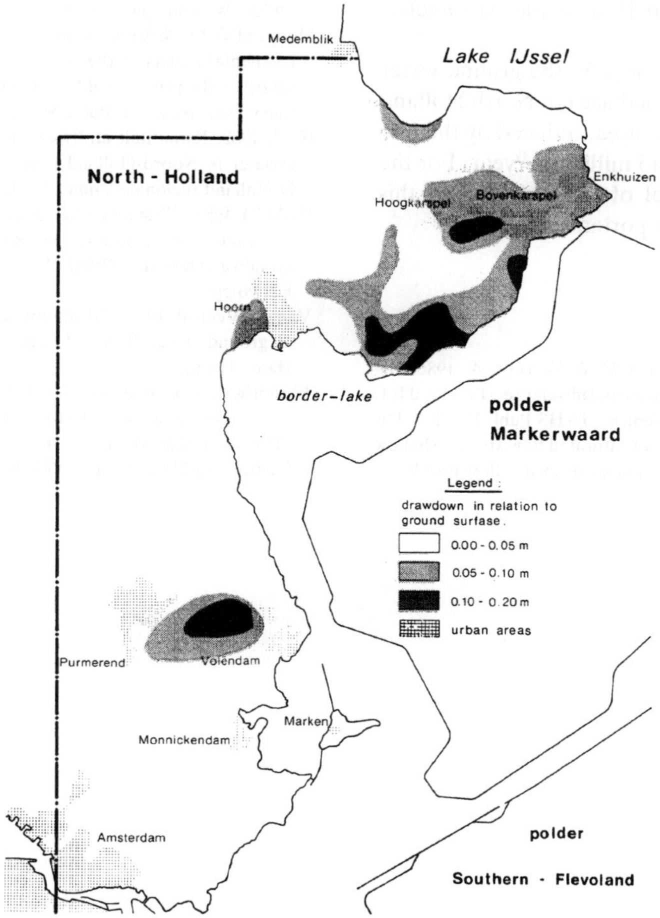

Fig. 9. Calculated maximum phreatic drawdown by reclamation of the Markerwaard polder.

compensated by the land subsidence was taken into account. Thirty summer half years were simulated for square areas of 1 by 1 km. Figure 9 shows the calculated average maximum drawdown during the summer half year. The largest phreatic drawdowns (0.1 to 0.2 m) will occur again in the areas near the coast.

These drawdown are only 10 to 15% of the piezometric drawdowns, because of the favourable soil moisture characteristics of these polders and the large thickness of the Holocene covering beds.

Conclusions

1. Piezometric drawdown in the upper aquifer in North Holland is the most important item for the geotechnical and other effects of the reclamation of the Markerwaard.
2. In the province of North Holland this drawdown takes place within 10 years and varies between less than 0,25 m at distances larger than 5 to 10 km from the coast of the Lake Marken and 1,0 to 1,25 m near the coast. The drawdown in the deeper aquifers is of the same magnitude.
3. Phreatic drawdown in the studied area of North

Holland is only 10 to 15% of the piezometric drawdown.

4. The calculated net change in the ground water flux between the studied area of North Holland and the Markerwaard area, induced by the reclamation is at about 15 million m^3/year. For the surface water control of this polder area this change is of minor importance.

References

Barends, F.B.J., Teunissen, J.A.M. & Verruijt, A. 1986. Assessment of the transient nature of subsidence – Proc. 3rd Int. Symp. Land Subsidence, Venice – IAHS Publ. 151: 107–116

De Laat, P.J.M. 1980. Model for saturated flow above a shallow water table, applied to a regional subsurface flow problem – Pudoc, Wageningen:1 26 pp

Mooy, H.G.M. & Vinkers, H.J. 1982. Agricultural and environmental impact of the reclamation of the polder Markerwaard on the province of North Holland – Heidemij, Consultancy Div., Arnhem, Publ. 662–81/3: 57 pp

ICW, 1982. Kwantiteit en kwaliteit van grond- en oppervlaktewater in Noord-Holland benoorden het IJ – Werkgroep N-Holland regionale studies 16, Wageningen: 188 pp

IWACO 1981. Fiesta berekeningen inzake de gevolgen van eventuele aanleg van de Markerwaard voor het huidige grondwaterregiem – IWACO Rotterdam, Intern Rept. 420-E2: 96 pp

Van der Veer, P. 1978. Calculation methods for two dimensional ground water flow – Comm. Rijkswaterstaat 28, Den Haag: 172 pp

Westerhoff, W.E. & de Mulder, E.F.J. 1986. Quarternary Geological Framework of North-Holland and the Markermeer (The Netherlands) – Proc. 3rd Int. Symp. Land Subsidence, Venice – IAHS Publ. 151: 877–883

Proceedings KNGMG Symposium 'Coastal Lowlands, Geology and Geotechnology', 1987: 283–291 (1989)

Secondary effects of the reclamation of the Markerwaard Polder

F.A.M. Claessen[1], A.J. Van Bruchem[2], G. Hannink[3], J.G. Hulsbergen[4] & E.F.J. De Mulder[5]

[1] Public Works Department, Rijkswaterstaat, Institute for Inland Water Management and Waste Water Treatment, Lelystad, The Netherlands; [2] Heidemij Consultancy Division, Hoofddorp, The Netherlands; [3] Delft Geotechnics, Delft, The Netherlands; [4] ABT Consulting Engineers, Velp, The Netherlands; [5] Geological Survey of The Netherlands, Haarlem, The Netherlands

Received 1 September 1987; accepted in revised form 26 April 1988

Keywords: environmental impact of reclamation, deltaic areas, land subsidence, geotechnics, damage to buildings, ecological effects

Abstract

The geohydrological effects of the reclamation of the Markerwaard polder in the Netherlands can be detrimental for urban areas, agriculture and nature reserves in the province of North Holland. Especially the piezometric and phreatic drawdown can result in crop yield reduction, drought damage in nature reserves, and land subsidence, causing damage to buildings and infrastructure. The possible effects are discussed. The methods of determination of these effects are briefly described successively. Damage estimates are presented.

Introduction

Construction of the Markerwaard polder in the Netherlands is expected to cause a piezometric drawdown in the province of North Holland. The piezometric drawdown in the uppermost Pleistocene aquifer will decrease the pore water pressure in the Holocene deposits and this will result in a phreatic drawdown (Claessen, 1988b).

These geohydrological effects will increase the effective stresses in the compressible Holocene deposits, leading to settlement and subsidence. This settlement in turn may cause damage to foundations of buildings and other constructions.

The afore-mentioned phreatic drawdown and land subsidence will also change the abiotic soil conditions. These alterations may have a negative impact upon crop yield as well as upon flora and avifauna in valuable nature conservation areas.

This article is a part of the integrated study about forecast and prevention of damage, in connection with the construction of the Markerwaard polder (Claessen, 1988a, this issue). Another preceding article (Claessen, 1988b, this issue) deals in detail with the study of the geohydrological effects of the reclamation.

Settlement and land subsidence

Prediction model

As a result of the decrease of the pore water pressure in the Holocene deposits, the effective stresses in the different layers will increase, as is shown

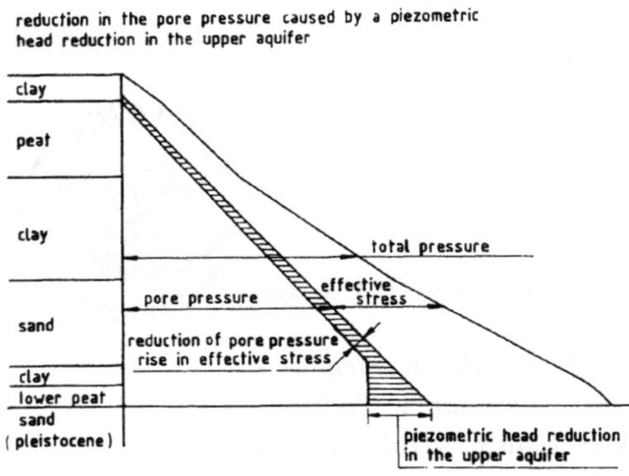

Fig. 1. Influence of piezometric drawdown upon vertical effective stress in Holocene soil layers.

schematically in Fig. 1. This increase in effective stresses in the Holocene layers will cause settlement of the compressible soil layers (clay and peat layers) so that land subsidence will occur.

To predict the geotechnical consequences it is necessary to determine the expected magnitude and rate of the settlements and land subsidence.

The magnitude of the settlement due to the changes in the groundwater regime depends on:

- the piezometric drawdown in the upper aquifer;
- the phreatic drawdown;
- the present hydrological conditions;
- the geotechnical properties of the Holocene deposits (e.g. permeability and compressibility).

The rate at which the settlement will occur depends on:

- the rate at which the piezometric head in the upper aquifer and the phreatic level will decline;
- the structure of the Holocene layers;
- the permeability and the compressibility of the various Holocene soil layers.

Soil investigations give adequate information on the structure of the Holocene deposits, the hydrological conditions, and the geotechnical properties of the various layers at one location, but there can be substantial variations within a large study area. For this reason the problem was approached area-wise, by dividing the study area into subareas which are, within certain limits, uniform as regards the structure of the Holocene deposits, the hydrological conditions and the geotechnical properties of the various Holocene layers.

The investigations show that 11 typical soil layers can be distinguished in the Holocene deposits. In each of these layers the relevant geotechnical properties, compressibility and permeability, are fairly uniform. Clear relationships were established between these parameters and the volumetric weight of these soil layers (Hannink & Talsma, 1986). Based upon the geotechnical properties of the soil layers and the hydrological conditions in the study area, 160 different soil profiles were distinguished. These different soil profiles were presented in a map. Figure 2 shows some of these profiles.

Computations of final land subsidence

A simplified formula to calculate final settlements which are considered to take place in about 27 years (10.000 days) was derived from the Koppejan formula (Koppejan, 1948):

$$S_f = (H/C_{10}{}^4\text{days})\ 1\text{n}\frac{P_o + \Delta P}{P_o}$$

where:

S_f = settlement of a soil layer in 27 years.

H = initial thickness of a soil layer,

$C_{10}{}^4$days $= {}^1/C_p + {}^4/C_s$, compression coëfficients determined from laboratory tests taking into account small increases in effective stress (Hannink & Talsma, 1986),

P_o = initial vertical effective stress,

Δp = increase in vertical effective stress.

Knowing the geotechnical properties of each soil layer, the final settlement of each profile has been calculated.

For each profile type the relationship was determined between the piezometric drawdown in the upper aquifer and the final settlement by calculating the settlement for piezometric head reductions of 0.25, 0.50, 0.75, 1.00 and 1.50 m. The phreatic drawdown for each profile type was input as a percentage of the piezometric head reduction. The results show that the settlement behaviour of the study area can be represented in 12 diagrams.

Statistical calculations were carried out to estimate the degree of accuracy of the settlement calculations. For a confidence interval of about 90% a

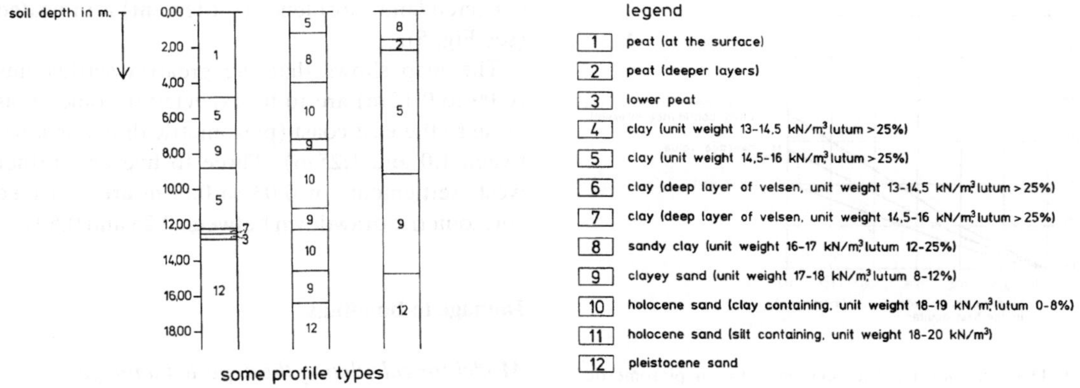

Fig. 2. Soil profiles representative for some Holocene deposits.

variation of 30% was estimated around the mean settlement values.

Rate of settlement

Besides the magnitude of the settlement the rate of settlement is important for possible damage to buildings and constructions. Therefore a number of calculations was carried out to determine the decrease in pore-water pressure in the Pleistocene sand deposits as a function of time and place during and after the reclamation of the Markerwaard polder (Barends et al., 1986, Hannink & Talsma, 1986). Figure 3 shows the results of a large number of calculations. About 90% of the final drop in pore-water pressure in the Pleistocene deposits will take place in 5 to 7 years after the start of reclamation, if the new waterlevels in the Markerwaard polder are established within 0.5 year. If this reclamation process extends over a period of 3 years instead of 0.5 years, about 90% of the final drop in pore-water pressure in the Pleistocene deposits will take place in 7 to 9 years after the start of reclamation.

Figure 3 also shows the results of the calculations of the rate of settlement in North-Holland. About 40 to 60% of the final settlement will take place during the first 3 years after the start of reclamation, depending on the rate of passing from the original lake level to the new phreatic level.

Areal extent of subsidence

On the basis of a map with the 160 soil profiles, the magnitude of the piezometric drawdown and the diagrams that demonstrate the relationship between piezometric drawdown and the final settlement (see Fig. 4), a map was produced on which the

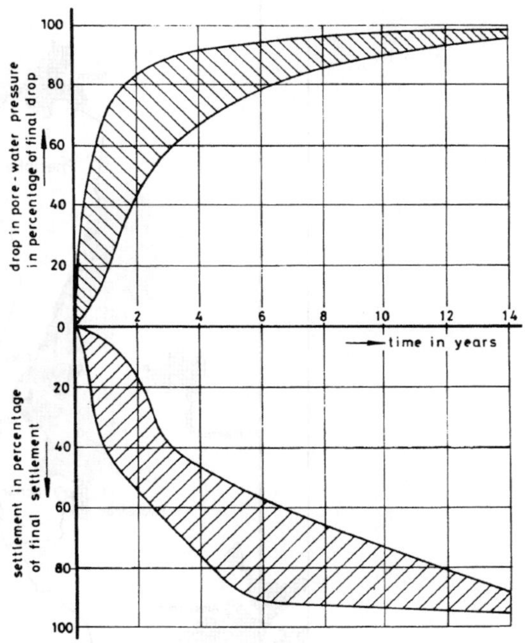

Fig. 3. Rate of fall in pore-water pressure in the Pleistocene deposits and the rate of settlement of the Holocene deposits.

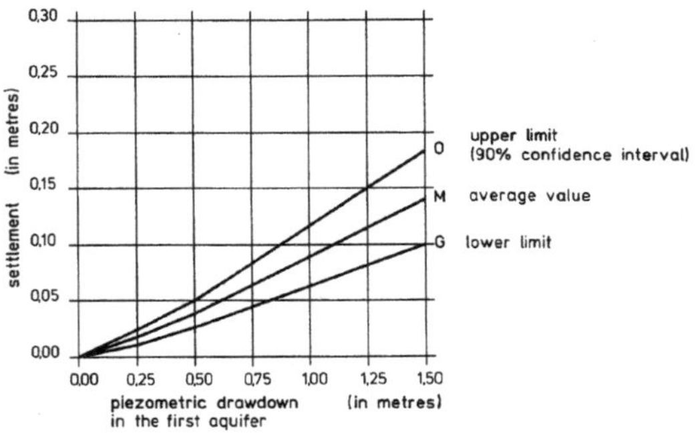

Fig. 4. Diagram showing the relation between piezometric drawdown and settlement for an average profile type.

expected land subsidence (settlement) is indicated (see Fig. 5).

The map shows that the greatest settlements (0.09 to 0.12 m) are to be expected in some areas close to the east coast (piezometric drawdown between 1.0 and 1.25 m). Three to five km further west, settlements of 0.03 to 0.06 m are expected (piezometric drawdown between 0.25 and 0.50 m).

Damage to buildings

Model for calculating the expected damage

The expected damage to buildings was calculated

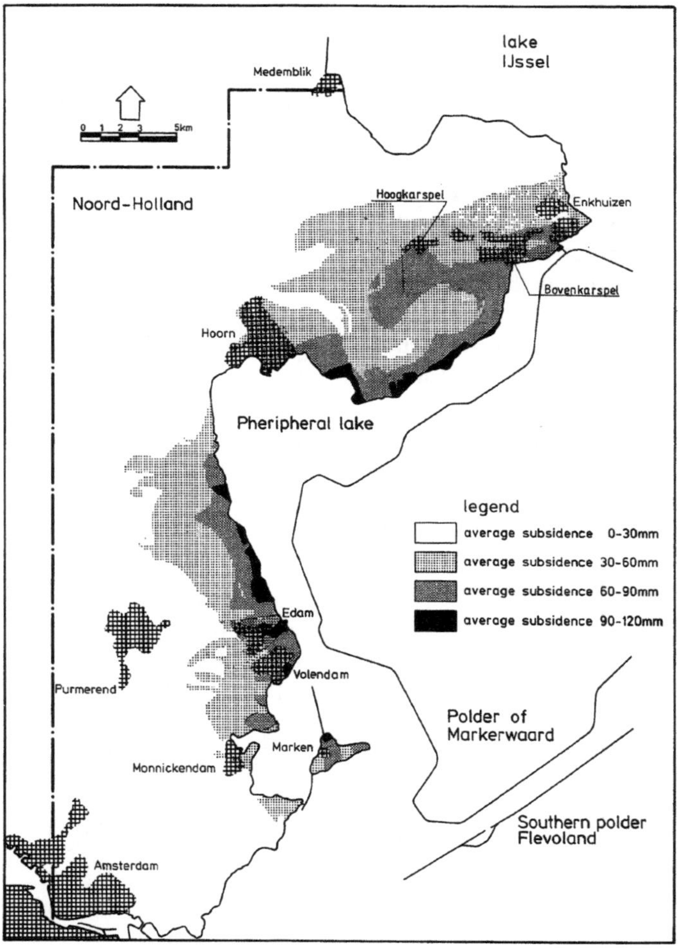

Fig. 5. Calculated subsidence caused by reclamation of the Markerwaard polder.

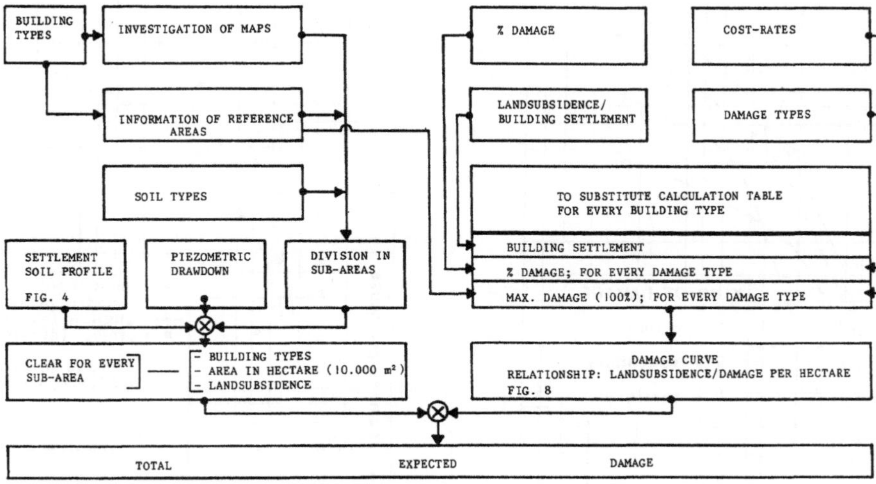

Fig. 6. Diagram of the model for calculating the expected damage.

using electronic and digital drawing equipment. A diagram of the calculation model is shown below (Fig. 6). Based on file-data, topographical- and soil-use maps, the built-up area was subdivided into areas of characteristic urbanisation. For the entire area 43 building types were distinguished in relation to the type of building, age of building, the method of construction and the type of foundation. Also areas exhibiting a high degree of homogeneity such as quarters of a town were modelled. For such 'reference areas' the relevant data about building types, building density and foundation types were collected and expressed in figures per hectare (10.000 m^2). Furthermore, for each sub-area the associated building type and soil type was mapped.

For each sub-area the types of building and all their related characteristics, the area in hectares and the land subsidence were obtained directly or were calculated from the above mentioned data.

Pile settlement

Foundation piles have been driven into the sand deposits which form the first aquifer. Due to settlement of the Holocene deposits around the pile shaft, conditions will arise which are likely to induce additional negative friction forces on the piles. To assess the consequences of these additional forces, calculations have been carried out for a number of single piles considered to be representative for the study area (Hannink & Talsma, 1986). According to the calculations the subsidence of a timber foundation pile may amount between 20 to 60% of the subsidence of the ground surface. The settlement of concrete piles will be negligible.

Standards for damage, leading to damage curves

To obtain a standard for the prediction of damage the damaging effects of a large number of drainage projects in the Netherlands have been evaluated. A total of about 5.000 buildings, situated within the region of influence of the soil water drainage, was considered. Of these about 12% had sustained damage. This analysis has resulted in a relationship between the average building settlement and the average constructional damage. A relationship between the total damage in an area and the total number of buildings, with or without damage was also derived. A distinction was made between shallow foundations and wooden pile foundations (Fig. 7).

Standards for damage were set up, for each type of building, based on literature information (Burland & Wroth, 1975 and Burland et al., 1978), the above mentioned cases of damage in the Nether-

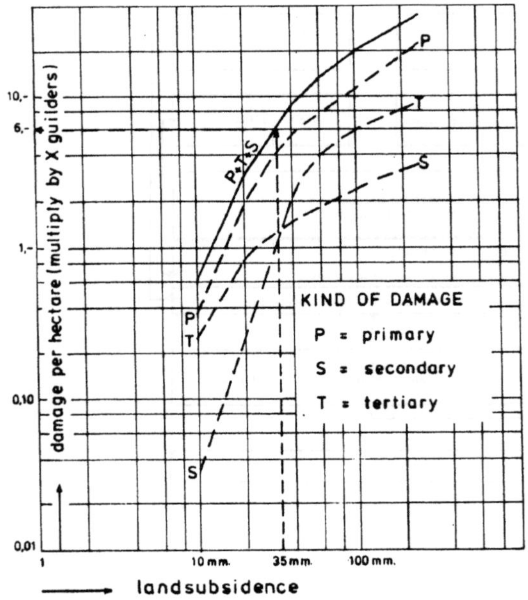

Fig. 7. Relationship between building settlement and damage according to different cases of damage in the Netherlands.

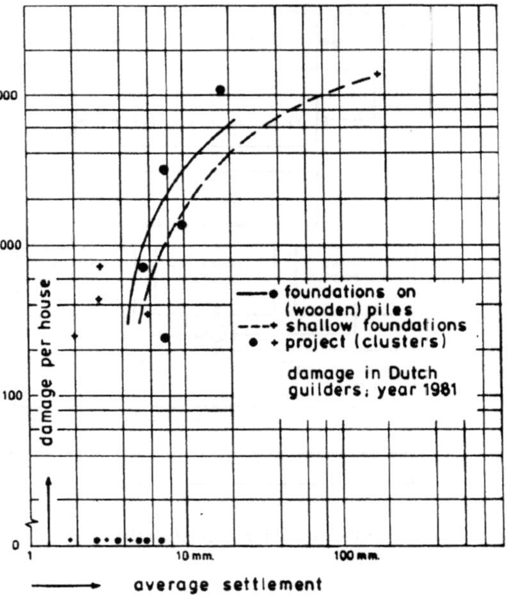

Fig. 8. Example of a damage curve.

lands, and a detailed estimate of the costs for repair. The standards for damage resulted in a damage-curve which, for each type of building, showed the expected relationship between land subsidence and damage per hectare (Fig. 8).

A distinction was made between primary damage (constructional damage, such as crack formation, burst pipes, subsidence of roads), secondary damage (damage resulting from decline in value, advice and legal assistance, increasing road maintenance and industrial damage) and tertiary damage (costs for the government, such as setting up and operating check points, administrative costs of recording and repairing the damage).

Calculation of expected damage

Based on the damage curves the expected damage can be calculated for each area. Adding up these figures gives an estimate of the total anticipated damage. To give a visual impression of the expected damage a damage map has been prepared (Fig. 9). The hatching is a measure for the damage per hectare.

Estimated damage without countermeasures

Without countermeasures, the expected average land subsidence in the urban areas, is about 35 mm. The total expected damage is about 800 million guilders plus or minus 40% (price level 1981). The built-up area (total $\pm 80\,km^2$) has an estimated value of 20 billion guilders. About 100.000 build-

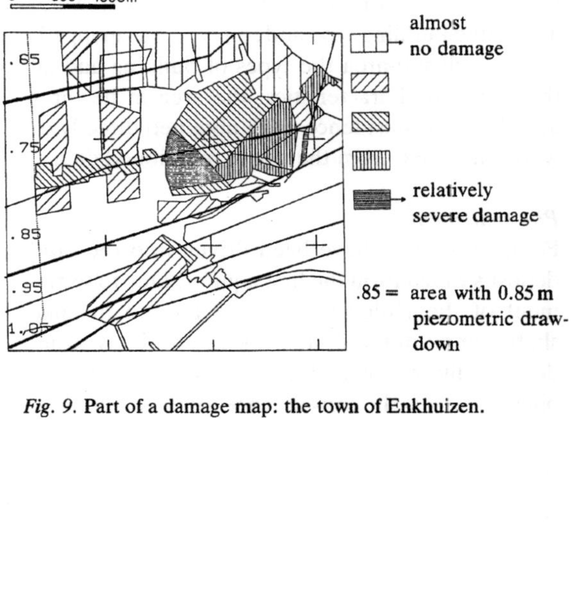

Fig. 9. Part of a damage map: the town of Enkhuizen.

ings are expected to be in the area at the moment of the reclamation. Of the total damage about 85% will occur to buildings and about 10% to factories. The total damage can be divided into 70% primary (constructional) damage, 15% secondary (non-constructional) damage and 15% tertiary (administrative) damage.

In general, it appears that the rate of settlement influences the magnitude of the damage. If the rate of settlement will be low, the building can deform, through creep, without cracking. Hence, damage can be limited considerably by lowering the rate of settlement, for instance by a temporary infiltration of water by means of injection wells.

Agricultural effects

The phreatic drawdown may give rise to an unfavourable moisture supply during the growing season, which may reduce the crop yield. The investigations showed that the maximum phreatic drawdown (0.10 to 0.20 m) may occur near the coast of the Lake. The decrease in the amount of moisture available for the crop, averaged over a long succession of years in the areas is a few millimetres as a maximum during the growing season (Mooy & Vinkers, 1982).

The effect of the phreatic drawdown and the decrease in the amount of moisture available on the crop depends on the present drainage situation and the type of crop.

For the crops which are dependent on ground water for moisture supply a compensation for the yield reduction was included, calculated on the basis of the marketprice of the crops concerned.

In determining the expected damage, additional expenses resulting from a more intensive sprinkling of the crops were taken into account. The damage estimations showed that the average damage to field and horticultural crops would not exceed 250.000 Dutch Guilders per year (price level 1981; Mooy & Vinkers, 1982).

Ecological effects

The phreatic drawdown will have also consequences for the air/moisture distribution in the soil. This can directly affect two vital processes for the plant's life: the oxygen and the moisture supply for the roots. An alteration of the oxygen content also influences mineralization and oxidation processes.

Because of this, the nutrient content of the soil can change, and so affect the nutrient supply to the plant. Alteration of these processes may influence all sorts of processes which take place in the plant (photosynthesis, respiration, transpiration, etc.) and which determine growth and development of the plant.

Some time after the drawdown of the phreatic groundwater is established, a new balance in the supply of oxygen, moisture and nutrients will be reached. Some species will better adapt to these new circumstances than others, so that a shift in species composition of the vegetation will occur, depending on the extent of the changes. If the composition of the vegetation is known, it is possible with the help of so-called indication values to predict which plant species will decrease or disappear, which species will increase in number, and what new species can be expected (Ellenberg, 1974).

Until the final drawdown of the phreatic level is completed, the above-mentioned processes will show variations. This can be considered as a temporary increase of environmental dynamics, to which certain species are known to be very sensitive, especially the rather rare and valuable vegetations (Londo, 1975).

Depending on the value attributed to the vegetation, it is possible to determine to what extent alterations in the species composition should be regarded as damage.

The predicted change in abiotic conditions (air/water content) in the topsoil and in plant composition can also affect the amount and composition of macrofauna in the soil (food for birds) and associated bird life. For instance, if at greater depth in the soil the air content increase, soil fauna migrates to greater depth too and is harder to catch by meadow birds.

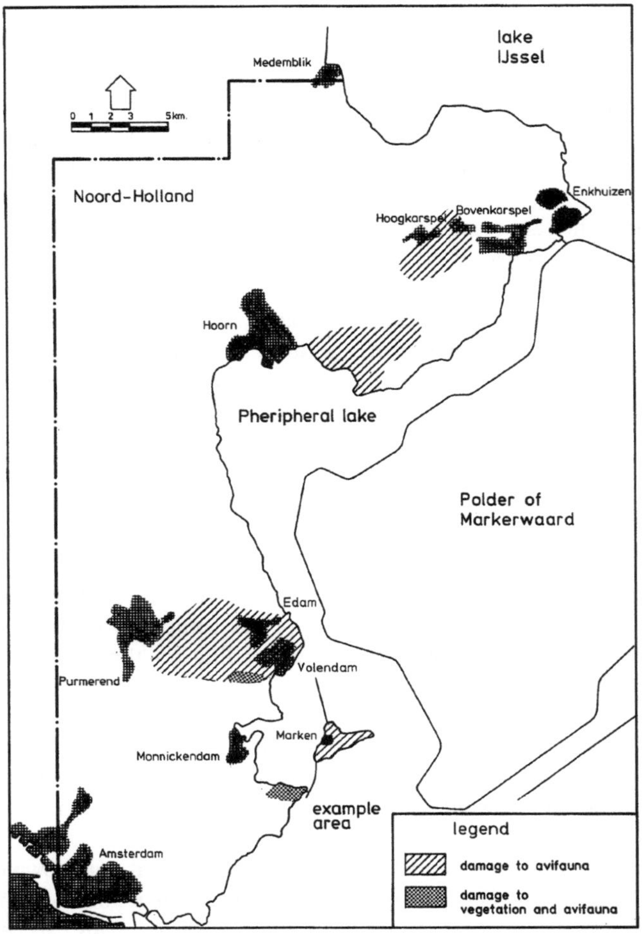

Fig. 10. Areas vulnerable for damage to plant and bird life by phreatic drawdown.

The results of the study show that in a limited part of the study-area a number of negative effects can be expected. This concerns in particular a small piece of peat land south-west of Volendam and a small area south of Marken where valuable marsh vegetation may deteriorate, which would threaten valuable species of marsh birds.

More-over, in other parts of the older area a decrease is to be expected in the number of valuable wet meadow plants and meadow birds such as the black-tailed godwit and the red-shank. The areas where negative effects are to be expected if no countermeasures are taken, are marked in Fig. 10.

Conclusions

- A relationship between piezometric drawdown and the final settlement was established. Near the east coast of the province of North-Holland the largest expected settlement amounts to 0.09 to 0.12 m. Three to five kilometre further inland, expected settlement is less than 0.03 to 0.06 m.
- The rate of settlement depends on the soil properties and the rate of piezometric drawdown in the upper Pleistocene aquifer. About 90% of the final settlement takes place within 5 to 15

years, 40 to 60% in the first 3 years.

- The settlement of timber piles is between 20 to 60% of the settlement of the ground surface. The settlement of concrete piles will be negligible.
- In urban areas the average expected subsidence of the ground surface is, without countermeasures, about 35 mm. The total estimated damage to buildings and infrastructure is about 800 million Dutch guilders.
- The amount of damage is dependent on the rate of settlement. If the rate of settlement is low, buildings can deform through creep without the occurrence of cracks. Hence damage can be limited by taking countermeasures that slow down the rate of settlement considerably.
- In only a very limited part of the province of North-Holland some negative effects are to be expected to valuable vegetation and wet-meadow birds by phreatic drawdown and land subsidence.
- The calculated phreatic drawdown during summer will have only small effects on the average crop yield (over a 30 year period). Damage estimates show an average damage to field and horticultural crops of less than 0,25 million Dutch guilders per year (additional expenses for extra sprinkling of the crops).
- The most important estimated negative effects of the reclamation of the Markerwaard are the damages to buildings and infrastructure by land subsidence.

References

Barends, F.B.J., Teunissen, J.A.M. & Verruijt, A. 1986. Assessment of the transient nature of subsidence-Proc. 3rd Int. Symp. Land Subsidence, Venice 1984 IASH Publ. 151: 107–116

Burland, J.B., Broms, B.B., De Mello, V.F.B. 1978. Behaviour of foundations and structures-BRE Curr. Pap. CP 51/78, England: 495–546

Burland, J.B. & Wroth, C.P. 1975. Settlement of buildings and associated damage-BRE Curr. Pap. CP 33/75, England: 44 pp

Claessen, F.A.M. 1988a. Study to forecast and to prevent damage resulting from reclamation of the Markerwaard polder. An integrated earth-scientific approach. In: W.J.M. van der Linden, S.A.P.L. Cloetingh, J.P.K. Kaasschieter, J. Vandenberghe, W.J.E. van de Graaf & J.A.M. van der Gun (eds): Coastal Lowlands: Geology and Geotechnology – Proc. KNGMG Symp. The Hague, 1987 – Kluwer Academic Publ. (Dordrecht): 267–271

Claessen, F.A.M. 1988b. Geohydrological effects of the reclamation of the Markerwaard polder. In: W.J.M. van der Linden, S.A.P.L. Cloetingh, J.P.K. Kaasschieter, J. Vandenberghe, W.J.E. van de Graaff & J.A.M. van der Gun (eds): Coastal Lowlands: Geology and Geotechnology – Proc. KNGMG Symp. The Hague, 1987 – Kluwer Academic Publ. (Dordrecht): 273–282

Ellenberg, H. 1974. Zeigerwerte der Gefärspflanzen Mitteleuropas – Scripta Geobotanica IX: 97 pp

Hannink, G. & Talsma, K.W. 1986. Geotechnical consequences to the environment of the Markerwaard polder, the Netherlands-Proc. 3rd Int. Symp. Land Subsidence, Venice 1984-IASH Publ. 151: 885–898

Koppejan, A.W. 1948. A formula combining the Terzaghi load-compression relationship and the Buisman secular time effect-Proc. 2nd. Int. Conf. Soil Mech. Found. Eng., 1948, Rotterdam, III: 32–37

Londo, G. 1975. Nederlandse lijst van hydro-, freato- and afreatofyten (List of hydro-, phreato- and aphreatophytes in the Netherlands) – R.I.N. Intern Rept, Leersum, The Netherlands: 52 pp

Mooy, H.G.M. & Vinkers, H.J. 1982. Agricultural and environmental impact of the reclamation of the Markerwaard polder on the province of North-Holland-Heidemij, Consult. Div., Arnhem. Intern Rept 662-81/3: 57 pp

Proceedings KNGMG Symposium 'Coastal Lowlands, Geology and Geotechnology', 1987: 293–300 (1989)

The Markerwaard reclamation project: geotechnical topics

Hannink, Geerhard
Delft Geotechnics, Box 69, 2600 AB Delft, The Netherlands

Received 1 September 1987; accepted in revised form 30 June 1988

Key words: settlement of soil, soil structure interaction

Abstract

In order to estimate the magnitude of induced settlement of the subsoil in the Province of North-Holland caused by the Markerwaard Reclamation Project, numerous soil profiles from borings were considered. Soil property data have been obtained from 600 compressibility tests and 250 permeability tests carried out in the laboratory. The parameters obtained have resulted in a reliable prediction of soil settlements in an area of 500 km^2.

The prediction of the rate of settlement has been based on calculations of the rate of piezometric drawdown in the upper Pleistocene aquifer, and in order to evaluate the settlement of buildings in North-Holland, a relationship has been developed between the predicted settlement of the Holocene deposits and the corresponding settlement of a pile foundation. This relationship has been developed by considering a single pile subjected to a change in negative skin friction, and shows that the settlement of a timber pile may be between 20 to 60% of the settlement of the ground surface.

Introduction

Due to a heavy storm in 1916, a large part of the Province of North-Holland was flooded and Amsterdam, the capital of the Netherlands, was endangered. That was the direct motive for a bill dealing with the reclamation of the ZuijderZee, which was offered to Parliament. This bill was accepted and the ZuijderZee (Reclamation) Act was published in 1918. The project, described in this act, consisted amongst others of the reclamation of four polders. So far three polders were reclaimed.

During the last years, however, environmental concern became more and more important. The reclamation of the Markerwaard Polder was reconsidered by government and consequently a comprehensive study was started. An important part of this study was devoted to the consequences of the reclamation which will result in an increase in the effective stress in the subsoil due to the associated lowering of the water table. This will cause the Holocene deposits and buidings in the Province of North-Holland to settle.

The following geotechnical aspects of this settlement have been considered:

- the determination of the geotechnical parameters of an area of 500 km^2;
- the reliability of the results of settlement calculations;
- the rate of settlement of the Holocene deposits;
- the relationship between the settlement of the Holocene deposits and the settlement of foundation piles.

These aspects form the subject of the present paper.

Subsoil in North-Holland

The subsoil in North-Holland consists of 10 to 20 m of Holocene compressible strata, overlying a sand deposit of Pleistocene age. All sediments of the Holocene in the study area belong to the Westland Formation. Apart from the layers found in the tidal gullies, the normal sequence of soil layers consists of the following main lithological units:

a lower peat: a thin layer of compact peat
b overlain by a clay deposit, followed by
c a sand deposit, with intermediate clay layers and lenses;
d Holland peat: a rather thick layer of peat
e sometimes covered by a layer of clay.

In the tidal gullies the units a and b are missing. The gullies have as a rule, been filled by sandy deposits.

Soil settlement calculations

In the Netherlands soil settlements are usually calculated by the Koppejan (1948) formula:

$$s = H\left(\frac{1}{C_p} + \frac{1}{C_s}\log\frac{t}{t_o}\right)\ln\frac{p_o + \Delta p}{p_o} \qquad (1)$$

where:

s = settlement of a soil layer [m],
H = initial thickness of a soil layer [m],
C_p, C_s = compression coefficients *independent* of effective stress [-],
t = time,
t_o = unit time (usually in days),
p_o = initial vertical effective stress [kN/m²],
Δp = increase in vertical effective stress [kN/m²].

The formula is often simplified in order to calculate final settlements which are considered to take place in about 27 years (10,000 days). The formula then reads:

$$s_f = (H/C_{10^4\ \mathrm{days}})\ln\frac{p_o + \Delta p}{p_o} \qquad (2)$$

where:

s_f = settlement of a soil layer in 10,000 days
$1/C_{10^4\mathrm{days}}$ = $1/C_p + 4/C_s$

Koppejan indicated that his formula may not necessarely be valid in the range of low pressures.

Determination of geotechnical parameters

The Holocene deposits in the study area have a uniform constitution. It is justified therefore, to assume that the soil layers in different areas which belong to the same lithological unit have similar geotechnical properties.

The two most important geotechnical properties in settlement calculations are compressibility and vertical permeability. The compressibility determines the magnitude of the settlement caused by an increase in effective stress in the compressible strata; the permeability of the different soil layers is important to the development of the pore-water pressure profile, and the rate of settlement.

To determine the magnitude of these two parameters, results were collected from previous laboratory tests on soil samples, obtained from borings carried out in the area. At the same time supplementary site and laboratory investigations were carried out in areas where insufficient data were available.

Figs 1 and 2 show the results of about 600 consolidation tests and 250 permeability tests. The curves show mean values for compressibility and permeability as a function of the vertical effective stress and have been used to calculate the settlement of 150 different soil profiles in the area. Fig. 1 shows that the $C_{10^4\ \mathrm{days}}$ values are indeed not constant at low pressures and therefore that the $1/C_p$ and $1/C_s$ values are also not constant.

Results of soil settlement calculations

The settlement of each soil profile, due to a decrease in pore-water pressure, can be calculated if both the present and future pore-water pressures are known. In this case:

- the present phreatic level is known from measurements and the calculations have been based on a present phreatic level of 0.20 m below polder water level (mean annually lowest phreatic

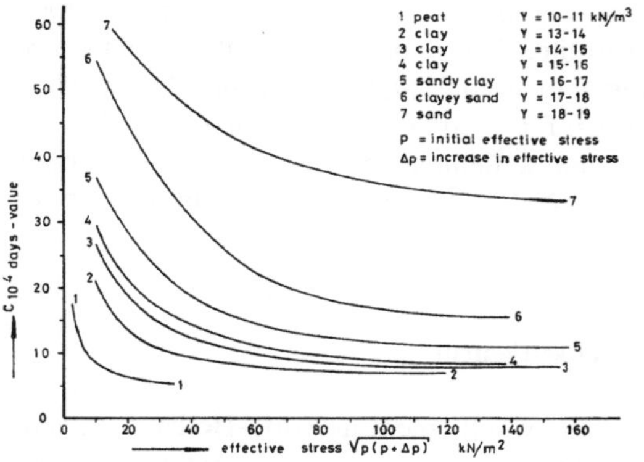

Fig. 1. Compressibility coefficient ($C_{10^4\ days}$) related to effective stress.

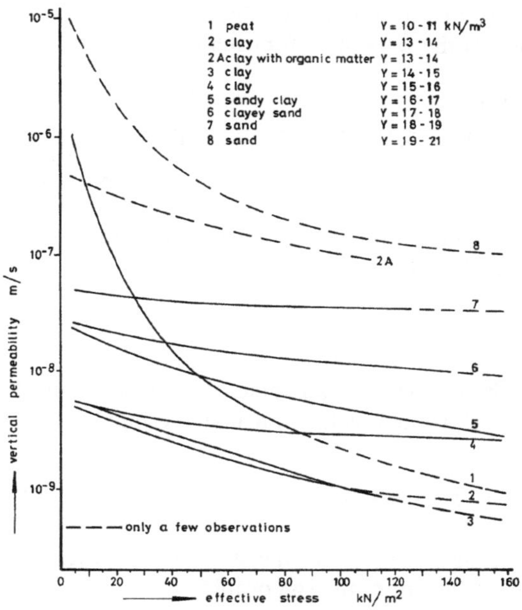

Fig. 2. Vertical permeability related to effective stress.

level); the expected drop in the phreatic level was determined using a hydrologic model;

- the pore-water pressure below the Holocene deposits also is known from measurements; the expected drop in the pore-water pressure (the piezometric drawdown in the upper Pleistocene aquifer) was calculated using a geohydrological model.

Knowing the geotechnical properties of each soil layer (Figs 1 and 2), the final settlement of each profile has been calculated. The maximum settlements amount to 9 to 12 cm, and are to be expected in some areas in the most eastern part of the Province of North-Holland.

Reliability of the results

The following formula has been used to assess the reliability of the results of the settlement calculations:

$$s_c = \sum_{i=1}^{N} s_i = \sum_{i=1}^{N} \alpha_i \frac{H_i}{C_i} \ln \frac{p_i + \Delta p_i}{p_i} \qquad (3)$$

where:

s_c = calculated expected mean value of the settlement of a soil profile,

s_i = calculated expected mean value of the settlement of a soil layer,

α_i = reliability factor for a soil layer.

The factor α was added to the simplified Koppejan formula (2), which indicates the degree of accuracy of the transposition of the soil data (borings and Cone Penetration Tests) into the various soil layers. The expected mean value of α in principle equals 1.0. The following coefficients of variation were determined by analysing the available data:

$V(H_i) = 0.20$
$V(p_i) = 0.20$
$V(\Delta p_i) = 0.25$
$V(C_i) = 0.04$
$V(\alpha_i) = 0.20$

The coefficient of variation of the compression coefficient is small due to the large amount of observations.

The standard deviation $\sigma(s_c)$ of s_c has been determined in a number of steps using equation 3:

a) The expected mean settlement was calculated using for each soil layer:

$$s_i = \alpha_i \frac{H_i}{C_i} \ln \frac{p_i + \Delta p_i}{p_i} = \alpha_i \frac{H_i}{C_i} \frac{\Delta p_i}{p_i} \quad (\Delta p_i \ll p_i) \qquad (4)$$

This results in a value for the coefficient of variation of the settlement for each soil layer of:

$$V_i(s_i) = \sqrt{V^2(H_i) + V^2(p_i) + V^2(\Delta p_i) + V^2(C_i) + V^2(\alpha_i)} = 0.43 \qquad (5)$$

and a value for the standard deviation of the settlement of a soil layer:

$$\sigma(s_i) = 0.43\, s_i \qquad (6)$$

b) The expected mean value of the settlement of the soil profile was then calculated using:

$$s_c = \sum_{i=1}^{N} s_i = s_1 + s_2 + \ldots + s_N, \qquad (7)$$

resulting in a value for the standard deviation for the settlement of the soil profile:

$$\sigma(s_c) = \sqrt{\sigma^2(s_1) + \sigma^2(s_2) + \ldots \sigma^2(s_N)}$$

$$= \sqrt{(V_1 \cdot s_i)^2 + \ldots + (V_N \cdot s_N)^2}$$

$$= V_i \sqrt{s_1^2 + s_2^2 + \ldots + s_N^2}$$

(provided $V_1 = V_2 = V_N$)

$$= 0.43 \sqrt{s_1^2 + s_2^2 + \ldots s_N^2} \qquad (8)$$

c) 150 different soil profiles were considered, the number of soil layers and the magnitude of the settlements of individual soil layers and of complete soil profiles leading to the following range of values for $\sqrt{s_1^2 + s_2^2 + \ldots + s_N^2}$:

$$0.32\, s_c < \sqrt{s_1^2 + s_2^2 + \ldots s_N^2} < 0.70\, s_c \qquad (9)$$

Combining equations (8) and (9) leads to:

$$0.14\, s_c < \sigma(s_c) < 0.30\, s_c, \qquad (10)$$

the mean value of $\sigma(s_c)$ being $0.19\, s_c$.

The final results of the settlement calculations are presented as follows:

$$s_{RE} = (1 \pm 0.30)\, s_c \qquad (11)$$

where s_{RE} is the range of settlements expected for a soil profile. This result is equivalent to a confidence interval of 70 to 95%, with a mean value of 90%.

Thus, for an area with a known soil profile the expected settlement may, in view of the above confidence, vary from 30% below to 30% above the calculated expected mean value.

Rate of settlement

The rate of settlement in North-Holland depends not only on the soil properties in that area but also on the rate of the piezometric drawdown in the upper Pleistocene aquifer. A number of calculations were carried out to determine the fall in pore-water pressure in the Pleistocene sand deposits as a function of time and place during and after the reclamation of the Markerwaard Polder.

Fig. 3 shows the results of a large number of calculations. About 90% of the final drop in pore-water pressure in the Pleistocene deposits will take place in 5 to 7 years after the start of reclamation, if the drawdown in the polder occurs in 0.5 year. If this drawdown extends over a period of 3 years instead of 0.5 year, about 90% of the final drop in pore-water pressure in the Pleistocene deposits will take place in 7 to 9 years after the start of reclamation.

Fig. 3 shows also the results of the calculations of the rate of settlement in North-Holland. About 40 to 60% of the final settlement will take place in the first 3 years after the start of reclamation, depending on the speed of the drawdown.

Pile settlement calculations

The construction of the Markerwaard will cause a piezometric drawdown in the sand layer into which foundation piles have already been driven. Therefore conditions will arise which are likely to induce additional negative friction forces on the piles due to the decrease of pore-water pressure in compressible layers around the pile shaft. To assess the consequences of these additional forces, calculations have been carried out for a number of single

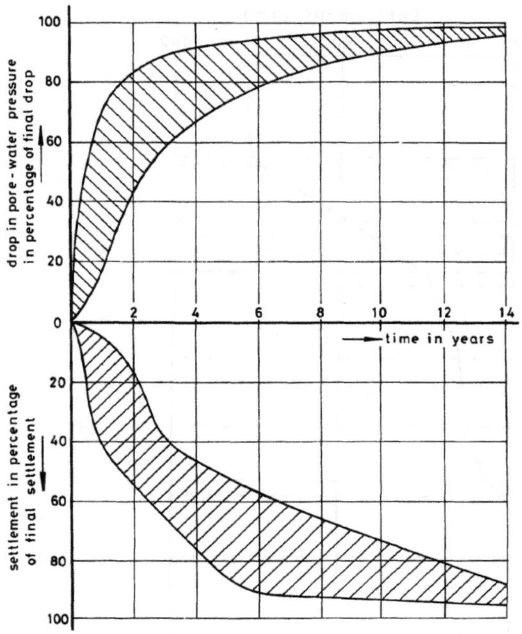

Fig. 3. Rate of fall in pore-water pressure in the Pleistocene deposits and the rate of settlement of the Holocene deposits.

piles considered to be representative for the study area.

From calculations it was found that the expected settlement of the compressible soil layers will decrease more or less linearly with depth (Fig. 4). The settlement of the Pleistocene sand layer due to the piezometric drawdown will be negligable. The resulting negative skin friction can now be calculated, if the pile is assumed to be incompressible (Heijnen 1972).

Calculations of the resulting negative skin friction were carried out where some relative displacement between pile and soil was necessary to mobilize the full friction. However, in the following simple model no relative displacement has been assumed to be necessary even to mobilize the full friction (Fig. 5). From Fig. 6:

$$Q_{nr} = (1 - \frac{2\, s_p}{s_{gs}})\, Q_{nm}, \qquad (12)$$

where:

Q_{nr} = resulting negative skin friction [kN],
s_{gs} = settlement of the ground surface [m],
s_p = settlement of the pile [m],
Q_{nm} = maximum negative skin friction (that can act on a rigidly embedded pile) [kN],
Q_h = load on the pile head [kN].

When the soil settles, the new equilibrium for a pile can be found by considering its load-settlement behaviour. In this example a timber pile has been considered. The ultimate bearing capacity of the sand layer at the base level of the pile can be determined from the results of a Cone Penetration Test (Van Mierlo & Koppejan 1952 and Van der Veen & Boersma 1957).

In order to establish the settlement of the pile at equilibrium the load on the pile which has caused the continuing settlement has to be assessed. Generally this situation occurs at 70 to 80% of the ultimate bearing capacity (Plantema 1948).

The expected load-settlement behaviour of the timber pile is reproduced in Fig. 7. The load on the pile-head (Q_h = 80 kN = constant) plus the (theoretical) maximum negative skin friction (when the friction acts downwards over the full length L in the compressible layers) on the pile (Q_{nm} = 100 kN) are drawn on the horizontal axis.

Q_{nm} can be determined from Cone Penetration Test results (Begemann 1969). As can be seen in Fig. 6, the resulting negative skin friction $Q_{nr} = 0$ when the settlement of the pile s_p is equal to half the settlement of the ground surface, s_{gs}. The settlement of the ground surface (equal to twice the settlement of the pile) is drawn at Q_h = 80 kN ($Q_{nr} = 0$) in Fig. 7. Load equilibrium of the pile exists at the intersection point of the load-settlement curve and the line which represents the resulting negative skin friction. By assuming several ground surface settlement values, a number of intersection points was found, and a relationship established between the settlement of the ground surface and the settlement of the pile (Fig. 8).

Results of pile settlement calculations

The results indicate that two different cases can occur. In the first case:

$$Q_h + Q_{nm} < 0.8\ (Q_p + Q_s) \qquad (13)$$

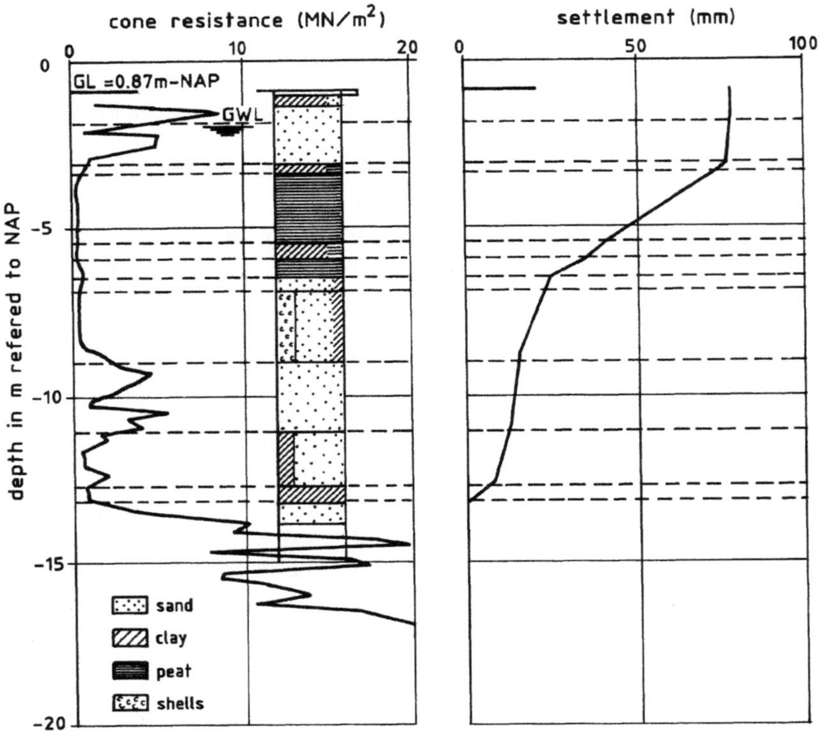

Fig. 4. Results of a CPT, of a 66 mm dia. Begemann boring and of calculated settlements due to a piezometric drawdown of 1 m in the Pleistocene deposits.

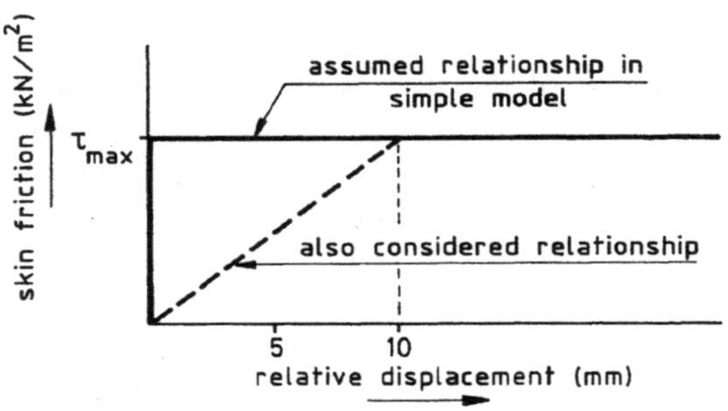

Fig. 5. Relationship between skin friction and relative displacement between pile and soil.

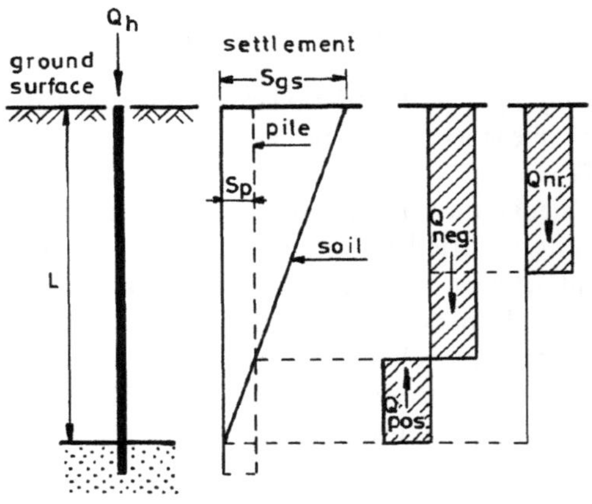

Fig. 6. Development of negative skin friction for a settling pile.

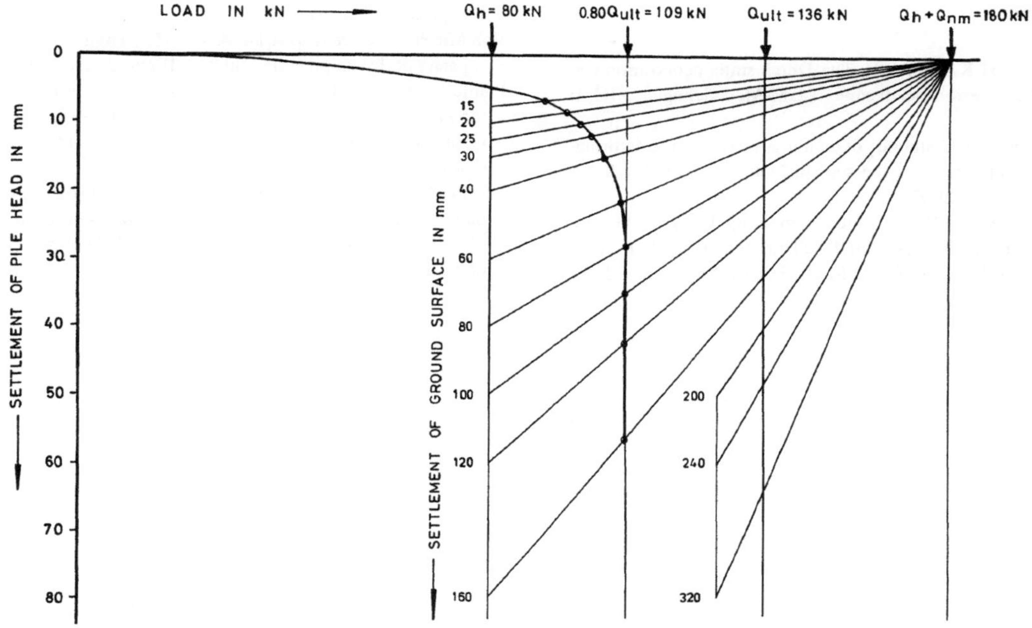

Fig. 7. Load-settlement behaviour of a timber pile.

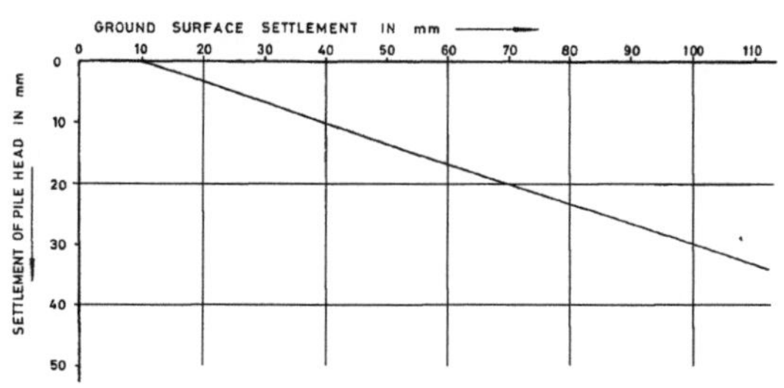

Fig. 8. Relationship between settlement of the ground surface and the settlement of the pile.

where:
Q_h = load on the pile head [kN],
Q_{nm} = maximum negative skin friction [kN],
Q_p = pile point resistance [kN],
Q_s = pile shaft resistance in the sand layer [kN].

In this case the sand layer will develop sufficient bearing capacity, without much settlement of the pile. In the second case:

$$Q_h + Q_{nm} > 0.8\ (Q_p + Q_s) \qquad (14)$$

The bearing capacity of the sand layer in this case is insufficient. Only when the pile has settled to a certain extent, is equilibrium in vertical direction possible.

In the study area concrete piles will normally react as described in the first case. However, timber piles will often react in the way described in the second case.

According to the calculations the settlement of a timber pile may be between 20 to 60% of the settlement of the ground surface.

Acknowledgements

This study forms part of a project undertaken by ABT Consulting Engineers, Delft Geotechnics, Geological Survey of the Netherlands, Heidemij Consultancy Division, and International Water Supply Consultants IWACO.

The author wishes to express his gratitude to Rijkswaterstaat (Dutch Public Works Department) for permission to present the results of the study made on their behalf.

References

Begemann, H.K.S.Ph. 1969 The Dutch static penetration test with the adhesion jacket cone – LGM-Mededelingen, XIII, 1: 1–86

Heijnen, W.J. 1972 Stability of buildings related to the deformation characteristics of the soil (in Dutch) – LGM-Mededelingen, XV, 2: 29–47

Koppejan, A.W. 1948 A formula combining the Terzaghi load-compression relationship and the Buisman secular time effect – Proc. 2nd Int. Conf. Soil Mech. Found. Eng., June 21–30, 1948, (Rotterdam) III: 32–37

Van Mierlo, W.C. & Koppejan, A.W. 1952 Length and bearing capacity of driven piles (in Dutch) – Bouw, January 19, 1952

Plantema, G. 1948 Results of a special loading test on a reinforced concrete pile, a so-called pile sounding – Proc. 2nd Int. Conf. Soil Mech. Found. Eng., June 21–30, 1948, (Rotterdam) IV: 112–118

Van der Veen, C. & Boersma, L. 1957 The bearing capacity of a pile predetermined by a cone penetration test – Proc. 4th Int. Conf. Soil Mech. Found. Eng., August 12–24, 1957, (London) II: 72–75

Proceedings KNGMG Symposium 'Coastal Lowlands, Geology and Geotechnology', 1987: 301–309 (1989)

The Markerwaard project: Countermeasures to prevent detrimental effects, a feasibility study

Hubertus M.C. Satijn
Geohydrological Consultant, IWACO B.V.,
P.O. Box 183, 3000 AD Rotterdam, The Netherlands

Received 9 September 1987; accepted in revised form 30 June 1988

Key words: reclamation works, land subsidence, Countermeasures, injection, infiltration, recirculation

Abstract

Without countermeasures to compensate the decline of the piezometric levels, settlement of the compressible Holocene clay and peat deposits will occur after reclamation of the Markerwaard. The resultant land subsidence may cause damage to buildings and infrastructure in the coastal area of the Province of North Holland. Decline of the piezometric levels can be completely or partly compensated by means of countermeasures. During the pre-feasibility study a wide range of countermeasures have been studied. A few solutions were selected to be studied in detail. These are mainly based on the principle of injection or infiltration of water in the Pleistocene aquifers underneath the mainland.

The systems studied in detail, including geohydrological and geotechnical calculations and cost estimates, are: the injection well system; the recirculation system consisting of extraction and injection wells; infiltration grooves and infiltration wells.

The injection well system proved to be the most promising countermeasure. Detailed geohydrological calculations are recommended to predict the local effects of the countermeasures.

Introduction

Without countermeasures, settlement of the compressible Holocene deposits will occur after reclamation of the Markerwaard. Due to resulting land subsidence, buildings and infrastructure may be damaged (Claessen 1984, Westerhoff & De Mulder 1984, Hannink & Talsma 1984, Van Bruchem 1984).

To prevent or compensate for the decline of the piezometric levels, which causes the detrimental effects, there are, in principle, three methods:

1. Compensation of the drawdown by injection of water, gas, or air into the aquifer or, by facilitating infiltration of water;
2. Creation of a resistance in the aquifers between the new polder and the main land by means of slurry walls or injection of chemicals, etc.;
3. Slowing down the land subsidence processes by phasing out the reclamation works over a number of years.

Compensation can be established in three degrees:

a) Complete compensation everywhere on the main land;
b) Complete compensation, restricted to the urban areas that are most vulnerable to damage;
c) Partial compensation in the most vulnerable areas.

During the pre-feasibility study, various measures, based on one of the aforementioned principles, have been studied; for example injection of air and water into the aquifers, injection of chemicals, construction of infiltration grooves and infiltration wells, construction of slurry walls, decreasing the existing groundwater exploitation on the main land, improvement of the infiltration by dredging activities.

All these measures were briefly studied and analysed provisionally with regard to their technical, practical, and economic feasibility.

Injection of air, gas or chemicals has been rejected as countermeasure, because the application is not well known and was considered to be technically not feasible. Slurry walls are rather costly. Unless they reach down to the third aquifer, (total depth 200 m), their effect is limited.

Decreasing groundwater abstraction on the main land has a very limited effect. Promotion of infiltration, by cleaning the existing ditches and canals, thus decreasing the entrance resistance, has also a limited effect. The remaining resistance of the Holocene layers is considerable and due to clogging after cleaning, the infiltration capacity will soon decrease.

A few solutions were selected to be studied in more detail:

a) Injection wells;
b) Infiltration grooves;
c) Infiltration wells;
d) Phasing out the reclamation works.

Countermeasures studied in detail

Injection wells/recirculation system

Injection of water by means of injection wells into the aquifers, is one of the most promising countermeasures (Fig. 1).The geohydrological effects of the countermeasures have been calculated with the finite element model FIESTA. This model was applied on a regional scale (IWACO B.V. 1982). Detailed calculations to predict the effects on a local scale have to be made in a later phase. For the regional calculations design parameters for injection wells have been applied, which are known from experiments in the Dutch coastal area for drinking water supply (KIWA 1979).

To achieve a compensation of 100%, these wells can be located along the border of the Marker lake (Fig. 2). Wells can be clustered near those urban areas that are most vulnerable to damage, to prevent detrimental effects related to subsidence (Fig. 3).

By varying the number of wells and the injection capacity, the degree of compensation can be regulated. Injection water can be taken from the peripheral lake and after purification pumped into a transport system towards the wells.

Injection water can also be pumped out of aquifers underneath the future Markerwaard or the deeper aquifer under the main land. In this way a recirculation system is established to compensate partially or fully for the declines of piezometric levels in the first aquifer due to the reclamation of the Markerwaard (Figs 4 and 5). In case of extraction from the deeper aquifers a significant resistance is required between upper and lower aquifers. Calibration of the geohydrological system showed that there are areas with a considerable resistance. The recirculation wells have to be situated on these areas.

Infiltration grooves

Increasing the infiltration capacity of the western peripheral lake by dredging infiltration grooves (depth 10 to 15 m), intersecting the covering clayey Holocene deposits, is another feasible countermeasure (Fig. 6).

By decreasing the hydraulic resistance of these deposits, the surface water from the lake will infiltrate more easily into the first aquifer thus compensating partly the declines of piezometric levels. The location of proposed infiltration grooves for a maximum compensation (90%) is presented in Fig. 7. Fig. 8 shows the location of infiltration grooves if only compensation in urban areas is required.

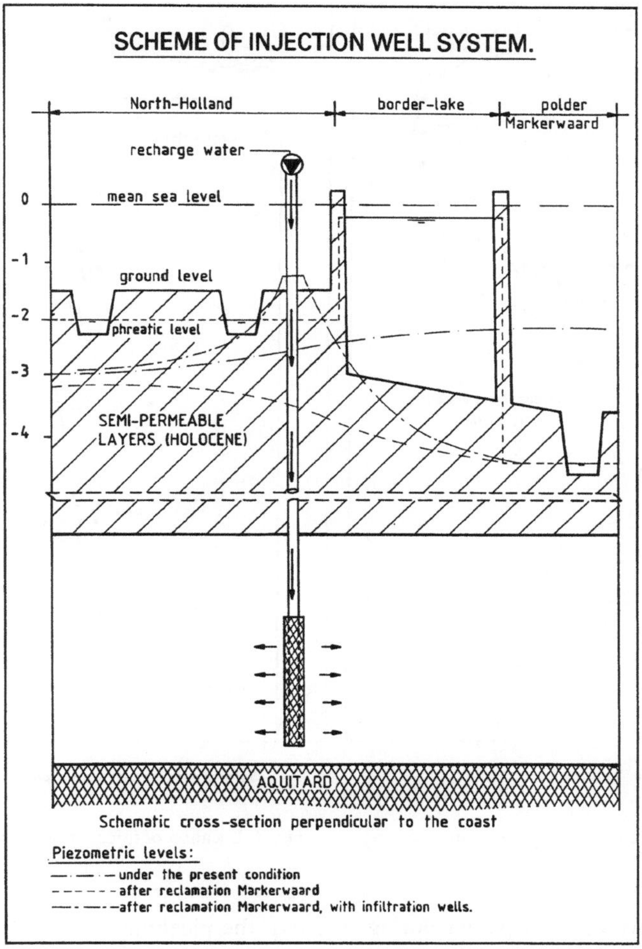

Fig. 1. Scheme of injection wells.

Infiltration wells

An infiltration well is a combination of a injection well and an infiltration groove. For injection wells purified water is required, resulting in high exploitation costs for this system.

Infiltration grooves may be clogged quickly due to the silt content of the surface water. Infiltration wells (Fig. 9) consist of wells in combination with sand filters, situated along the new polder dike. Surface water, infiltrating through the sand bed, will be filtered. After passing the sand filters, the water will flow by gravity into the infiltration wells. In this way a possible clogging of the infiltration wells is minimized.

Phasing out the reclamation works

Damage, due to land subsidence, is closely related to the speed of the processes. By slowing down these processes the damage to buildings and infrastructure will be reduced considerably. Phasing out the reclamation works over several years,

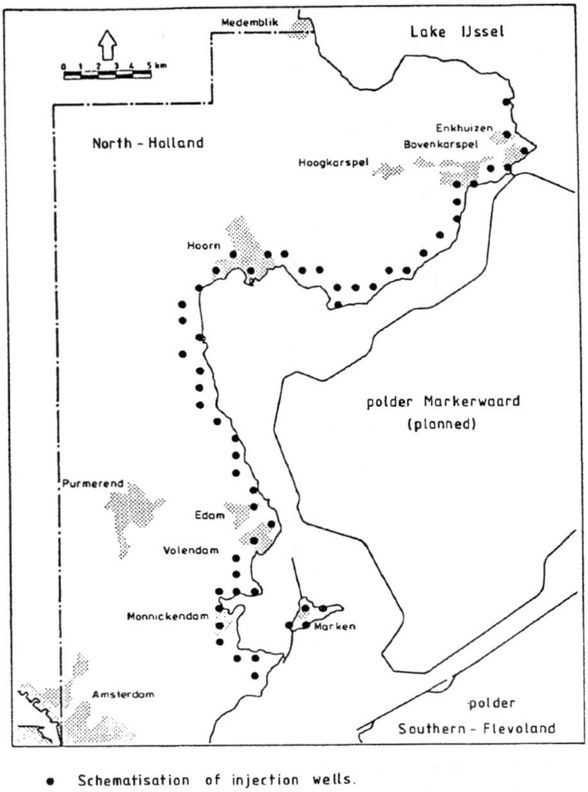

Fig. 2. Location of injection wells for complete compensation.

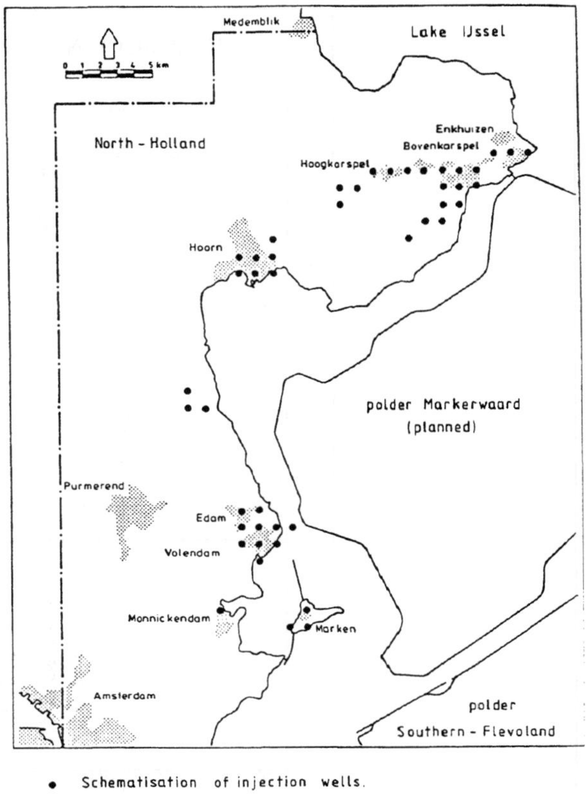

Fig. 3. Location of injection wells near urban areas.

which means pumping dry the new polder during three or five years instead of during half a year, can reduce the detrimental effects.

A gradual settlement of the Holocene deposits can also be achieved by decreasing the quantity of water in the injection wells system for example, over a period of 30 years.

Calculations

Methodology

The impact of the countermeasures on the geohydrological system and on the detrimental effects has been calculated. The model FIESTA was used to forecast the regional geohydrological effects, c.q. the piezometric drawdowns (Claessen 1984). The residual damage for partial compensatory measures has been calculated using a model in which relations were established between the drawdown of the piezometric levels, land subsidence and damage to buildings and infrastructure (Claessen 1984).

Results of the calculations

Injection wells

An annual amount of 40 million m^3 water is required to compensate fully for the piezometric drawdowns under the mainland. In this case about 215 injection wells are required. The water will be injected into the first aquifer, the second aquifer

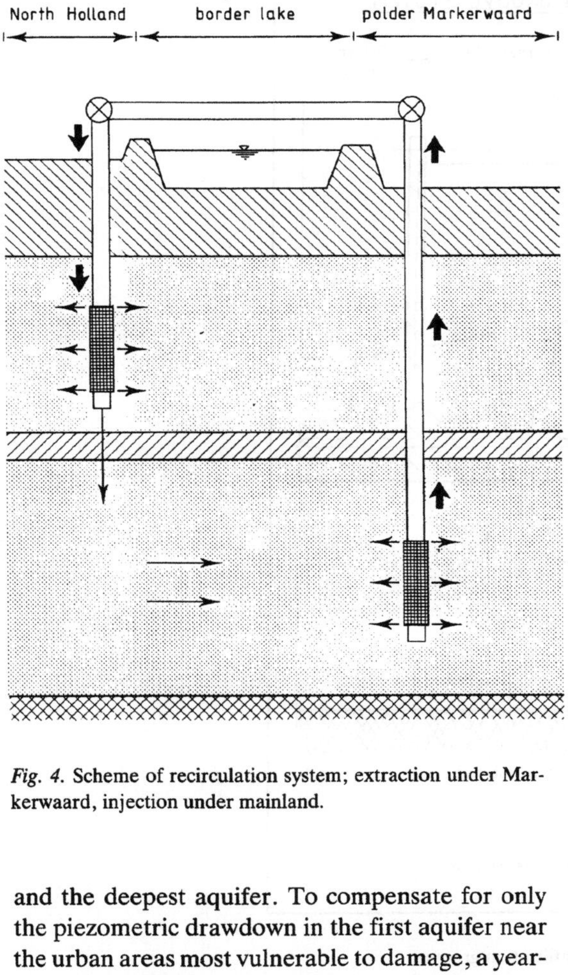

Fig. 4. Scheme of recirculation system; extraction under Markerwaard, injection under mainland.

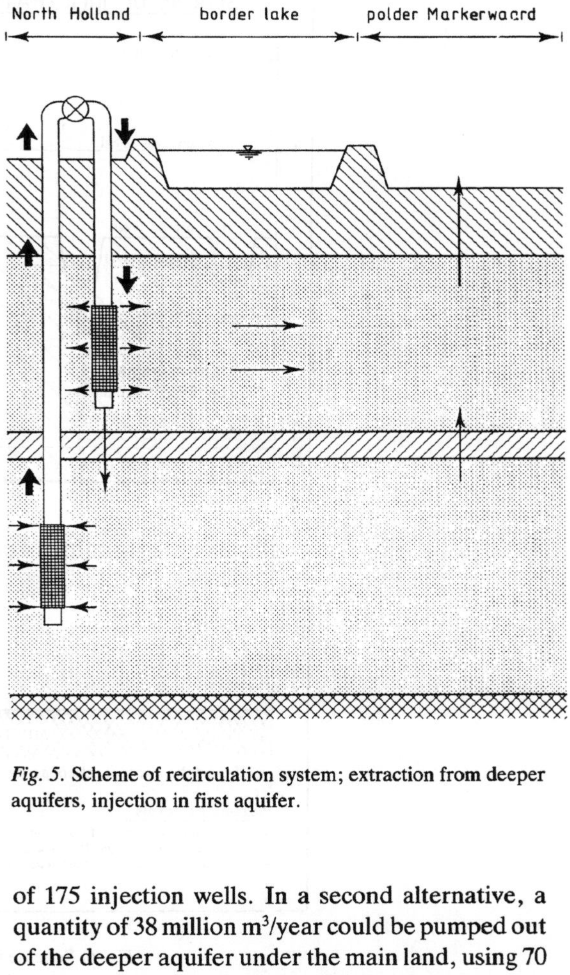

Fig. 5. Scheme of recirculation system; extraction from deeper aquifers, injection in first aquifer.

and the deepest aquifer. To compensate for only the piezometric drawdown in the first aquifer near the urban areas most vulnerable to damage, a yearly amount of 20 million m^3 pre-treated water and about 125 injection wells are required.

Recirculation system

A yearly amount of 120 million m^3 water has to be exploited from 150 wells under the future Markerwaard. This quantity of water has to be transported to 300 injection wells along the coast where it will be injected into one of the aquifers. Treatment of this water is not considered necessary.

In case compensation is only desired near the urban areas, two alternatives are feasible. Firstly a quantity of 27 million m^3/year could be exploited by means of 50 wells under the Markerwaard. This quantity is injected near the urban areas by means of 175 injection wells. In a second alternative, a quantity of 38 million m^3/year could be pumped out of the deeper aquifer under the main land, using 70 wells. This water could be pumped directly into the shallow aquifer by means of 250 injection wells.

Infiltration grooves

For optimum compensation (90%) the infiltration grooves should have an infiltration surface of 8 to 24 km^2. This wide range of the required surface area is related to the uncertainty about the entrance resistance of the bottom of the groove.

Based on experience the resistance has been varied in the model calculations from 100 to 300 days, resulting in an area of respectively 8 and 24 km^2. In the case that the groove has a length of 50 km this will result in a width between 160 and 500 m (see Fig. 7).

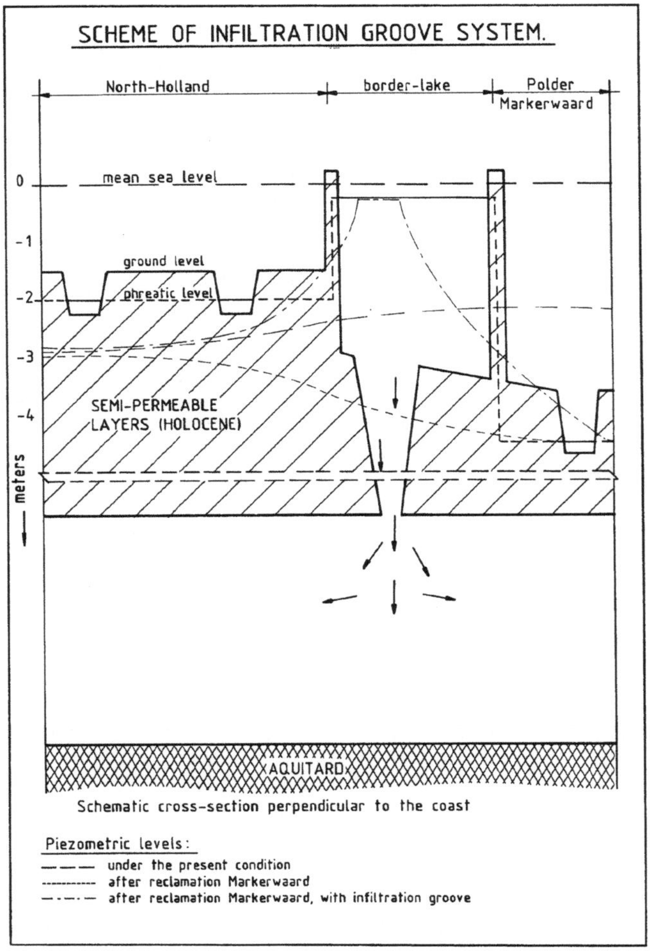

Fig. 6. Scheme of infiltration grooves.

Infiltration wells

As shown by pre-liminary calculations about 600 infiltration wells, situated along the new western dike of the Markerwaard, are required (Fig. 10). Maximum compensation (80 to 90%) can be achieved by infiltrating about 36 million m^3/year into the aquifers (Hebbink et al. 1986).

Phasing out the reclamation works

In case the Markerwaard will be pumped dry in a period of three to five years instead of in 0,5 year, the calculated damage to buildings and infra-structure will be reduced to about 50%.

Cost estimates

Cost estimates have been made for the counter-measures which have been studied in detail. For the scope of this article the cost estimates of the injection wells and the infiltration grooves will be presented. Investment costs, as well as exploitation costs have been considered.

The costs are presented at present values, based on a real interest of 6% and a depreciation period of 50 years. An overview of the costs including the costs of residual damage, is presented in Table 1.

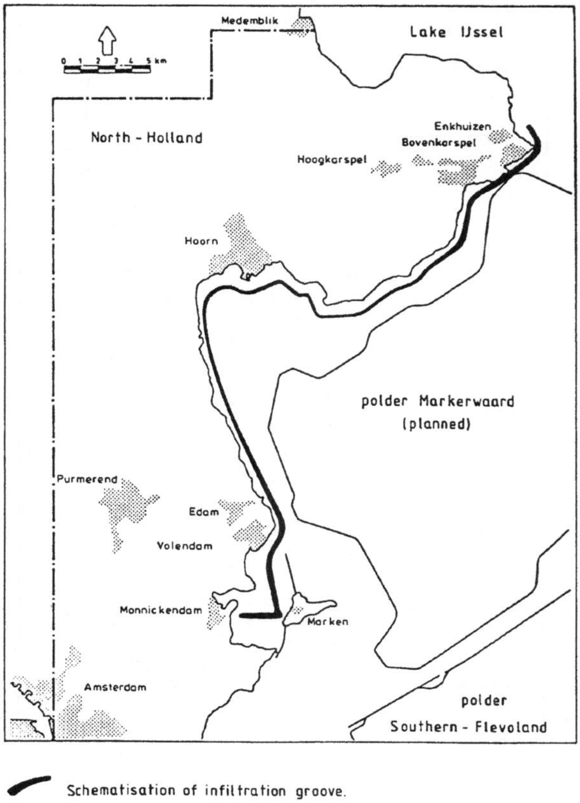

Fig. 7. Location of infiltration grooves for maximum compensation.

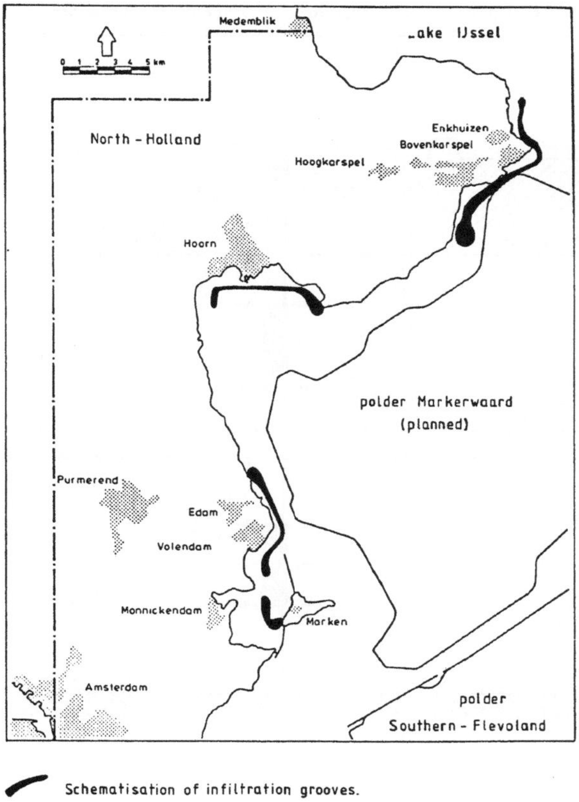

Fig. 8. Location of infiltration grooves near urban areas.

Evaluation and discussion

The injection well system and recirculation system have the advantage that full compensation can be achieved. Lay-out and exploitation are very flexible.

It is not necessary to exploit these systems for ever. The exploitation can be phased out gradually over a period of 30 to 50 years. In this way the damage will be reduced considerably due to the low rate of land subsidence. In this case it is necessary to monitor the geohydrological and consolidation processes. Given the results of the monitoring program, the quantity of injection water and the number of injection wells can be adapted. Although pilot schemes with these systems are required experience in the drinking water sector in the Netherlands is quite promising.

A recirculation system has the advantage that probably no purification is required. Extra attention should be given, however, to the qualitative aspects; deep groundwater with a high chloride content pumped into the shallow aquifer, might cause unacceptable changes. A detailed study is required to investigate the consequences.

Infiltration grooves and infiltration wells are less flexible options. Full compensation is more difficult to realize: some residual damage might be possible. The effect of clogging on the bottom of the grooves is not well known. The required infiltration area and thus the investment costs vary over a wide range. The process is more difficult to control.

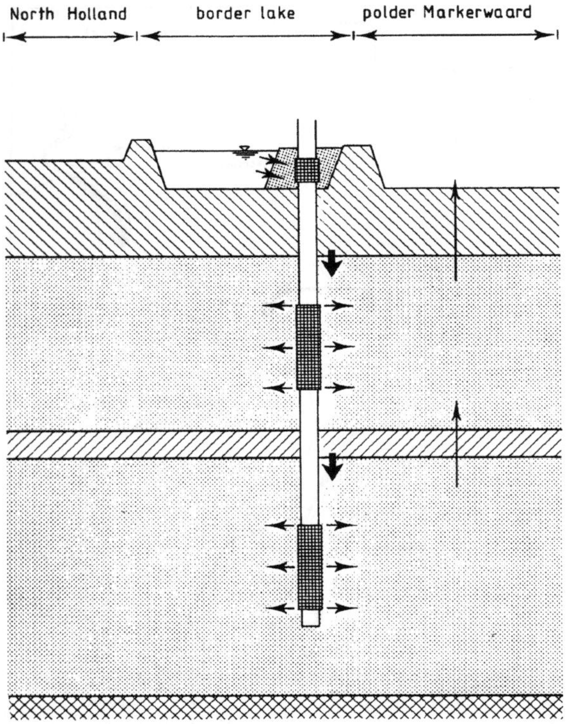

Fig. 9. Scheme of infiltration wells.

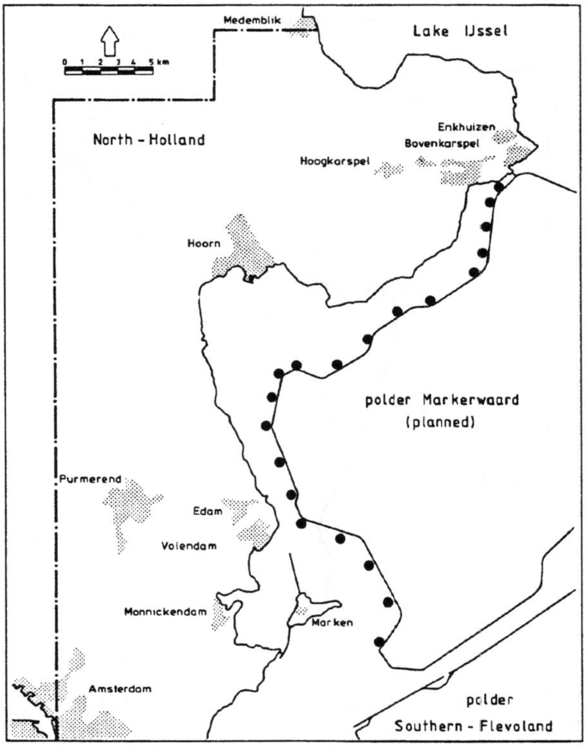

Fig. 10. Location of infiltration wells.

Phasing out the reclamation works over several years implies a considerable residual damage (50%). Delay of construction of the polder implies extra costs due to loss of interest. This process is not easy to control. It seems to be impossible to apply this method on its own.

Conclusions

- It is feasible to take countermeasures to prevent or compensate for the declines in the piezometric level fully or partially.
- Injection well systems and recirculation systems seem to be more flexible and easier to control. Both systems could be gradually phased out over a period of 30 to 50 years.
- Infiltration wells do not guarantee a full compensation (only 90%) and are less flexible, but compared to infiltration grooves there are probably fewer problems with clogging. The exploitation costs for both systems are considerably less compared to those for injection well systems.
- A combination of several countermeasures might prove to be the most feasible solution to partly or completely compensate the piezometric drawdowns.
- Further pilot schemes of injection wells and infiltration wells and detailed cost estimates have to show which countermeasure will be selected.

Table 1. Costs of counter measures and residual damage.

	Rate of recharge $10^6\,m^3$/year	Investment costs* in 10^6 Dfl	Exploitation costs* in 10^6 Dfl/year	Cash value* in 10^6 Dfl	Residual* damage in 10^6 Dfl
A. Measures for permanent full compensation					
1. Injection wells	40	160	15	400	0
2. Recirculation system	120	190	10	320	0
B. Measures for permanent partial compensation (vulnerable areas only)					
1. Injection wells	20	100	10–12	260–290	50
2. Recirculation system					
– withdrawal in the Markerwaard polder	27	116	3	160	50
– withdrawal near vulnerable areas	38	28	4.5	80	50
3. Infiltration grooves (for c-value of bottom of the groove resp. 100 and 300 days)	35	160–720	1	375–735	50

* Price level 1981.

References

Claessen, F.A.M. 1984 An integrated study to forecast and to prevent detrimental effects in the province of North Holland resulting from a change in the groundwater regime after the reclamation of the Markerwaard Polder-Proc. 3rd Int. Symp. Land Subsidence, March 19–25, 1984 Venice: 869–877

Hannink, G. & Talsma, K.W. 1984 Geotechnical consequences to the environment, by construction of the Markerwaard Polder-Proc. 3rd Int. Symp. Land Subsidence, March 19–25, 1984 Venice: 885–899

Hebbink, A.J. et al. 1986 Infiltratieputten in de randmeren van de Markerwaard – H_2O, 19, 21: 512–518

IWACO B.V. 1982 Compenserende maatregelen ter voorkoming van potentiaaldalingen van het grondwater onder Noord-Holland als gevolg van aanleg van de Markerwaard – Intern. Rept.

KIWA 1979 Concept eindrapport Persputten. Persputten Onderzoek – Intern Rept.

Van Bruchem, A.J. 1984 Geohydrological aspects of the reclamation of the Markerwaard polder. Geotechnical investigations-Proc. 3rd Int. Symp. Land Subsidence, March 19–25, 1984 Venice: 899–907

Westerhoff, W.E. & De Mulder, E.F.J. 1984 The Markerwaard reclamation project. Quarternary geological framework of the study area (North Holland and western part of the lake IJssel, the Netherlands) – Proc. 3rd Int. Symp. Land Subsidence, March 19–25, 1984 Venice: 877–885

Proceedings KNGMG Symposium 'Coastal Lowlands, Geology and Geotechnology', 1987: 311–320 (1989)

The geology and geotechnology of the Keta basin with particular reference to coastal protection

Nicholas Kwasi Kumapley
University of Science & Technology, Private Mail Bag, Kumasi, Ghana

Received 31 August 1987; accepted in revised form 29 February 1988

Key words: Coastal erosion, Quaternary geology, geotechnology, Keta basin, soft sediments

Abstract

While most of the coast of the Bight of Benin in West Africa is in a state of active erosion, the problem is most acute in the south-eastern part of Ghana, along a 70 km-long narrow sandbar extending from the border with the Republic of Togo in the east to the Volta river estuary in the west. The erosion problem along the sandbar is aggravated by the fact that a series of lagoons immediately to the north of the sandbar are also subject to extensive periodic flooding. On the bank of the Keta Lagoon, the largest of these lagoons, is the town of Keta which is estimated, from oral historical accounts, to have lost over 4 km of land through active coastal erosion since 1860, although records of high tide contours carefully kept since 1907, show that the maximum land loss through coastal erosion within the past eighty years probably does not exceed 1 km in the Keta area. Presently, a strip of land, approximately 150 km wide, separates the sea from the lagoon at Kedzi, the narrowest point along the sandbar, about 3 km to the east of Keta.

Water and oil exploration activities in the basin since the early 1950's and middle 1960's respectively, have helped to establish the geology of the area. Substantial thicknesses of sand, gravel and clay of Quaternary age, overlie undifferentiated basement rocks that are suspected to be the Dahomeyan gneisses and schists which outcrop to the north of the basin.

The relatively unfavourable geology of the area has made the design of effective coastal defense measures difficult. This paper presents a review of previous unsuccessful attempts to protect the Keta coastline, dating back to the early 1950's, within the context of the geology and geotechnology of the area.

Introduction

The Keta Basin, the largest of the coastal sedimentary basins of Ghana, (see Fig. 1) occupies the extreme south-eastern corner of Ghana and is bounded on the west by the Volta river estuary, on the east and south by the Gulf of Guinea and on the north by the international highway connecting Accra and Lome, the capital cities of Ghana and Togo respectively. The basin, which, according to Ajayi (1980) and Kesse (1985), represents the westernmost extremity of a coastal sedimentary basin extending westward from the Niger Delta, covers a total area of 2200 km^2, of which about 15% is made up of creeks and lagoons, the biggest of which, the Keta Lagoon, is aproximately 50 km long and 20 km wide.

While four main rivers, including the Volta, and numerous other streams drain into the Keta basin, the absence of any major natural outlet from the basin, coupled with the relatively flat topography and the poor natural drainage characteristics, have led to periodic flooding of the towns, villages and agricultural lands within the basin. As a result,

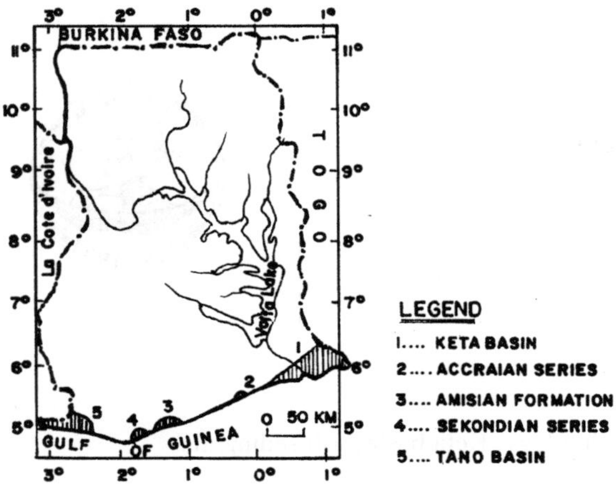

Fig. 1. Location of coastal sedimentary basins in Ghana.

most of the population of about a quarter of a million (1984 census) occupies a narrow sandbar – about 70 km long, 6 km at its widest portion and 150 m wide at its narrowest – which forms the southern boundary of the basin, resulting in population densities which are about three times the national average. Unfortunately the sandbar is also subject to active coastal erosion in places, particularly in the vicinity of Keta, the largest town in the area. The plight of the inhabitants of the area is compounded by the fact that both the coastal erosion and the inundation problems often occur within the same period, namely from July to November.

This paper discusses previous attempts at, and recommendation relating to, finding solutions to the problem of coastal erosion along the sandbar forming the southern boundary of the Keta basin within the context of the geology and geotechnology of the area. Detailed discussion of the flooding problem is considered outside the scope of this paper. For a detailed discussion on the problem and the various solutions proposed, reference may be made to the reports of three independent studies of the problem (Anon, 1968; Anon, 1970; Anon, 1982).

Geology of the Keta basin

Although it is one of the major sedimentary basins of West Africa, and in spite of its hydrocarbon potential, the geology of the Keta basin has not been systematically studied until the early 1950's when Cox, (1952), following studies of cores obtained from water supply boreholes, established the existence of Eocene and Cretaceous sediments within the basin. Subsequent activities relating to hydrocarbon prospecting within the basin, beginning in the mid-1960's, facilitated a reconstruction of the geologic structure of the Keta basin by Khan (1970) and Akpati (1975) and culminated in the production of what is probably the first geologic map of the basin, shown in Fig. 2, with a typical North-South geologic section given in Fig. 3.

Based on the study of the logs of two oil wells drilled at Atiavi (depth 1560 m) and Anloga (depth 2135 m), Khan (1970) proposed the stratigraphic sequence shown in Table 1, for the basin. This sequence was essentially also confirmed by Akpati (1975) who, in discussing the geologic structure and evolution of the basin, based on evidence derived from analysis of the results of gravity surveys and well data, concluded that 'basement faulting and subsidencce have played a major part in the evolution of the Keta basin'. Although the basement rocks within the basin have not been positively identified, both Akpati (1975) and Kesse (1985) believe that the basin is also underlain by the acid and basic gneissses and schists of the Dahomeyan (Lower Precambrian) system which outcrop on the northern fringes of the basin.

The tectonic structure of the Keta basin has also been discussed in some detail by Kirton (1967), Khan (1970) and Akpati (1975), who hold that the NE-SW trending fault forming the northern boundary of the basin extends off-shore. However, validity of attempts by Mascle & Sibuet (1974) and Francheteau & Le Pichon (1972) to associate this fault with the Romanche Fracture Zone (Arens et al., 1971) has been questioned by Blundell (1976).

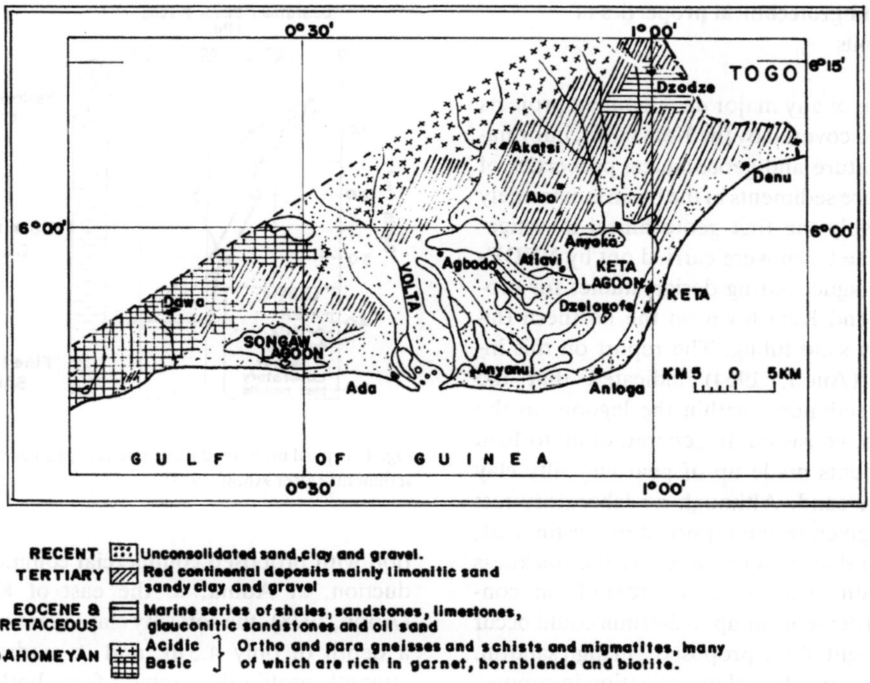

Fig. 2. Geologic map of the Keta basin (after Akpati, 1975).

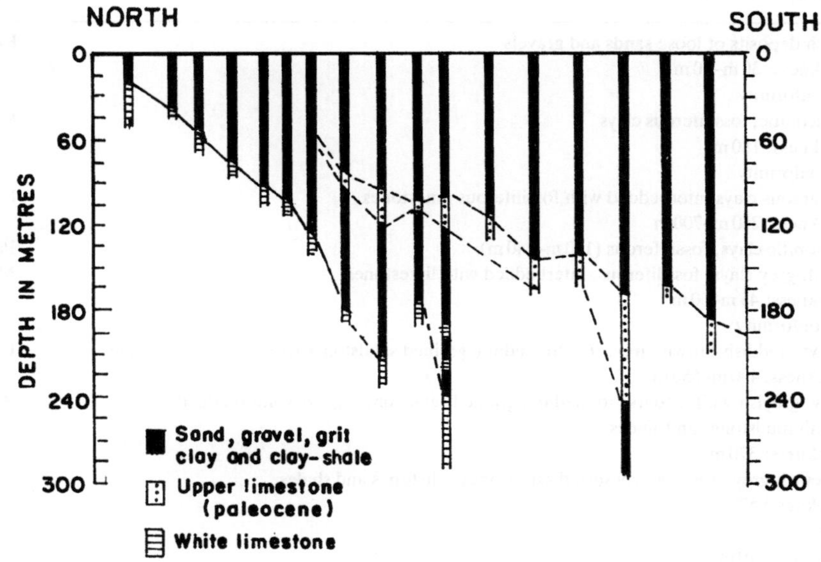

Fig. 3. Typical N-S stratigraphic section across the Keta lagoon (after Akpati, 1975).

The nature and geotechnical properties of the surficial soils

In the absence of any major development projects within the area covered by the Keta basin, information on the nature and geotechnical properties of the near-surface sediments in the basin is extremely scanty. Probably the first geotechnical investigations within the basin were carried out by NEDECO of The Hague, during design studies for proposals to extend Keta town on the northern (lagoon) side by sand-filling. The report on this investigationn (Anon, 1964) indicated that the near-surface sediments within the lagoon, to the north of Keta, consisted, in general, of up to 10 m of soft sediments made up of peat and silty clay overlying fine sand. Although no laboratory test results were given in the report, it was estimated, (Anon, 1964) that, in an area where the thickness of the soft sediments was of the order of 3 m, consolidation settlements of up to 300 mm could occur under the weight of the proposed 2 m high sandfill.

In another geotechnical investigation in connection with proposed commercial common salt production, at Adina, to the east of Keta Town, (Anon, 1978), the soft silty clay layer was proved to a depth of only 4.5 m, and its undrained shear strength profile determined from both field vane

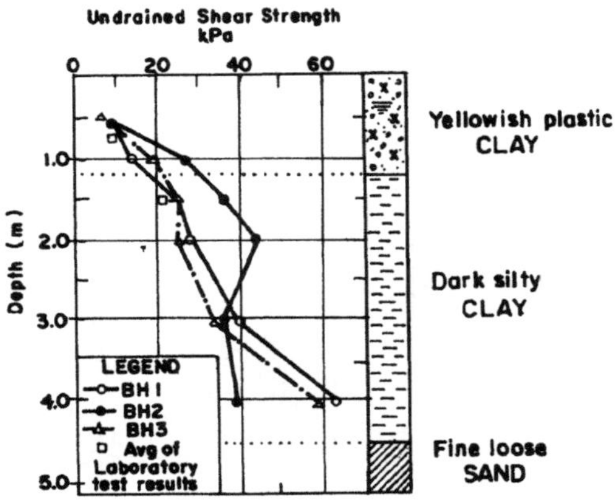

Fig. 4. Typical undrained shear strength profiles of near-surface sediments (after Anon, 1978).

Table 1. Stratigraphic sequence for the Keta basin (after Khan, 1970).

Unit No.	Lithology & Thickness	Age
I	Beach deposits of loose sands and gravels Thickness: 30 m–60 m Unconformity	Recent
II	Glauconitic, fossiliferous clays Thickness: 180 m Unconformity	Miocene
III	Calcareous clays interbedded with fossiliferous limestones Thickness: 250 m–700 m	Eocene
IV	Bentonitic clays, fossiliferous (120 m–240 m)	Paleocene
V	Bluish-grey clays, fossiliferous, interbedded with limestones Thickness: 45 m–60 m Unconformity	Maestrichtian
VI	Brown, reddish-brown, grey, fine to medium grained sandstones with subordinate shales Thickness: 400 m–550 m	Campanian
VII	Grey, greyish-white, coarse to medium grained sandstones, gravels interbedded with mudstones and shales Thickness: 370 m	Albian
VIII	Greenish-grey, grey, poorly sorted sandstones; siltstones and shales Thickness: 579 m Dolerite Thickness: 70 m	Aptian
IX	Dark grey, Micaceous, often varve-like shales and siltstones, fossiliferous Thickness: 610 m	Devonian

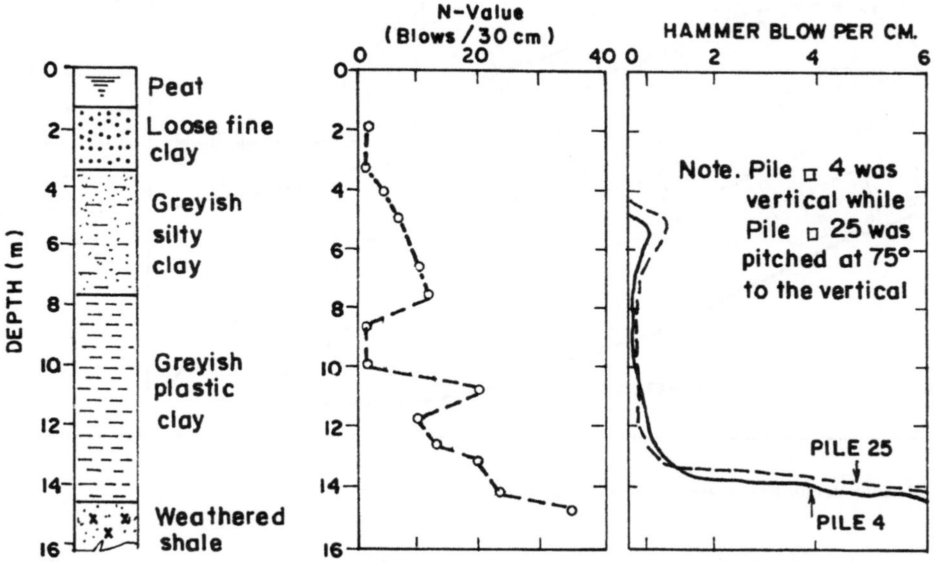

Fig. 5. Typical variation of SPT blow-count and pile penetration resistance with depth (after Andam, 1987).

and laboratory tests is given in Fig. 4.

Perhaps the most detailed geotechnical investigations within the Keta basin were carried out at Srogboe to the west of Keta, in connection with the design of a 30 m-long single span reinforced-concrete road bridge across the only permanent drainage channel linking the Keta Lagoon to the Volta river estuary. This investigation, which involved the sinking of five boreholes to depths of about 20 m, showed that the fine sand layer was only 7 m thick at this site and is underlain to a depth of 20 m by clay and weathered shale. The standard penetration blow-counts (N-values) within the sand layer never exceeded two within the top 3 m and were generally less than ten. A typical borehole log and other pertinent information on the SPT blow-counts and the penetration resistance of two 350 mm square precast concrete piles are given in Fig. 5. The problems encountered during the construction of the pile foundations of this bridge have been discussed by Andam (1987).

The soils underlying the Keta basin may therefore be generally characterised as soft, highly compressible organic or inorganic clays overlying fine sand to great depth. The water table within the basin is usually close to or above the ground surface, and generally fluctuates in sympathy with the mean sea level, especially along the sandbar. Similar sub-soil conditions have also been reported for the Niger Delta by Ajayi (1980) and Balogun (1980) and for the continental shelf along the Gulf of Guinea by Sullivan & Squire (1980).

The impact of the geology and geotechnology of the Keta basin on the past and present attempts at finding solutions to the problem of coastal erosion in the area will be examined in the following sections. Basically the aspects of the geology of the basin which are of relevance to the solution of the coastal erosion problem are:

a) the fact that the near-surface sediments of the basin consist of between 30 m and 60 m of beach deposits comprising mainly loose sands and gravels and that these deposits increase in thicknesses to as much as 140 m in the Keta area where the coastal erosion problem is most serious.
b) the complete absence of competent rock outcrops within the basin.

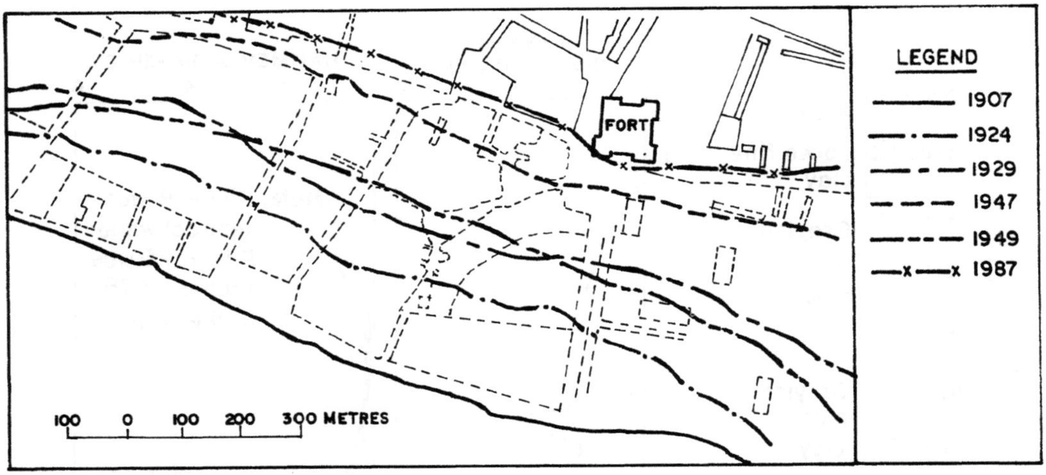

Fig. 6. High tide contours for Keta town for the period 1907–1987.

Coastal erosion along the sand bar

The problem of coastal erosion along the sandbar forming the southern boundary of the Keta basin and previous attempts at coping with it have been exhaustively reviewed by Anon (1976). Even though the problem occurs to varying degrees along the whole sandbar – and, indeed, along most of the coastline of Ghana – it is most serious along a 9 km stretch of coastline extending from Dzelukope to Kedzi, particularly along the shoreline fronting the central part of Keta town which also happens to be located at the narrowest part of the sandbar, and which, before the construction of a deep sea port in Tema in 1955, was one of the main ports along the Ghana coast.

Engineering studies have been carried out by various experts, including Coode (1929), Batley (1950), Anon (1964) and Anon (1976), since the problem was first identified in 1907. Accounts of the actual extent of coastal recession in front of Keta town vary widely from as much as 4 km since 1860, based on oral historical sources, to 240–300 m between 1907 and 1929 (Coode, 1929). Most of the forts built along the Ghana coast in the 17th and 18th centuries were located either on high ground for obvious military advantage or as close to the sea as possible to facilitate easy loading of slaves on to surf boats transporting them to ships anchored off-shore. Given the generally flat topography of the Keta area, there would appear to be no major military advantage to the Danes in constructing Fort Prinzenstein anywhere else except on the Keta coast in 1782, particularly in a period when the construction of such a massive structure could only be carried out using human labour. It would appear logical to assume, therefore, that the coastline in the Keta area in 1782 must have been close to the present location of the fort. A similar opinion was expressed by Batley (1950), who reported that he was informed by an 80-year old chief that, in his youth, the fort was so close to the sea that slaves were loaded directly from it on to boats. While the estimate of about 4 km of coastal recession since 1860 may be exaggerated, it is believed that the large discrepancy between this and Coode's (1929) apparently more reliable estimate of the extent of coastal erosion may be attributed, in part, to the possibility of significant coastal accretion in the Keta area between about 1860 and 1907. Indeed, records of high tide contours between 1907 and 1987 (Fig. 6) showed evidence of substantial coastal accretion between 1947 and 1949 and also indicated that a more realistic estimate of the maximum coastal recession in front of Keta town since the mid-1800's is probably close to 1 km. The pre-

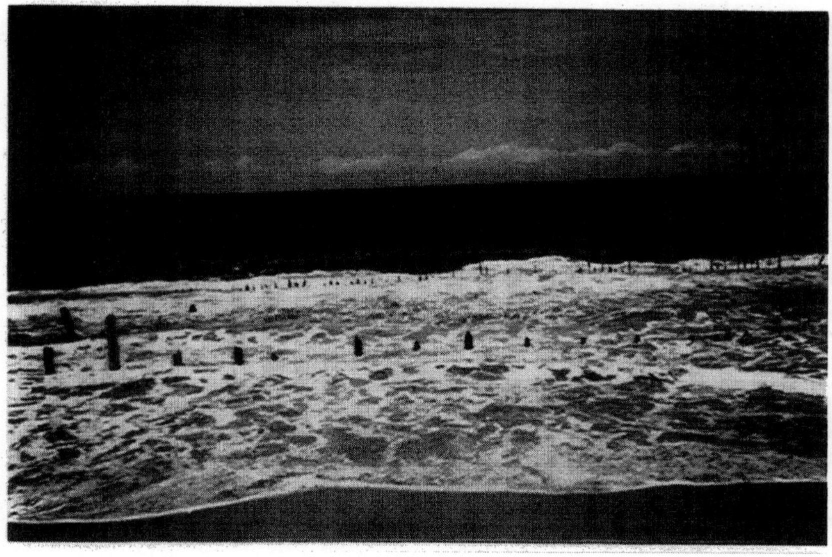

Fig. 7. View showing remnants of previous coastal protection walls in the area fronting Keta town (April 1987).

Fig. 8. View showing the onset of destruction of fort Prinzenstein (April 1987).

sent situation in the vicinity of the fort is depicted in Figs 7, 8 and 9 which show the onset of the destruction of the fort and the extent of property damage.

Remedial measures

The possible mechanisms responsible for the active coastal erosion along sections of the sandbar, in the

Fig. 9. General view showing destruction of property in Keta town (April 1987).

vicinity of Keta town, in particular, have been discussed in some detail during a study of the problem by the Delft Hydraulics Laboratory of the Netherlands (Anon, 1980), while previous attempts made towards finding an engineering solution to the problem are discussed by Anon (1976) and summarised in Table 2. These attempts have however been hampered by uncertainties regarding the predominant mechanism responsible for the morphology of the coastline of the Keta Basin, particularly the influence of sub-marine canyons, one of which is located off Cape Saint Paul (Anon, 1976). Results of previous studies (Anon, 1980) have led to the conclusion that onshore-offshore sediment transport, rather than longshore transport probably controls erosion and accretion phenomena along the coastal areas of the basin.

On the basis of the available information, the following options have been identified (Anon, 1980) as probably the most cost-effective coastal protection measures for the Keta area:

- the construction of experimental groynes in the most seriously affected areas, to be followed by a field study of their effectiveness before a final decision is taken about expanding the system to the estimated number of 30 groynes required to protect the sections of Keta town under active erosion
- sand nourishment of the beach using sand reclaimed from the lagoon near Keta
- the construction of a longshore rubble mound beach wall.

The rubble mound solution had also been suggested earlier as an alternative to the use of sea gabions by Anon (1976) after a critical review of the available options on remedial measures, given the relatively unfavourable foundation conditions prevailing in Keta, especially the absence of competent formation close to the surface.

As noted earlier, the Keta basin itself is completely devoid of any rock outcrops. Consequently rock for the construction of either the experimental groyne system or the rubble mound will have to be quarried from an area north of the basin where the Dahomeyan basement rocks outcrop and where suitable sources of quartz hornblende gneiss and granite have been located. However, based on the existing road system in the area, exploitation of these rock outcrops will involve haul distances of between 50 km and 90 km, thereby adversely af-

Table 2. Previous remedial measures carried out for coastal protection at Keta (Modified after Anon, 1976).

Period	Description of work	Remarks
1949–1950	Experimental groyning with local materials such as coconut tree and palm tree trunks.	Effectiveness of installation was difficult to assess since installation was made during a period of general accretion in the Keta area. Material too light and unsuitable for the wave action experienced in the area; installation destroyed.
1951–1957	Extensive heavier groyning system, using heavy constructional timber.	Undermined by heavy wave action and destroyed.
1957–1962	Steel sheet piling to cover a frontage of about 1800 m.	Partially successful in protecting Keta town; within 3 to 4 years of its completion, the protection failed; two miles of coastal stretch east of Keta town was seriously eroded resulting in heavy loss of houses and property. Due to extensive corrosion and impact of breaking waves, steel sheet piles gave way in many places, and installation may be considered as having failed far ahead of its design life of 40 years.
1971 & 1973	Temporary timber revetment works to protect a road length of about 300 m or so.	In 1971, temporary timber revetment works, supported by sand bags saved the situation and maintained the road link of vital importance between Keta and Denu. In 1973, however, the heavy wave action resulted in the need to divert the road leading to demolishing of school and residential buildings.
1975	Temporary stone revetment to protect the affected coastal stretch at Abutiakofe, a suburb of Keta to the west of the Fort.	An effective temporary measure feasible under the prevailing circumstances to prevent further intrusion of sea. Construction completed in August 1975. It has so far protected the coastal stretch at Abutiakofe, Keta.

fecting the economics of the proposed project. The construction of a shorter and more direct haul road across the Keta Lagoon from Abor to Keta is under active consideration, and its feasibility is currently being evaluated through geotechnical investigations along the proposed alignment.

The sand nourishment solution has been previously considered and rejected as unfeasible by Anon (1976) and also in a different context by Anon (1964) in an investigation relating to the development of a flood relief and resettlement plan for Keta. While it has been established that the sand in the lagoon close to Keta was of grading suitable for dredging and pumping for sand filling purposes, this particular option is not highly favoured because of the difficulty of obtaining sufficient sand from the lagoon to initiate and sustain the process. The possibility of dredging sand from the sea for the purpose has not been seriously considered in the past but is currently under active consideration as one of the viable options.

Concluding remarks

The failure of previous attempts at finding engineering solutions to the problems of coastal erosion in the Keta area may be attributed, in part, to the unfavourable geology of the area, although the difficulty of identifying the predominant mechanism responsible for sediment transport along the sandbar has also meant that these attempts at coastal protection have not been based on sound engineering principles. While geological investigations and drilling in connection with prospecting for hydrocarbons and water in the area have facilitated the reconstruction of a fairly accurate geology of the basin, available information is still too scanty to permit a reasonably accurate geotechnical characterisation of the near-surface deposits. More detailed knowledge of the engineering properties of these deposits and of the predominant sediment transport mechanism will increase the available options on coastal protection in the Keta area, as well as facilitate the proper design of associated flood control structures such as the proposed dyke and the Kedzi outfall as part of an integrated engineering solution to the twin problems of flooding and coastal erosion in the Keta basin.

References

Ajayi, L.A. 1980 Geotechnical properties of a deep organic clay stratum underlying Lagos area of Nigeria – Proc. 7th Reg. Conf. for Africa on Soil Mech. and Found. Eng. 1, Accra, June 1980: 75–82.

Akpati, B.N. 1975 Geological structure and evolution of Keta Basin – Rept No. 75/3. Geol. Surv. Ghana, Accra, Aug. 1975: 31 pp.

Andam, K.A. 1987 Non-destructive pile load test in a sedimentary basin – Proc. Int. Conf. on Found. and Tunnels, London, March 1987: 8 pp.

Anonymous 1964 Flood relief and resettlement plan for Keta Town, Chana – NEDECO Rept., The Hague, Holand, November 1964: 27 pp.

Anonymous 1968 Feasibility study for the Avu-Keta Project – WAKUTI Rept., West Germany, December 1968: 86 pp.

Anonymous 1970 Flood control in the Avu-Keta, Accra – Nathan Consortium for Sector Studies Rept., Accra, Feb. 1970: 26 pp.

Anonymous 1976 Coastal erosion and proposed protection works at Keta – Rept. CE/2, 1, Hydrological Div. Architect. Eng. Serv. Corp. (AESC) Accra, 1976: 20 pp.

Anonymous 1978 Sub-soil investigation report on the proposed Adina Salt Project – Unpubl. Rept. Earth Science Consultancy, Accra, 1978: 6 pp.

Anonymous 1980 Coastal erosion and protective measures at Keta: Report on Investigation – Rept. R1255, Delft Hydraulics Laboratory, May 1980: 30 pp.

Anonymous 1982 Outfall Keta Lagoon: Report on model investigation – Rept. M1612, Delft Hydraulics Laboratory, Feb. 1982: 60 pp.

Arens, G., Delteil, J.R., Valery, P., Damotte, B., Montabert, L. & P. Patriat 1971 The Continental margin off the Ivory Coast and Ghana. In: F.M. Delany (ed.): The Geology of the East Atlantic Continental Margin. – Rept. 70/16, Inst. Geol. Sci.: 65–78.

Balogun, L.A. 1980 Foundation design for some tall buildings in Lagos Area – Proc. 7th Reg. Conf. for Africa on Soil Mech. and Found. Eng. 2, Accra, June 1980: 627–649.

Batley, W. 1950 Keta coastal erosion – Rept. Public Works Dept. Gold Coast, (Ghana), Accra, Jan. 1950: 13 pp.

Blundell, D.J. 1976 Active faults in West Africa – Earth Planet. Sci. Lett. 31, 287–290.

Coode, A.T. 1929 Keta coast erosion – Rept. Messrs Coode, Wilson, Mitchell & Vaughan-Lee, Accra, 1929–30: 13 pp.

Cox, L.R. 1952 Cretaceous and Eocene fossils from the Gold Coast – Bull. Geol. Surv. Gold Coast 17, (Ghana), Accra: 68 pp.

Francheteau, J. & X, Le Pichon 1972 Marginal fracture zones as structural framework of continental margins in South Atlantic Ocean – Am. Assoc. Petrol. Geol. Bull. 56: 991–993.

Kesse, G.O. 1985 The mineral and rock resources of Ghana – A.A. Balkema Publishers, Rotterdam: 610 pp.

Khan, M.H. 1970 Cretaceous and Tertiary rocks of Ghana, with historical account of oil exploration – Bull. Geol. Survey of Ghana, 40, Accra, 1970: 55 pp.

Kirton, M. 1967 Exploration of petroleum and gas in Ghana – Unpubl. Rept. Geol. Survey of Ghana, Accra: 24 pp.

Mascle, J. & J.C. Sibuet 1974 New pole for early opening of South Atlantic – Nature 252: 464–465.

Sullivan, R.A. & J.M. Squire 1980 Geotechnical properties of West and North African continental shelf sediments – Proc. 7th Reg. Conf. for Africa on Soil Mech. and Found. Eng. 1, Accra, June 1980: 43–54.

Proceedings KNGMG Symposium 'Coastal Lowlands, Geology and Geotechnology', 1987: 321–335 (1989)

The geotechnics of the Coastal Lowlands of the United Arab Emirates

P.M. Maurenbrecher[1] & M. Van der Harst[2]
Technical University Delft, P.O. Box 5028, 2600 GA Delft, [1] Faculty of Mining Engineering, The Netherlands; [2] Fugro Geotechnical Engineers, P.O. Box 63, 2260 AB Leidschendam, The Netherlands

Received 2 September 1987

Key words: Coast, engineering geology, geotechnics, nearshore, penetration tests, soil profiles, United Arab Emirates

Abstract

The coastal region of the United Arab Emirates has been investigated extensively during the building boom between 1975 and 1985. This paper makes use of the data accumulated during that period by classifying geotechnical ground profiles prevalent along the coastline of the U.A.E. Correlations are given for these profiles between the Cone Penetration Test (CPT) and other geotechnical parameters such as the Standard Penetration Test (SPT) and laboratory tests.

The coastline of the U.A.E. is influenced by the topography of the coastal plain and by the bathymetry and geography of the Arabian Gulf. Sand dunes, apparently derived from the Rub al Khali, owe their morphology to ancient weather patterns. They extend from the mountain outwash plains to the coast, except for sections where either present and ancient drainage systems or coastal accretion give rise to sabkha (salina) tidal flat and flood plain formation. The high salt and sulphate content of sabkhas together with underlying weakly cemented detrital and aeolian calcarenites, often with solution cavities, create special geotechnical problems.

Case histories are given of investigations for structures and utilities often situated on the sabkha plains and extending out to sea. This paper outlines typical site investigation procedures and techniques that are carried out for this region.

Introduction

With the exception of one town the principal centres of the United Arab Emirates are all situated on the coastline; along the west coast (Arabian Gulf) Ras al Khaimah, Umm Al Quwain, Ajman-Sharjah-Dubai-Jebel Ali, Abu Dhabi, Ruweis; on the east coast (Gulf of Oman) Dibba, Khor Fakkan and Fujeirah (Fig. 1). Hydro-carbon discoveries have led to considerable development of these centres from small compact towns to major cities, vastly exceeding the urban limits of only twenty years ago. All west coast centres are situated on tidal creeks and hence much of the construction occurs on recent infilled tidal flats and on landfill. At Ajman, Umm al Quwain and Ras Al Khaima the tidal inlets form estuaries for wadis that cross a barrier of desert dunes covering the area between the coast and the outwash plains of the Omani mountains to the east. In other places remnants of such estuaries occur as 'tidal creeks' at Dubai and Sharjah, whereas at Abu Dhabi accretional tidal

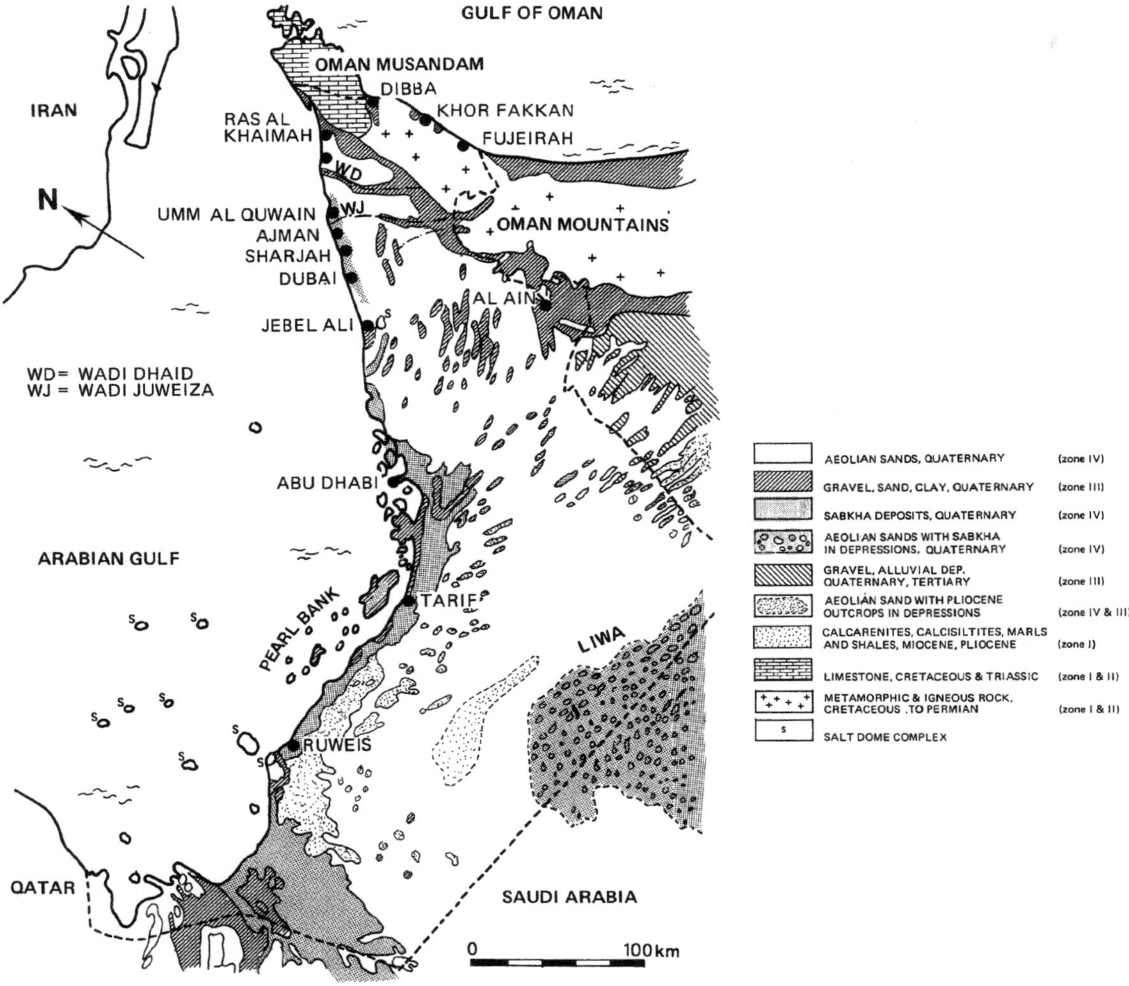

Fig. 1. Drift geology of the United Arab Emirates from USGS & ARAMCO (1963).

flats were built up behind nearshore core islands (Purser & Evans 1973). They are thus better protected from the erosive fetch of the NW Shamal winds to which the more northerly coastlines are exposed.

Elsewhere development has occurred essentially for the oilfield areas, both located offshore, near coastal areas and on the outwash planes of the Omani mountains and the sand sea areas extending inland from the sabhka flats of southern Abu Dhabi towards the Rub al Khali. The oilfields required the development of a number of islands such as Das, Arzanah and Delma as well as coastal areas at Ruweis (Abu Dhabi), Jebel Ali (Dubai), Hamriyah (Sharjah) and recently at Fujeirah to ensure an oil outlet on the geographically more favourable eastcoast. Significantly, the oil wealth has spurred agricultural development inland, on the outwash plains of the Omani mountains at Ras Al Khaimah, Dhaid (Emirate of Sharjah), Al Ain /Buraimi Oasis (Emirate of Abu Dhabi and Sultanate of Oman) and on the Rub Al Khali's deflation sabkha plains northerly limit, the Liwa oasis complex.

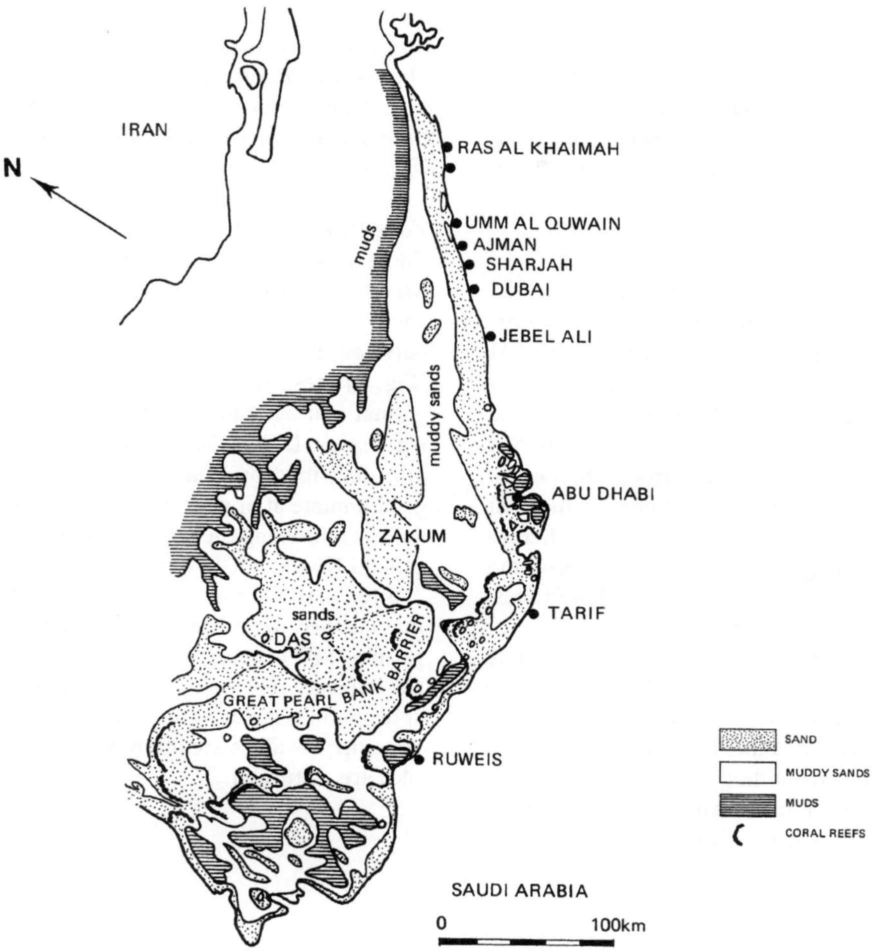

Fig. 2. Drift seabed geology of Arabian Gulf section of the United Arab Emirates based on Dennis (1978) and Wagner & Van der Togt (1973).

This paper concentrates on the west coast centres where thanks to rapid development the low-lying areas have special geotechnical problems that are associated with tidal sabhka flats and a rise in ground water levels.

Engineering geological setting

The United Arab Emirates and Sultanate of Oman occupy the northeastern end of the Arabian Peninsula. The U.A.E. western coastline extends 600 km from Qatar to Oman Musandam along the Arabian Gulf. The eastern coastline extends 75 km along the Gulf of Oman. The geology and the principal geography of the region are shown in Fig. 1, based on USGS & ARAMCO (1963) and in Fig. 2, based on Dennis' (1978) version of Wagner & Van der Togt's (1973) sediment distribution map, for the onshore and offshore parts respectively of the Arabian Gulf of the U.A.E.

Onshore region

The land soil profiles and the type of geotechnical investigation that can be carried out are best described in terms of physiographic 'Soil engineering zones' developed by Fookes & Knill (1969). They consist of four zones:

Zone I. Rock pediment

This zone forms coastal features in the drowned landscape of the Musandam Peninsula and as foothill spurs extending to the coast north of Ras Al Khaimah and along the east coast from Dibba to Qidfa (just south of Khor Fakkan, Fig. 1). The rocks are Permian to Cretaceous in age. North of a major fault zone that extends from Idn (40 km south of Ras Al Khaimah) to Dibba on the east coast the rock is predominantly limestone whereas south of the faulting there are mainly gabbros and peridotites with outcrops of Tertiary limestones, surrounded by the outwash plains.

In the western region of the U.A.E. carbonate rocks, marls and shales are found at or near the surface, of Miocene and Pliocene age. These outcrops form locally pediment features rising 40 to 50 m above the surrounding flat lands.

Site investigation in this zone is by rock coring. The highly weathered and fissured condition of the peridotites may require drilling using double tube core barrels with lining to ensure good recovery. The top weathering zone of the Miocene and Pliocene strata can be penetrated to a limited extent by cone penetration testing and by percussion drilling with tube sampling.

Zone II. Fan head deposits

These consist of gravels with sands and cobbles which among the east coast are often cemented into conglomerates. Where the coastal plane narrows between the Oman Mountains and the west coast (Fig. 3) these sediments are interbedded with lagoonal and aeolian deposits of Zone III and IV, whereas at Dubai they occur at depth (Fig. 4) as also described by Warren (1985). These deposits predominate at or near the surface in the mountain valleys and along the narrow coastal strip north of Ras Al Khaimah and along the east coast.

Site investigation is by percussion boring, often making use of chisselling. Standard penetration testing (SPT) is used to determine the strength of the deposits. The SPT driving head, consisting of a split spoon sampler, is usually substituted by a solid cone tip. N-values are often very high (>75). Samples are disturbed.

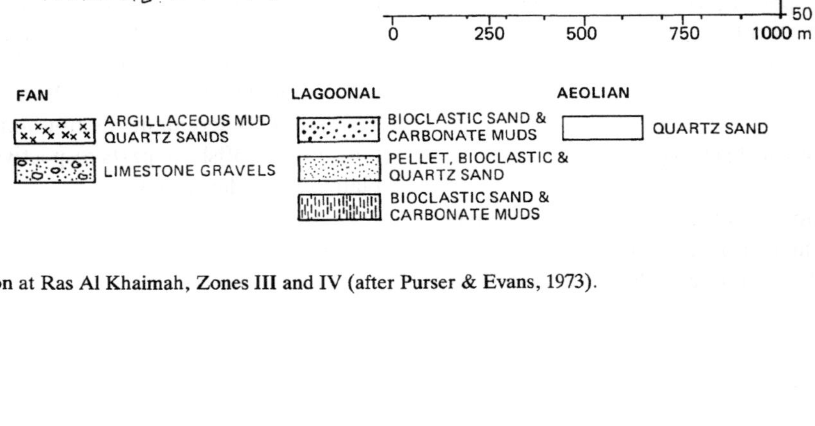

Fig. 3. Crossection at Ras Al Khaimah, Zones III and IV (after Purser & Evans, 1973).

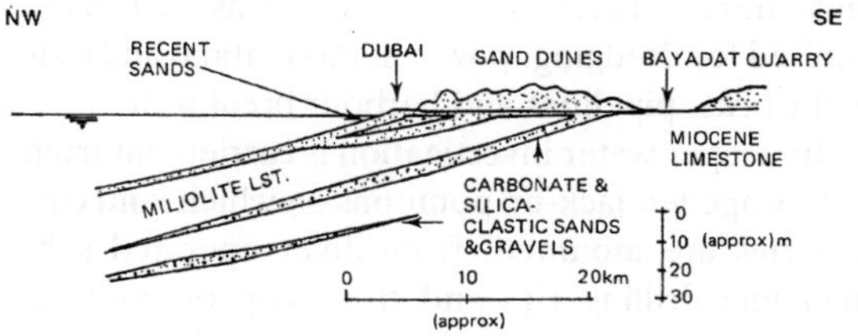

Fig. 4. Crossection at Dubai, Zones III and IV (after Tarrant, 1978).

Zone III. Alluvial plains

These consist of extensive outwash plains and extend as a narrow coastal strip between the Oman Mountain's fan head deposits and the coast southward from Qidfa. The coastal strip increases in width along the Batin coast of Oman, and forms a continuous green strip 1 to 3 km wide along the shore line. On the west coast the outwash plain extends up to 20 km from the Oman Mountains on its east flank to the sand dune belt on its western flank. The outwash plain surfaces are exposed within the sand dunes as elongated flat deflation plains, trending north-south in the northern Emirates and trending increasingly east-west in the southern Emirate of Abu Dhabi. The gravel deposits become finer in size with increasing distance from the mountains and are more and more interbedded with water and wind born dune sands.

Site investigation is by percussion boring with split spoon standard penetration testing (SPT), while samples are cut by either driven samplers or by coring (double tube core barrels). The zone can be further investigated by the static cone penetration test (CPT) though the test depth is limited between 2 and 7 m due to the strength of the soils.

Zone IV. Recent deposits

This zone forms the most extensive feature in the U.A.E. and consists of dunes and salt flats, playas and sabkhas. Fookes & Collis (1976, cf Fig. 5) subdivided this zone into subzones A to D which correspond with features along the west coast, as suggested by the Tarrant (1978) profile (Fig. 4).

The site investigation is similar to that of Zone III. CPT penetration depths up to 25 m have been attained in this zone.

Marine zone

Superficial sea bed deposits are shown in Fig. 2, a simplified version of Wagner & Van der Togt's (1973) map. The map serves as a useful framework to define coastal sedimentation processes as described by Purser & Evans (1973).

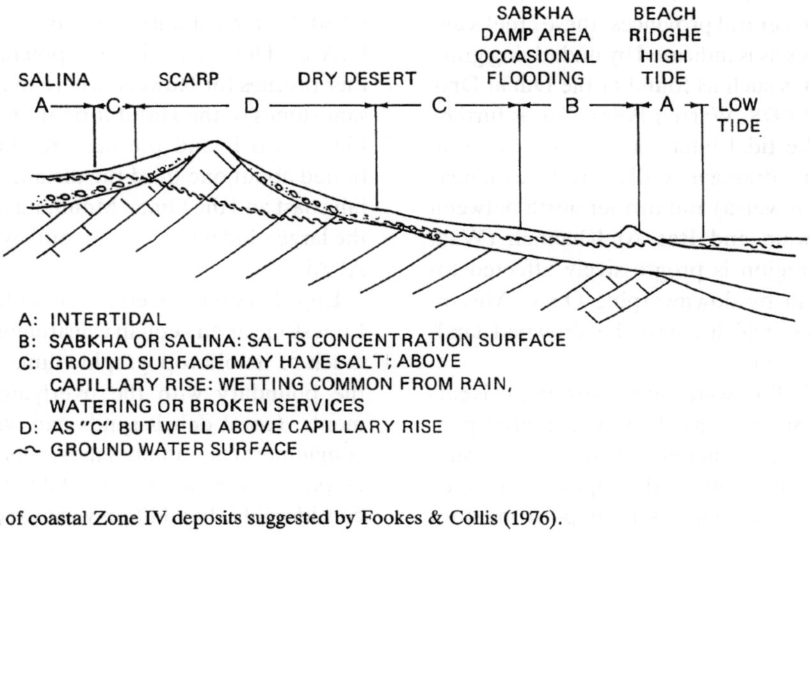

Fig. 5. Subdivision of coastal Zone IV deposits suggested by Fookes & Collis (1976).

The west coast can be divided into three sedimentological provinces, a western, a central and a northeastern province.

The principal influences affecting coastal deposition are the northwestern 'Shamal' winds and the 'fetch shadows' of the Qatar peninsula and the structurally high Pearl Bank.

The western province consists of lagoonal and inter-tidal flats which rapidly give way to a coast that is more open to longer fetch, such as at Jebel Dhanna/Ruweis.

In the central province the Pearl Bank axis rises towards Abu Dhabi and its obliqueness reduces the fetch with the coast towards the west. The increasing protection allows extensive accretional processes to develop such as coral banks and littoral sand banks/dunes and formation of inter-tidal flats and supra-tidal sabkhas. Such processes tend to follow a sequence and start with the growth of corals over relatively shallow Miliolite outcrops. In turn these features trap coral sand debris causing bars to form. Finer deposits settle behind the bars, whereas more coral debris accumulates on the seaward side. Abu Dhabi island was formed in this way.

The northeastern province is open to the complete fetch of the Arabian Gulf, causing lateral accretion and longshore drift which forms beaches and dunes. A section 75 km long extends from Ras Ghanada to Dubai which has no tidal inlets. Tidal inlets reappear from Dubai northwards. Unlike in the western and central provinces, these creeks are ancient estuaries as is indicated by underlying gravel conglomerates such as found at the Dubai Drydocks (Tarrant 1978, Warren 1985). This is further born out by the tidal inlets under the increasing influence of land drainage wadis as is the case near Ajman (Wadi Juweiza) and further north between Umm Al Quwain and Ras Al Khaimah (Wadi Dhaid). This region is progressively affected towards the north by downwarping Oman Musandam mountains, which cause the drowned landscape of this region.

Nearshore shallow water site investigation is carried out at its simplest by floating anchored pontoon using a wireline percussion rig. To measure strengths SPTs are made and sampling is done by wireline hammer. Coring is seldom performed for nearshore work, as investigative depths are usually limited for dredging, power station intake and outfall works, pipelines and harbour breakwaters.

In deeper water investigation is carried out from self propelled jack-up pontoons on which land coring rigs are mounted, from diver operated submersible drilling rigs and from ship or pontoon mounted rigs with wave compensating devices. Investigations are for the oil industry, such as for pipelines, piled jacket platform structures, oil-rig mounted jackup rigs, oil well conductor pipes and mooring terminals for tanker loading. These investigations are often preceded by shallow continuous seismic reflection surveys to determine best positions to drill the boreholes. Principal considerations, prior to investigations, are water depths, currents, seasonal weather conditions and the type of anchoring system for the investigation vessel. Often 'gravity' concrete block anchors placed by divers in oil fields are specified to avoid damaging pipelines.

Soil description profiles

Land

Lithological sequences shown in Figs. 3 and 4 represent conditions along the northern coast and the coast from Dubai southwards, respectively. More detailed general sequences are given by Epps (1980) for typical soil profiles of various areas in the U.A.E. These have been supplemented with further profiles for Ruweis, where the mudstones and sandstones of the Hofuf formation are exposed. In Fig. 6 two inland profiles are shown, one positioned about one third the distance from the Sharjah coast and the Oman Mountains and the other in the large sand dunes on the south west perimeter of Al Ain.

Epps (1980) referred to the underlying Miliolite Limestone as an aeolian calcarenite which retained much of its old topographic dune features so that the boundary with the overlying younger sediments, (coral debris sands and weakly cemented conglomerates, sabkha sands, silts and, sometimes clays, recent beach or inland dune sands) can vary considerably. In addition various sea level rises led

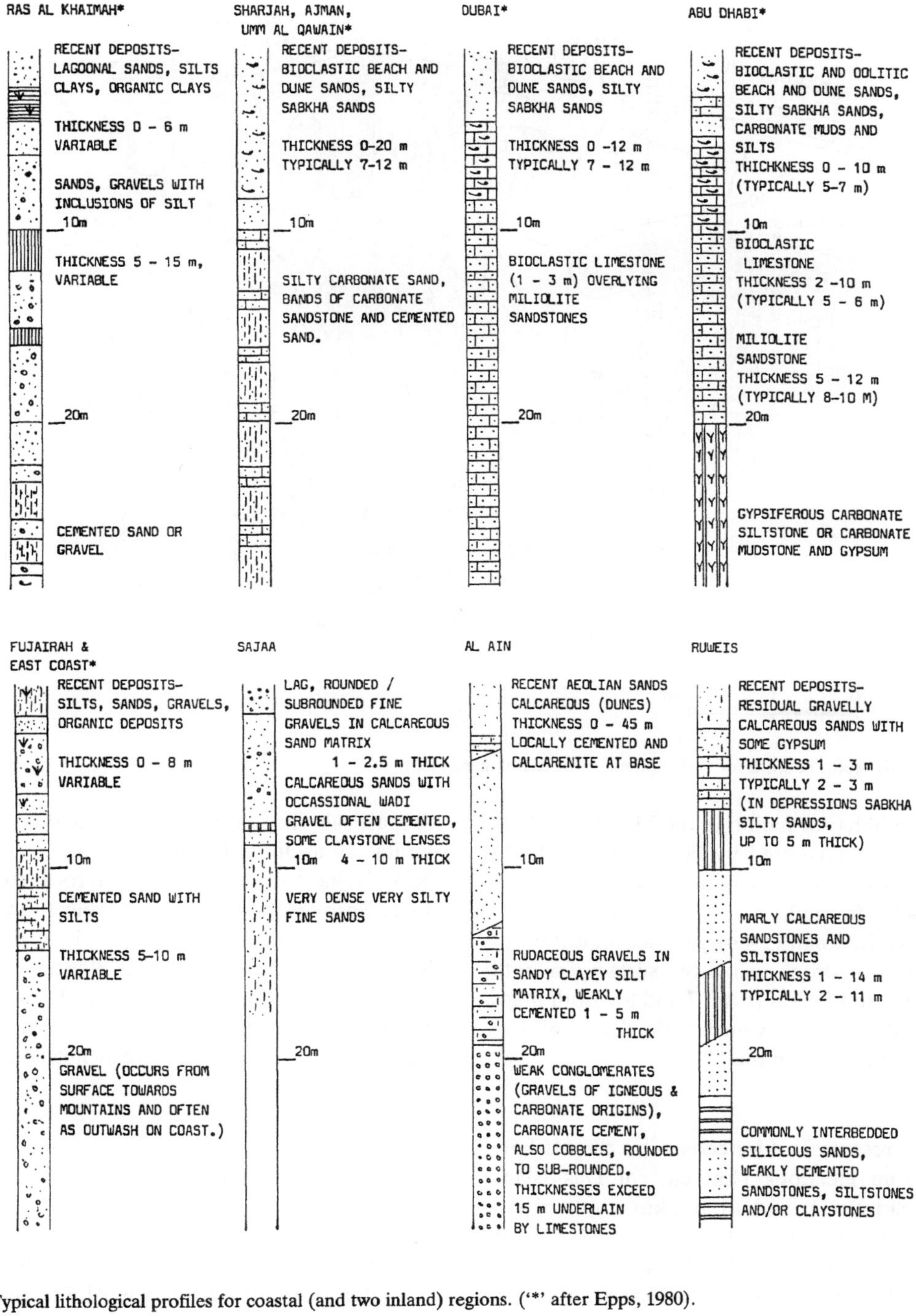

Fig. 6. Typical lithological profiles for coastal (and two inland) regions. ('*' after Epps, 1980).

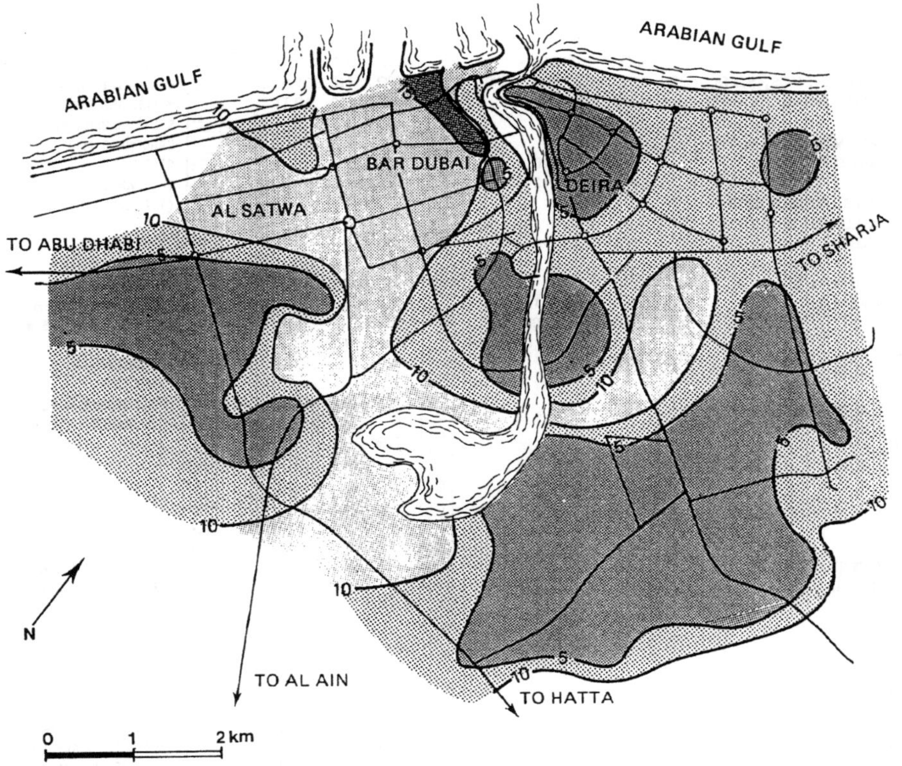

Fig. 7. Isopleth depths bedrock for Dubai (after Harris & Rahman, 1983).

to palaeosabkha formation in the Miliolites (Warren 1985). As suggested by Fookes et al. (1985) sometimes these palaeosabkhas resolved and thus cavities were formed.

Harris & Rahman (1984) gave an indication of the depth to bedrock below the surface for both Abu Dhabi and Dubai (Figs. 7 and 8). Both maps correspond reasonably well with the depths of penetration obtained with the CPT.

Marine

Figure 9 presents profiles chosen for the offshore region from locations 1 km east of Das Island (Broadhead 1971), and from 2 km south of Arzanah Island, nearshore Abu Al Abyadh (Tarif), from the Zakum Field, the SW Fateh Field, (± 15 km NNE of Sir Bu Nu'air, 100 km W of Dubai) and from nearshore Jebel Ali (Dubai).

The islands of Das, Arzanah and Sir Bu Nu'air are all salt diapirs which affected the overlying strata, as at Jebel Dhana. Another diapir, near Ruweis, has forced Archaen aged rocks to the surface. Despite this the strata are typically marine, in many instances exhibiting a cap rock, which may be a duricrust produced during the Pleistocene regressions. The duricrust can cause jack-up platform spud-can to punch through, resulting in a sudden tilt of the rig as the spud-can penetrates the less firm sands beneath.

Soil strength profiles

The CPT profiles give a good indication of the variation in strength for the coastal deposits. A selection of profiles from Ras Al Khaimah in the northeast to Ruweis in the southwest is given in Fig. 10 for the principal centres along the coast. All

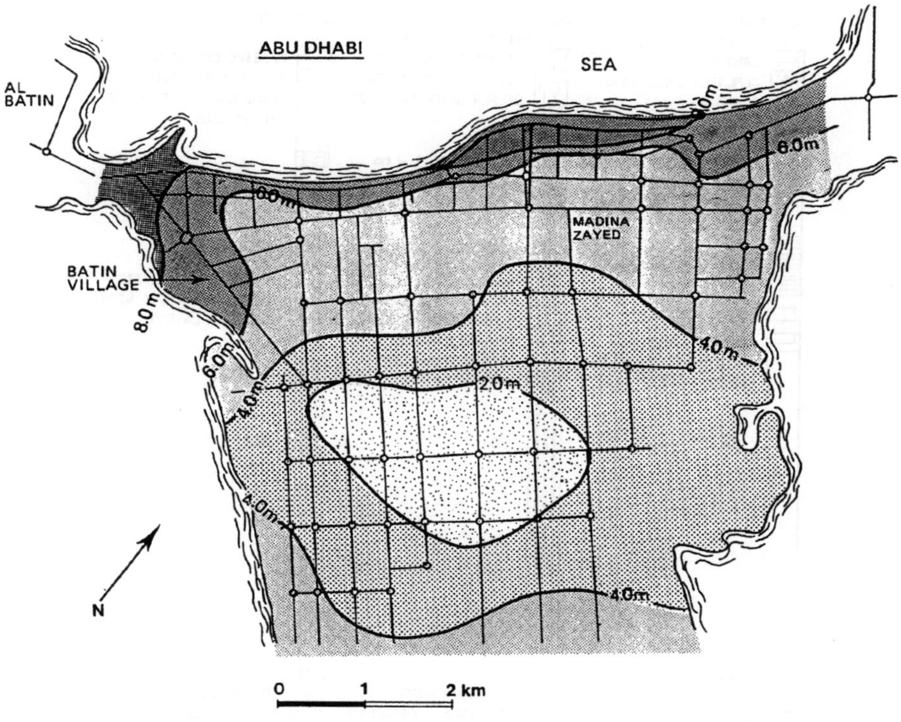

Fig. 8. Isopleth depths bedrock for Abu Dhabi (after Harris & Rahman, 1983).

profiles indicate sharp variations in density and cementation of the Zone IV deposits; many indicate a surface crust that is underlain by sometimes very weak sediments that are typical of sabkha formations. The local variation per profile can be substantial as shown in the example of a site in Sharjah (Fig. 11), where shallow dunes overlie sabkhas. Also there may be considerable variation in a particular horizon (coral debris sands) as shown in Fig. 12 for Abu Dhabi.

In a separate study comparisons are made between the SPT and CPT values. Results are given in Fig. 13 for typical sabkha deposits, five sites in Abu Dhabi and one in Ruweis. Although the correlation can be considered good by virtue of standard borehole sampling/ testing sequence, the SPTs fail to measure the strength of the surface crust and often miss relatively lower density layers such as the two SPT profiles from site AD2. In this instance a pile founded at 8 m moved excessively during its loading test, the cause indicated by the CPT, C-AD2, carried out subsequent to the boreholes with SPTs, S1-AD2 and S1-AD2, pre-construction investigations.

Chemical profiles

The combination of extremely hot weather (see Fig. 14 for Dubai) and frequent and extreme variations in daily humidities produces an environment suited to chemical attack of structures both below and above the surface. Lifespans of many buildings are no more than ten years. The buildings are often built to specifications suited for temperate climates and the building boom led to poor quality control. Steel sheet piling can rust by electrolysis within two years but instances of concrete rot in piles have not been recorded although they are often subjected to salt groundwater tidal fluctuations.

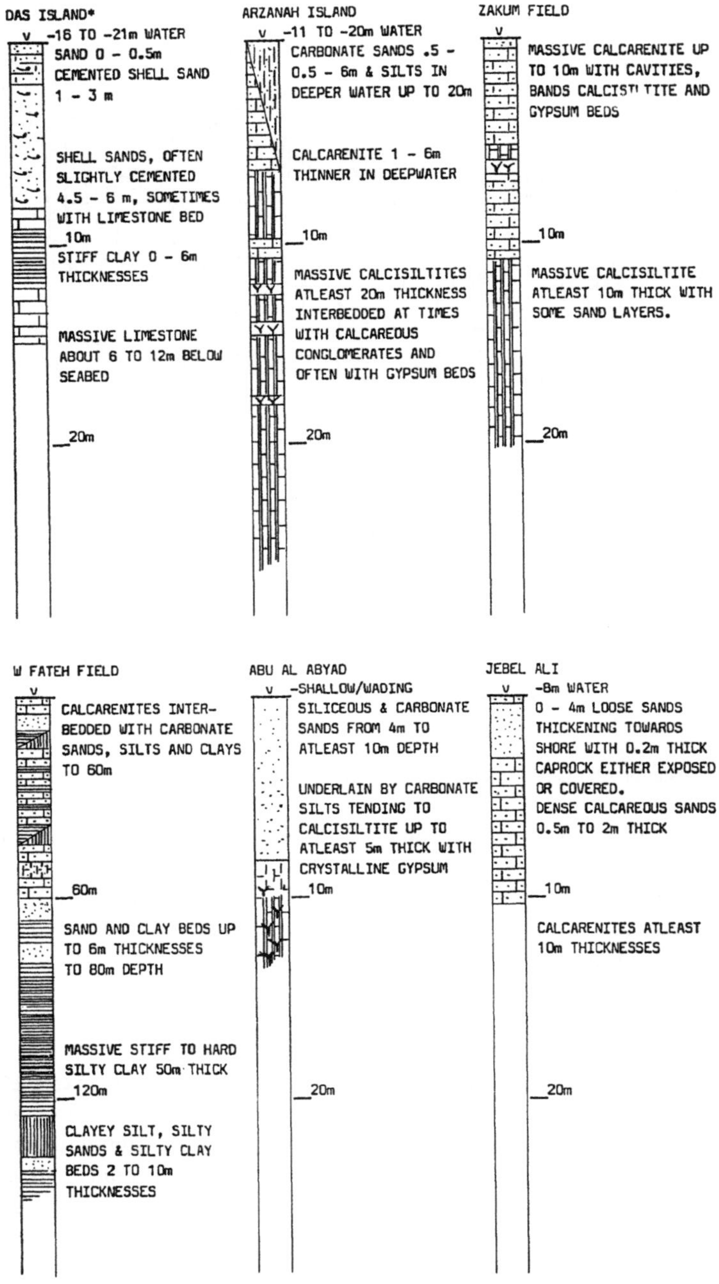

Fig. 9. Selection lithological seabed profiles offshore U.A.E. ('*' after Broadhead, 1971).

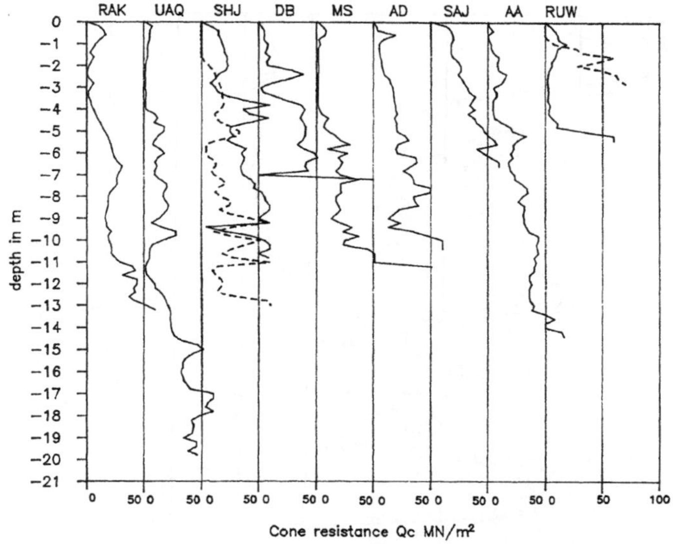

Fig. 10. CPT soil strength profiles along coast of U.A.E., (AA: Al Ain, AD: Abu Dhabi, DB: Dubai, MS: Musafa- near Abu Dhabi, RAK: Al Hamra Island- near Ras al Khaimah, RUW: Ruweis, SHJ: Sharjah, SAJ: Sajaa- between Sharjah and Dhaid, UAQ: Umm al Quwain).

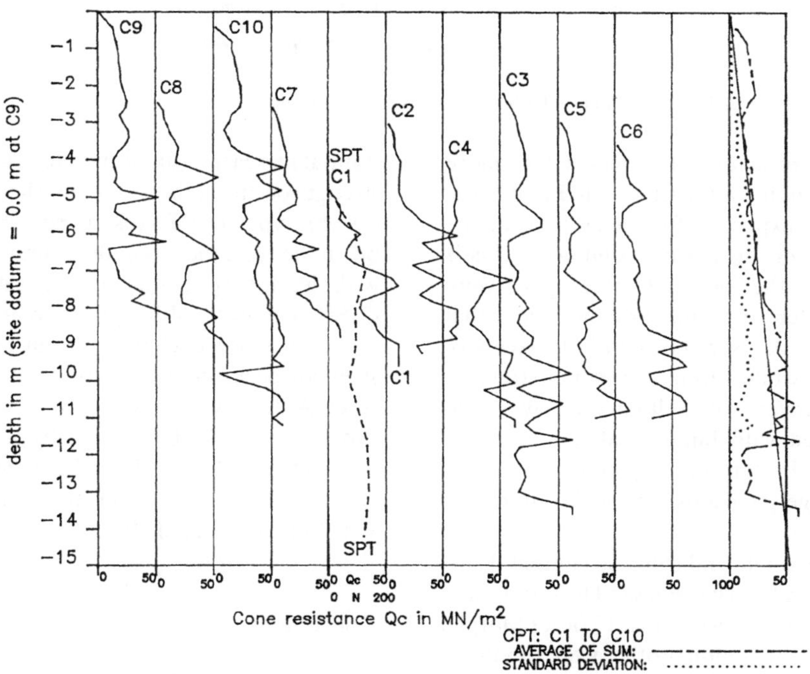

Fig. 11. Local variations in CPT soil strength profiles for a site in Sharjah, the total horizontal distance is about 200 m – not to scale.

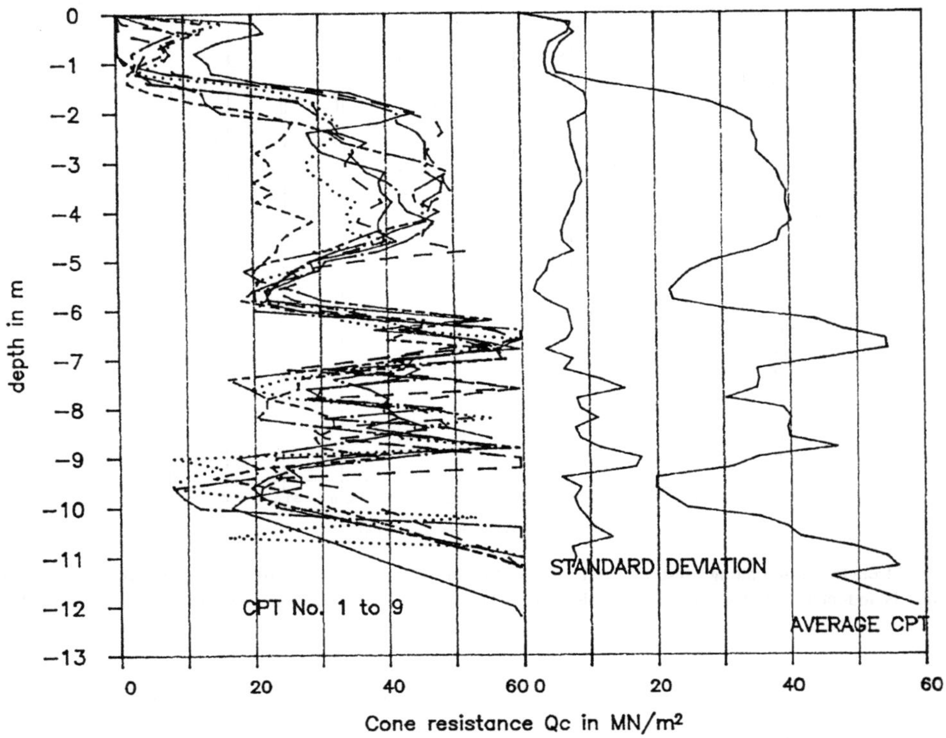

Fig. 12. Local variations in CPT soil strength profiles for a building site in Abu Dhabi.

One piling contractor, who switched to more lucrative demolition contracting, noted that foundation pile tops exposed after demolition of the superstructure were in perfect condition. This is probably due to the fact that driven piles were usually cast under strict quality control and that insitu concreted piles were not subject to excessive heat and drying while curing. The super structure concreting, however, with salt covered steel reinforcement mixed with impure sulphate contaminated water and often blazenly improperly compacted, was largely responsible for the short lifespan of many building.

Profiles for sulphate and chloride content distribution with depth from coastal sand pits, typical of the Gulf region, are shown in Fig. 15 (after Morgan 1982). Sabkha cementation profiles were simulated in the laboratory from different solution concentrations of calcium carbonate and calcium sulphate by Akili & Torrance (1981). They showed that substantial increases (about two- to four-fold) in CPT strength occurs at the water table boundary as is the case for typical CPT sabkha profiles for Abu Dhabi. In the tests the strengths continued to increase with depth despite a decrease in either $CaCO_3$ or $CaSO_4$ contents, although in some instances the rate of increase leveled of or even a limited drop in strength occurred below the peak concentration level. A study was made on changes in the chemical profile as a result of hydraulically placed marine fill at Jebel Ali. Restoration of a previously lowered water table resulted in profiles similar to that of Fig. 15 (Fookes et al. 1985).

Conclusions

The United Arab Emirates represent varied desert coastal lowland conditions from drowned landscapes to tidal accretional sabkha areas. The devel-

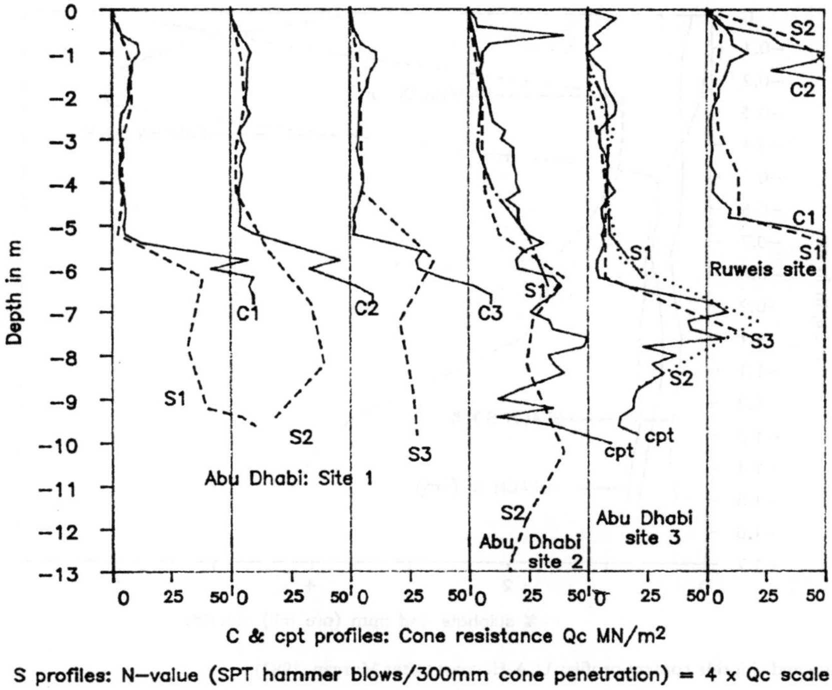

Fig. 13. Comparison CPT with down hole SPT profiles for sites in Abu Dhabi and Ruweis.

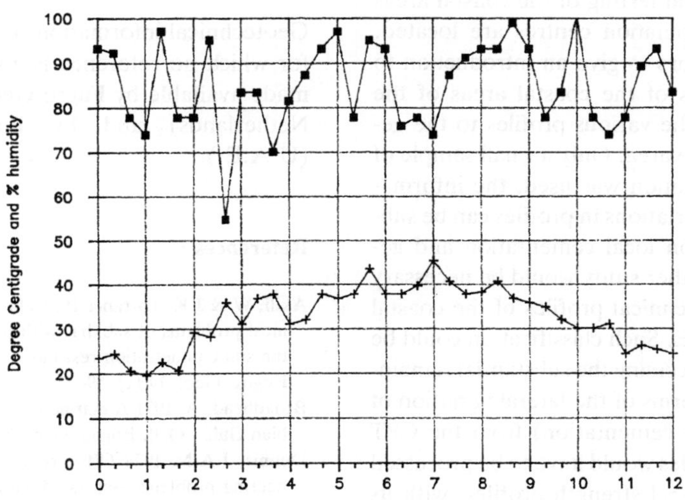

Fig. 14. Temperature and humidity variation for Dubai (after Fookes, 1978).

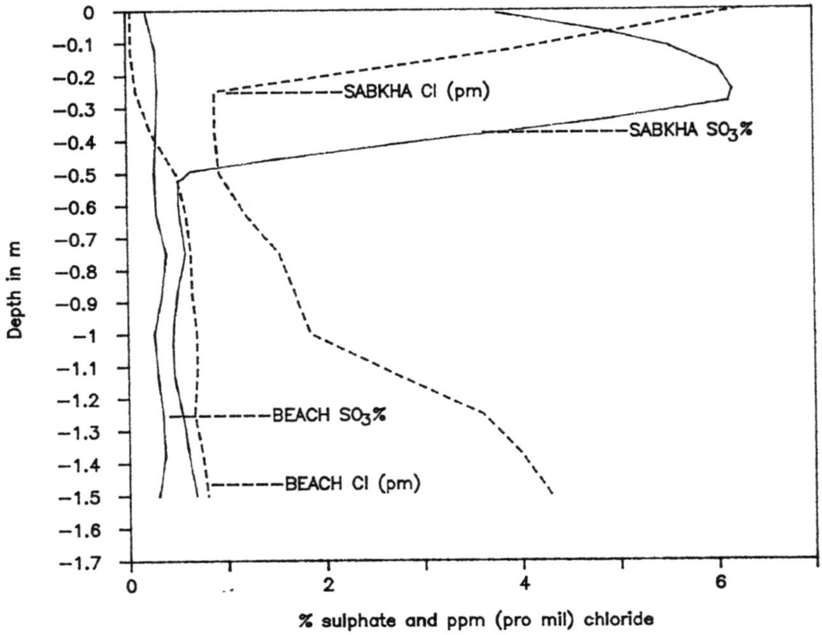

Fig. 15. Soil sulphate and chloride content profiles U.A.E. coast (after Morgan, 1982).

opment boom of the last 20 years has resulted in extensive geotechnical testing of the coastal areas where the main population centres are located. This paper endeavours to give an introduction to geotechnical profiles of the coastal areas of the U.A.E. and relate the various profiles to the geological setting of the area. Only a small sample of the available information was used; the information indicates that variations in profiles can be substantial depending on local cementation and geological setting. Further study would be necessary to classify the geotechnical profiles of the coastal regions of the U.A.E. Such classification could be best achieved by processing the substantial remaining information in terms of the lateral variation of strength (and hence cementation) from the CPT profiles. These profiles would have to be processed to give 'average lateral strength profiles' with its associated 'standard deviation' profiles. This would then allow classification of profiles and hence mapping in terms of type profiles with overlays showing the variation in terms of standard deviation.

Acknowledgement

Geotechnical information contained in this paper, for which no reference is given, has kindly been made available by Fugro Geotechnics B.V., (The Netherlands) and by Fugro Middle East, (U.A.E.).

References

Akili, W. & J.K. Torrance 1981 The development and geotechnical problems of sabkha, with preliminary experiments on the static penetration resistance on cemented sands – Q.Jl. Engng. Geol. 14 (1): 59–73

Broadhead, A. 1971 A marine foundation problem in the Arabian Gulf – Q.Jl. Engng. Geol. 3 (2): 73–84

Dennis, J.A.N. 1978 Offshore structures. Proc. Conf. Engineering problems associated with ground conditions in the Middle East – Q.Jl. Engng. Geol. 11 (1): 79–90

Epps, R.J. 1980 Geotechnical practice and ground conditions in coastal regions of the United Arab Emirates – Ground Eng. 13 (5): 19–25

Fookes, P.G. 1978 Middle East – Inherent ground problems. Proc. Conf. Engineering problems associated with ground conditions in the Middle Easst – Q.Jl. Engng. Geol. 11 (1): 33–49

Fookes, P.J. & L. Collis 1976 Cracking and the Middle East – Concrete 10 (2): 14–19

Fookes, P.G. & J.L. Knill. 1969 The application of engineering geology in the regional development of northern and central Iran – Eng. Geol. 3: 81 pp

Fookes, P.G., W.J. French & S.M.M. Rice 1985 The influence of ground and groundwater geochemistry on construction in the Middle East – Q.Jl. Engng. Geol. 18 (2): 101–128

Harris, R.W. & S.Z. Rahman 1984 The nature and characteristics of the soils in the regions of Dubai and Abu Dhabi. Compendium of papers presented at the symposium on the nature and characteristics of soils in the U.A.E. – U.N. Assistance to Ministry Public Works and Housing, U.A.E., Dubai: 3.1–3.17

Morgan, D.H. 1982 When problems arise call in the analyst. In: Special report, Gulf construction – Khaleej Times, Dubai: 65–66

Purser, B.H. & G. Evans 1973 Regional sedimentation along the Trucial Coast, SE Persian Gulf. In: B.H. Purser (ed.): The Persian Gulf, Holocene carbonate sedimentation and diagenesis – Springer (Berlin): 211–232

Tarrant, L.F.C. 1978 Dock and harbour problems. Proc. Conf. Engineering problems associated with ground conditions in the Middle East – Q.J. Eng. Geol. 11 (1): 91–97

USGS & ARAMCO 1963 Geology of the Arabian Peninsula, Map 1 : 2,000,000 Sponsorship: The Kingdom of Saudi Arabia – Ministry of Petroleum (Washington)

Wagner, C.W. & C. Van der Togt 1973 Holocene sediment types & their distribution in Southern Persian Gulf. In: B.H. Purser (ed.): The Persian Gulf, Holocene carbonate sedimentation and diagenesis – Springer (Berlin)

Warren, C.D. 1985 Dubai Dry Dock: engineering significance of dock floor geology – Q.Jl. Engng. Geol. 18 (4): 391–411

Proceedings KNGMG Symposium 'Coastal Lowlands, Geology and Geotechnology', 1987: 337–348 (1989)

Geodata management system, a computerized data base for geotechnical engineering

Jaap J.A. Hartevelt
Fugro, Geotechnical Engineers,
P.O. Box 63, 2260 AB Leidschendam, The Netherlands

Received 10 September 1987; accepted in revised form 3 February 1988

Key words: Computerized geotechnical data base, correlation studies, automatic 3-D data presentation

Abstract

This paper describes the Geodata Management System, a data base system for the management of large quantities of geotechnical data. Borelogs, in-situ tests, laboratory tests and geophysical data can be handled by the system. The advantages of such a system for the planning of urban developments and industrial projects are discussed. They include:

- Convenient storage of all the relevant geotechnical and geological parameters, including borehole, laboratory and in-situ testing data.
- Capability to model and present the data in two or three dimensions, in a variety of fence diagrams, contour maps and graphs allowing correlations between borehole and cone penetration tests (CPT) locations.

The creation of comprehensive geotechnical data base systems has become essential for the planning of urban developments and industrial projects. The large number of boreholes and CPTs made in densely populated lowland areas, result in excessive paper files of soil data. Poor accessibility of these files often prohibits their efficient reference and may reduce the quality of the consultation of valuable data.

On the other hand data bases may be consulted for many different types of projects and access to the original soil data is easily achieved. This allows the engineer to make correlations with new data and re-evaluate earlier engineering assessments. Definition of geological formations or geotechnical soil units and addition of engineering parameters to the data base will lead to thematic maps and diagrams related to foundation design.

Introduction

In densely populated coastal lowlands like the Netherlands building activities have always been very intense. We are continuously remodelling our natural environment by expanding our cities and industrial areas, the building of new roads and waterways and the reclamation of new land from the sea to satisfy our hunger for land caused by these same activities.

The complex geological history of the Quaternary has created an intricate pattern of sandy, clayey and peat deposits often distributed in rather unpredictable channel patterns. Building activities in this complex deltaic environment necessarily have to be preceded by thorough soils investigations and therefore the number of investigated sites runs in many millions. These site investigations have generated large quantities of soils data which have been collected in various archives of geotechnical consultants, municipalities and other governmental agencies. It is estimated that the number of borings and cone penetration tests (CPTs) made during the last 50 years in an average large city in the Netherlands would be in the order of 100,000 and that the number of laboratory test data would be a multiple of this. Hence it will not be an easy task to consult existing soil data files. Often we know neither what data exist nor if it is worthwhile to search for it. A computerized data base would solve these problems, provided it is designed to handle these specific geotechnical and geological data and has the additional software to present the data in an integrated collection in two or three dimensions. To this purpose Fugro developed the Geodata Management System (GMS). After the introduction of the GMS (Hartevelt & Geise 1984) the changing requirements in geotechnical reporting and the rapid developments of computer hardware, stimulated the complete redevelopment of the system. The software, originally developed for a PDP-11 mini-computer, now runs on a VAX. Furthermore the improvements specifically allowed the extensive use of micro-computers for data collection, data preparation and data input during the field work and laboratory stage. The data base has now the capability to store the complete set of parameters produced by modern geotechnical laboratory and in-situ testing programs. The development of menu driven programs has made the system more user friendly. The new generation of GMS, as it is in use today is described below.

Computer hardware

The GMS, as in operation at Fugro, is developed for a VAX mini computer. The hardware as shown on Fig. 1 consists of a central processor, disk drives, a number of alpha-numerical terminals and graphic work stations, hardcopy unit, plotter and printer. The graphic work station (Fig. 2) includes a digitizing table with a cursor, a keyboard, a menu tablet for standard graphic manipulations and two graphic screens (colour and black and white) for graphical and non-graphical information.

Data collection and preparation is usually performed on micro computers which are used in the field for automatic registration and processing of in-situ tests or laboratory tests. The data are subsequently transmitted to the VAX computer. There is a general tendency to use the more cost effective micro computers for all data preparation and input.

Computer software

The GMS software as shown in Fig. 3 consists of a

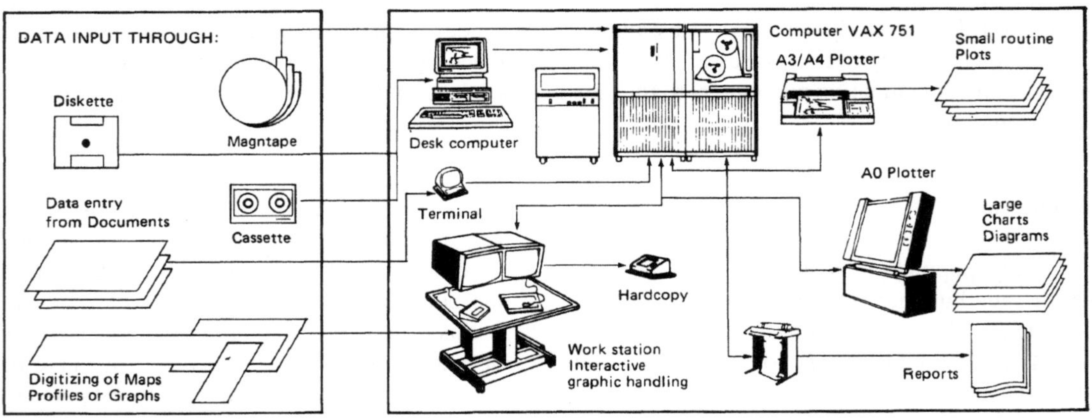

Fig. 1. Computer hardware.

Fig. 2. Interactive graphic manipulation at workstation.

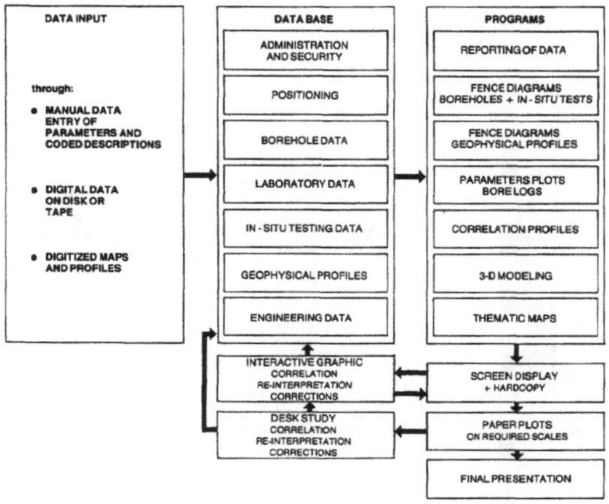

Fig. 3. Dataflow in GMS.

hierarchically structured data base, and a set of menu driven programs for data input, data manipulation and retrieval. The hierarchic structure of the data base relates to the essentially hierarchic order observed in site investigations, where the project is the basic entity to which all other entities relate in a hierarchic order like (1) project, (2) location, (3) borehole, (4) recovered sample, (5) laboratory test, (6) test parameter (Fig. 4). The same logic applies to CPT locations and geophysical profiles. Each entity comprises a group of attributes with data values. For example the entity PROJECT has the attributes like: project no., name of project, area, client, consultant and year. The entity LOCATION has attributes like: location name, coordinates, terrain level or water depth and date.

Data storage

The procedures for data storage depend very much on the way the data can be made available. Older numerical data (e.g. lab. testing results) which have to be collected from paper files and reports, have to be tabulated and borehole descriptions have to be coded for subsequent alpha-numeric input. The actual input can be performed on a micro computer and subsequently be transmitted to the VAX, or alternatively directly be stored in the VAX. Older CPT data, not yet available in a digital from on tape or diskettes, have to be digitized and subsequently stored.

Data from more recent projects can be handled in a more efficient way. Laboratory and in-situ data (CPTs) are recorded on diskettes of the micro computer and borehole descriptions can be prepared on diskettes in the field.

Transmitting the data to the VAX is all that remains to be done. The main elements to be stored include:

- position: geographic coordinates, TM coordinates, local grids.
- borehole data: coded description of lithologic logs interpreted from visual inspection, terrain level or water depth.
- laboratory data: data from geological, geotechnical, chemical tests, etc.
- geophysical data: digital data from profiles.

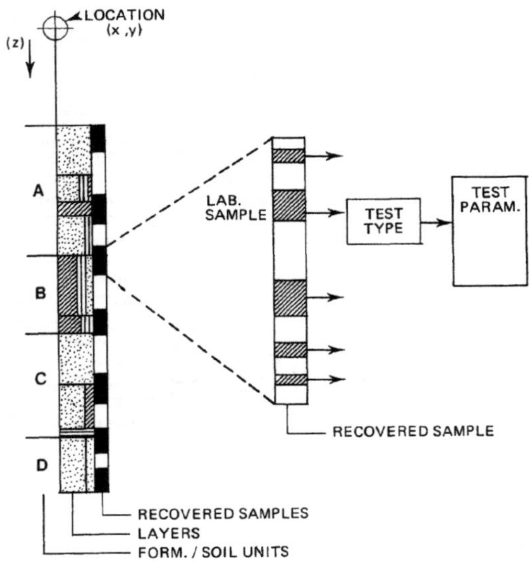

Fig. 4. Hierarchic relationship of data.

Standardization of data input

The data base is organized to accept all relevant geological, geophysical, geotechnical and chemical (pollutants) data in an internationally accepted form. Soils descriptions and coding are directly linked to the Fugro systematic classification triangle for soils (Beringen 1979). Because of the logic behind it, the coding system is extremely easy to use and at the same time results in a completely standardized input of borelog descriptions.

Geotechnical parameters are standardized according the list of symbols, units and definitions of the Int. Soc. for Soil Mechanics and Foundation Engineering (ISSMFE 1977) and pollutants are stored according to the standard list of pollutants published by the Netherlands Ministry of Environment (Ministerie VROM 1983).

Data retrieval and presentation

The GMS can be used in various ways. The most simple way is to use it as an archive of soils investigation data and just extract the data in the same form as they have been stored: lists of laboratory parameters and individual borelogs and CPTs. More useful however, is to use the system in regional studies (e.g. for town planning, industrial sites and infrastructure planning), where there is a need for data compilation, correlation studies and the construction of a geological/geotechnical framework. The GMS software allows the presentation of all or a selection of the data existing within a geographically defined area. Within this newly defined project area, data can be manipulated and the required presentations can be made.

Data presentation techniques

The various two and three dimensional presentation techniques commonly used in geological and geotechnical data presentations have been built into the GMS software package to enable the geologist or geotechnical engineer to perform correlation studies using all the data in the data base. The individual presentation techniques are discussed below:

Fence diagrams
The fence diagram in pseudo-isometric projection is one of the most extensively used presentation methods for correlation studies in geology and geotechnics. The fence diagram is especially useful in establishing reliable correlations of layers and soil units between all boreholes and CPTs in the study area. The important features are an undisturbed map surface combined with a three dimensional view from any chosen angle, which allows correlations in any direction. Horizontal and vertical scales can be chosen independently in these projections. Software has been developed for presentations of borehole lithological logs (Fig. 5), CPT graphs (Fig. 6) and geophysical profiles (Fig. 7).

Correlation profiles
Borelogs and CPT graphs can be presented as correlation profiles (Fig. 8) which are especially useful to show the variations in the soil profile along road alignments and pipelines. Two options are available:

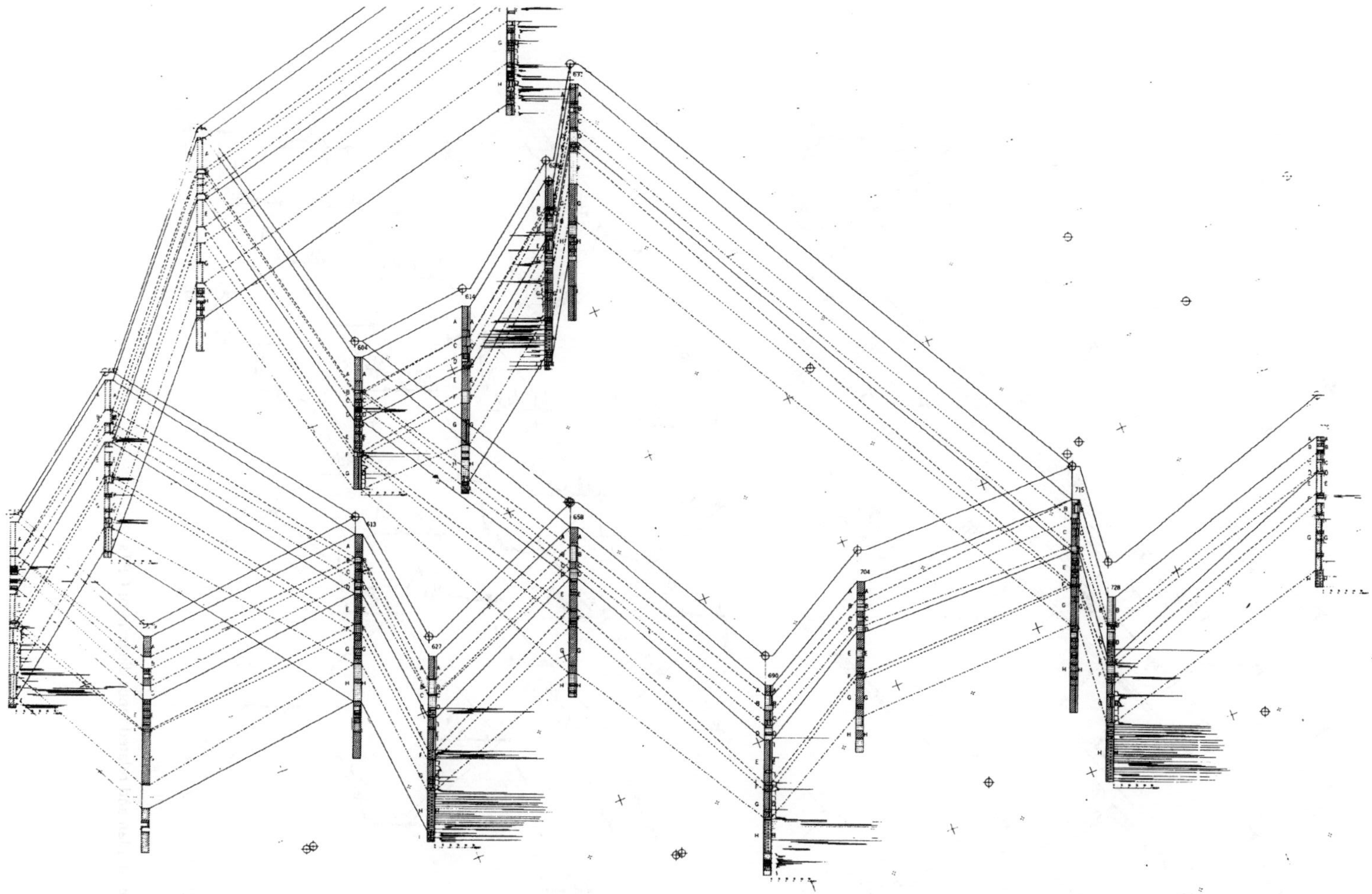

Fig. 5. Fence diagram of borelogs and CPTs.

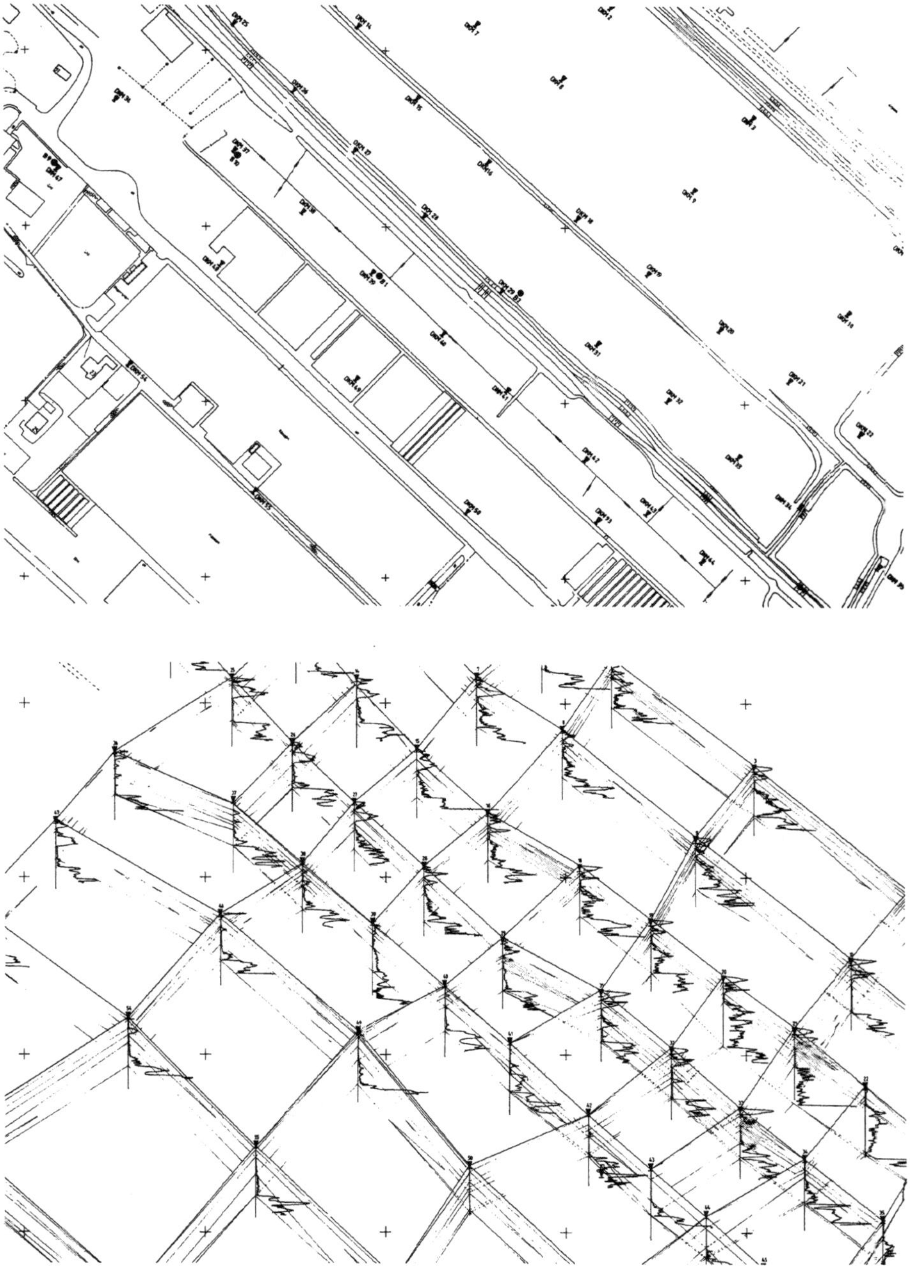

Fig. 6. Town extension plan and related fence diagram of CPTs.

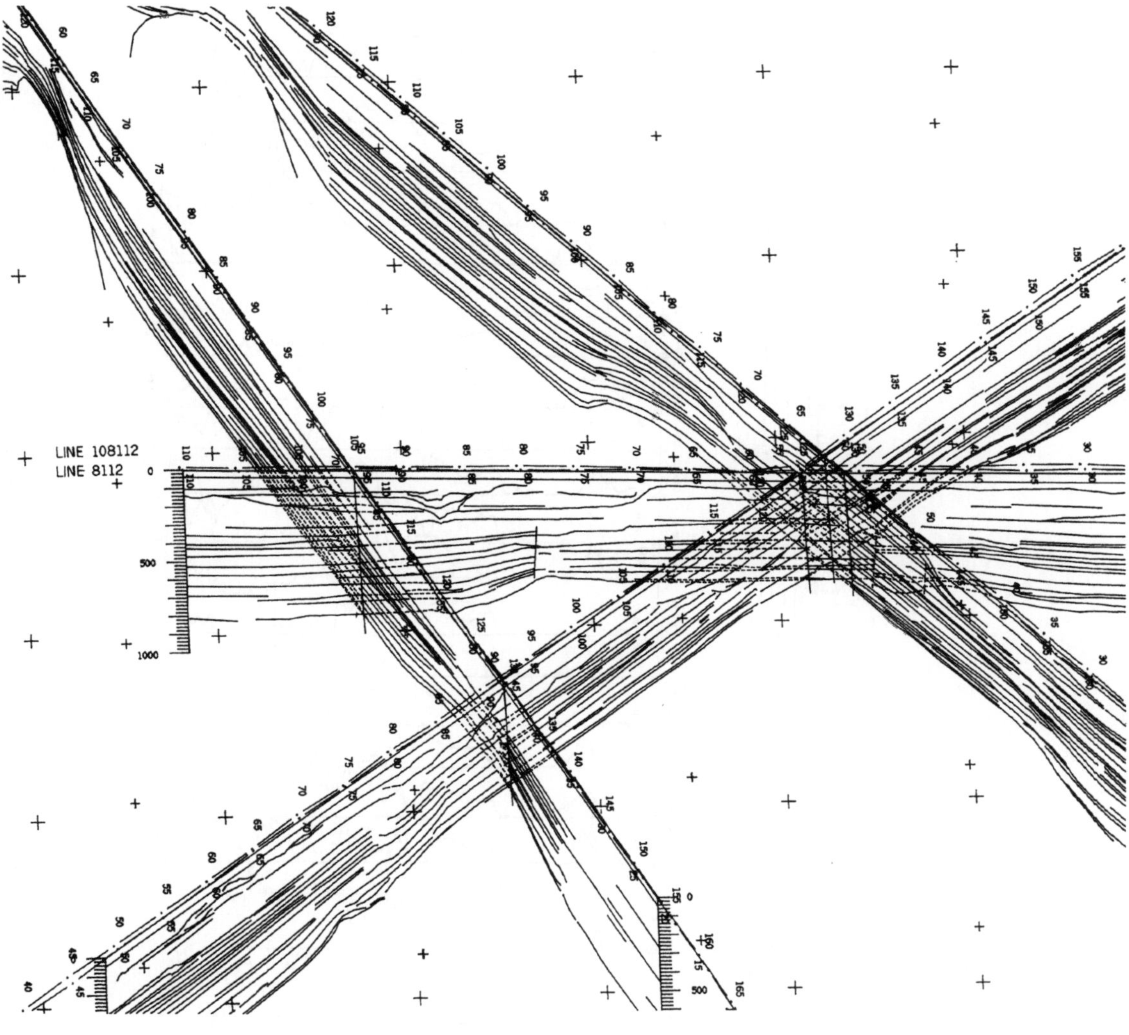

Fig. 7. Fence diagram of seismic profiles offshore.

1. a correlation profile over the actual locations showing real distances to scale;
2. a correlation profile through the projections of borings and CPTs on a chosen alignment.

Contouring and 3-D modelling

Lateral variations of geological or geotechnical data are ideally presented on contour maps. Automatic contour programs are available to construct maps from the numerical data linked to locations (e.g. boreholes and CPTs) or from continuous data (e.g. echosounding and geophysical profiles). Depth contour maps and isopach maps can be constructed and volumetric calculations can be made for resource studies. Contours can also be constructed of the lateral parameter variation within a selected layer or soil unit. Contoured surfaces can be converted to 3-D displays in isometric projection of the morphology of surfaces as observed from any angle of view (Fig. 9).

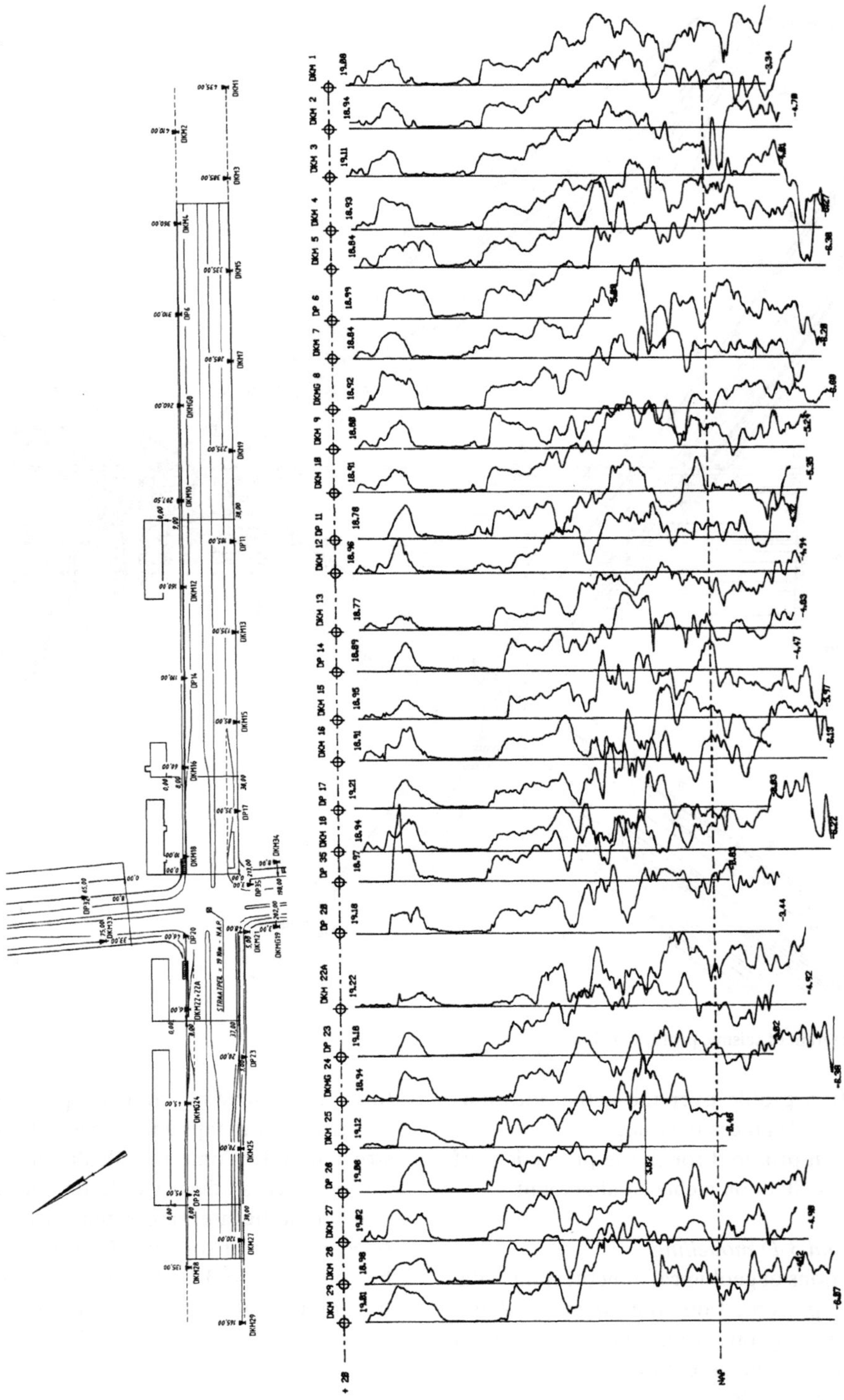

Fig. 8. Correlation profile of CPTs along a road alignment.

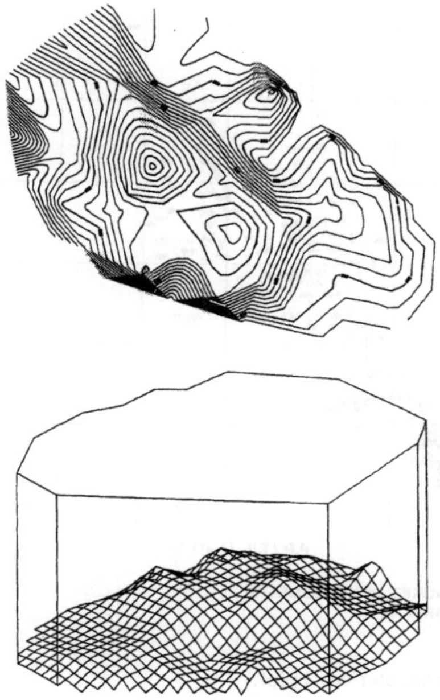

Fig. 9. Contour map and related block diagram.

Thematic maps
Engineering assessments and calculations using soil parameters stored in the database generate new engineering parameters which can be used for the construction of thematic contour maps (e.g. pile foundation level maps, settlement maps and sand and gravel resource maps, Fig. 10).

Parameter plots
Vertical and horizontal variations of geotechnical or geological properties can be rapidly observed from these plots. Parameters can be selected from specific boreholes and soil units and plots can be combined to compare characteristic parameters (Figs. 11 and 12). Plots can be presented next to the related borelog to show the correlation with the soil profile.

Data manipulation and correlation procedures

The actual use of the GMS in the phase of planning and preliminary foundation design, will be illustrated below.

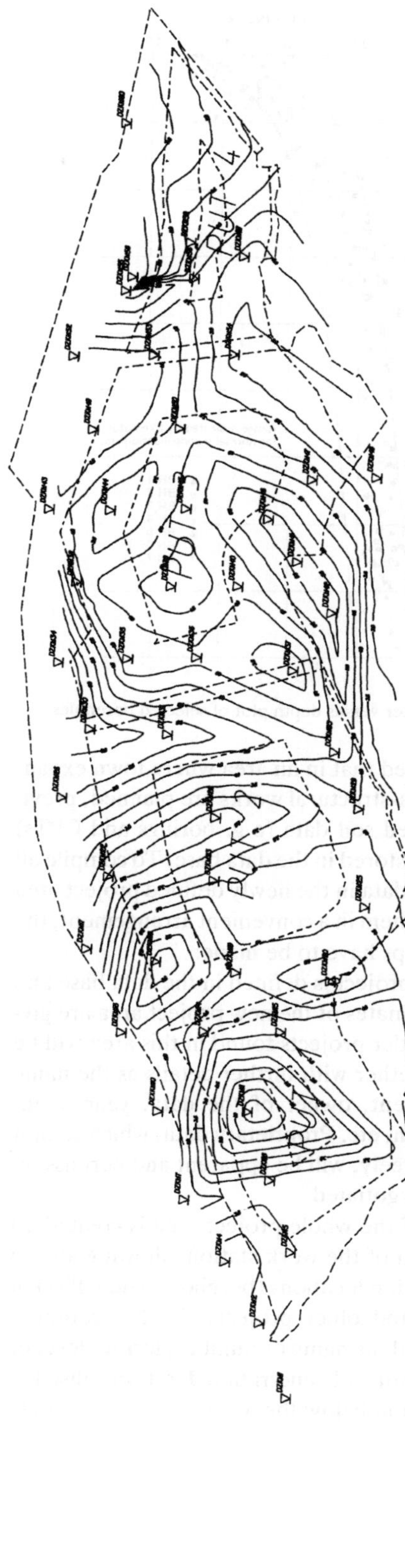

Fig. 10. Thematic contour map of depth to sand and gravel deposits with planned exploration pits.

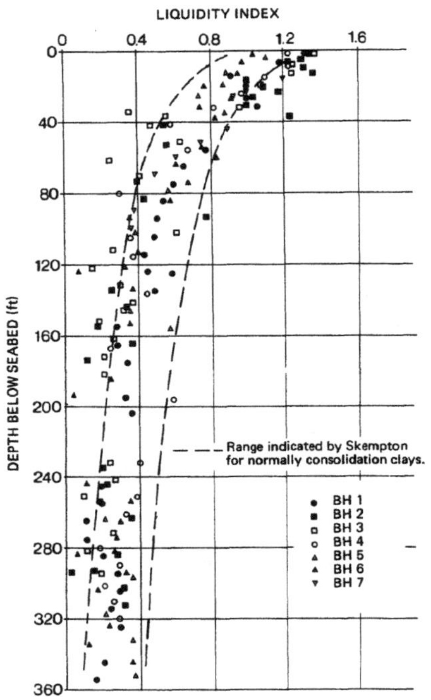

Fig. 11. Parameter versus depth plot of selected boreholes.

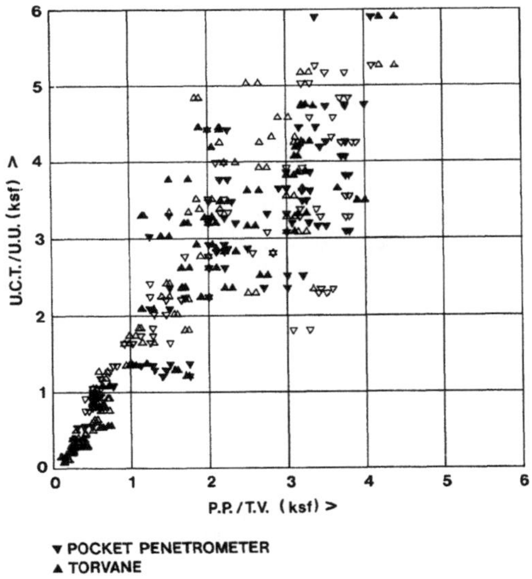

Fig. 12. Parameter cross plot.

It is assumed that in an area where town expansion and infra-structural works are planned, previously collected soil data (e.g. borings and CPTs) exist and are stored in the data base. To compile all relevant soil data in the newly defined project area and present them in a convenient arrangement, the following steps have to be made:

1. The new project is defined in the data base and the coordinates of the new project area are given. All older projects found in this area will be listed together with further details as the name of the client, owner of the data, year of investigation, etc. Proprietary data, which cannot be used freely, will be apparent and permission may be negotiated.
2. A map of the whole project area is created on the screen of the work station, showing all site investigation locations (boreholes and CPTs) of the selected older projects. Each location is shown with its name or number, terrain level or water depth and penetration depth visualised as a projection below the location. From the data arrangement shown on the screen it is apparent which data presentation would serve the purpose best.
3. A pseudo-isometric projection of all borelogs and CPTs appears after view point definition and the choice of horizontal and vertical scales. If soil units in previous projects are defined in a consistent way, the correlation lines can be created between the borelogs and/or CPTs. These lines can be generated through a user command simply by indicating on the screen those locations that should be connected. Correlation lines are then constructed automatically between all corresponding soil units thus creating the fence diagram. Finally coordinates (TM and geographic) are indicated along the map edges and a paper plot can be made. Topographic features as street plans or outlines of important building activities, which have been digitized separately, can be merged with the plot of the fence diagram, if required.
4. The paper plot of the fence diagram can be reviewed and the data coverage be evaluated. This may lead to the conclusion that additional CPTs or borings have to be made, which subsequently have to be added to the data base.

Possibly the soil unit definitions have to be revised for the new project area. The software has provisions to redefine/rename soil units to adapt them for a new project area. After all corrections have been made the final fence diagram can be plotted.

5. Data for alignment studies (e.g. roads or pipelines) can best be presented as profiles through locations along the alignment or as projections of boreholes and CPTs onto the alignment. Through the GMS software and simple manipulations at the work station these profiles can be created.
6. The final correlations and definition of soil units allows the construction of contour maps (e.g. depth contours and isopachs) and 3-D models of relevant surfaces.
7. Thematic contour maps can be generated after additional assessments and calculations have been made and the derived engineering data have been stored in the data base.

Applications

The system has been used in various foundation studies on land as well as offshore. Two examples to such studies will be briefly described.

Town expansion plan

For a town expansion in the Netherlands about 10 km^2 of polderland were investigated to assess the overall foundation characteristics of the subsurface. To this purpose about 100 CPT's and some boreholes were made regularly distributed over the area (Fig. 6, top). After storage of the soil investigation data in the GMS data base, a fence diagram (Fig. 6, bottom) was generated showing the spatial distribution of the sand, clay and peat deposits and their geotechnical characteristics.

From this fence diagram, areas could be delineated of similar foundation characteristics and pile foundation depth, allowing estimates of foundation costs and final planning of land use. Considerable savings could be attained in foundation costs by avoiding zones of thick peat and clay deposits.

Offshore foundation planning

For the extensive exploration of a large oilfield offshore South Asia, soil investigations (boreholes and CPT's) have been performed at about 150 locations by a number of consultants over an area of 1500 km^2. These investigations, which took place over a 10 year period, were followed by the installation of a still increasing number of platforms and pipelines. Because of the large quantity of data, and the ongoing platform installation activity in the area, it was decided to create a geotechnical/geological data base for the top 150 m below seabed, being the foundation zone. Once the data base was available, fence diagrams (Fig. 5) could be generated displaying the geotechnical framework of the area. Contour maps of specific foundations levels were made. The study formed the basis for planning and preliminary foundation design of future platforms and pipelines.

Conclusions

The GMS has been developed to solve the problem of comprehensive and efficient storage of data produced in site investigations. Emphasis has been on the storage, retrieval and presentation of the original and detailed data such as the actual borelogs and CPTs to enable the geologist or geotechnical engineer to reinterpret the data after updating the data base with the latest information. Special attention has been given to a set of sophisticated programs to facilitate 3-D correlation and presentation of soil data. The versatility of the system makes it ideally suitable for:

- use as geotechnical archive;
- combined presentations of soils data and topographical maps for efficient planning, management and maintenance of infrastructural elements like roads, canals, pipelines and dikes and the foundation of buildings;
- use in extensive long term projects in which regular updating of maps and diagrams with the latest soil data is essential.

References

ISSMFE 1977 Proc. 9th Int. Conf. on Soil mechanics and foundation engineering, Tokyo – Rep. Subcomm. Symbols, Units, Definitions, III: 153–171

Beringen, F.L. 1979 Geotechniek als basis voor het stalen offshore platform (Post academiale cursus berekening van vaste offshore vakwerk constructies) Internal Rept.: 37 pp

Hartevelt, J.J.A. & Geise, J.M. 1984 Use of a Geodata Management System for Offshore Geotechnical Investigations – Oceanology Intern. Conf. Brighton, Pap. 2.5: 10 pp

Ministerie VROM 1983 Leidraad Bodemsanering Afl. 3 (Interim wet Bodemsanering) – Staatsuitgevery (Den Haag), II: 1–6

Proceedings KNGMG Symposium 'Coastal Lowlands, Geology and Geotechnology', 1987: 349–354 (1989)

Some thoughts on hydrocarbon exploration in the Paris Basin

A. Koning[1]
[1] *Consulting geologist, 15 rue Marguerite de Navarre, 78540 Vernouillet, France*

Received 5 May 1987; accepted in revised form 18 January 1988

Key words: Paris Basin, initial graben hydrocarbon exploration

Abstract

Paris Basin exploration started in earnest in the '50s with Dogger, Neocomian and Rhaetian reservoirs as main targets. This search waned in the '70s but was reactivated in the '80s when unexpected large accumulations of hydrocarbons were found in the Keuper and, additionally, in the Dogger.

This revival could well continue, especially when the following two inter-related hypotheses are considered:

1) An initial Late Carboniferous graben system triggered the subsequent Mesozoic origin of the Paris Basin through a visco-elastic relaxation of the crust. The postulated graben configuration from the Lorraine-Saar Trough in the east, over the 'gravity troughs' in the centre, to the Laval-Contres Trough in the west, resulted from offsets by late Hercynian wrench faults. The grabens could contain thousands of metres of Permo-Carboniferous sediments with good source rock potential. Attractive places to explore for these sediments would exist in the central and western parts of the Paris Basin rather than in the Lorraine Trough.

2) Most if not all Keuper oils are postulated to have been generated in the underlying graben sediments; otherwise a problematic 'per descensum' migration from Lias source rocks has to be envisaged, hydrodynamically difficult to comprehend.

Introduction

Much has already been said on the hydrocarbon habitat of the Paris Basin, but the whole story may not have been fully told yet.

A look at the exploration results, as evidenced by the production figures, shows a steady rise in output since the late '50s, until a production peak of some 11,000 b/d was reached in 1964. Thereafter a sharp decline followed till a virtual standstill of exploration in the '70s. Notwithstanding the rise of the oil price, exploration was considered unattractive due to the very small size of the newly discovered fields, combined with a staggering high number of dry wells. The latter were due to weak structural expression of Mesozoic prospects in general, incorrectly applied static corrections, as well as the lack of deep seismic penetration.

Suddenly, early in the '80s, unexpected large accumulations, of more than 50.10^6 barrels of recoverable reserves, were found at Chaunoy in Keuper sandstones and at Villeperdue in Dogger carbonates. Oil production at end 1986 rose to as much as three times the earlier peak of 1964! It triggered an enormous scramble for acreage which is still going on.

This exploration revival is bound to continue in light of the following two hypotheses, viz.:

1) an initial Palaeozoic graben system lies at the origin of the Mesozoic Paris Basin,

2) most if not all oil found in the Keuper reservoirs has been generated in the Palaeozoics of the underlying central graben system.

The initial graben

The Lorraine-Saar Trough

Some years after the discovery of the huge Groningen gas field in The Netherlands, late in the '50s, it was realized that the gas had been generated from Carboniferous coals underneath the Permian reservoirs. A comparable framework exists in the Lorraine-Saar Trough: the Permo-Carboniferous fill (with thick coal seams) is known from outcrop in the Saar and dips westwards to great depths in Lorraine below the Mesozoic of the Paris Basin (Donsimoni, 1981) (Fig. 1).

A first round of exploration in the Lorraine-Saar Trough took place from the mid-'50s till the mid-'60s. The deepest well drilled so far in the Paris Basin, Gironville-101, was drilled in that period down to a total depth of 5,683 m in the Visean below some thousands of metres of Westphalian with a cumulative thickness of 100 metres of coal (Hervouet, 1966). A second round of exploration started in the early '80s, but yet only relatively small gas accumulations have been discovered (in the Muschelkalk of the Trois-Fontaines area). It is thought that the overall regional rise towards the east of the main Triassic reservoirs (as a result of an early Alpine uplift of the whole Rhenish Massif-Vosges frame) and the presence of numerous faults in the Carboniferous and overlying beds favoured the escape of the gas. The same would hold true for oil which could have been generated from excellent Westphalian source rocks. So far only a minor accumulation of such oil was found at Forcelles in the Triassic.

The western continuation of the Lorraine-Saar trough is unknown, although the 1/2,500,000 tectonic map of Europe suggests a sudden termination in the Lorraine subsurface. Such a termination is not at all evident because of the lack of well information. Only Lhuitre-1 drilled through thick Autunian and possibly some Stephanian; it could herald the northern rim of a continuation of the Trough.

Troughs in the central Paris Basin

In the early '70s the 'Bureau de Recherches Géologiques et Minières' (BRGM) organized a symposium on Paris Basin tectonics. One of the papers discussed the pronounced negative Bouguer anomalies north and south of Paris, known as the 'gravity troughs', suggested to reflect grabens at depth, filled with thousands of metres of sediments, which, therefore, could be similar to the Permo-Carboniferous of the Lorraine-Saar Trough (Gérard, 1971) (Fig. 1).

Some years later, however, the well Grisy-1 was drilled on a Mesozoic structure at the centre of the 'gravity trough' south of Paris. This well was reported to have bottomed in Pre-Triassic (?) 'granitic basement'. Consequently, the graben interpretation was converted into one of a granitic batholith. Therefore, the 'Grisy batholith' figured in the 3-volume synthesis on the Paris Basin, issued by the BRGM for the 26th International Geological Congress in Paris in 1980 (BRGM, 1980).

The above data from Grisy-1 can, however, be doubted seriously as:

- the 'granitic basement' definition was based on cuttings only;
- detailed petrographic studies of the relevant cuttings showed the presence of both sedimentary and igneous rocks;
- radiometric dating of these cuttings suggested possibly Permian, but Devonian not excluded at all;
- a continuous dipmeter survey firmly suggests a sedimentary section over the so-called 'granitic basement'.

It seems more likely that the well Grisy-1 bottomed in conglomerates, probably strongly silicified, as suggested by the high resistivities of the formation. Hydrothermal silicification is well-known from the Permian of the Saar Trough where locally also rather thick volcanics are encountered (Falke, 1975). Pre-Triassic (?) effusives are known from the 'gravity trough' south of Paris (Heurtebise-1 and Vulaines-1). Cores of comparable age from the well Nantouillet-1, in the 'gravity trough' north of Paris, consist of albitic sandstones, slightly metamorphosed, probably due to metasomatic alteration related to Permian volcanism.

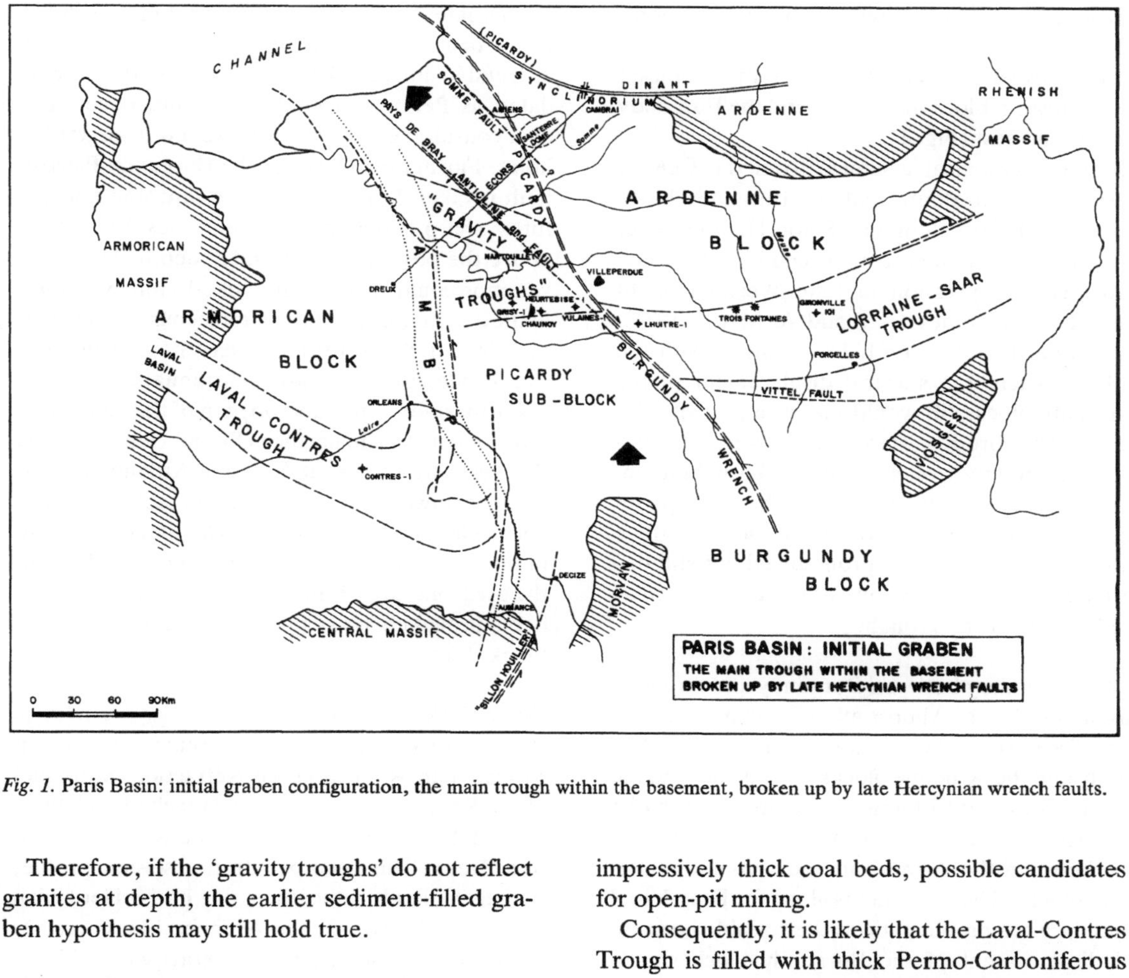

Fig. 1. Paris Basin: initial graben configuration, the main trough within the basement, broken up by late Hercynian wrench faults.

Therefore, if the 'gravity troughs' do not reflect granites at depth, the earlier sediment-filled graben hypothesis may still hold true.

The Laval-Contres Trough in the southwest

An exploration well, Contres-1, 65 km southwest of Orléans, bottomed at a total depth of 2,623 m in a thick Permian section. The corresponding depocentre of yet unknown magnitude probably connects westwards with the Laval Basin of the Central Armorican depression where Dinantian and some Stephanian are exposed (Weber, 1973). Other wells in the general Contres region penetrated coal-bearing Autuno-Stephanian (Sapin, 1967) (Fig. 1).

It is of interest that coals have been exploited more to the southeast, at Decize and in the Aumance depression. Recent reconnaissance wells of the BRGM in the area southeast of Decize showed impressively thick coal beds, possible candidates for open-pit mining.

Consequently, it is likely that the Laval-Contres Trough is filled with thick Permo-Carboniferous sediments. The Trough might be related to the 'gravity troughs' in the centre of the Paris Basin, which similarly might be connected with the Lorraine-Saar Trough in the east; the whole graben framework being offset by late Hercynian wrench-faulting.

Late Hercynian wrench fault tectonics

1. The boundary between Armorican and Burgundy basement blocks. A very conspicuous positive magnetic anomaly, known as the AMBP ('Anomalie Magnétique du Bassin de Paris') runs in an approximate N-S direction from the Central Massif to the Channel. It has been interpreted as a rift, possibly of Devonian (or ? younger) age, with basic

intrusions at depth being responsible for the anomaly (Autran et al., 1986). It would represent the suture between two Hercynian basement blocks: the Armorican block in the west and the Burgundy block in the east (Fig. 1).

Surface geological evidence from the Central Massif leaves no doubt that the Burgundy block moved northward along the 'Sillon Houiller' over at least tens of kilometres, succumbing to a late Hercynian compression in Late Westphalian to Early Permian times (Arthaud & Matte, 1977). The 'Sillon Houiller' probably connects with the AMBP, where the sinistral displacement of the Burgundy block still would measure tens of kilometres. The abrupt western termination of the pronounced 'gravity trough' south of Paris hints indeed at a major fault truncation. The corresponding counterparts on the Armorican block of these 'gravity troughs' are less pronounced but still discernable in two gravity lows at the eastern end of the Laval-Contres Trough.

2. The boundary between Burgundy and Ardenne basement blocks. Almost all tectonic maps of the Paris Basin show the geosuture between the Burgundy and the Ardenne blocks to coincide with the Pays de Bray and Vittel surface faults, although far less evidence exists for this boundary than for the one between the Armorican and Burgundy basement blocks. The movements along the Bray-Vittel boundary are problematic as the sizeable horizontal displacement of the Burgundy block should have caused strong collision tectonics with the Ardenne block, although the available evidence only suggests some minor strike-slip movements along the Bray fault.

The Bouguer anomaly map of northern France, however, shows the western part of the Paris Basin distinctly divided into two gravity provinces, which could be separated by a major break in the lithosphere. This suture would originate in Burgundy and terminate in the Channel. A sliver of the Burgundy block would have continued its northward movement as the Picardy sub-block. In this interpretation the greater part of the boundary between the Burgundy and Ardenne blocks would then coincide with a Burgundy-Picardy wrench fault (Fig. 1).

This late Hercynian wrench-faulting coincided with extrusives being poured to the surface along its trend in northern France. Corresponding rocks, dated as Permo-Carboniferous, were penetrated over considerable thicknesses in a few stratigraphic holes. Further evidence for the Burgundy-Picardy fault is taken from early Alpine rejuvenation features: sharp curvatures of Cretaceous folds, right above the possible fault trend at depth. This rejuvenation could also have provoked a renewed magmatic activity, bulging the Cretaceous sequence in the circular Santerre dome, east of Amiens, the site, incidentally, of historic earthquakes.

Furthermore, the subcrop map of the Hercynian erosional surface in northern France clearly reveals the westward continuation of the Ardenne tectonics in northwesternmost France, notably of the Dinant (Picardy) synclinorium. The marked change in the axial trend of the synclinorium from ENE in the Ardenne to NW in northern France may have been provoked by a punch of the Picardy sub-block, forging its way northward.

The ECORS geotraverse

As part of the 'Etude de la Croûte terrestre en France par Réflection et Réfraction Sismique' (ECORS), a geotraverse was launched end 1983/early 1984 in northern France between Cambrai and Dreux. This deep-penetration seismic profile, with as main objective the study of the Hercynian framework, starts on the Ardenne block, crosses the Picardy sub-block with its 'gravity trough' and terminates on the Armorican block beyond the AMBP (Fig. 1).

The full report on these ECORS data is still awaiting publication but preliminary interpretations (Cazes et al., 1985) would seem to be at variance with the model of this article, although a closer examination seems warranted. The geotraverse crosses a major disturbance, described by the ECORS interpreters as the Somme fault, but which more likely coincides with the Burgundy-Picardy wrench fault. Furthermore, the section suggests a graben feature of some magnitude to the south of the Bray anticline, possibly the 'gravity trough'. In this area ECORS seismic hints at the presence of a truncation surface (already deep in the basement according to ECORS-staff) but very similar to the seismic picture of the Stephanian-Westphalian unconformity in the Lorraine Trough, well known from well data.

Hydrocarbon prospects related to the initial Paris Basin graben

Late Hercynian wrench fault tectonics could thus have caused the break-up of the initial Paris Basin main graben: the Lorraine-Saar Trough cut off by the Burgundy-Picardy wrench fault, and the continuation on the Picardy sub-block in the 'gravity troughs' cut, in turn, by the wrench faults of the AMBP.

After extension ceased in the graben in late Hercynian times, the crust started to relax visco-elastically (Beaumont, 1978; Brunet & Le Pichon, 1982). The resulting expanding depression on its top became then gradually filled with Late Permian and Mesozoic sediments. The actual shape of the western part of the Paris Basin does indeed reflect a bipolar subsidence: one over the 'gravity troughs', the other over the Laval-Contres depocentre.

As mentioned earlier, the early Alpine regional uplift of the Basin in the east tilted the Triassic reservoirs into a general westward dipping, monoclinal position, diminishing the chances for closed structural traps. The western part of the Basin, especially over the 'gravity trough' north of Paris, remained in a high position throughout the Mesozoic, as did the western part of the Laval-Contres Trough, with Late Permian reservoirs on truncated Permo-Carboniferous potential source rocks. Hydrocarbon exploration of the graben and overlying reservoirs stands therefore a better chance of success there than exploring in the Lorraine-Saar Trough in the east.

The Keuper oil migration hypothesis

The Keuper-on-basement model

As discussed earlier, the well Grisy-1 probably did not bottom in Hercynian basement. Subsequent wells in 'gravity trough' position again were reported to have hit basement. Also here, firm evidence is lacking as only cuttings have been considered to define basement. In this respect it should be recalled that exploration permits in France are granted for one initial period and two subsequent periods after reduction of acreage, each of 3–5 years. The sizeable Chaunoy field was only discovered three years before the final expiry of the permit. Therefore, the operator was left with no choice but to drill as many wells as possible in order to carve out further potential accumulations from the acreage-to-be-surrendered. Consequently, each well, after having penetrated (and, eventually, tested) the Dogger, Rhaetian and Keuper reservoirs, was abandoned/suspended without any cores being taken at total depth, either in Keuper or in so-called basement.

Petrographical analyses of these new 'basement' cuttings did not produce any unambiguous answers. Permian volcanism and related subsequent hydrothermal activities were wide-spread phenomena throughout western and southern Europe and would have been absolutely normal in the central graben of the Paris Basin. Their possible effects on sedimentary rocks, such as silicification, albitisation, etc., could well simulate basement lithologies. It is therefore felt that the model of Hercynian basement directly below Keuper sediments in the 'gravity troughs' needs more solid evidence, preferably from a number of adequately cored wells.

The Keuper-on-basement model lies at the root of a difficult-to-grasp 'per descensum' hydrocarbon migration path. As the Triassic is devoid of any source rocks, Keuper sandstone accumulations could only have been fed from the excellent Hettangian source rocks above. Such migration would have taken place either through the omnipresent Upper Keuper marls (otherwise known as good caprocks) or laterally through non-sealing faults with source rock and reservoir in direct contact. The latter scheme would require fault throws of at least 60 metres which seem rather excluded in light of the very weak Mesozoic tectonics in the central Paris Basin, with a possible exception for faults on trend with the southeastern continuation of the Bray fault and for the Burgundy-Picardy wrench fault.

It also has been suggested that a number of faults originating in the Keuper and terminating in the descensum' migration into the Keuper reservoirs. Still, such a theory needs a better foundation, especially as pressure differences such as would be required to draw the oil down to deeper levels, are not known in the region of Keuper accumulations.

The graben theory
An initial graben system in the Paris Basin would likely be filled with Permo-Carboniferous sediments. From the Lorraine-Saar Trough excellent Westphalian oil source rocks and coal are known. Furthermore, Stephanian and Autunian could be present with similar source rocks as well.

The graben system would be bound by faults along tilted fault blocks. The early Alpine compression would have caused conjugate faults within the trough. In such a tectonic framework sufficient fairways would be available for graben-generated oil to fill the Keuper reservoirs in weak undulations over remobilized faults at depth. A 'per descensum' migration could then be dispensed with.

Geochemical data, however, do not seem to support the theory of oil generation in the graben. In fact, oils from Dogger, Rhaetian and Keuper pools as well as extracts from Lias source rocks possess similar geochemical fingerprints (Espitalié, in press). Therefore, it is difficult to believe the Lias not having generated all the oils discovered so far in the central Paris Basin. Naturally, the above does not 'ipso facto' exclude deeper source rocks having generated similar oils.

The first imperative for future exploration in the central Paris Basin would be to confirm the presence of the graben system, mainly by seismic and other geophysical methods, e.g. magneto-tellurics. Subsequently, drilling would be required to confirm the lithological nature of the graben fill together with its source rock and reservoir potential. Only then the full story of the hydrocarbon habitat of the Paris Basin would become clear together with an appreciation of the chances to find deeper oil and gas accumulations.

Conclusion

The concept of an initial Late Palaeozoic graben would add new dimensions for hydrocarbon exploration in the Paris Basin, especially as present-day exploration is dominated by only one source interval, the Lias, which is only mature in the deepest parts of the Basin.

References

Arthaud, F. & Ph. Matte. 1977. Late-Paleozoic strike-slip faulting in southern Europe and northern Africa: results of a right-lateral shear zone between the Appalachians and the Urals – Geol. Soc. Amer. Bull. 83: 1305–1320

Autran, A., C. Castaing, N. Debeglia, A. Guillen & C. Weber. 1986. Nouvelles contraintes géophysiques pour l'interprétation de l'anomalie magnétique du Bassin de Paris: hypothèse d'un rift paléozoïque refermé au Carbonifère – Soc. Géol. Fr. Bull. (8)II: 125–141

Beaumont, C. 1978. The evolution of sedimentary basins on a visco-elastic lithosphere: theory and examples – R. Ast. Soc. Geophys. J. 55: 471–497

BRGM. 1980. Synthèse géologique du Bassin de Paris – Mém. 101: Stratigraphie et paléogéographie: 468 pp

Brunet, M.-F. & X. Le Pichon. 1982. Subsidence of the Paris Basin – J. Geophys. Res. 87: 8547–8560

Cazes, M., G. Torreilles, C. Bois, B. Damotte, A. Galdeano, A. Hirn, A. Mascle, Ph. Matte, Ph. Van Ngoc & J.-F. Raoult. 1985. Structure de la croûte hercynienne du Nord de la France: premiers résultats du profil ECORS – Soc. Géol. Fr. Bull. (8)I: 925–941

Donsimoni, M. 1981. Le bassin houiller lorrain – Mém. BRGM 117: 100 pp

Espitalié, J. (in press). Organic chemistry of the Paris Basin – Proc. 3rd Conf. Petr. Geol. NW Europe

Falke, H. 1975. Problems of the continental Permian in the Federal Republic of Germany. In: Falke, H. (ed.): The continental Permian in central-, west- and south Europe – NATO Advanced Study Institutes (C)22: 38–52

Gerard, A. 1971. Apports de la gravimétrie à la connaissance de la tectonique profonde du Bassin de Paris – Bull. BRGM (2)I: 75–88

Hervouet, M. 1966. Beitrag zur Kenntnis der Geologie und der Erdölführung Lothringens – Z. Dt. Geol. Ges. 117: 225–242

Sapin, S. 1967. Principaux résultats géologiques des travaux d'exploration réalisés par la SNPA dans le sud-ouest du Bassin de Paris – Soc. Géol. Fr. Bull. (7)IX: 327–355

Weber, C. 1973. Le socle anté-triasique sous la partie sud du Bassin de Paris d'après les données géophysiques – Bull. BRGM (2)II: 219–343

Proceedings KNGMG Symposium 'Coastal Lowlands, Geology and Geotechnology', 1987: 355–361 (1989)

Natural radioactive heavy minerals in sediments along the Dutch coast

R.J. De Meijer[1], L.W. Put[1], R.D. Schuiling[2], J.H. De Reus[3] & J. Wiersma[4]
[1] Kernfysisch Versneller Instituut, Rijksuniversiteit Groningen, Zernikelaan 25, 9747 AA Groningen, The Netherlands; [2] Instituut voor Aardwetenschappen, Rijksuniversiteit Utrecht, Postbus 80021, 3508 TA Utrecht, The Netherlands; [3] Rijkswaterstaat, Dienst Getijdewateren, Postbus 207, 9750 AE Haren, The Netherlands; [4] Rijkswaterstaat, Directie Noordzee, Postbus 5807, 2280 HV Rijswijk, The Netherlands

Received 14 August 1987; accepted in revised form 29 February 1988

Key words: Natural radioactivity, heavy minerals, sand transport, provenance of minerals

Abstract

The radioactivity of heavy-mineral sands, due to the inclusion in their crystals of U and Th, has proven to be a fast and reliable method for mapping surface occurrences of these minerals in dune and beach sands along the Dutch coast. Gamma-spectrometric and granulometric analysis revealed that the radioactivity can be correlated with the smaller grain sizes. A correlation was found between U and Th abundances and elements like Zr, P and e.g. La. This indicates a correlation between U, Th and K and the mineralogical composition, thus leading to 'radiometric finger printing'. The method described may be of interest not only in prospecting for heavy minerals but also when unraveling the provenance and the depositional history of the coastal sands. Concentrations of heavy-mineral sands were not only found at the foot of dunes but also in the dunes and at places where islands merged, indicating that heavy-minerals are being concentrated not only thanks to erosion but that this also happens during deposition.

Introduction

Heavy-mineral sand-deposits are the main sources for the world supply of rare metals as zirconium and titanium. Concentrations of these minerals are known to occur in beach and dune sands. They are found e.g. along the coasts of Kerala in India, Australia, Malaysia, Brasil, the Gulf of Mexico and the North Sea. In these sands minerals are present as ilmenite, magnetite, garnet, epidote, zircon, monazite and rutile. These deposits usually may visually be recognized by a darker colouring of the sands.

Since heavy-mineral concentrations are often found in beach ridges it is generally accepted that the concentration took place by the selective removal of the lighter minerals such as quartz during erosion along a sandy coast. The presence of heavy minerals in beaches and dunes can therefore be used as an indicator for coastal development processes, i.c. erosion.

In the search for heavy-mineral concentrations usually samples are taken at various locations and the abundance of the heavy minerals is determined in the laboratory by floating off the lighter components of the sand in heavy liquids as bromoform. Investigations (Baak, 1936; Crommelin & Slotboom, 1945; Eisma, 1968; Lamcke, 1937; Van der

Kleyn, 1975; Wasmund, 1938) in and along the North Sea have been carried out in this way and represent a data base that was obtained through elaborate work and only by rather coarse sampling.

It is known that the radionuclides ^{238}U, ^{235}U and ^{232}Th may become incorporated in igneous materials when they are originally formed from the molten state. A mineral is likely to be enhanced in radioactive nuclide content if one of its predominant ions is of similar size to that of U and Th. Based on isomorphism, radioactivity is expected in minerals containing zirconium, calcium and rare earth elements like cerium and lanthanum. It is known that U and Th are found in the minerals such as apatite and zircon and that the mineral monazite contains a relatively large quantity of thorium.

The radioactivity of the minerals allows the identification of deposits by radiometric means. Fast and detailed information of surface deposits were obtained in the southeast part of the USA by registration of the emitted gamma radiation using detectors mounted onto a plane (Force et al., 1982; Grosz et al., 1983). Discrete sources of ^{232}Th were used by Kamel & Johnson (1962) to determine the drift along the Californian coast. Also along the North Sea coast heavy minerals were located by radiometric means. Bonka (1980) observed dark radioactive sands along the beaches of some German Frisian Islands.

The aim of this investigation was to find out whether radioactive heavy mineral sands also occur along the Dutch coast, to study the correlation of the radioactivity with physical parameters as grain size and magnetic susceptibility, and to investigate the feasibility of a detailed radiometric mapping of the coast in order to obtain information on the physical processes that play a role in transport of sand along the coast. This latter topic may help to obtain a better understanding of the processes that play a role in the dynamic changes of the Dutch coast and may contribute to the present research program 'Kustgenese'.

In previous publications (De Meijer et al., 1985; Schuiling et al., 1985) we have shown that the heavy-mineral enriched sands have a bimodal grain-size distribution. Moreover it was indicated that the specific radioactivity for ^{222}Rn and ^{232}Th increased exponentially with decreasing grain size: a three times smaller grain size corresponded with three orders of magnitude larger activity. Measuring the variation of the radioactivity with magnetic susceptibility showed a strongly varying U and Th content, not only in absolute magnitude but also in ratio. These results indicate the difference in preference of the various minerals for U and Th. Another result was that the average grain sizes of various minerals in a sample decrease linearly with the density of the mineral. This indicates that the various minerals were in hydraulic equivalence during the sedimentation process. Whether this process occurred in air or in water is still unknown since a quantitative analysis, including corrections for the shape of the minerals, has not yet been made. Stolk (1985) presented an overview of concentration processes in beach deposits; see also Slingerland (1977) and Stapor (1973).

Experimental technique

Figure 1 shows a map of the Netherlands with an indication of locations mentioned in the text.

Radiometric maps have been made of the beaches and adjacent dune formations along the North Sea coast from IJmuiden to the island of Borkum (FRG). The measurements were carried out with a portable $4 \times 4 \times 5\,cm^3$ NaI crystal, mounted on a photomultiplier tube and connected to accessory electronics, including a single-channel analyser and a scaler display (Scintrex GIS5). Measurements were usually made, at a height of 1 m (hip height) above the surface, every 100 m along the foot of the dunes; at every kilometre a transect was made from the waterline into the dunes in steps of about 25 m.

Occasionally, usually in areas with a higher count rate, more detailed measurements were made, including contact readings. The sensitivity of the instrument for cosmic-rays was determined in measurements over (fresh) water with a depth of at least 5 m. It was assumed that the residual count rate could be attributed to cosmic-ray background.

At a few locations profiles down to a depth of about 1 m were made using a radiation monitor

Fig. 1. The Netherlands.

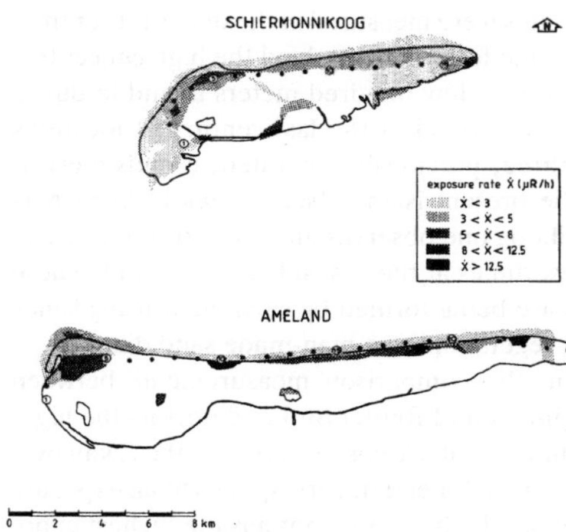

Fig. 2. Gamma-ray exposure ($\mu R.h^{-1}$) on the beach and in the dunes of the islands Ameland and Schiermonnikoog.

with a GM-tube mounted on a stick (Wallac GMP-256).

Samples of sand were collected from the surface, from greater depth and from the sea bottom to be investigated in detail in the laboratory. Samples were dried at 100° C. Bulk samples, sieved fractions and isomagnetic separated fractions were investigated on their U, Th and K content by gamma-ray spectrometry with a Ge detector.

The conversion of total count rate to exposure rate was deduced from a series of calibration measurements at various locations on Ameland with the Scintrex and a Reuter Stokes detectors at the same spot and the same height above the ground.

The ^{238}U activity is deduced from the decay of ^{214}Bi assuming equilibrium in all preceding decay stages. A more direct determination of the ^{238}U activity can only proceed via the $E_\gamma = 1.001$ MeV gamma rays in the decay of ^{234m}Pa. The results are only deducible with a relatively large uncertainty. Using the 1.001 MeV gamma ray of ^{234m}Pa yielded consistently a ^{238}U concentration that was a factor two higher than if deduced via the other methods. This discrepancy needs further investigation.

The mineralogical composition of the sands was determined visually under a polarization microscope and/or by Guinier X-ray diffraction.

Results

The results of radiometric mapping indicate rather low concentrations on the islands of Terschelling, Schiermonnikoog, Rottumerplaat and Rottumeroog and on the mainland, except for a spot near Bergen (Rottumerplaat and Rottumeroog are located between Schiermonnikoog and Borkum, FRG; c.f. Fig. 1). High values were found on the islands of Texel, Vlieland and Ameland. In general it can be said that each island has its own characteristic pattern. Moreover it was observed that the concentrations were lowest at the shore and increased landwards, often reaching a maximum at the foot of the dunes, probably caused by diminishing grainsize. At various locations even higher values were found in the dunes. Since most detailed information has been gathered on the adjacent islands Schiermonnikoog and Ameland, the results presented in this paper will mainly concern these two islands.

Figure 2 shows the generalized results of the measurements of the total counts on these islands. The exposure rates are obtained from a conversion procedure described in the previous section on experimental technique. One notices that the values for Ameland are much higher than those found on Schiermonnikoog. On both islands the highest count rates were measured in the dunes rather than at the dune foot. On Ameland the high concentrations occur a few hundred meters inland in dunes that were formed in the last century at locations were three, previously separated, islands merged into the present island (Isbary, 1936). At Schiermonnikoog one observes an elevated count rate in the new dunes at the east side of the island. These dunes are being formed by catching drifting sands in the vegetation on a man-made sand dike.

From the comparison measurements between the Scintrex and Reuter Stokes detectors the highest count rate of 620 cps, observed in the Zwanewaterduinen at Ameland, corresponds to an exposure rate of 25 $\mu R \cdot h^{-1}$. This spot has the highest exposure rate due to natural radiation in the Netherlands known to us. A full year stay at that spot would result in an effective dose equivalent of 1.3 $mSv \cdot a^{-1}$, corresponding to about 50% of the average annual dose of the Dutch population due to natural radiation in the living environment. At beaches and dunes of the other islands and along the coast of the mainland maximum count-rates of 80–250 cps were obtained. It should be noted that, on the dikes protecting the southern part of the island, count rates were obtained which were two times higher than at the hot spot in the Zwanewaterduinen. The high exposure rate on the dikes is attributed to slags embedded in the concrete blocks as part of the dike construction.

Depth profiles on the beach show bands of dark minerals varying in thickness from a few millimetres to a few centimetres. The bands become more diffuse at larger depth and are usually unobservable below 0.5 m. In the older dunes no clear band structure was observed. Figure 3 shows the count-rate variation with depth measured with the Wallac radiation monitor at two locations in the Zwanewaterduinen at Ameland. The left part of the figure shows a broad maximum, 10–40 cm below the surface; at the right-hand part of the figure the maximum occurs at a depth of 30–50 cm.

Gamma-ray spectra indicate that the radioactivity in the sands is predominantly (a small Cs contamination was observed in some surface samples) due to the naturally present radionuclides U, Th and K and their radioactive decay products. The

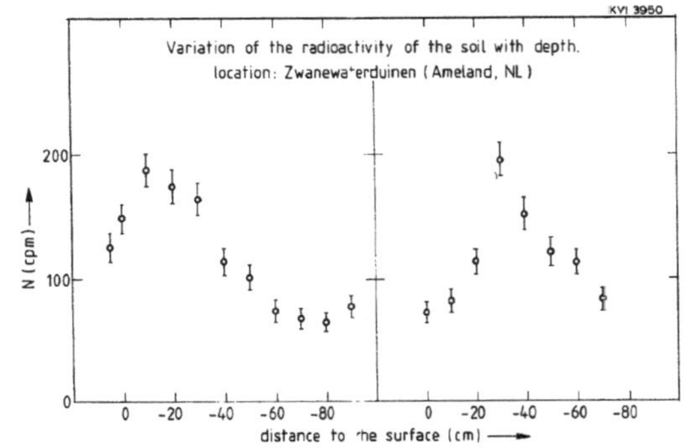

Fig. 3. Variation in radioactivity of the soil with depth at two locations in the dunes of Ameland. The measurements were made with a Wallac radiation monitor.

results are presented in Table 1 for selected samples from dunes and beaches usually at places with high activity, bulk samples from two locations elsewhere in coastal deposits (Domburg & Bennebroek, c.f. Fig. 1), and samples obtained from the sea bottom between the islands Ameland and Terschelling. From the table one sees that the U/Th ratio of the samples from Borkum, Ameland and various locations in the tidal inlet between Terschelling and Ameland is consistent with a value near or slightly smaller than unity; the ^{40}K values vary between 250 and 400 $Bq \cdot kg^{-1}$. The samples from Bergen have an U/Th ratio of about 1.2 but have a low ^{40}K content. The samples of Domburg and Bennebroek are samples from one of the islands in Zeeland and an old dune formation south of Haarlem, respectively. The Domburg sample has an U/Th ratio of 1.3 and a ^{40}K content of about 100 $Bq \cdot kg^{-1}$; the U/Th ratio of the Bennebroek sample has radiometric characteristics comparable to the samples in the northern part of the country.

The samples collected from the sea bottom of the tidal inlet between Terschelling and Ameland all have a similar ^{40}K content but show variations in the U and Th activities up to a factor of four. From these scattered data it is not yet clear whether these differences are due to present or ancient selection processes.

Figure 4 shows concentrations of various radioactive and stable elements obtained from a chem-

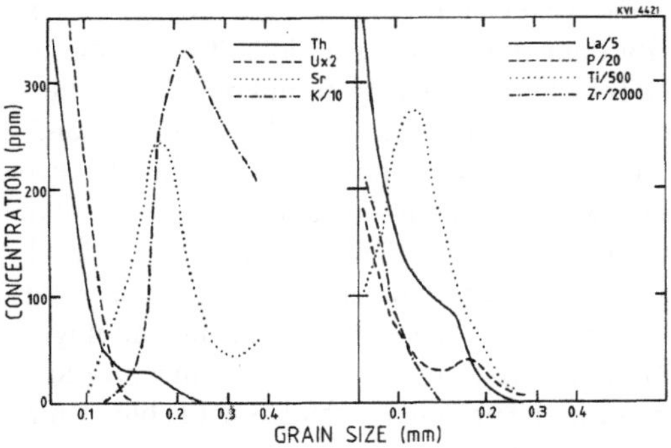

Fig. 4. Concentrations of various elements as determined by chemical analysis in the Ameland P19 sample. The multiplicative numbers indicate the factor by which the original concentration was multiplied (×) or divided (/) to bring them to the presented scale.

ical analysis. In this figure La is chosen as a representative of the rare earth elements. From this figure it follows that in the smaller fractions Zr dominates (400,000 ppm) pointing to the occurrence of zircon in the fine fractions, and that K is mainly present in the larger grains, obviously indicating the occurrence of light minerals as K-feldspar and muscovite (Schuiling et al., 1985). Moreover one notices that the distribution of U follows that of Zr, and that Th and La follow P thus reflecting the replacement in zircon and monazite, respectively. The P anomaly near the grain size of 0.2 mm corresponds with the maximum of the Sr concentration. The similarity of Sr and P values near $\varnothing = 0.2$ mm may be due to e.g. apatite.

Conclusions and discussion

Based on relations between the natural radioactivity and the mineralogical composition, the ratio of U and Th and the activity of ^{40}K may probably be used to identify the origin of the sands. Such a possibility was recently found in heavy mineral sands collected at Florida beaches along the Gulf of Mexico (De Meijer, personal observation). Here variations of an order of magnitude were observed apparently reflecting the differences in mineral composition of the sands of nearby rivers.

If the previous paragraph it was indicated that sands may be qualified by means of their U/Th ratio and their K concentration ('radiometric finger printing'). In retrospect this result should not be too surprising. U and Th are mainly present in the

Table 1. U, Th and K contents in Bq.kg^{-1} for various sand samples collected on beaches and in dunes and samples at various depths collected from the sea bottom of the tidal inlet between Terschelling and Ameland; *) bulk samples.

Location	^{238}U	^{235}U	^{232}Th	^{40}K	^{238}U/^{232}Th
Beaches and dunes					
Borkum (FRG)	238 ± 8	22 ± 8	238 ± 8	340 ± 30	1.00 ± 0.05
Ameland P19	167 ± 8	23 ± 3	174 ± 8	360 ± 30	0.96 ± 0.06
Ameland P23	13.3 ± 1.5	3.0 ± 1.5	20.0 ± 1.9	380 ± 30	0.67 ± 0.10
Bergen P32	384 ± 10	20 ± 2	314 ± 15	11 ± 11	1.22 ± 0.07
Bergen P32.6	249 ± 3	13 ± 2	205 ± 7	45 ± 8	1.21 ± 0.03
Domburg P15.1*)	6.5 ± 0.7	0.5 ± 0.4	4.9 ± 0.5	109 ± 8	1.3 ± 0.2
Bennebroek*)	7.9 ± 1.4	1.2 ± 1.3	8.8 ± 1.4	534 ± 14	0.9 ± 0.2
*Sea bottom**)					
depth (m)					
20.1	5 ± 2	0.5 ± 0.7	7.3 ± 0.6	373 ± 11	0.7 ± 0.3
8.8	5.4 ± 0.6	0.6 ± 0.2	4.4 ± 0.5	250 ± 9	1.2 ± 0.2
5.4	5.6 ± 0.9	ND	5.7 ± 0.9	267 ± 12	1.0 ± 0.2
4.4	9 ± 2	0.5 ± 0.3	15.3 ± 1.2	264 ± 11	0.59 ± 0.14
6.8	4.8 ± 0.4	0.4 ± 0.3	5.0 ± 0.8	367 ± 11	0.96 ± 0.17
7.4	6.8 ± 0.7	0.2 ± 0.5	5.6 ± 0.8	321 ± 11	1.21 ± 0.17
3.4	9.7 ± 2.0	1.0 ± 0.2	11.7 ± 1.4	364 ± 8	0.8 ± 0.2

heavy minerals whereas K is associated with the light minerals. Heavy minerals are usually a smaller fraction of the sand; their concentration varies depending on the selectivity in the transport processes. Thus the absolute concentration is strongly dependent on the degree of enrichment of heavy minerals, whereas the K-concentration only varies slowly with the enrichment factor. The U/Th ratio reflects the heavy-mineral composition and of course is more or less independent on the enrichment factor.

It is interesting to know that near Bergen a border line exists between sand deposits of Scandinavian and Middle European origin (e.g. Eisma, 1968). From these results it seems that the samples from Bergen and from Domburg are of Middle European origin and that the others, including possibly the sample from Bennebroek, contain sand of predominantly Scandinavian origin.

The present results indicate an empirical method to identify the origin of sands in a rather fast and sensitive way. From this rather limited number of samples it is, however, too early to draw definite conclusions, but a more systematic study of this aspect is presently considered.

The difference in heavy-mineral concentrations between the islands Ameland and Schiermonnikoog seems at first sight in agreement with the general hypothesis (Crommelin & Slotboom, 1945; Lamcke, 1937; Wasmund, 1938; Koning, 1947) that eroding coasts show higher concentration factors of heavy minerals than stable coasts: nowadays Ameland is eroding, Schiermonnikoog is stable to slightly prograding at its western part. A closer look at the situation shows, however, that the high concentrations occur in the dunes at places where former islands have merged. Also on Schiermonnikoog higher activities are found in new dune formations. Aside from the 'hot spots' it can be concluded that the bulk material on Ameland has higher content than the one on Schiermonnikoog. This is surprising since the two islands are only separated by a narrow tidal inlet. Veenstra & Winkelmolen (1976) also stressed this difference, mainly based on higher garnet percentages and higher rollability of beach sands from Ameland. Similarly, in our mapping we find that the bulk sands on Texel and Vlieland have higher radioactivity than those on the mainland or on the islands Terschelling, Schiermonnikoog, Rottumerplaat and Rottumeroog. At Terschelling, sandwiched between Vlieland and Ameland the two ends show areas with higher activity. In the front of the islands, at a depth of about 6 m below sea level, there is also a difference in the percentage of the heavy minerals (Vakgroep Mijntechnologie, 1982). The variation in U and Th concentrations observed in the present investigation (Table 1) may be related to this difference.

In the chemical analysis of one sand sample a correlation was found between the grainsize distribution and the concentration of La and P and of U and Zr. The first correlation corroborates with the presence of monazite $(Ce,La,Y,Th)PO_4$ in the sample; the latter result agrees with the known inclusion of U in zircon. The results may be used to estimate the heavy-mineral concentrations in sands from their radioactivity (see also Schuiling et al., 1985).

So far this project has raised more questions than it has given answers. In our opinion the method described has the potential of revealing information on the origin of the islands, the origin of the sands and the transport mechanisms in the sand household around the islands and along the coast in general. Plans are being developed to investigate the possible sources of heavy minerals in front of the island and to obtain more insight in the dynamics of transport by following the time dependence of the heavy-mineral concentrations at certain areas of the beach by radiometric means. Also in this way we aim to contribute to a better insight in coastal processes and to a more effective defence of the coast of a country of which the sea is both its best ally and its strongest foe.

Acknowledgements

We acknowledge the help of drs J.G. Ackers and A. van der Wijk in earlier stages of this investigation. The assistance in the analysis of various samples by F.J. Aldenkamp, R. van der Made and O. Voorwinde and of P. Guinee, T. Top and R. Verburg in the mapping is greatly appreciated.

References

Baak, J.A. 1936. Regional petrology of the southern North Sea – Diss. Univ. Wageningen: 127 pp

Bonka, H. 1980. Erhöhte natürliche Strahlenexposition durch Schwermineral-anreicherung an der Küste Norddeutschlands – Atomkernenergie-Kerntechnik, 35, 1: 5–11

Crommelin, R.D. & G. Slotboom. 1945. Een voorkomen van granaatzandlagen op het strand van Goeree – Tijdschr. Kon. Ned. Aardr. Gen., 62: 143–147

De Meijer, R.J., L.W. Put, R. Bergman, G. Landeweer, H.J. Riezebos, R.D. Schuiling, M.J. Scholten & A. Veldhuizen. 1985. Local variations of outdoor radon concentrations in the Netherlands and physical properties of sand with enhanced natural radioactivity – The Science of the Total Environment, 45: 101–109

Eisma, D. 1968. Composition, origin and distribution of Dutch coastal sands between Hoek van Holland and the island Vlieland – Thesis Univ. Groningen, also: Neth. J. Sea Res. 4, 1968: 123–267

Force, E.R., A.E. Grosz, P.J. Loferski & A.H. Maybin. 1982. Aeroradioactivity maps in heavy-mineral exploration Charleston, South Carolina – U.S. Geol. Surv. Prof. Pap., 1218: 19 pp

Grosz, A. & K.L. Kosanke. 1983. Application of total-count aeroradiometric maps to the exploration for heavy-mineral deposits in the coastal plain of Virginia – U.S. Geol. Surv. Prof. Pap., 1263: 20 pp

Isbary, G. 1936. Das Inselgebiet von Ameland bis Rottumeroog – Archiv der Deutschen Seewarte, 56, 3: 55 pp

Kamel, A.M. & J.W. Johnson. 1962. Tracing coastal sediment movement by naturally radioactive minerals – Proc. 8th Coast. Eng. Conf.: 324–330

Koning, A. 1947. Sedimenten met een hoog gehalte aan zware mineralen op het strand van Vlieland – Natuurwet. Tijdschr., 29: 197–202

Lamcke, K. 1937. Natürliche Anreicherungen von Schwermineralien in Küstengebieten – Geologie der Meere und Binnengewässer, 1: 106–125

Schuiling, R.D., R.J. de Meijer, H.J. Riezebos & M.J. Scholten. 1985. Grain size distribution of different minerals in a sediment as a function of their specific density – Geol. Mijnbouw 64: 199–203

Slingerland, R.L. 1977. The effects of entrainment on the hydraulic equivalence relationship of light and heavy minerals in sand – J. Sediment. Petrol. 47, 2: 753–770

Stapor, F.W. 1973. Heavy mineral concentrating processes and density/shape/size equilibria in the marine and coastal dune sands of the Apalachicola, Florida, region – J. Sediment. Petrol. 43, 2: 396–407

Stolk, A. 1985. Koncentratie van zware mineralen in strandafzettingen – Inst. Earth. Sc., Utrecht State Univ., Internal Rept.: 41 pp

Vakgroep Mijntechnologie. 1982. Project Noordzee – Memorandum 82–01, Delft Univ. Technology, Delft: 56 pp

Van der Kleyn, P.H. 1975. Enkele aspecten van de verspreiding van winbare mineralen in het oppervlakte sediment van de Noordzee – Jaarboek Mijnbouwk. Ver., Delft, 45: 175–183

Veenstra, H.J. & A.M. Winkelmolen. 1976. Size, shape and density sorting around two barrier islands along the north coast of Holland – Geol. Mijnbouw 55: 87–104

Wasmund, E. 1938. Die Schwermineralllagerstätten der deutschen Küsten – Geol. Rundsch. 29: 237–300

Proceedings KNGMG Symposium 'Coastal Lowlands, Geology and Geotechnology', 1987: 363–370 (1989)

Evaluation of landsat imagery for Coastal-Lowland uranium exploration

Ren, You-liang[1]
[1] *Inst. of estuarine and Coastal Res., East China Normal University, Shanghai, People's Rep. of China*

Received 22 June 1987; accepted in revised form 4 December 1987

Key words: uranium mineralization, remote sensing, lineament

Abstract

Computerenhanced Landsat images of northwestern Sri Lanka have been used to define facies changes and linear features along a belt of structural basins. This was possible because of the relation between the vegetation density and the local substrate. A major lineament coincident with an alignment of basin axial zones was discovered based on the convergence of evidence derived from both spacial and spectral data manipulations. This basin axial lineament is believed to be related to uranium mineralization in the following ways. 1) It marks a regional facies change, which is necessary for the efficient trapping of dissolved uranium. 2) It marks an abrupt change in regional slope and in hydrodynamic gradient. 3) It marks an abrupt change in structural domain. 4) It marks an abrupt change in geochemical gradient. The abovementioned conditions favour a continuing concentration of uranium through long geological time.

Introduction

A geochemical exploration programme for uranium covering the entire island of Sri Lanka commenced in 1979 (Abeysinghe & Fernando, 1986). The designation of sample points at this stage was based on the distribution of the drainage basins. The bulk stream sediment samples and their heavy mineral concentrates collected during the preliminary survey were analysed for the leachable uranium by fluorimetry. Forty-eight concentrates collected from the Puttalam district had uranium values ranging from 5.6 to 460 ppm with a mean of 66 ppm, which was considered one of the most promising anomalous areas of the Island. In addition, scintillometric readings were in good correlation with the uranium anomalies. Furthermore, the approximate alignment of the anomalous locations in this area in a northsouth line suggests that the

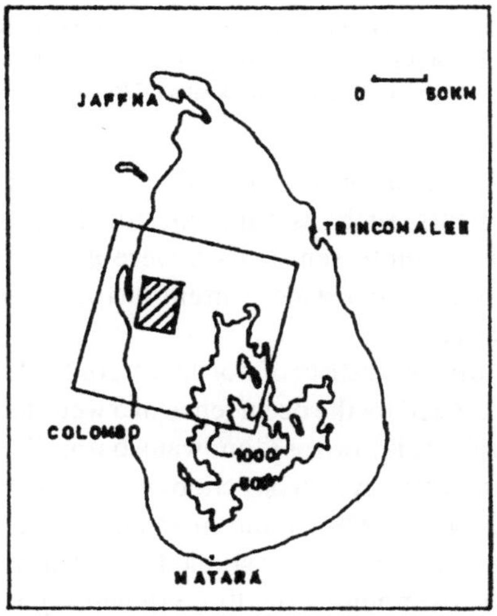

Fig. 1. Index map showing the coverage of the Landsat MSS scene (square) and the (hachured) area of the report.

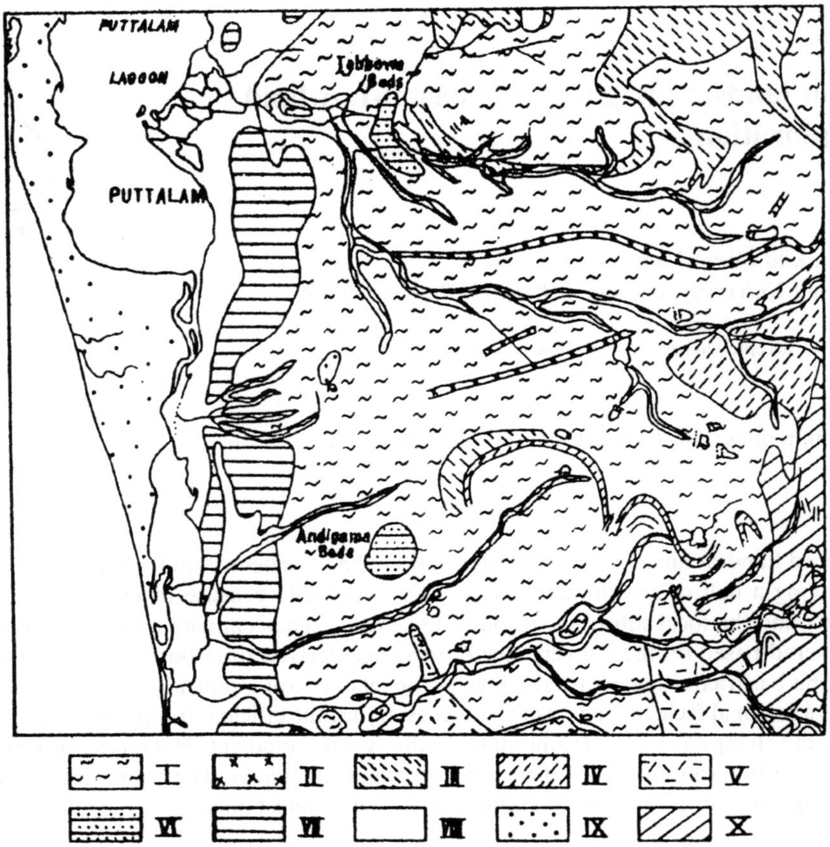

Fig. 2. Sketch map showing the geologic setting of the study area (after Cooray, 1978). Precambrian Vijayan Complex: I Granitic gneiss; II Granite; III Biotite gneiss; IV hornblende gneiss; V Charnockite; Mesozoic: VI Gondwana beds; Cainozoic: VII Red-brown earth or sand; VIII Alluvial and lagoonal deposits; IX Beach and dune sand; Precambrian Highland Series: X Undifferentiated metasediments.

uranium mineralization is most likely concentrated in lineaments such as fault-controled grabens. Therefore, remote sensor data were used in this follow-up study to pay more attention to such structural features.

The Puttalam district of northwestern Sri Lanka is characterized by the occurrence of a well-defined belt of grabens known as Gondwana basins that are aligned in a roughly N-S direction subparallel to the coast (Cooray, 1978). Landsat images integrated with stereo airphotos were studied, which provided a synoptic overview of the three-dimensional geomorphic features that consist of a contiguous set of genetically related component-facies, such as a fluvial, deltaic, lagoonal, and shelf system (Figs. 1, 2). The uranium in sandstone deposits is generally considered to have been leached from uraniferous granitic or other source rocks, to have been transported in solution in hexavalent form, but to have been precipitated in quadrivalent form in a reducing environment. The surface showings of mineralization are commonly associated with red iron-stained altered ground. The coastal lowlands of northwestern Sri Lanka are dry yet heavily vegetated; special approaches needed to be developed to deal with vegetation cover. Landsat images were computer enhanced, which enables the possible delineation of altered ground and the differentia-

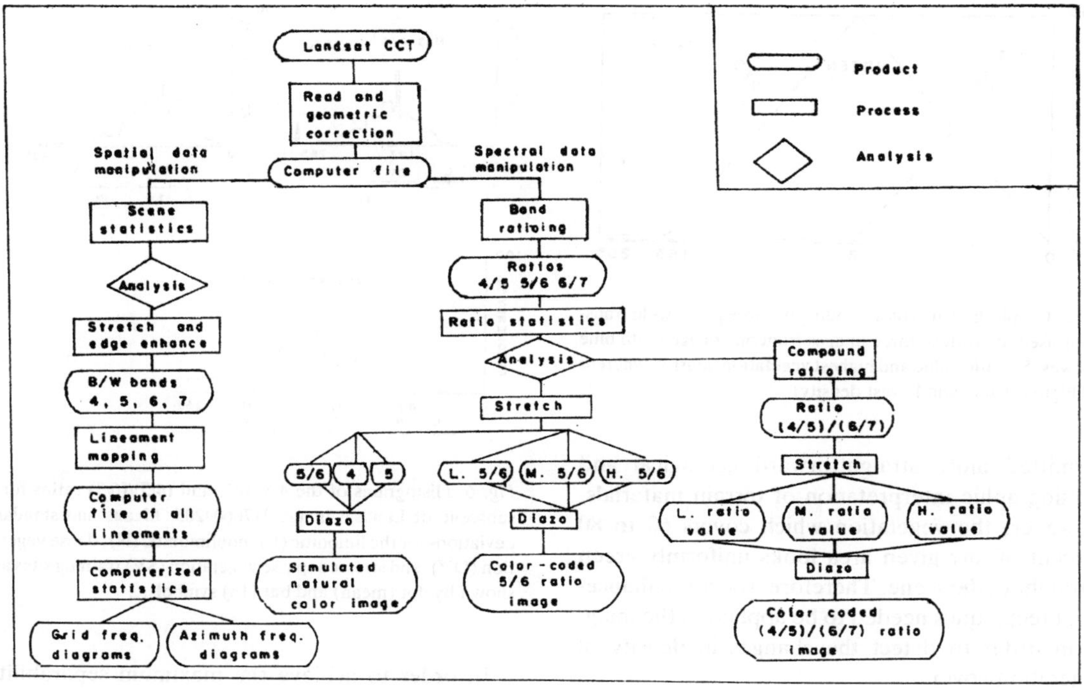

Fig. 3. Flow chart of digital image processing performed in this study.

tion of metal-stressed sparse vegetation from unstressed dense vegetation. Other subjects of interest in remote sensing investigations included the possible detection of surface expressions of buried channel sands and associated rock types that make up Jurassic basin-fill as well as computerized statistical analyses of image lineaments that might have influenced the structural framework of an uranium-bearing flow system.

Digital image processing

The digital image processing method used in this study is presented in a simplified flow chart in Fig. 3. As a first-step exploratory survey of the main features of this region, a simulated approximately natural colour image was produced by adding a calculated blue component (missing from Landsat data) to the green and red components in a colour composite. This simulated normal colour image

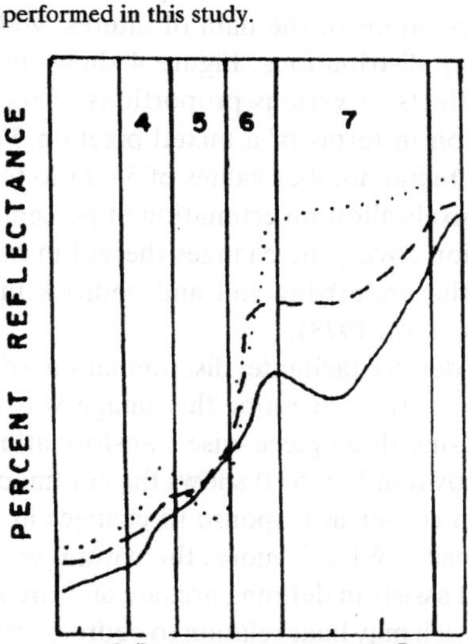

Fig. 4. Graph showing the combined effects of various proportions of rock and vegetation. The solid line: 10% vegetation and 90% rock and soil; the dashed line: 50% vegetation and 50% rock and soil; the dotted line: 90% vegetation and 10% rock and soil (after Raines, 1978).

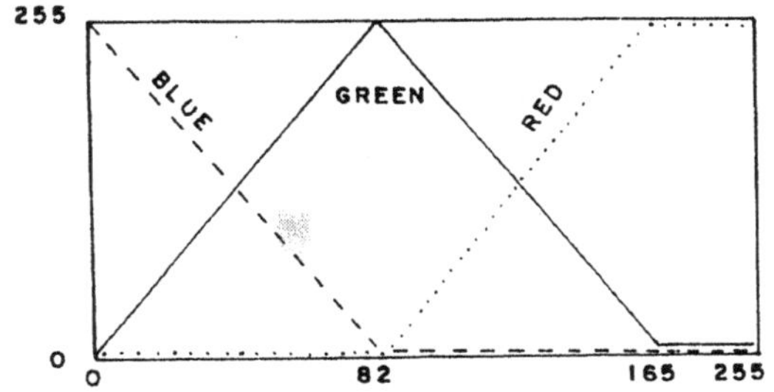

Fig. 5. Graph showing colour-coding of three piece-wide transforms used to produce three images from one image (with blue as lowest 5/6 ratio value and highest vegetation density, whereas red highest value and lowest density).

permitted more straightforward geological and physiographic interpretation of terrain materials. However, the vegetation which covers 60 to 80 percent of any given area looks uniformly green throughout the scene. Therefore, specific enhancement techniques needed to be applied to the images in order to detect the changes in density of vegetation cover.

The choice of enhancement parameters is based on the positions of the data of interest within the frequency distributions. Figure 4 shows the combined effects of various proportions of rocks and vegetation in terms of a mixed pixel on 5/6 band ratios. Digital number values of 5/6 ratio for these mixed pixels allow an estimation of percentages of vegetation cover, the changes thereof in turn may reflect the underlying soil and bedrock lithology (Raines et al., 1978).

In order to facilitate discrimination of subtle changes in the 5/6 ratio, this image was colour-coded using three piece-wise transformation functions shown in Fig. 5. It shows the continuous variation in colour in response to changes in vegetation density. What is more, the natural vegetation contrast assists in defining arcuate and circular features which may bear relation to bedrock structures and strata-bound hydrothermal alteration. Because hot, hydrothermal groundwater inhibited vegetation in granitic and metamorphic terrain, zoning of linear features and diapiric folding beds concentric with one another are shown in their entirety and hints of concealed intrusions are implied (Figs. 9, 10).

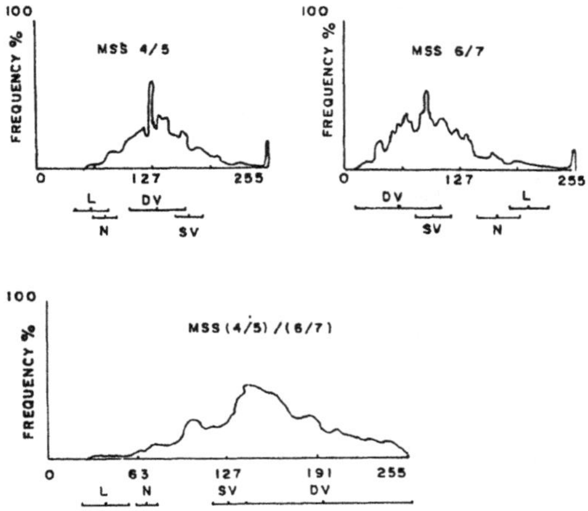

Fig. 6. Histograms of the 4/5, 6/7, and (4/5)/(6/7) ratios for a subscene of Landsat image 3176604205. Means and standard deviations for the limonitic (L), nonlimonitic (N), dense vegetation (DV), and stressed sparse vegetation (SV) training sites are shown by dot (mean) and bar (1σ) symbols.

In order to achieve the maximum separability among limonitic rocks, stressed and unstressed vegetation, a compound ratio transformation (Segal, 1983) was applied to the data set. Because the limonitic materials show both low 4/5 and high 6/7 ratio values, a compound ratio of (4/5)/(6/7) should force the exposed limonitic materials to the lowest part of the histogram. On the other hand, the unstressed dense vegetation has the highest compound ratio value because of its extremely low 6/7 ratio value. In between, the exposed nonlimonitic rocks and the stressed sparse vegetation occupy the medium low and medium high positions respectively (Fig. 6).

As mentioned above, vegetation was found to be adjusted in a very sensitive way to soil composition and bedrock lithology. Perhaps the most significant result derived from manipulation of both spectral (ratioing) and spacial data (lineaments) was the discovery of a basin-axis lineament, which is about 50 km long extending from the north of Tabbowa through Antigama to the south of Pallama. Its geological significance is fully discussed in the Regional Exploration Criteria section.

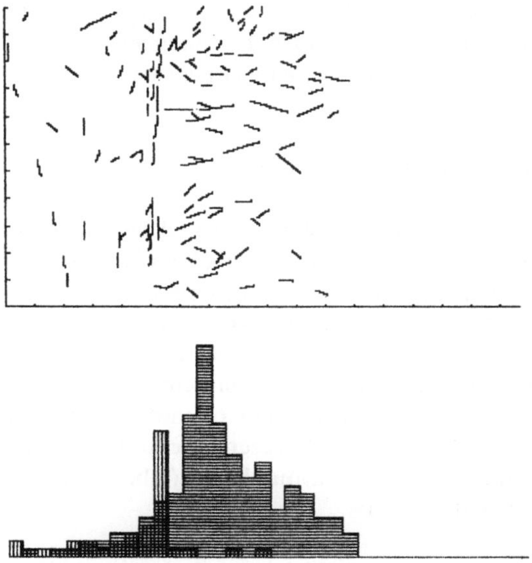

Fig. 7. Histograms showing the frequency distribution of NW + NE lineaments and near NS lineaments. The basin axial zone marks an abrupt change in structural domain from east to west.

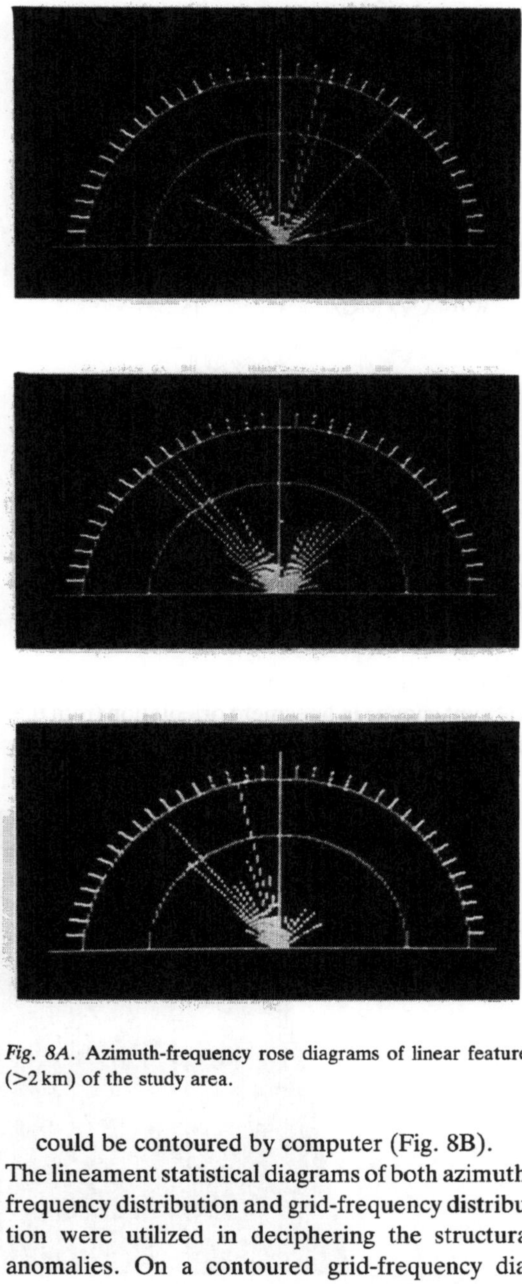

Fig. 8A. Azimuth-frequency rose diagrams of linear features (>2 km) of the study area.

Lineament statistical analysis

The primary purpose of lineament mapping and statistics was to evaluate the enhanced Landsat images for elucidating the geologic settings of potential uranium deposits on both regional and local scales. In the sedimentary region of basin fill, uranium was assumed to be deposited where facies changes or faulting-induced cross-stratal flow forms uranium-enriched roll fronts, which might be indicated by the structural framwork of the aquifer system detectable on Landsat scale. The computerized analyses of mapped linear features proceed as follows:

1. The end points of all linear features were digitized and input to a computer (Fig. 7).
2. The computer constructed an azimuth-frequency rose diagram of linear feature distribution.
3. Statistically significant azimuth-frequency maxima occurred at NNW, NS-NNE, and ENE intervals (Fig. 8A).
4. The grid-frequency of linear feature concentration at any intervals and of their intersections could be contoured by computer (Fig. 8B).

The lineament statistical diagrams of both azimuth-frequency distribution and grid-frequency distribution were utilized in deciphering the structural anomalies. On a contoured grid-frequency diagram, the structural anomalies are the alignment of the high concentration zones passing along the longer axes of the girdles. Fig. 8B shows high girdles oriented in a pattern corresponding to the surface trace of the rift axis as represented by the

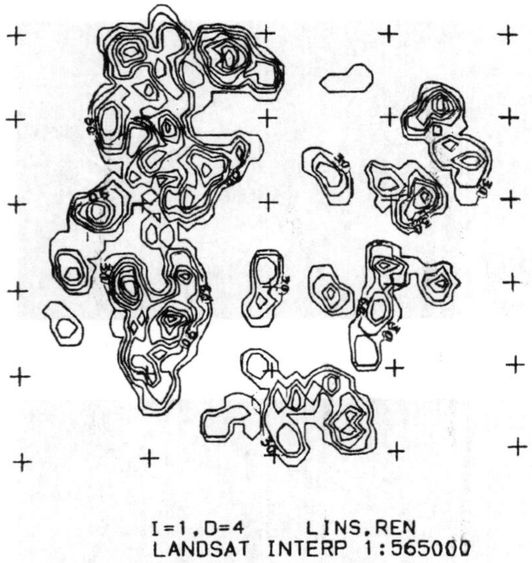

Fig. 8B. Grid-frequency diagram of lineament intersection incidence (NS versus NW + NE).

N-S alignment of a series of grabens, which marks an abrupt change in lineament orientation from the eastern domain (NNW and ENE dominant) to the western domain (NS and ENE dominant). Furthermore, a number of minor anomalies are obtained from each domain on either side of the axis.

Regional exploration criteria

Integration of the remotely sensed data and the auxiliary data presented so far allows prediction of uranium mineralization in this terrigenous coastal-lowland setting. The Jurassic continental sandstones qualified as a favourable host for uranium deposits. The greatest uranium enrichment tends to occur in the axial zones of these Gondwana basins based on widely accepted criteria for sedimentary uranium favourability (Abeysinghe & Fernando, 1986).

Abrupt changes in sedimentary facies
A basin-axis lineament was discovered on the colour-coded ratiocomposite image as a marked change in colour (vegetation density), which can be used for recognition of the facies changes in the subsurface. Exposures and subsurface data revealed the fact that the basin-axis lineament rough-

Fig. 9. Aerial photo of concentric circular features east of Tabbowa basin. Because the strata-bound hydrothermal solutions inhibit the growth of vegetation, the altered ground stands out among heavy tropic jungles.

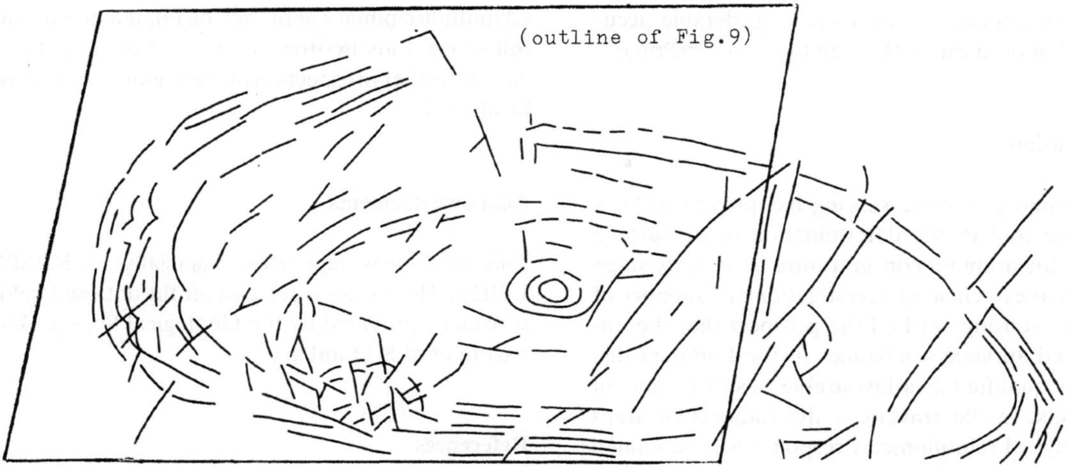

Fig. 10. Interpretation of Figure 9 and adjacent area.

ly divides the lacustrine sediments into two end-member subfacies: 1) prograding sub-deltas of coarsening-up sequence; and 2) fine-grained horizontal-bedding lake-basin squence. Such regional facies change is necessary for the efficient trapping of dissolved uranium.

Abrupt change in regional slope

The basin-axis lineament marks a regional change in slope. Uranium which entered the surface drainage may have been retained in once ponded areas. The environment of uranium deposition appeared to be where the more active streams from the eastern flanks of the basin joined the more sluggish streams of the Jurassic basin axis.

Abrupt change in regional hydrodynamic gradient

The Jurassic basin axis in this coastal-lowland setting may have served as a groundwater hydrologic boundary. The ancient coastal lake surface represented the water table at the time of or soon after the deposition of terrigenous sediments. Water recharging at the updip outcrops and moving downdip must ultimately move upward and be discharged to the water table. The basin sandstones might well be the host rocks for uranium.

Abrupt change in structural domain

The NNW and ENE trending lineaments which dominate the eastern domain may have aided the flow of uranium-bearing groundwater into the basin from the east, and then, where a NS structural barrier was encountered the groundwater flow might be inhibited, thus favouring uranium deposition.

Abrupt change in geochemical gradient

As evidenced by the drill cores and surface exposures of the Andigama, Tabbowa, and Pallama basins, carbonaceous shales and sandstones containing detritus trash are common, as indicative of a regionally reducing subsurface environment during both syndepositional and postdepositional groundwater flow history. For a roll-type uranium deposit, the geochemical gradient usually occurs at the margin of the alteration front. In this regard, the basin axis marks a regional redox interface. Moreover, the basin axial zone served as a natural electrochemical cell which connected the high Eh conditions of the oxygenated surface weathering environment with the low Eh conditions at depth. The uranium-bearing groundwaters might have permeated downward and channelled into reducing environments wherein uranium was precipitated. This process might have resulted in a continuing concentration and hence considerable accumulation of uranium through long geologic time.

Conclusions

In summary, remote sensing has proved to be a definite tool in the determination of favourable zones for uranium concentration at an early stage of a survey. Ground investigations in support of remote sensing verified the prospect that the unexposed Jurassic sandstones in the Puttalam district is qualified as a favourable host for uranium deposits. While traversing the radioactive areas readings of scintillometer response were obtained at various stations ranging from 10 up to 150 counts per second (cps) above a background of 60 cps for Jurassic sandstones and shales and their contact with granites and granitic gneisses. Parallel with and west of this major Jurassic basin fill, similar grabens were discovered, the filling of which demonstrates stratigraphic leaking of Miocene sandy limestones into Jurassic sandstones as indicative of a multistage reopening of the NS-trending fracture system. Radioactive anomalies at various concentration levels were detected from these rocks as well.

The results of the study suggest that an integrated multidisciplinary approach be implemented in a follow-up study in order to pinpoint the target areas for further prospection at the regional as well as local level.

Acknowledgements

This study was supported financially by RRSP/UNDP. The author is grateful for the data and field assistance provided by the Geological Survey Department of Sri Lanka.

References

Abeysinghe, P.B. & M.R.D. Fernando 1986 Uranium mineralization in Sri Lanka. In: The L.J.D. Fernando Felicitation Volume – Geol. Soc. Sri Lanka: 45–58.

Cooray, P.G. 1978 Geology of Sri Lanka. In: Proc. 3rd Reg. Conf. Geology and Mineral Resources of southeast Asia: 701–710.

Raines, G.L., Offield, T.W. & E.S. Santos 1978 Remote-sensing and subsurface definition of facies and structure related to uranium deposits, Powder River Basin, Wyoming – Econ. Geol. 73: 1706–1723.

Segal, D.B. 1983 Use of Landsat MSS data for the definition of limonitic exposures in heavily vegetated areas – Econ. Geol. 78: 711–722.

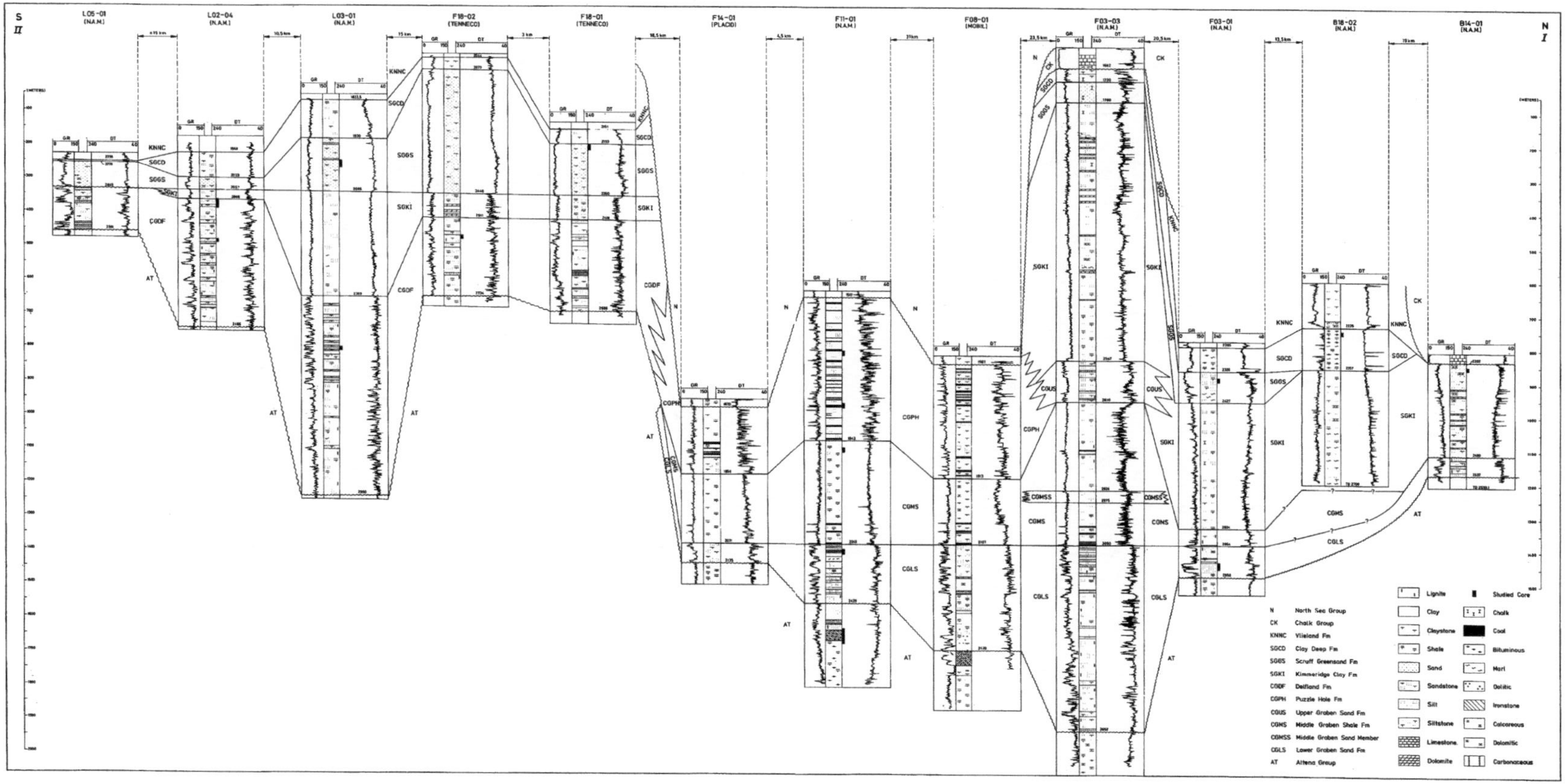

Encl. 1. Section I-II. North-South log correlation through the Central North Sea Graben. In the northernmost part of the section (B14-1) the Kimmeridge Clay is present. This formation is overlain in B18-2 and F3-1 by the bituminous Clay Deep. In the F3 area the entire Callovian-Ryazanian sequence comprising the Lower Graben Sand, Middle Graben Shale, Upper Graben Sand, Kimmeridge Clay, Scruff Greensand, and Clay Deep may be encountered. In the lower part of the Middle Graben Shale the newly established Middle Graben Sand Member is present. Southward, the Upper Graben Sand grades into the Puzzle Hole, i.e., in F8-1 and F11-1. South of F14-1 the Puzzle Hole in turn grades into the Delfland Fm. In addition, most of the other formations become thinner or even wedge out. Hence, the southern area is characterized by a generally well-developed Delfland Fm. overlain by either Kimmeridge Clay, Scruff Greensand, or Clay Deep formations.

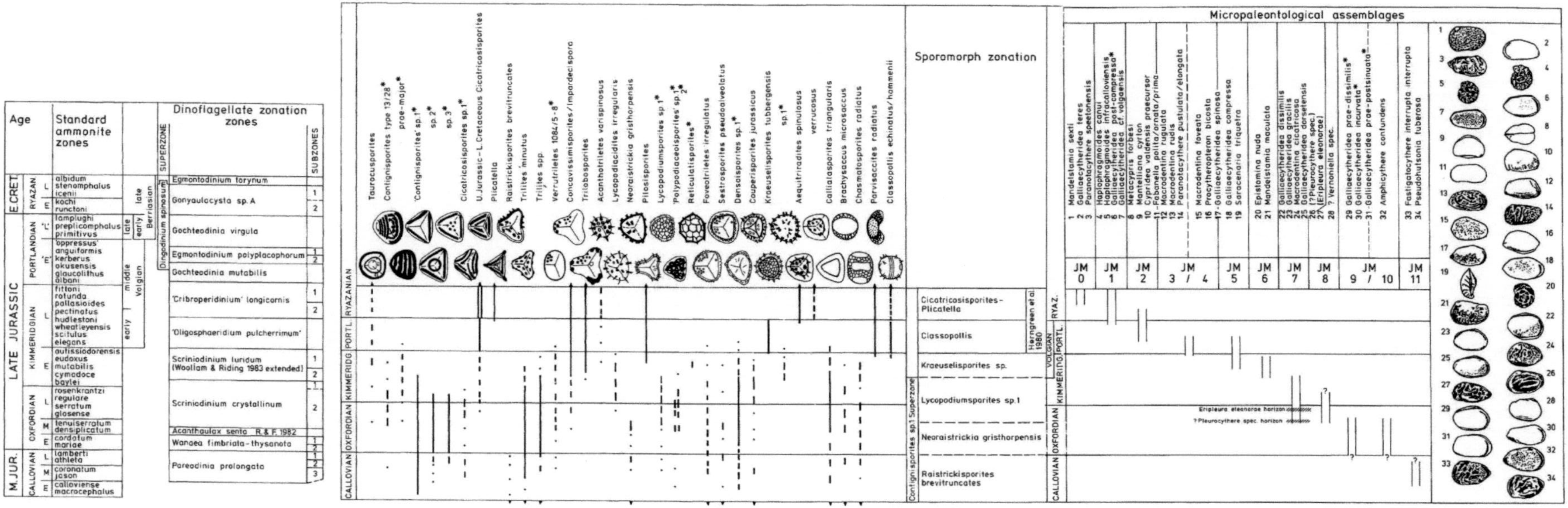

Encl. 2. Biostratigraphic zonation based on dinoflagellates, sporomorphs, ostracods and foraminifera. The limits of the sporomorph and micropaleontological zones are not known as precisely as those of the dinoflagellates and represent a (sub)stage accuracy.

MIX
Papier aus verantwortungsvollen Quellen
Paper from responsible sources
FSC® C105338

If you have any concerns about our products,
you can contact us on
ProductSafety@springernature.com

In case Publisher is established outside the EU,
the EU authorized representative is:
Springer Nature Customer Service Center GmbH
Europaplatz 3, 69115 Heidelberg, Germany

Printed by Libri Plureos GmbH
in Hamburg, Germany